Rumen Microbiology: From Evolution to Revolution

Anil Kumar Puniya • Rameshwar Singh
Devki Nandan Kamra
Editors

Rumen Microbiology: From Evolution to Revolution

Springer

Editors
Anil Kumar Puniya
College of Dairy Science and Technology
Guru Angad Dev Veterinary and Animal Sciences University
Ludhiana, Punjab, India

Rameshwar Singh
Directorate of Knowledge Management in Agriculture
Indian Council of Agricultural Research
New Delhi, Delhi, India

Devki Nandan Kamra
Division of Animal Nutrition
ICAR-Indian Veterinary Research Institute
Izatnagar, Uttar Pradesh, India

ISBN 978-81-322-2400-6 ISBN 978-81-322-2401-3 (eBook)
DOI 10.1007/978-81-322-2401-3

Library of Congress Control Number: 2015944078

Springer New Delhi Heidelberg New York Dordrecht London

Printed on acid-free paper

Springer (India) Pvt. Ltd. is part of Springer Science+Business Media (www.springer.com)

Dedicated to Animal Science Researchers

Preface

The rumen microbiology has been on the forefront of modern livestock productivity-related researches. In general, bacteria, protozoa, fungi and viruses exist in close proximity in the rumen, where bacteria predominate accounting for nearly 95 % of the total microbial community. This book provides an in-depth description of different groups of microorganisms that reside in the rumen, and our goal is to make the book well informative to scholars, researchers and teachers of animal and veterinary sciences, especially rumen microbiology. For this, throughout the text, we have focused on specific areas related to the biology and complex interactions of the strict anaerobes present in the rumen, with an aim of improving the animal productivity and the integration of significant key issues of this under-explored area of animal science.

This book presents a series of chapters on the current knowledge with an emphasis on the interactions of host animal with the microbes in the rumen. The whole content is divided into 6 units and 24 chapters. The introductory chapters cover an overview of rumen microbiology; rumen microbial ecosystem of domesticated ruminants; domesticated rare animals (yak, mithun and camel), rumen microbial diversity; wild ruminants; and structure and function of a nonruminant gut, a porcine model (i.e. Chaps. 1, 2, 3, 4 and 5). Following this, Chaps. 6, 7, 8, 9 and 10 summarize the involvement of microorganisms in the rumen such as rumen bacteria, fungi, protozoa, viruses (bacteriophages, archaeaphages) and methanogens.

The rumen manipulation is discussed in Chaps. 11, 12, 13, 14 and 15. Chapter 11 deals with plant secondary metabolites, while microbial feed additives and utilization of organic acids to manipulate ruminal fermentation and improve ruminant productivity are described in Chaps. 12 and 13, respectively. The selective inhibition of harmful rumen microbes and 'omics' approaches to understand and manipulate rumen microbial function are covered in the next two chapters. A unit of exploration and exploitation of rumen microbes deals with the current knowledge of rumen metagenomics, rumen: an underutilized niche for industrially important enzymes, ruminal fermentations to produce liquid and gaseous fuels, in Chaps. 16, 17 and 18, respectively. Finally, the commercial application of rumen microbial enzymes (Chap. 19) and molecular characterization of *Euryarcheal* community within an anaerobic digester (Chap. 20) are described. The next unit of intestinal disorders and rumen microbes covers three chapters on acidosis in cattle, urea/ammonia metabolism in the rumen and toxicity in ruminants and nitrate/ nitrite toxicity and possibilities of their use in ruminant diet. The last unit of future prospects of rumen microbiology describes the revolution in rumen microbiology.

In summary, the book shaped out to be a very systematic collection of the knowledge available in the area of rumen microbiology up to the entire satisfaction of editors. We are sure this compilation, by virtue of its content and continuity, will popularize itself among rumen microbiologists, students and researchers of related areas. The editors fully acknowledge the contributions of experts and also greatly appreciate the support of Springer.

Punjab, India — Anil Kumar Puniya
New Delhi, India — Rameshwar Singh
Izatnagar, India — Devki Nandan Kamra

Acknowledgements

This book on 'rumen microbiology' came into reality through the substantial contributions of a number of researchers, especially the contributors of the chapters. The great determined efforts done by all of them to complete this significant task are truly acknowledged by the editors.

The editors feel privileged to express their heartfelt gratitude to Dr. A. K. Srivastava, Director, ICAR-National Dairy Research Institute, Karnal; Dr. A. S. Nanda, Vice-Chancellor, Guru Angad Dev Veterinary and Animal Sciences University, Ludhiana and Dr. R. K. Singh, Director, ICAR-Indian Veterinary Research Institute, Izatnagar, who extended their full support and greatly encouraged for this academic assignment. We also owe our profound thanks to Dr. G. R. Patil, Joint Director (Academic), and Dr. R. K. Malik, Joint Director (Research), ICAR-National Dairy Research Institute, Karnal, for their constructive comments throughout the editing of this book.

We extend our thanks to all of our professional colleagues in the institute and elsewhere, who supported us directly or indirectly in giving a final shape to this book. Dr. Bhuvnesh Shrivastava and Dr. Monica Puniya are especially acknowledged for their significant involvement in the compilation of this book.

We are sincerely thankful to Dr. Prakash Sinha and Dr. Harmesh Sahay, who provided their technical help in compiling and editing of the chapters. In addition, the scholars who assisted me, viz. Prasanta Kumar Choudhury and Vipul Kashayap, deserve full appreciation.

We are also grateful to Springer for inviting us to edit this book.

Words would not suffice to thank Mr. Dharun Vijay and Ishaan Vijay for their affectionate concern throughout the compilation of this work.

Contents

Contributors

Neeta Agarwal Animal Nutrition Division, ICAR-Indian Veterinary Research Institute, Izatnagar, India

Rustam Aminov National Veterinary Institute, Technical University of Denmark, Frederiksberg C, Denmark

K. Arunasri Centre for Cellular and Molecular Biology (CCMB), Hyderabad, India

Saurabh Bansal Department of Biotechnology and Bioinformatics, Jaypee University of Information Technology, Solan, India

Serena Calabrò Department of Veterinary Medicine and Animal Production, University of Napoli Federico II, Naples, Italy

M.D. Carro Departamento de Producción Agraria, E.T.S.I. Agrónomos, Universidad Politécnica de Madrid, Madrid, Spain

Laura M. Cersosimo Department of Animal Science, College of Agriculture and Life Sciences, University of Vermont, Burlington, VT, USA

Subhojit Chakraborty Lignocellulose Biotechnology Lab, University of Delhi South Campus, New Delhi, India

Hotam Singh Chaudhary Department of Biotechnology, Madhav Institute of Technology and Science, Gwalior, India

L.C. Chaudhary Animal Nutrition Division, ICAR-Indian Veterinary Research Institute, Izatnagar, India

Y. Chen Department of Agricultural, Food, and Nutritional Science, University of Alberta, Edmonton, Canada

Prasanta Kumar Choudhury Dairy Microbiology Division, ICAR-National Dairy Research Institute, Karnal, India

Gustavo Durante Cruz Cargill Animal Nutrition, Elk River, MN, USA

Sumit Singh Dagar Micorbial Science Division, Agarkhar Research Institute, Pune, India

A. Dey Rumen Microbiome Laboratory, Division of Animal Nutrition and Feed Technology, Central Institute for Research on Buffaloes, Hisar, India

Tejpal Dhewa Bhaskaraycharya College of Applied Sciences, University of Delhi, New Delhi, India

Katerina Fliegerova Institute of Animal Physiology and Genetics, Czech Academy of Sciences, v.v.i., Prague, Czech Republic

Rosalind A. Gilbert Department of Agriculture and Fisheries, EcoSciences Precinct, Dutton Park, QLD, Australia

Centre for Animal Science, Queensland Alliance for Agriculture and Food Innovation, University of Queensland, St Lucia, QLD, Australia

Gunjan Goel Department of Biotechnology and Bioinformatics, Jaypee University of Information Technology, Solan, India

L.L. Guan Department of Agricultural, Food, and Nutritional Science, University of Alberta, Edmonton, Canada

Phil J. Hobbs Anaerobic Analytics Ltd., Okehampton, UK

Suzanne L. Ishaq Department of Animal Science, College of Agriculture and Life Sciences, University of Vermont, Burlington, VT, USA

Kavish Kumar Jain Lignocellulose Biotechnology Lab, University of Delhi South Campus, New Delhi, India

Rajashree Jena Dairy Microbiology Division, ICAR-National Dairy Research Institute, Karnal, India

Kerstin Kaerger National Reference Centre for Invasive Mycoses, Leibniz Institute for Natural Product Research and Infection Biology – Hans Knöll Institute (HKI), Jena, Germany

D.N. Kamra Animal Nutrition Division, ICAR-Indian Veterinary Research Institute, Izatnagar, India

Paul Kirk Royal Botanic Gardens, Kew, Richmond, UK

S. Kishore Rothamsted Research, North Wyke, Okehampton, UK

Athol V. Klieve Department of Agriculture and Fisheries, EcoSciences Precinct, Dutton Park, QLD, Australia

Centre for Animal Science, Queensland Alliance for Agriculture and Food Innovation, University of Queensland, St Lucia, QLD, Australia

School of Agriculture and Food Sciences, University of Queensland, Gatton, QLD, Australia

R.C. Kuhad Lignocellulose Biotechnology Lab, University of Delhi South Campus, New Delhi, India

Nikhil Kumar Basic and Applied Sciences, National Institute of Food Technology Entrepreneurship and Management, Kundli, India

Ravinder Kumar Dairy Microbiology Division, ICAR-National Dairy Research Institute, Karnal, India

Sandeep Kumar Lignocellulose Biotechnology Lab, University of Delhi South Campus, New Delhi, India

Sanjay Kumar Agriculture Systems and Microbial Genomics Laboratory, Department of Clinical Studies, School of Veterinary Medicine, University of Pennsylvania, Philadelphia, PA, USA

Sanjeev Kumar Department of Life Science, Assam University, Silchar, India

Robert W. Li United States Department of Agriculture, Agriculture Research Service, Animal Genomics and Improvement Laboratory, Beltsville, MD, USA

Ravinder Nagpal Division of Laboratories for Probiotic Research, Juntendo University Graduate School of Medicine, Hongo, Bunkyo-ku, Tokyo, Japan

Rodrigo Dias Lauritano Pacheco Universidade Federal do Mato Grosso, Cuiabá, MT, Brazil

Amlan Kumar Patra Department of Animal Nutrition, West Bengal University of Animal and Fishery Sciences, Kolkata, India

S.S. Paul Rumen Microbiome Laboratory, Division of Animal Nutrition and Feed Technology, Central Institute for Research on Buffaloes, Hisar, India

Dipti W. Pitta Agriculture Systems and Microbial Genomics Laboratory, Department of Clinical Studies, School of Veterinary Medicine, University of Pennsylvania, Philadelphia, PA, USA

Mamatha Potu Institute of Biological and Environmental and Rural Sciences (IBERS), Aberystwyth University, Aberystwyth, UK

Sonali Prusty Dairy Cattle Nutrition Division, ICAR-National Dairy Research Institute, Karnal, India

Anil Kumar Puniya College of Dairy Science and Technology, Guru Angad Dev Veterinary and Animal Sciences University, Ludhiana, Punjab, India

Monica Puniya Dairy Microbiology Division, ICAR-National Dairy Research Institute, Karnal, India

Mamta Raghav Department of Biotechnology and Bioinformatics, Jaypee University of Information Technology, Solan, India

Sreenivas Rao Ravella Institute of Biological and Environmental and Rural Sciences (IBERS), Aberystwyth University, Aberystwyth, UK

Rothamsted Research, North Wyke, Okehampton, UK

Harmesh Sahay Dairy Microbiology Division, ICAR-National Dairy Research Institute, Karnal, India

Abdelfattah Zeidan Mohamed Salem Facultad de Medicina Veterinaria y Zootecnia, Universidad Autónoma del Estado de México, Estado de México, México

S. Shivaji Centre for Cellular and Molecular Biology (CCMB), Hyderabad, India

Bhuvnesh Shrivastava Dairy Microbiology Division, ICAR-National Dairy Research Institute, Karnal, India

Rameshwar Singh Directorate of Knowledge Management in Agriculture, Indian Council of Agricultural Research, KAB-1 (PUSA), New Delhi, India

Kiyoshi Tajima National Institute of Livestock and Grassland Science, National Agriculture and Food Research Organization, Tsukuba, Japan

E.M. Ungerfeld Instituto de Investigaciones Agropecuarias (INIA), Carillanca, Temuco, Chile

Yutaka Uyeno Frontier Agriscience and Technology Center, Shinshu University, Nagano, Japan

Kerstin Voigt Jena Microbial Resource Collection, Leibniz Institute for Natural Product Research and Infection Biology – Hans Knöll Institute (HKI), Jena, Germany

Department of Microbiology and Molecular Biology, Institute for Microbiology, Friedrich Schiller University, Jena, Germany

Paul J. Weimer US Dairy Forage Research Center, US Department of Agriculture – Agricultural Research Service, Madison, WI, USA

Department of Bacteriology, University of Wisconsin Madison, Madison, WI, USA

Andre-Denis G. Wright Department of Animal Science, College of Agriculture and Life Sciences, University of Vermont, Burlington, VT, USA

M. Zhou Department of Agricultural, Food, and Nutritional Science, University of Alberta, Edmonton, Canada

About the Editors

Anil Kumar Puniya is Dean, College of Dairy Science and Technology, Guru Angad Dev Veterinary and Animal Sciences University, Ludhiana. Earlier, he served as principal scientist at ICAR-National Dairy Research Institute, Karnal, and contributed significantly in rumen and dairy microbiology. He received his PhD from NDRI, in 1994, during which he was awarded DAAD fellowship in Germany. He developed expertise in direct-fed microbial for enhanced animal productivity besides probiotics for controlling lifestyle diseases.

He published over 125 research papers/reviews/book chapters of impact and five lab manuals. He acted as chairman and keynote/invited speaker in scientific events, nationally and internationally. He has been involved in 18 externally funded and institutional projects as PI/Co-PI and supervised a number of master and doctoral scholars. He remained paper coordinator of Food and Dairy Microbiology for e-contents (i.e. e-PG Pathshala) of UGC under National Mission on Education. He also developed e-course of Introductory Dairy Microbiology for BTech (DT) under NAIP of ICAR.

He was awarded DSE Grant to attend the 'International Conference of Tropical Veterinary Medicine' in 1995 at Germany and DST and CSIR funding to present a paper at 'BioMicroWorld 2007' in Spain. He chaired sessions of 'Food Microbiology' during 'BioMicroWorld 2009' in Portugal and 'Rumen Microbiology' during 'ANINUE-2012' in Thailand. Besides, he visited Aberystwyth University, UK, twice under 'Exchange of Scientists Programme' of INSA with the Royal Society of London (2007) and as 'DBT-CREST' Awardee (2013). He visited Taiwan in 2010 during an exposure visit of DST.

He is a recipient of Young Scientist Award in 1996 from the Association of Microbiologist of India, Kautilya Gold Medal in 1998–1999 from Nature Conservators, Best Young Scientist Award in 2004 from Eureka Forbes, K.K. Iya Award in 2006, Honorable Mention in 2010 from the American Society for Microbiology and certificate of appreciation from ABRCMS, 2012–2013 and 2014, USA. He is a member of the International Scientific Advisory Board of Formatex Research Centre, Spain.

He underwent training for scientists and technologists in 2007 at DST and in molecular biology at IARI and IMTECH besides e-course development at NAARM. He was 'co-organizing secretary' of the conference of AMI, 2004, and convener of the National Environment Awareness Campaign of MoEF at NDRI, 2007. He is editorial board member and referee of several reputed journals worldwide. He is a life member of a number of renowned scientific bodies and fellow of ANSI and NADSI.

Devki Nandan Kamra is presently the ICAR national professor at the Animal Nutrition Division of Indian Veterinary Research Institute, Izatnagar, and has a research and teaching experience of 36 years. Previously, he had been the director of the Centre of Advanced Faculty Training in Animal Nutrition at IVRI, Izatnagar, from 2010 to 2013 and had an additional charge of joint director (research) from July 2012 to February 2013.

He has completed several research projects funded by the Department of Biotechnology, New Delhi; National Agricultural Technology Programme, ICAR, New Delhi; ICAR Cess Fund; International Atomic Energy Agency/ Food and Agricultural Organization (IAEA-FAO Joint Division), Austria; and National Agricultural Innovative Programme (NAIP), New Delhi. His research interests are microbial diversity of the rumen, metagenomics and rumen manipulation for increasing livestock productivity, microbial feed additives for improving nutrient utilization and inhibition of enteric methane emission by the ruminants. He has guided four MVSc and ten PhD students for their research leading to respective degrees. He has written five books on animal nutrition, which are very well accepted by the students, teachers and researchers. He has published more than 130 research papers in journals of high repute and contributed several chapters in a book edited by experts in animal nutrition.

He has been conferred the *Best Teacher Award* of IVRI (Deemed University), *Award of Merit* and *Award of Honour* of IVRI, *Rafi Ahmed Kidwai Award*, *Fellowship of the National Academy of Agricultural Sciences*, *Recognition Award of the National Academy of Agricultural Sciences*, *Bharat Ratna Dr. C. Subramaniam Award* for *Outstanding Teacher*, *CLFMA Lifetime Achievement Award* and *ANA Dr. D.V.R. Prakash Rao Outstanding Researcher Award*. Two of his students have received prestigious *Jawaharlal Nehru Award* for Outstanding P.G. Research and one student received *Dr. N. D. Kehar Award* for Outstanding P.G. Research.

Dr. Rameshwar Singh is presently serving as project director at ICAR – Directorate of Knowledge Management in Agriculture at the Indian Council of Agricultural Research, New Delhi. In this role, Dr. Singh is responsible for communication, information, publication of research journals and books and other outreach activities of the ICAR and its 100 odd research institutions at the national level. He has helped in further opening up ICAR through social media and e-publications.

He has over 36 years of research and teaching experience in dairy and rumen microbiology. He has completed a number of projects on biodegradation of lignin, animal and human probiotics, improvement and preservation of feeds and fodders through microbial fermentation, indigenous fermented milks and dairy starter cultures. His effort in expansion and strengthening of the National Collection of Dairy Cultures (NCDC) has been noteworthy. He also led the All India Network Project on Dairy Microbes as nodal officer. Dr. Singh has been previously associated with the Indian Veterinary Research

Institute, Izatnagar, and National Dairy Research Institute, Karnal. He has guided a number of postgraduate students for the master's and doctoral programmes and has published highly cited research papers and review articles.

Dr. Singh is life member of the Association of Microbiologists of India, Indian Dairy Association, Indian Science Congress and Animal Nutrition Association. He has also been awarded the fellowship of National Academy of Dairy Sciences (India).

Part I

Overview of Rumen and Ruminants

1 Rumen Microbiology: An Overview

Prasanta Kumar Choudhury, Abdelfattah Zeidan Mohamed Salem, Rajashree Jena, Sanjeev Kumar, Rameshwar Singh, and Anil Kumar Puniya

Abstract

The systematic exploration of microbial ecosystem of the rumen was commenced by the father of rumen microbiology, Robert Hungate, in 1950s. His contributions toward the development of anaerobic culture techniques have illustrated the ways to explore the complex microbial structures of the rumen and other anaerobic ecosystems. The understanding of rumen microbiology has strengthened an awareness to improve the feed utilization and manipulation of microbial compositions. Microbes and their interactions in interspecies H_2 transfers were first studied in the rumen ecosystems and attracted pioneers to investigate the alternate abatement strategies of methane production along with enhanced animal productivity. The discovery of alternate hydrogenotrophs and industrially important novel microbes and the management of rumen disorders via microbial manipulations make this community an interesting research platform for different microbial theories. The discovery of anaerobic fungi as a part of rumen flora by Orpin during the 1970s disproved their mistaken identity as flagellated protozoa and the concept that all fungi are aerobic organisms. Upcoming biotechnological strategies and deciphering on microbial com-

P.K. Choudhury • R. Jena
Dairy Microbiology Division, ICAR-National Dairy Research Institute, Karnal 132001, India

A.Z.M. Salem
Facultad de Medicina Veterinaria y Zootecnia, Universidad Autónoma del Estado de México, Estado de México, Mexico

S. Kumar
Department of Life Science, Assam University, Silchar 788011, India

R. Singh
Directorate of Knowledge Management in Agriculture, Indian Council of Agricultural Research, KAB-1 (PUSA), New Delhi 110012, India

A.K. Puniya (✉)
College of Dairy Science and Technology, Guru Angad Dev Veterinary and Animal Sciences University, Ludhiana, Punjab, India
e-mail: akpuniya@gmail.com

A.K. Puniya et al. (eds.), *Rumen Microbiology: From Evolution to Revolution*, DOI 10.1007/978-81-322-2401-3_1

munity using molecular tools, novel gene discovery, community-community interactions, and phylogenetic relationships have opened new avenues of microbial ecology in rumen ecosystem. The present chapter deals with the microbial ecosystems of rumen and their interactions.

Keywords
Rumen • Bacteria • Anaerobic fungi • Methanogens • Microbial interactions

1.1 Introduction

Rumen, the four-chambered stomach of grazing animals, harbors a complex bionetwork, where all forms of primitive starting from archaea to protozoa exist in close proximity (Choudhury et al. 2012). Some of these microbes interact with each other in a synergistic relationship to extract energy while producing highly active lignocellulolytic enzymes supporting digestion of the host. This yields volatile fatty acids (VFAs, acetate, butyrate, propionate), formic acid, H_2, CO_2, and CH_4 (Krause et al. 2003). For many years, nutritionists, microbiologists, and physiologists have been studying the rumen with the aim of maximizing productivity and improving overall host health via manipulating the rumen and its microbial ecosystem. Without rumen microorganisms, or when the rumen ecosystem is disturbed, there can be a cascade of detrimental effects on animal health and productivity. Altering the rumen ecosystem to reduce methane emissions is a significant scientific challenge that must take the effects on animal health, farm management, and animal productivity into account. Understanding of complex microbial communities' function and microbial interactions within their niches represents a major challenge for rumen microbiologists even today. Advances in molecular biology and genomics offer new opportunities to conduct a holistic examination of the structure and function of rumen microbial communities (Sirohi et al. 2012a). Analyzing the role of individual group of microbes, microbial structure, composition, and interaction may provide functional dimensions of the rumen network that will help to achieve a major goal of rumen microbiology.

1.2 Rumen Microbial Ecosystem

Herbivores feed on plant structural carbohydrates such as lignin, cellulose, and hemicellulose, which they are not able to utilize themselves because of their inability to produce respective hydrolytic enzymes. Therefore, symbiotic microorganisms are established in their alimentary tracts that can hydrolyze these compounds to generate energy for themselves, as well as for the host animal. Since strict anaerobic conditions (−150 to −350 mv) prevail in the rumen (Clarke 1977), the natural rumen microflora consists of strict anaerobes. Woese's classification represented all microbes in the rumen ecosystem and distinguished them into three domains: Bacteria (bacteria), Archaea (methanogens), and Eucarya (protozoa and fungi). The rumen is open to the external environment, and there is a continuous flow of material into and out of the rumen. It has a dry matter content of 10–13 %, and temperature is typically regulated between 38 and 41 °C. The physiological pH range is between 5.5 and 6.9, and it is one of the most variable factors in the rumen environment. However, the buffering capacity of the rumen content is quite high (Hungate 1966; Church 1969; Clarke 1977; Dehority 2003). The details of different rumen physical, chemical, and microbiological parameters are discussed in Table 1.1.

1.2.1 Bacteria

Rumen harbors different types of bacteria, which are most actively involved in the plant fiber degradation, as revealed by the fact that bacteria associ-

Table 1.1 Physical, chemical, and microbiological characteristics of rumen ecosystem

Physical properties	
Dry matter (%)	10–18
Osmolality	250–350 mOsmol/Kg^{-1}
pH	5.5–6.9 (Mean 6.4)
Redox potential	−350 to −400 mV
Temperature	38–41 °C
Chemical properties	
Amino acids and oligopeptides	<1 mmolL^{-1} present 2–3 h post feeding
Ammonia	2–12 mmolL^{-1}
Dietary (cellulose, hemicelluloses, pectin) component	Always present
Endogenous (mucopolysaccharides)	Always present
Gas phase (%)	CO_2 65; CH_4 27, N_2 7; O_2 0.6, H_2 0.2
Growth factors	Good supply; branched-chain fatty acids, long-chain fatty acids, purines, pyrimidines, other unknown
Lignin	Always present
Minerals	High Na; generally good supply
Nonvolatile acids (mmol L^{-1})	Lactate <10
Soluble carbohydrates	<1 mmolL^{-1} present 2–3 h post feeding
Trace elements/vitamins	Always present; good supply of B vitamins
Volatile fatty acids (mmol L^{-1})	Acetate 60–90, propionate 15–30, butyrate 10–25, branched chain and higher 2–5
Microbiological properties	
Anaerobic fungi	10^{3-5} g^{-1} (6 genera)
Bacteria	10^{10-11} g^{-1} (>200 species)
Bacteriophage	10^{7-9} g^{-1} particles ml^{-1}
Ciliate protozoa	10^{4-6} g^{-1} (25 genera)

Source: Adapted from Mackie et al. (1999), Castro-Montoya et al. (2011)

ated with feed particles account for nearly 50–75 % of the total microbial population (Minato et al. 1966). The adhesion of microbes to solid substances is an important factor in successful competition and survival in the rumen, as well as in the digestion of solid feed (Minato et al. 1966; Mitsumori and Minato 1997). In addition, most of the rumen endoglucanase and xylanase activities are contributed by bacteria, and therefore, fiber-associated bacteria are pivotal to the rumen digestion. These are divided into four groups on the basis of their association as free floating in the liquid phase, attached to the feed particles (firmly/loose), rumen epithelium, protozoa, and fungi. Besides, these microbes have also diversified with respect to their functions. These carry out in degradation of plant ingredients. These have been classified as fiber (i.e., cellulose, hemicelluloses, and pectin)-degrading bacteria, lactic acid utilizers, acetogens, starch utilizers, etc. (Table 1.2). Starch and sugar digesters make up a significant part of the rumen bacterial population. Generally, high-producing dairy cows are fed diets containing more than 30 % starch and sugars; therefore, these bacteria are greatly needed. Even if a cow is on an all-straw diet, the fiber digesters still never account for more than 25 % of the rumen bacterial population. Some microbes utilize the gaseous end products of the rumen to form methane and are grouped under archaea, while others are involved in converting the reducing power to useful end products as acetate, propionate, H_2S, and butyrate. Always, there is an interest for non-methanogenic sinks to release H_2 produced during fermentation (Joblin 1996, 1999). In the rumen, CH_4 is produced by methanogens using H_2 to reduce CO_2. The non-methanogenic hydrogenotrophic bacteria residing in the rumen include acetogens that reduce CO_2 to form acetate by the Wood-Ljungdahl pathway (i.e., reductive aceto-

Table 1.2 Microbes involved in various rumen functions

Microbial types	Important genera and species
Bacteria	
Acetogens	*Acetitomaculum ruminis, Eubacterium limosum*
Acid utilizers	*Megasphaera elsdeni, Wolinella succinogenes, Veillonella gazogene, Micrococcus lactolytica, Oxalobacter formigenes, Desulfovibrio desulfuricans, Desulfotomaculum ruminis, Succiniclasticum ruminis*
Cellulolytic	*Fibrobacter succinogenes, Butyrivibrio fibrisolvens, Ruminococcus flavefaciens, Ruminococcus albus, Clostridium cellobioparum, Clostridium longisporum, Clostridium lochheadii, Eubacterium cellulosolvens*
Hemicellulolytic	*Prevotella ruminicola, Eubacterium xylanophilum, Eubacterium uniformis*
Lipolytic	*Anaerovibrio lipolytica*
Pectinolytic	*Treponema saccharophilum, Lachnospira multiparus*
Proteolytic	*Prevotella ruminicola, Ruminobacter amylophilus, Clostridium bifermentans*
Amylolytic	*Streptococcus bovis, Ruminobacter amylophilus, Prevotella ruminicola*
Saccharolytic	*Succinivibrio dextrinosolvens, Succnivibrio amylolytica, Selenomonas ruminantium, Lactobacillus acidophilus, Lactobacillus casei, Lactobacillus fermentum, Lactobacillus plantarum, Lactobacillus brevis, Lactobacillus helveticus, Bifidobacterium globosum, Bifidobacterium longum, Bifidobacterium thermophilum, Bifidobacterium ruminale, Bifidobacterium ruminantium*
Tanninolytic	*Streptococcus caprinus, Eubacterium oxidoreducens*
Ureolytic	*Megasphaera elsdenii*
Bacteriophages	*Methanobacterium* phage Ψ M1, *Methanobacterium* phage Ψ M10, *Methanobacterium* phage Ψ M100, *Methanothermobacter* phage Ψ M100, *Methanobacterium* phage ΨM2
Fungi	*Piromyces communis, Piromyces mae, Piromyces minutus, Piromyces dumbonicus, Piromyces rhizinflatus, Piromyces spiralis, Piromyces citronii, Piromyces polycephalus, Anaeromyces mucronatus, Anaeromyces elegans, Caecomyces communis, Caecomyces equi, Caecomyces sympodialis, Cyllamyces aberensis, Cyllamyces icaris, Neocallimastix frontalis, Neocallimastix patriciarum, Neocallimastix hurleyensis, Neocallimastix variabilis, Orpinomyces joynii, Orpinomyces intercalaris*
Methanogens	*Methanobacterium formicicum, Methanobacterium bryantii, Methanobrevibacter ruminantium, Methanobrevibacter smithii, Methanomicrobium mobile, Methanosarcina barkeri, Methanoculleus olentangyi*
Protozoa	*Entodinium bovis, Entodinium bubalum, Entodinium bursa, Entodinium caudatum, Entodinium chatterjeei, Entodinium parvum, Entodinium longinucleatum, Entodinium dubardi, Entodinium exiguum, Epidinium caudatum, Isotricha prostoma, Isotricha intestinalis, Dasytricha ruminantium, Diplodinium dendatum, Diplodinium indicum, Oligoisotricha bubali, Polyplastron multivesiculatum, Eremoplastron asiaticus, Eremoplastron bubalus*

Source: Pfister et al. (1998), Luo et al. (2001), Kamra (2005), Janssen and Kirs (2008), Wright and Klieve (2011), Sirohi et al. (2012b), Choudhury et al. (2012), Kumar et al. (2014)

genesis), sulfate-reducing bacteria (SRB) that reduce sulfate to H_2S (Morvan et al. 1996), and bacteria that use H_2 to reduce fumarate to succinate (Asanuma and Hino 2000). Succinate, the product of fumarate reduction, is decarboxylated to propionate, a valuable animal nutrient (Wollin et al. 1997). Dissimilatory sulfate-reducing bacteria produce a toxic end product as H_2S from the reducing power.

The type of feed also controls the bacterial populations inside the rumen. A high-fiber-containing feed increases the population of *Fibrobacter succinogenes*, *Ruminococcus albus*, and *Ruminococcus flavefaciens*. The H_2 and VFA production are also enhanced by high-fiber diet. Minato et al. (1989) clarified the effect of feeding ammonia-treated straw on rumen bacterial flora, showing that it stimulates the growth of *Eubacterium ruminantium*. Orpin et al. (1985) described seasonal variations of rumen bacteria in wild reindeer by demonstrating an increased cellulolytic *Butyrivibrio fibrisolvens* under severe dietary conditions in winter. Interestingly, non-cellulolytic bacteria were also detected as members of the fiber-associating consortia in the rumen (Koike et al. 2003; Larue et al. 2005). Tannin-degrading bacteria have also been isolated (Goel et al. 2005a, b, 2011) from ruminants of wild origin identified as *Selenomonas ruminantium* (Odenyo and Osuji 1998) as these consume tanniferous plants. *Streptococcus bovis* is present only when large amounts of starch or sugars are fed and pH is low. It produces lactic acid, a stronger acid than many of the other VFAs produced in the rumen. When conditions are favorable for *Streptococcus bovis*, it will grow explosively. This type of growth causes rumen acidosis. *Megasphaera elsdenii* requires lactic acid to grow that helps to clean up the rumen a bit and raise rumen pH, sustaining the growth of the acid-intolerant fiber digesters in the rumen.

1.2.2 Anaerobic Fungi

The rumen is a storehouse of anaerobic fungi that are not found elsewhere and play a very dominant role in the degradation of lignocellulosic components of the feed particles. These constitute the smallest in numbers and represent only about 20 % (Rezaeian et al. 2004) of the rumen microbial biomass but are considered to be the most efficient fiber degrader (Akin et al. 1988). According to the recent classification system (Hibbett et al. 2007), anaerobic fungi have been described under the order *Neocallimastigales*, class *Neocallimastigomycetes* in the phylum *Neocallimastigomycota* (Dagar et al. 2011; Sirohi et al. 2013a; Gruninger et al. 2014). The thallus may be monocentric (single reproductive body, i.e., sporangium from single zoospore) or polycentric (numerous sporangia from single zoospore) (Ho 1995). These forms are determined in the earliest stages of growth, soon after zoosporogenesis. The zoospores may be posteriorly uniflagellated or polyflagellated in both forms. The life cycle of anaerobic fungi is asexual, and no sexual stage has been described, as only mitotic nuclear divisions have been observed (Heath et al. 1986). It alters between a motile zoospore, encysted zoospore, and vegetative zoosporangial stage. Once released from zoosporangium, the motile zoospores move by chemotaxis to colonize the plant fibrous material and shed its flagella to get transformed into cyst (encystment). The cyst germinates by producing a germ tube, which ultimately gives rise to a rhizoidal system. In monocentric forms, the development may further be classified as endogenous or exogenous (Sirohi et al. 2012b).

These fungi differ from their aerobic counterparts in having hydrogenosomes as a site for energy production, instead of mitochondria (Yarlett et al. 1986). These fungi are not exclusively found in the rumen of the animals, but throughout the entire digestive tract (Davies et al. 1993). Moreover, these have also been isolated from the saliva and feces (Lowe et al. 1987), suggesting the presence of a resistant structure tolerant to the oxygen and desiccation. Brookman et al. (2000) identified the formation of spores in the polycentric anaerobic gut fungi, which enhance their ability to survive and extend a possibility toward their use as direct-fed microbial (Tripathi et al. 2007a; Saxena et al. 2010; Puniya et al. 2014). Anaerobic fungi have been isolated from a number of ruminant and nonruminant herbivorous mammals (Ljungdahl 2008; Paul et al. 2010). These contribute significantly to overall metabolism of host with their high cellulolytic activity and play a greater role in degradation of lignified plant tissues (Akin and Borneman 1990; Dey et al. 2004; Thareja et al. 2006; Sirohi et al. 2012b) with the help of a wide range of hydrolytic enzymes (Tripathi et al. 2007b; Shelke et al. 2009; Nagpal et al. 2009, 2010, 2011) including cellulases (Barichievich and Calza 1990), hemicellulases (Novotna et al. 2010; Mountfort and Asher 1989), proteases (Michel et al. 1993), amylases, amyloglycosidases (Paul et al. 2004), feruloyl and p-coumaroyl esterases (Borneman et al. 1990), various disaccharidases (Chen et al. 1995), and pectinases (Kopecny and Hodrova 1995). To date, only six genera (Table 1.2) have been described (Griffith et al. 2009; Sirohi et al. 2012b). However, certain uncharacterized isolates have also been reported (Phillips and Gordon 1988; Ho 1995).

Since these fungi are obligatory anaerobic, these obtain energy by fermentation of carbohydrates (Gordon and Phillips 1998; Trinci et al. 1994), a process in which the energy source acts as both the electron acceptor and the electron donor. These lack mitochondria (Yarlett et al. 1986), cytochromes, and other biochemical features of the oxidative phosphorylation pathway. Instead of mitochondria, hydrogenosomes are present, which couple the metabolism of glucose to cellular energy production. In the beginning, hydrogenosomes were thought to be novel inventions to cope with anaerobic conditions (Cavalier-Smith 1987), but now it has become quite clear that hydrogenosomes are either specialized or derived mitochondria (Embley et al. 1997). These organelles are not unique to anaerobic fungi and were discovered in parasitic trichomonads (Lindmark and Müller 1973). Since then, these have been found in a wide variety of anaerobic eukaryotes. Typical mitochondrial features like genome, electron transport chain, and cristae are missing from fungal hydrogenosomes and contain unusual enzymes like hydrogenase and pyruvate/ferredoxin oxidoreductase instead of usual mitochondrial pyruvate dehydrogenase (van der Giezen 2002). The common properties of hydrogenosomes and mitochondria include the existence of a transmembrane pH gradient and an alkaline lumen. Just like mitochondria, hydrogenosomes also serve as the site for conversion of pyruvate to acetyl CoA and ATP production, whereas the presence of other metabolic processes like amino acid metabolism and β-oxidation is presently unknown. Hydrogenosomes contain hydrogenase and produce H_2, CO_2, and acetate as metabolic waste products (Brul and Stumm 1994; Theodorou et al. 1996).

1.2.3 Protozoa

Rumen protozoa constitute about 50 % of the viable biomass in the rumen and are dependent. Majority are ciliates and few flagellates and are very motile. Unicellular organisms with 20–200 μm size are not attached to feed particles. They engulf bacteria and feed particles and digest carbohydrates, proteins, and fats (Williams and Coleman 1992). Two groups, i.e., holotrich and entodiniomorphid protozoa, have been studied inside the rumen (Table 1.2). Morphological studies have identified more than 250 species of ciliates living in the various ruminants (Williams and Coleman 1992; Imai 1998). These ciliates play an important role in fiber digestion and the modulation of the fermentation profiles. The rumen protozoa produce fermentation end products similar to those made by the bacteria, particularly acetate, butyrate, and H_2. Rumen methane bacteria actually attach and live on the surface of rumen protozoa for immediate access to H_2. They utilize large amounts of starch at one time and can store it in their bodies. This may help to slow down the production of acids that lower rumen pH, benefiting the rumen. Rumen protozoa multiply very slowly in the rumen over 15–24 h as opposed to the bacteria. For this reason, the rumen protozoa hide out in the slower-moving fiber mat of the rumen, so that these are not washed out before these have a chance to multiply. Low-roughage diets reduce the retention of fiber in the rumen and may decrease the number of protozoa in a cow's rumen.

The enzymatic profile of holotrich protozoa indicates that these have amylase, invertase, pectin esterase, and polygalacturonase in sufficiently large quantities for using starch, pectin, and soluble sugars as energy source (Bailey and Howard 1963; Williams 1979). The enzymes responsible for cellulose and hemicellulose degradation have also been reported in the holotrich protozoa, but the levels are very low compared to those present in the entodiniomorphid protozoa (Williams and Coleman 1985). The rumen ciliates are proteolytic producing ammonia and amino acids as end products (Warner 1956). Their N metabolism is based largely on the digestion of engulfed bacteria (Coleman 1975), although all rumen ciliates contain enzymes capable of digesting plant proteins (Coleman 1983). The utilization of these protein substrates is inefficient, as a significant proportion of the total amount engulfed by the protozoa is excreted as amino acids (Coleman 1975). Degradation of protein or amino acids of microbial origin reduces the net yield of microbial amino acids available for intestinal digestion and hence, the net efficiency of microbial growth. Further, a large proportion of the protein incorporated in the biomass of protozoa in the rumen may not be available for digestion in the intestines. Direct enumeration of protozoa in the rumen fluid entering the omasum (Weller and Pilgrim 1974), concentration of a protozoa marker (i.e., phosphatidylcholine) in abomasal digesta, and measurement of the turnover rates of the protozoa (Leng 1982) all indicate a low outflow rate of ciliates from the rumen in relation to their concentration in the rumen. The synthesis of protozoa nitrogenous compounds and their breakdown in the rumen, therefore, contribute considerably to the N recycling within the rumen, as the apparent turnover of protozoa protein was calculated to be 0.67 of the total protein synthesized by protozoa (Leng 1982).

1.2.4 Methanogens

Methanogenic archaea, considered as the earliest organisms originated in our planet, are phyllogenetically widespread and represent three of the seven classes in the phylum *Euryarchaeota* (Sirohi et al. 2010; Singh and Sirohi 2012). These rumen microorganisms utilize the H_2 and CO_2 produced by the protozoa, fungi, and bacteria from the catabolism of hexoses to produce CH_4 and generate ATP (Albers et al. 2007; Ferry and Kastead 2007), which benefits the donors by providing an electron sink for reducing equivalents to minimize the partial pressure of H_2 (Wolin and Miller 1988; Lange et al. 2005) inside the rumen. These are comprised of about 70 species among 31 genera as reviewed in "Bergey's Manual of Systematic Bacteriology" (Whitman et al. 2001) and "The Prokaryotes." These microorganisms specifically possess three coenzymes, i.e., coenzyme F-420, involved in electron transfer; coenzyme M, involved in transfer of methyl groups; and a low molecular weight, oxygen-labile, heat-stable coenzyme B, involved in the final reaction of CH_4 production. Methanogens are found in a symbiotic association with rumen bacteria (Wolin and Miller 1988) and protozoa (Lange et al. 2005). The establishment and maintenance of the stable population of methanogens is affected by the type of diet and level and frequency of feeding (Kumar et al. 2012, 2013a, b; Sirohi et al. 2013b). The symbiotic association of hydrophobic methanogens with H_2 producers is usually realized by the attachment or by floc formation (Thiele et al. 1988; Lange et al. 2005). Among these ciliates, protozoa are the only ones for which such interaction can be microscopically demonstrated (Vogels et al. 1980). The symbiotic relation between methanogens and ciliates may generate up to 37 % of rumen CH_4 emission (Finlay et al 1994). In the rumen, H_2 is produced during plant cell wall degradation as an intermediate compound by cellulolytic bacteria (i.e., *Ruminococcus albus, Ruminococcus flavifaciens*) and anaerobic fungi (Stewart and Bryant 1988). It never gets accumulated, as it is rapidly utilized by methanogens in ruminants (Wolin and Miller 1988) to produce CH_4 and generate ATP (Albers et al. 2007; Ferry and Kastead 2007). Details of different genera are given in Table 1.2.

1.2.5 Bacteriophages

Bacteriophages are one of the most important component of the rumen microbial community and are present typically at >10^9 particles per mL. These are specific for different bacteria present in the rumen. These are also considered to be obligate pathogens for the bacteria, as bacteriophages are capable of lysing bacteria. These phages help in bacterial mass turnover in the rumen, which may be considered not so useful for the animals on different feeding schedules (Klieve and Swain 1993), but by lysing the bacterial cells, the bacterial protein is easily made available to the animals as a source of amino acids. The specificity of the bacteriophages for a particular rumen bacterium may be exploited for removal or killing by lysis of unwanted rumen bacteria from the ecosystem like *Streptococcus bovis* and methanogens (Klieve et al. 1999; Bach et al. 2002). A little information is available on the genetic blueprint and gene functionality of archaeal, particularly methanogenic, phages, but more are being discovered using electron microscopy (Ackermann 2007) and in vitro techniques (Stanton 2007). McAllister and Newbold (2008) reported siphophages that can infect methanogens (*Methanobacterium*, *Methanobrevibacter*, and *Methanococcus* spp.) (Table 1.2), although these phages have not been isolated from the rumen. Despite the lack of knowledge, ruminant phage and their enzymes that are involved in lysing host cell represent significant opportunities for controlling both methanogen population and other community members in the rumen. In addition, it is possible to identify key host enzyme targets that are susceptible to inhibition by phage proteins understanding the life cycles of phage.

1.3 Microbial Interactions

The bionetwork inside the rumen is responsible for the complete degradation of organic matter ingested by the animals. There is no single organism that has the capacity for a complete degradation of complex substrates in the rumen (Bladen et al. 1961). A complex succession of microorganisms takes part in the cooperative catabolism of substrates in the rumen and production of fermentative end products. The complex feedstuffs are broken down into simple sugars, transported, and fermented by other members of the microbial population that in turn produce branched-chain fatty acids, vitamins, or other cofactors for other bacteria responsible for degrading feedstuffs (Allison et al. 1961; Bryant 1973). The community analysis approach to the breakdown of nutrients by a microbial consortium led to the discovery of the concept of cross feeding of nutrients across and within environmental niches and can support the overall health of the host animal. The steady supply of food and constant removal of digested feed material and end products along with the abovementioned rumen conditions allow a dense population of microorganisms to grow inside the rumen (Hungate 1966). Some bacteria and protozoa (i.e., holotrichs) are closely associated with the rumen wall (McAllister et al. 1994; Dehority 2003). The rumen wall-associated bacteria scavenge oxygen and break down urea into CO_2 and NH_3 that enter into the rumen via feed, saliva, and blood (McAllister et al. 1994). Further, some of those degrade the sloughing off of epithelial cells. The NH_3 is a source of nitrogen to the rumen bacteria, while CO_2 is an essential growth factor for the fiber-digesting bacteria. The extramural bacteria (99 % of the rumen bacteria) can be either free floating or attached to protozoa (i.e., symbiotic methanogens) or feed particles (Koike and Kobayashi 2009). About 75 % of the bacteria are associated with feed particles (Craig et al. 1987). Out of the feed particle-associated bacteria, approximately 80–90 % is responsible for fiber digestion, 75 % for protein digestion, and 70 % for starch digestion (Brock et al. 1982; Minato et al. 1993). The free-floating bacteria or bacteria associated with liquid material inside the rumen are about 30 % of the total bacteria inside the rumen (Legay-Carmier and Bauchart 1989). These can either become attached to the feed particles or simply washed away into the lower gastrointestinal tract and then become available as a good source of protein for the host animal.

Production of B vitamins by rumen bacteria is important to ensure bacterial growth and, particularly, fiber degradation and is also responsible for ensuring animal health and well-being (Scott and Dehority 1965). For example, a strain of the prominent rumen cellulolytic bacterium (*Ruminococcus albus*) was found to require phenylpropanoic acid (PPA) to digest cellulose (Stack and Hungate 1984). Rumen microbes produce PPA from the fermentation of phenolic compounds, and the lack of PPA in cultures meant that *R. albus* could not adhere to cellulose. The PPA was thought to act as a "glue" that holds the cellulolytic enzyme complex in a stable conformation, allowing for cellulose degradation (Stack and Hungate 1984). As researchers delineated bacterial functions and nutritional requirements, it became clear that there are significant interactions between microbial populations, both within an environmental niche and between niches. Application of this concept spreads into analysis of other microbial habitats (Schultz and Breznak 1979) and has shaped the view of how microbes interact with the world around those. From the time of these, it has been discovered that quorum signals play a role in other natural microbial environments (Khan et al. 1995; Miller and Bassler 2001; Sperandio et al. 2003) and that these biochemical messenger molecules can cause changes in microbial growth and fermentation. Furthermore, it now appears that a 2-way communication between the host and microbial population exists, meaning that the endocrine status of the host can have a direct impact on the gut microbial population (Sperandio et al. 2003; Lyte and Freestone 2009; Freestone and Lyte 2010; Bailey et al. 2011). This has, in turn, led to the developing concept of the "microbial organ" playing a critical role in animal production and health (Lyte 2010) and the theory that probiotics or the rumen microbial population could be used as a drug delivery vehicle to improve animal performance for higher productivity.

The increase in internal H_2 concentration has several harmful effects in lowering the rumen pH and overall inhibits the biodegradation by creating an unfavorable equilibrium (Ragsdale and Pierce 2008). But, some of the groups involved in the transfer of produced H_2 as reducing power. The process of syntrophic H_2 transfer was first hypothesized by Hungate (1966). Methanogens, a group of archaeal microbes, combine this molecular H_2 with CO_2 to produce CH_4. Both from environment and nutrient utilization point of view, methanogenesis is not a suitable sink as the end product is a greenhouse gas with 23 times global warming (Eggleston 2006) potential than CO_2 with a loss of 8–13 % digestible energy ingested (Moss et al. 2000). *Ruminococcus albus*, a major cellulose-digesting species in the rumen, produces ethanol, acetate, H_2, and CO_2 in the absence of methanogen but produces more acetate and less or no ethanol in the presence of methanogens (Iannotti et al. 1973). Because acetate production results in ATP synthesis, the interaction results in increased growth and cellulose digestion. This startling concept demonstrated that although excessive rumen CH_4 production was negative from the perspective of the host animal and the environment, a certain level of CH_4 production is necessary to remove reducing the equivalents, thereby, allowing the overall rumen fermentation to proceed. As a result, rumen fermentations will likely always continue to produce at least some CH_4. Major precursors of methanogens are acetate, H_2 and CO_2, the question of how propionate and butyrate would become CH_4 in anaerobic ecosystems other than the rumen arose. Several alternate approaches from diet to animal management have been employed since long, but most of them failed as being costly and noxious to host. Accordingly, microbial interventions that employ biological routes are ideal approaches to sequester these gaseous end products from the rumen.

1.4 Future of Rumen Microbiology

The rumen microbial systems have now been examined in detail for some decades by microbiologists, nutritionist, and biotechnologists. Always, the focus is to enhance the animal productivity by manipulating the dietary composition or by altering the microbial ecosystems. The upcoming strategies in discovery of novel gene/ microbes with genomic approaches will pave a manipulating way toward the developments of in-depth knowledge on microbial sequence signature. Application of highly active enzymes (i.e., endoglucanase, xylanase, esterases, etc.) from the rumen source for commercial applications will provide a new dimension in agro-industries. Genetic engineering, by adding a desired activity to an established rumen microorganism, may overcome some of the problems. But there still remain the few problems of how long a new microorganism, once established, will remain active. The control of rumen methanogenesis by inhibiting the active sites of different enzymes by the in silico prediction tools from the genome sequences will limit the research inputs in terms of manpower, energy, and time. Bioinformatics approaches used for the inhibitor prediction against the F_{420}-dependent NADP oxidoreductase enzyme that catalyzes an important electron transfer step in the methanogenesis from *Methanobrevibacter smithii* reveal that lovastatin and compactin had high affinity to the enzyme and can act as the potential inhibitors (Sharma et al. 2011). Both in silico approaches and in vitro enzyme assays may be useful for screening chemical inhibitors of methanogenesis. Most recent report of Lee et al. (2013) that only very few rumen methanogens are cultured as pure isolates and 13 genome projects are completed as yet (Morgavi et al. 2013). Most of these genome sequences are from the genus *Methanobrevibacter*, which is considered to be dominating rumen methanogen, as per the global data set of rumen microbes (Jeyanathan et al. 2011; Kumar et al. 2012). This will provide knowledge on how microbes and their metabolic products affect the gastrointestinal physiology and immune responses in the host and also will provide genetics of syntrophic interactions, e.g., interspecies hydrogen transfer, microbe-microbe interactions relevant to host nutrition and health, optimization of fiber utilization, and probiotic and prebiotic (synbiotic) developments.

1.5 Conclusions

The rumen microbial ecosystem has a significant impact on understanding the digestive physiology and also provides a test bed for other microbial environments in technology development and validation. This will also have a potential application in industry for a cost-effective ruminant feed production and management of enteric fermentation. Recent developments in microbial ecology have been adapted again and validated in the rumen, providing the ability to understand the gastrointestinal consortium at a level never dreamed of in the times of Hungate. Application of rumen microbial ecosystems will contribute much to the understanding of the ecology of rumen microorganism including the potential to elucidate and overcome the biochemical, ecological, or physiological processes that are rate limiting for rumen fiber degradation. By linking these data with animal diets and in vitro analysis, it may be possible to define key microbes and their role in the fiber degradation. The learning novel pathways will also lead to the knowledge about the currently non-culturable bacteria to culture. The modern genomics-based strategies to elucidate the microbial community structure, novel gene discoveries, and recombinant DNA technology will provide information for eco-friendly animal production and management.

References

Ackermann HW (2007) 5500 phages examined in the electron microscope. Arch Virol 152:227–243

Akin DE, Borneman WS (1990) Role of rumen fungi in fiber degradation. J Dairy Sci 73:3023–3032

Akin DE, Borneman WS, Windham WR (1988) Rumen fungi: morphological types from Georgia cattle and the attack on forage cell walls. Biosystems 2:385–391

Albers SV, Konings WN, Driessen JM (2007) Solute transport. In: Cavicchioli R (ed) Archaea: molecular and cellular biology. ASM Press, Washington, DC, pp 354–368

Allison MJ, Bryant MP, Doetsch RN (1961) Studies on the metabolic function of branched-chain volatile fatty acids, growth factors for *Ruminocci*. J Gen Microbiol 5:869–879

Asanuma N, Hino T (2000) Activity and properties of fumarate reductase in ruminal bacteria. J Gen Appl Microbiol 49:119–125

Bach SJ, Mcallister TA, Veira DM et al (2002) Transmission and control of *Escherichia coli* O157: H7 – a review. Can J Anim Sci 82:475–490

Bailey RW, Howard BH (1963) The biochemistry of rumen protozoa: the maltases of *Dasytricha ruminantium*, *Epidinium ecaudatum* (Crawley) and *Entodinium caudatum*. Biochem J 86:446

Bailey MT, Dowd SE, Galley JD et al (2011) Exposure to a social stressor alters the structure of the intestinal microbiota: implications for stressor-induced immunomodulation. Brain Behav Immun 25:397–407

Barichievich EM, Calza RE (1990) Supernatant protein and cellulase activities of the anaerobic ruminal fungus *Neocallimastix frontalis* EB188. Appl Environ Microbiol 56:43–48

Bladen HA, Bryant MP, Doetsch RN (1961) A study of bacterial species from the rumen which produce ammonia from protein hydrolyzate. Appl Microbiol 9:175–180

Borneman WS, Hartley RD, Himmelsbach DS et al (1990) Assay for trans-p-coumaroyl esterase using a specific substrate from plant cell walls. Anal Biochem 190:129–133

Brock FM, Forsberg CW, Buchanan-Smith JG (1982) Proteolytic activity of rumen microorganisms and effects of proteinase inhibitors. Appl Environ Microbiol 44:561–569

Brookman JL, Ozkose E, Rogers S et al (2000) Identification of spores in the polycentric anaerobic gut fungi which enhance their ability to survive. FEMS Microbiol Ecol 31:261–267

Brul S, Stumm CK (1994) Symbionts and organelles in anaerobic protozoa and fungi. Trends Ecol Evol 9:319–324

Bryant MP (1973) Nutritional requirements of the predominant rumen cellulolytic bacteria. Fed Proc 32:1809–1813

Castro-Montoya JM, Makkar HPS, Becker K (2011) Chemical composition of rumen microbial fraction and fermentation parameters as affected by tannins and saponins using an in vitro rumen fermentation system rumen. Can J Anim Sci 91:433–448

Cavalier-Smith T (1987) Eukaryotes with no mitochondria. Nature 326:332–333

Chen H, Li XL, Ljungdahl LG (1995) Biomass degrading enzymes from anaerobic rumen fungi. SAAS Bull Biochem Biotechnol 8:1–6

Choudhury PK, Sirohi SK, Puniya AK et al (2012) Harnessing the diversity of rumen microbes using molecular approaches. In: Sirohi SK, Walli TK, Singh B, Singh N (eds) Livestock Green House Gases: emission and options for mitigation. Satish Serial Publishing House, Delhi, Section-B, Chapter-6, pp 65–82. ISBN 978-93-81226-52-065

Church DC (1969) Digestive physiology and nutrition of ruminants. OSU. Book Stores, Corvallis

Clarke RTJ (1977) Protozoa in the rumen ecosystem. In: Clarke RTJ, Bauchop T (eds) Microbial ecology of the gut. Academic, London, pp 251–275

Coleman GS (1975) The interrelationship between rumen ciliate protozoa and bacteria. In: McDonald IW, Warner ACI (eds) Digestion and metabolism in the ruminant. University of New England Publishing Unit, Armidale, pp 149–164

Coleman GS (1983) The cellulolytic activity of thirteen species of rumen entodiniomorphid protozoa. J Protozool 30:36A

Craig WM, Broderick GA, Ricker DB (1987) Quantitation of microorganisms associated with the particulate phase of ruminal ingesta. J Nutr 117:56–62

Dagar SS, Kumar S, Mudgil P, Singh R, Puniya AK (2011) D1/D2 domain of large subunit rDNA for differentiation of *Orpinomyces* spp. Appl Environ Microbiol 77(18):6722–6725

Davies DR, Theodorou MK, Lawrence MI (1993) Distribution of anaerobic fungi in the digestive tract of cattle and their survival in faeces. J Gen Microbiol 139:1395–1400

Dehority BA (2003) Rumen microbiology. Nottingham University Press, Nottingham

Dey A, Sehgal JP, Puniya AK, Singh K (2004) Influence of anaerobic fungal culture (*Orpinomyces* sp) administration on growth rate, ruminal fermentation and nutrient digestion in calves. Asian Aust J Anim Sci 17:820–824

Eggleston H (2006) 2006 IPCC guidelines for national greenhouse gas inventories. Forestry 5:1–12

Embley TM, Horner DA, Hirt RP (1997) Anaerobic eukaryote evolution: hydrogenosomes as biochemically modified mitochondria? Trends Ecol Evol 12:437–441

Ferry JG, Kastead KA (2007) Methanogenesis. In: Cavicchioli R (ed) Archaea: molecular and cellular biology. ASM Press, Washington, DC, pp 288–314

Finlay BJ, Esteban G, Clarke KJ et al (1994) Some rumen ciliates have endosymbiotic methanogens. FEMS Microbiol Lett 117:157–162

Freestone P, Lyte M (2010) Stress and microbial endocrinology: prospects for ruminant nutrition. Animal 4:1248–1257

Goel G, Puniya AK, Singh K (2005a) A study on tannic acid resistance in ruminal streptococcal isolates. J Basic Microbiol 45:243–245

Goel G, Puniya AK, Singh K (2005b) Xylanolytic activity of ruminal *Streptococcus bovis* in presence of tannic acid. Ann Microbiol 4:295–298

Goel G, Kumar A, Beniwal V, Raghav M, Puniya AK, Singh K (2011) Degradation of tannic acid and purification and characterization of tannase from *Enterococcus faecalis*. Int Biodeterior Biodegrad 65:1061–1065

Gordon GLR, Phillips MW (1998) The role of anaerobic gut fungi in ruminants. Nutr Res Rev 11:1–36

Griffith GW, Ozkose E, Theodorou MK et al (2009) Diversity of anaerobic fungal populations in cattle revealed by selective enrichment culture using different carbon sources. Fungal Ecol 2:87–97

Gruninger RJ, Puniya AK, Callaghanc TM, Edwardsc JE, Youssef N, Dagar SS, Fliegerova K, Griffith GW, Forster R, Tsang A, McAllister T, Elshahed MS (2014) Anaerobic fungi (Phylum *Neocallimastigomycota*): advances in understanding of their taxonomy, life cycle, ecology, role, and biotechnological potential. FEMS Microbiol Ecol. doi:10.1111/1574-6941.12383

Heath IB, Kaminskyj SG, Bauchop T (1986) Basal body loss during fungal zoospore encystment: evidence against centriole autonomy. J Cell Sci 83:135–140

Hibbett DS, Binder M, Bischoff JF et al (2007) A higher-level phylogenetic classification of the fungi. Mycol Res 111:509–547

Ho YWBD (1995) Classification of anaerobic gut fungi from herbivores with emphasis on rumen fungi from Malaysia. Mycologia 87:655–677

Hungate RE (1966) The rumen and its microbes. Academic, New York, p 41

Iannotti EL, Kafkewitz D, Wolin MJ et al (1973) Glucose fermentation products of *Ruminococcus albus* grown in continuous culture with *Vibrio succinogenes*: changes caused by interspecies transfer of H_2. J Bacteriol 114:1231–1240

Imai S (1998) Phylogenetic taxonomy of rumen ciliate protozoa based on their morphology and distribution. J Appl Anim Res 13:17–36

Janssen PH, Kirs M (2008) Structure of the archaeal community of the rumen. Appl Environ Microbiol 74:3619–3625

Jeyanathan J, Kirs M, Rominus RS et al (2011) Methanogen community structure in the rumens of farmed sheep, cattle and red deer fed different diets. FEMS Microbiol Ecol 74:311–326

Joblin KN (1996) Options for reducing methane emissions from ruminants in New Zealand and Australia. In: Bouma WJ, Pearman GI, Manning MR (eds) Greenhouse: coping with climate change. CSIRO Publishing, Collingwood, pp 437–449

Joblin KN (1999) Ruminal acetogens and their potential to lower ruminant methane emissions. Aust J Agr Res 50:1307–1313

Kamra DN (2005) Rumen microbial ecosystem. Curr Sci 89:124–135

Khan S, Spudich JL, McCray JA et al (1995) Chemotactic signal integration in bacteria. Proc Natl Acad Sci U S A 92:9757–9976

Klieve AV, Swain RA (1993) Estimation of ruminal bacteriophage numbers by pulsed field gel electrophoresis and laser densitometry. Appl Environ Microbiol 59:2299–2303

Klieve AV, Heck GL, Prance MA (1999) Genetic homogeneity and phage susceptibility of ruminal strains of *Streptococcus bovis* isolated in Australia. Lett Appl Microbiol 29:108–112

Koike S, Kobayashi Y (2009) Fibrolytic rumen bacteria: their ecology and function. Asian Aust J Anim Sci 22:131–138

Koike S, Yoshitani S, Kobayashi Y (2003) Phylogenetic analysis of fiber-associated rumen bacterial community and PCR detection of uncultured bacteria. FEMS Microbiol Lett 229:23–30

Kopecny J, Hodrova B (1995) Pectinolytic enzymes of anaerobic fungi. Lett Appl Microbiol 20:312–316

Krause D, Denman AE, Mackie RI (2003) Opportunities to improve fiber degradation in the rumen: microbiology, ecology, and genomics. FEMS Microbiol Rev 27:663–693

Kumar S, Dagar SS, Puniya AK (2012) Isolation and characterization of methanogens from rumen of Murrah buffalo. Ann Microbiol 62(1):345–350

Kumar S, Dagar SS, Puniya AK, Upadhyay RC (2013a) Changes in methane emission, rumen fermentation in response to diet and microbial interactions. Res Vet Sci 94(2):263–268

Kumar S, Dagar SS, Sirohi SK, Upadhyay RC, Puniya AK (2013b) Microbial profiles, in vitro gas production and dry matter digestibility based on various ratios of roughage to concentrate. Ann Microbiol 63(2):541–545

Kumar S, Choudhury PK, Carro MD et al (2014) New aspects and strategies for methane mitigation from ruminants. Appl Microbiol Biotechnol 98:31–44

Lange M, Westermann P, Ahring BK (2005) Archaea in protozoa and metazoan. Appl Microbiol Biotechnol 66:465–474

Larue R, Yu Z, Parisi VA (2005) Novel microbial diversity adherent to plant biomass in the herbivore gastrointestinal tract, as revealed by ribosomal intergenic spacer analysis and rrs gene sequencing. Environ Microbiol 7:530–543

Lee JH, Kumar S, Lee GH et al (2013) *Methanobrevibacter boviskoreani* sp. nov., isolated from the rumen of Korean native cattle. Int J Syst Evol Microbiol 63:4196-4201

Legay-Carmier F, Bauchart D (1989) Distribution of bacteria in the rumen contents of dairy cows given a diet supplemented with soya-bean oil. Br J Nutr 61:725–740

Leng RA (1982) Modification of rumen fermentation. In: Hacker JB (ed) Nutritional limits to animal production from pastures. CAB, Farnham Royal, pp 427–453

Lindmark DG, Müller M (1973) Hydrogenosome, a cytoplasmic organelle of the anaerobic flagellate *Tritrichomonas foetus*, and its role in pyruvate metabolism. J Biol Chem 248:7724–7728

Ljungdahl LG (2008) The cellulase/hemicellulase system of the anaerobic fungus *Orpinomyces* PC-2 and aspects of its applied use. Ann N Y Acad Sci 1125:308–321

Lowe SE, Theodorou MK, Trinci APJ (1987) Isolation of anaerobic fungi from saliva and faeces of sheep. J Gen Microbiol 133:1829–1834

Luo Y, Pfister P, Leisinger T (2001) The genome of archaeal prophage ψm100 encodes the lytic enzyme responsible for autolysis of *Methanothermobacter wolfeii*. J Bacteriol 183:5788–5792

Lyte M (2010) The microbial organ in the gut as a driver of homeostasis and disease. Med Hypotheses 74:634–638

Lyte M, Freestone P (2009) Microbial endocrinology comes of age. Microbe 4:169–175

Mackie RI, McSweeney CS, Aminov RI (1999) Rumen. In: Encyclopedia of life sciences. Nature Publishing Company, London

McAllister TA, Newbold CJ (2008) Redirecting rumen fermentation to reduce methanogenesis. Aust J Exp Agric 48:7–13

McAllister TA, Bae HD, Jones GA et al (1994) Microbial attachment and feed digestion in the rumen. J Anim Sci 72:3004–3018

Michel V, Fonty G, Millet L et al (1993) In vitro study of the proteolytic activity of rumen anaerobic fungi. FEMS Microbiol Lett 110:5–9

Miller MB, Bassler BL (2001) Quorum sensing in bacteria. Annu Rev Microbiol 55:165–199

Minato H, Endo A, Higuchi M et al (1966) Ecological treatise on the rumen fermentation. I. The fractionation of bacteria attached to the rumen digesta solids. J Gen Appl Microbiol 12:39–52

Minato H, Ishizaki S, Adachi Y et al (1989) Effect on rumen microbial populations of ammonia treatment of rice straw forage for steers. J Gen Appl Microbiol 35:113–124

Minato H, Mitsumori M, Cheng KJ (1993) Attachment of microorganisms to solid substrates in the rumen. In: Proceedings of the MIE bioforum on genetics, biochemistry and ecology of lignocellulose degradation. Institut Pasteur, Paris, pp 139–145

Mitsumori M, Minato H (1997) Cellulose-binding proteins from rumen microorganisms. In: Onodera R, Itabashi H, Ushida K, Yano H, Sasaki Y (eds) Rumen microbes and digestive physiology in ruminants. Japan Scientific Societies Press/S. Karger, Tokyo/Basel, pp 35–45

Morgavi DP, Kelly WJ, Janssen PH et al (2013) Rumen microbial (meta)genomics and its application to ruminant production. Animal 7:184–201

Morvan B, Bonnemoy F, Fonty G et al (1996) Quantitative determination of H_2-utilizing acetogenic and sulfate-reducing bacteria and methanogenic archaea from digestive tract of different mammals. Curr Microbiol 32:129–133

Moss AR, Jouany JP, Newbold J (2000) Methane production by ruminants: its contribution to global warming. Ann Zootech 49:231–253

Mountfort DO, Asher RA (1989) Production of xylanase by the ruminal anaerobic fungus *Neocallimastix frontalis*. Appl Environ Microbiol 55:1016–1022

Nagpal R, Puniya AK, Singh K (2009) In-vitro activities of immobilized anaerobic rumen fungus *Caecomyces* sp for its use as direct-fed microbials. J Anim Feed Sci 18:758–768

Nagpal R, Puniya AK, Singh K, Sehgal JP (2010) Influence of bacteria and protozoa from the rumen of buffalo on in-vitro activities of anaerobic fungus *Caecomyces* sp isolated from the feces of elephant. J Yeast Fungal Res 1(8):152–156

Nagpal R, Puniya A, Sehgal J et al (2011) In vitro fibrolytic potential of anaerobic rumen fungi from ruminants and non-ruminant herbivores. Mycoscience 52:31–38

Novotna Z, Prochazka J, Simunek J et al (2010) Xylanases of anaerobic fungus *Anaeromyces mucronatus*. Folia Microbiol (Praha) 55:363–367

Odenyo AA, Osuji PO (1998) Tannin-tolerant ruminal bacteria from East African ruminants. Can J Microbiol 44:905–909

Orpin CG, Mathiesen SD, Greenwood Y et al (1985) Seasonal changes in the ruminal microflora of the high-arctic Svalbard reindeer (*Rangifer tarandus platyrhynchus*). Appl Environ Microbiol 50:144–151

Paul SS, Kamra DN, Sastry VR et al (2004) Effect of anaerobic fungi on in vitro feed digestion by mixed rumen microflora of buffalo. Reprod Nutr Dev 44:313–319

Paul SS, Deb SM, Punia BS et al (2010) Fibrolytic potential of anaerobic fungi (*Piromyces* sp.) isolated from wild cattle and blue bulls in pure culture and effect of their addition on in vitro fermentation of wheat straw and methane emission by rumen fluid of buffaloes. J Sci Food Agric 9:1218–1226

Pfister P, Wasserfallen A, Stettler R et al (1998) Molecular analysis of *Methanobacterium* phage ΨM2. Mol Microbiol 30:233–244

Phillips MW, Gordon GL (1988) Sugar and polysaccharide fermentation by rumen anaerobic fungi from Australia, Britain and New Zealand. Biosystems 21:377–383

Puniya AK, Salem AZM, Kumar S, Dagar SS, Griffith GW, Puniya M, Ravella SR, Kumar N, Dhewa T, Kumar R (2014) Role of live microbial feed supplements with reference to anaerobic fungi in ruminant productivity. J Integr Agric. doi:10.1016/S2095-3119(14)60837-6

Ragsdale SW, Pierce E (2008) Acetogenesis and the Wood-Ljungdahl pathway of CO (2) fixation. Biochim Biophys Acta 1784:1873–1898

Rezaeian M, Beakes GW, Parker DS (2004) Distribution and estimation of anaerobic zoosporic fungi along the digestive tracts of sheep. Mycol Res 108:1227–1233

Saxena S, Sehgal JP, Puniya AK, Singh K (2010) Effect of administration of rumen fungi on production performance of lactating buffaloes. Benefic Microbes 1(2):183–188

Schultz JE, Breznak JA (1979) Cross-feeding of lactate between *Streptococcus lactis* and *Bacteroides* sp. isolated from termite hindguts. Appl Environ Microbiol 37:1206–1210

Scott HW, Dehority BA (1965) Vitamin requirements of several cellulolytic rumen bacteria. J Bacteriol 89:1169–1175

Sharma A, Chaudhary PP, Sirohi SK et al (2011) Structure modeling and prediction of NADP oxidoreductase enzyme from *Methanobrevibacter smithii*. Bioinformation 6:15–19

Shelke S, Chhabra A, Puniya A et al (2009) In vitro degradation of sugarcane bagasse based ruminant rations using anaerobic fungi. Ann Microbiol 59:415–418

Singh N, Sirohi SK (2012) Biochemistry, bioenergetics and genetics of methanogenesis in rumen methanogenic archaea. In: Sirohi SK, Walli TK, Singh B, Singh N Livestock Green House Gases: Emission and options for mitigation. Satish Serial Publishing House, Delhi, Section-B, Chapter-8, pp 103–120. ISBN 978-93-81226-52-065

Sirohi SK, Pandey N, Singh B, Puniya AK (2010) Rumen methanogens: a review. Indian J Microbiol 50:253–262

Sirohi SK, Choudhury PK, Dagar SS et al (2012a) Isolation, characterization and fibre degradation potential of anaerobic rumen fungi from cattle. Ann Microbiol 63:1187–1194

Sirohi SK, Singh N, Dagar SS, Puniya AK (2012b) Molecular tools for deciphering the microbial community structure and diversity in rumen ecosystem. Appl Microbiol Biotechnol 95(5):1135–1154

Sirohi SK, Chaudhary PP, Singh N, Singh D, Puniya AK (2013a) The 16S rRNA and mcrA gene based comparative diversity of methanogens in cattle fed on high fibre based diet. Gene 523(2):161–166

Sirohi SK, Choudhury PK, Puniya AK, Singh D, Dagar SS, Singh N (2013b) Ribosomal ITS1 sequence-based diversity analysis of anaerobic rumen fungi in cattle fed on high fiber diet. Ann Microbiol 63(4):1571–1577

Sperandio V, Torres AG, Jarvis B et al (2003) Bacteria-host communication: the language of hormones. Proc Natl Acad Sci U S A 100:8951–8956

Stack RJ, Hungate RE (1984) Effect of 3 phenylpropanoic acid on capsule and cellulases of *Ruminococcus albus* 8. Appl Environ Microbiol 48:218–223

Stanton TB (2007) Prophage-like gene transfer agents: novel mechanisms of gene exchange for

Methanococcus, Desulfovibrio, Brachyspira, and *Rhodobacter* species. Anaerobe 13:43–49

Stewart CS, Bryant MP (1988) The rumen bacteria. In: Hobson PN (ed) The rumen microbial ecosystem. Elsevier, New York, pp 21–75

Thareja A, Puniya AK, Goel G, Nagpal R, Sehgal JP, Singh PK, Singh K (2006) In-vitro degradation of wheat straw by anaerobic fungi from small ruminants. Arch Anim Nutr 60:412–417

Theodorou MK, Mennim G, Davies DR et al (1996) Anaerobic fungi in the digestive tract of mammalian herbivores and their potential for exploitation. Proc Nutr Soc 55:913–926

Thiele JH, Chartrain M, Zeikus JG (1988) Control of interspecies electron flow during anaerobic digestion: role of floc formation in syntrophic methanogenesis. Appl Environ Microbiol 54:10–19

Trinci APJ, Davies DR, Gull K et al (1994) Anaerobic fungi in herbivorous animals. Mycol Res 98:129–152

Tripathi VK, Sehgal JP, Puniya AK, Singh K (2007a) Effect of administration of anaerobic fungi isolated from cattle and wild blue bull (*Boselaphus tragocamelus*) on growth rate and fibre utilization in buffalo calves. Arch Anim Nutr 61:416–423

Tripathi VK, Sehgal JP, Puniya AK, Singh K (2007b) Hydrolytic activities of anaerobic fungi from wild blue bull (*Boselaphus tragocamelus*). Anaerobe 13:36–39

Van der Giezen M (2002) Strange fungi with even stranger insides. Mycologist 16:129–131

Vogels GD, Hoppe WF, Stumm CK (1980) Association of methanogenic bacteria with rumen ciliates. Appl Environ Microbiol 40:608–612

Warner ACI (1956) Proteolysis by rumen microorganisms. J Gen Microbiol 14:749

Weller RA, Pilgrim AF (1974) Passage of protozoa and volatile fatty acids from the rumen of the sheep and from a continuous in vitro fermentation system. Br J Nutr 32:341–351

Whitman WB, Boone DR, Koga Y et al (2001) Taxonomy of methanogenic archaea. In: Garrity GM (ed) Bergey's manual of systematic bacteriology. Springer Verlag, New York, pp 211–213

Williams AG (1979) The selectivity of carbohydrate assimilation in the anaerobic rumen ciliate *Dasytricha ruminantium*. J Appl Bacteriol 47:511–520

Williams AG, Coleman GS (1985) Hemicellulose degrading enzymes in rumen ciliate protozoa. Curr Microbiol 12:85–90

Williams AG, Coleman GS (1992) The rumen protozoa. Springer, New York, pp 4–83

Wolin MJ, Miller TL (1988) Microbe-microbe interactions. In: Hobson PN (ed) The rumen microbial ecosystems. Elsevier, London, pp 343–459

Wollin MJ, Miller TL, Stewart CS (1997) Microbe-microbe interactions. In: Hobson PN, Stewart CS (eds) The rumen microbial ecosystem. Chapman and Hall, London, pp 467–488

Wright ADG, Klieve AV (2011) Does the complexity of the rumen microbial ecology preclude methane mitigation. Anim Feed Sci Technol 166–167:248–253

Yarlett N, Orpin CG, Munn EA et al (1986) Hydrogenosomes in the rumen fungus *Neocallimastix patriciarum*. Biochem J 236:729–739

Rumen Microbial Ecosystem of Domesticated Ruminants

2

Neeta Agarwal, Devki Nandan Kamra, and L.C. Chaudhary

Abstract

The research has indicated that ruminants and their gut microbiota both have co-evolved while adapting to their climatic and botanical environments. The microorganisms present inside the rumen of livestock have a profound influence on the conversion of feed into end products, which can impact on both the animal and environment. The present chapter deals with the different types of microbes present inside the rumen of domesticated animals and their activity along with future prospects in the area of rumen microbial ecosystems.

Keywords

Ruminants • Rumen microbes • Bacteria • Protozoa • Archaea • Fungi

2.1 Introduction

Rumen is a specialized chamber of gastrointestinal tract, where a complex microbial ecosystem thrives and extracts nutrients from a variety of poor quality lignocellulosic agricultural by products consumed by the animals. These nutrients are utilized by the microbes as well the host animal. Since the nutrient densities in most of the roughage feeds consumed by the ruminants are poor, therefore, to meet out nutrient demands, animals have to consume large quantities of such feeds. The important segments of the digestive tract are mouth cavity, oesophagus, complex stomach (rumen, reticulum, omasum, abomasum and pylorus) and intestines (duodenum, small intestine, caecum, colon and rectum). The other glands associated with the digestion are spleen, pancreas and liver.

The mouth of the ruminants is a cone-shaped tubular structure with a narrow anterior opening and wide posterior end. It contains jaws, cheeks, lips, teeth, salivary glands and pharynx. The lips are short, thick and almost immobile in the ruminants. A long, muscular and mobile tongue with pointed tip is the main organ of prehension. The main salivary glands in bovines are parotid, mandibular or submaxillary and sublingual glands. The pharynx is a short and wide muscular passage between the mouth and the oesophagus. The mouth cavity opens into the oesophagus through the pharynx. It is a

N. Agarwal • D.N. Kamra (✉) • L.C. Chaudhary
Animal Nutrition Division, ICAR-Indian Veterinary Research Institute, Izatnagar 243 122, India
e-mail: dnkamra@rediffmail.com

A.K. Puniya et al. (eds.), *Rumen Microbiology: From Evolution to Revolution*, DOI 10.1007/978-81-322-2401-3_2

muscular tube and opens into the rumen perforating through the diaphragm. The size varies from 60 to 100 cm depending on the size of the animals. The main function of the oesophagus is it acts as a passage for food and water to the stomach and also during regurgitation.

Unlike monogastric, the stomach in the ruminants is divided into four compartments, i.e. rumen, reticulum, omasum and abomasum or true stomach or glandular stomach. The average proportion of the rumen, reticulum, omasum and abomasum of the complex stomach is about 80 %, 5 %, 3 % and 9 %, respectively. The capacity of various parts of GIT depends on the size of the animals. The volume of the rumen increases at a faster rate with the start of the intake of solid feeds and fodders. The complex stomach forms about 55 % of the digestive tract in adult ruminants.

2.2 Ruminant's Stomach

2.2.1 Rumen

The rumen is the largest compartment of the complex stomach occupying left half of the abdominal cavity. It has enormous space filled with digested and partially digested feed material. The feed is swallowed and regurgitated time and again. A combination of mechanical mastication and enzymatic action on the hard fibre-containing feeds results in breakdown into smaller pieces. Various pillars divide the rumen into different sacs. The pillars are folds of ruminal wall thickened with additional muscle fibres. The rumen is separated from the reticulum by the rumino-reticular fold. The mucosal surface of the rumen is mostly dark brown except the margins of the ruminal pillars where it is pale in colour. The oesophageal groove starts at the cardia and terminates at the reticulo-omasal orifice. In young calves, it forms a temporary tube by a reflex action at the time of suckling or drinking milk from the pail and provides a passage directly to the abomasum bypassing the rumino-reticulum.

2.2.2 Reticulum

The name has been derived from its honeycomb-like interior formed by numerous polygonal cells made of the folds of mucous membrane which are about 5 mm in length. The size of cells gradually diminishes towards the oesophageal groove, and these are almost absent towards the oesophageal groove as well as on the edge of the rumino-reticular fold. A peculiar claw-shaped curved papillae known as unguliform papillae are found at the reticulo-omasal orifice.

2.2.3 Omasum

It lies to the right of the rumen and is lined with leaf-shaped folds of the mucous membrane known as omasal laminae. There are different sizes of laminae, covered with numerous cornified papillae and are meant for pressing and grinding of feed particles. A groove between reticulo-omasal orifice and omaso-abomasal inlet is called sulcus omasi. It is covered with low folds and short papillae and provides a direct channel from the reticulum to the abomasum for the passage of liquid and fine feeds (chyme). The oval omaso-abomasal orifice is surrounded by a thick muscular omasal pillar and mucosal folds of abomasum on each side of the orifice which functions as valves for preventing regurgitation of the abomasal contents. The omasum helps in the absorption of water, and the variation in omasum dry matter was reported by Holtenius and Bjornhag (1989) as it was 0.6 %, 1.2 % and 3.3 % of the body weight in the sheep, goats and cows, respectively. The liquid leaving the reticulo-rumen passed quickly through the omasum to the abomasum in sheep (43 %), while it was lower in goats (17 %). There was a difference in delayed water absorption in omasum of sheep (15 %), goats (18 %) and cows (19 %).

2.2.4 Abomasum or True Stomach

This is the only compartment (also called the true stomach) with a glandular lining. Hydrochloric acid and digestive enzymes, needed for the breakdown of feeds, are secreted into the abomasums, which are comparable to the stomach of nonruminants. It may be differentiated into a capacious fundus partly attached with the reticulum, and the terminal small and narrower pyloric region joins the duodenum. The sphincters of the pylorus are very strong, and the passage of digesta through the pylorus is controlled by the pyloric valve. A round prominent structure on the upper part of valve is called torus pyloricus. Cardiac glands are found near the omaso-abomasal orifice. The secretions from these glands are thin and pH is always acidic due to production of HCl.

2.2.5 Intestine

The intestine is a very long coiled musculo-membranous tubular part of the alimentary canal with several modifications for specific functions. It is differentiated into the small intestine including duodenum and the large intestine consisting of caecum, colon and rectum.

2.3 Digestion of Feeds in the Rumen

Most of the feeds consumed by the ruminants are digested in the rumino-reticulum. The digestion in the mouth is mainly mechanical, while in the rumino-reticulum, the digestion is both mechanical and microbiological. There is no significance of host enzymes in the pre-abomasal compartments. The enzyme present in the ruminant saliva is a mild lipase of weak activity capable to hydrolyze some short-chain fatty acids like butyric acid and to some extent caproic acid. Its role is primarily limited to the digestion of short-chain lipids in pre-ruminant stage. The ruminants are capable to digest highly fibrous feeds with the help of fermentative digestion carried out by the microorganisms (bacteria, protozoa and fungi) present in the rumen and reticulum. The digestion is influenced by several factors like retention time of feeds in the rumino-reticulum, mastication and mixing of saliva, flow rate of digesta to omasum and lower part of the digestive tract, buffering action of saliva and continuous removal of the metabolites like volatile fatty acids (VFA), ammonia, carbon dioxide, and methane. Majority of VFA and ammonia are absorbed through the walls of the pre-abomasal compartments mainly by diffusion, and the gases CO_2 and CH_4 are removed during eructation. The highest gas production occurs 1–2 h of post-feeding depending on the chemical composition of the diets. The gases produced during fermentation are also absorbed into the respiratory tract. About one-third of ruminal gases are eructed through the mouth or nose, and the rest are expired from the trachea. A part of the fermentation products is utilized by the ruminal microbes for their nutritional supply.

The ingesta passing into the omasum contains unfermented feed residue, unabsorbed fermentation products and large microbial biomass. The processes of salivation, fermentation, absorption and flow of digesta are continuous while that of rumination and eructation are periodic. The contents in the rumino-reticulum are retained for a long time (30–50 h) depending on the nature of diets.

Cellulose, the major component of plant tissue, is degraded by the rumen microbes only, since the required enzymes (cellulases) are not secreted by the host. Therefore, for degradation of plant tissue, the animal is absolutely dependent on the rumen microbes. Approximately 35–55 % of the cellulose and hemicellulose are digested by the rumen microbes. More than 60 % of starch is also digested by the rumen microbes, depending on the amount fed and its rate of passage through the rumen. Almost all the sugars are digested in the rumen. Playne (1978) reported that the digestibility of dry matter of grass hay was higher in cattle (49.6 %) than that in sheep (34.6 %). This difference was neither due to intake nor due to diet. However, neutral-detergent fibre (NDF) was digested better by the cattle than by the sheep; for each kg of dry matter consumed, the cattle digested 415 g NDF and sheep 279 g NDF. This was because the cattle digested 60 % and 35 % more hemicellulose and cellulose than the sheep. Aerts et al. (1984) conducted 82 comparative digestion trials, involving 26 maize silages, 24 grass silages, 18 grass hays and 14 other feedstuffs, and reported that on an average, crude fibre of maize silage, ether extract and crude fibre of grass silage, N-free extracts and organic matter of grass hay were better digested by cows, whereas sheep had better ability to digest crude protein of grassland products. The mean net energy utilization was about 3 % higher in cows than sheep for each of the three forages. The organic matter digestibility of maize silage and grassland products was about 65 % in both cows and sheep. Forages of lower quality were better digested by cows. But Arman and Hopcraft (1975) conducted a series of digestibility trials on sheep, cattle and antelope and reported that there were no significant differences in the overall mean digestibilities of diets in cattle and sheep. They concluded that digestibility of nutrients in different animals differed with feed, feeding habits and digestive physiology of the animals.

Huston et al. (1986) conducted four experiments to study the factors affecting digestibility of forages in cattle, sheep, goats and white-tailed deer. The digestibility of dry matter was moderately higher with high-fibre diet in cattle than in deer. Digestibilities of the diet in sheep and goats were intermediate. They suggested that rate of digestion was related more to diet than to the animal species. The goats digested less per cent of feed consumed than cows and sheep. This difference was related to a faster turnover and shorter retention time in goats. Chaudhary et al. (1987) reported no difference in nutrient utilization ability of pregnant heifers of cattle (Jersey and Haryana) and Murrah buffaloes. Chanthakhoun et al. (2012) reported higher ability of swamp buffalo than cattle in digesting low-quality roughages, and this might be due to the higher number cellulolytic bacteria (*Ruminococcus albus*) in buffalo rumen. Jetana et al. (2009) demonstrated that Brahman cattle were better fibre utilizer than swamp buffaloes when fed on pineapple waste silage.

The carbohydrates are fermented into VFA (mainly acetic, propionic and butyric acids) which meets out about 50 % of the energy requirements of the ruminants. When large amounts of roughage are fed, the formation of acetic acid is increased (60–70 %), propionate is lowered (15–20 %), but butyric (5–15 %) acid is least affected. However, with the increase in concentrate in an animal diet, the proportion of acetic acid decreases, while the propionic acid increases. This change in molar proportion of VFA is normally associated with a reduction in milk fat.

The crude proteins (CP) consumed by the ruminants are recycled in the rumen with help of microbial consortia. Some of the protein escapes from breakdown in the rumen. The CP is converted into ammonia and amino acids. About 40–75 % of the CP of feed is degraded in the rumen. This breakdown depends on many factors (solubility of the protein, resistance to breakdown, rate of feed passage through the rumen). Some nonprotein nitrogen sources such as urea, ammonium salts, nitrates, etc., can also be utilized by the rumen microbes as a source of protein. Rumen microbes convert ammonia and organic acids into amino acids that are converted into microbial protein. A significant portion of these metabolites are utilized by the microbes for the synthesis of protein to be digested in the intestine by the host animals. There were differences between cattle and sheep in the blood urea and sulphate concentration and relatively lower excretion of N, P and Ca by the cattle suggested that the rumen microorganisms of the cattle were able to utilize these nutrients better than the sheep (Playne 1978).

Degradation and transformation of dietary lipids occur in the rumen through lipolysis and biohydrogenation. The dietary fats are mainly converted into saturated fatty acids like palmitic and stearic acids. Almost an entire amount of carotenes entering to the rumen are passed into the intestine where a part of carotenes are converted to vitamin A in the epithelial cells. All the vitamin B complex is synthesized in the rumen. Vitamins C and K are also produced in the rumen during microbial fermentation of feeds in the rumen.

The feeding behaviour of ovine and bovine is different, for example, sheep are exclusively grazers while goats are browsers. Cattle and buffaloes also belong to the category of grazers. Though there is difference in feeding behaviour, the structure of the gastrointestinal tract is quite similar in all the four animals. The difference in size and capacity of various parts of the gastrointestinal tract is mainly dependent on the size and weight of the animal. The digestive capacity of different feeds also depends on the rumen microbial system of the breed.

2.4 Microbial Ecosystem

2.4.1 Bacteria

The bacterial community present in the rumen is highly diverse, but the number of active bacteria depends upon the animal species, type and chemical composition of diet, frequency of feeding and many more identified or unidentified factors. Under Indian feeding system of ruminants, the bacteria in the rumen of domesticated animals are primarily Gram-negative, strictly anaerobic, buffered around a pH of 6.5–6.9, partially halophiles and optimized at a temperature of 39 °C. These bacteria are represented by cellulose, hemicellulose and starch degraders, lipid, protein and urea utilizers, tannin and saponin detoxifiers. In addition to that, methanogenic archaea are also an important component of the ecosystem. At any time in the rumen of animals, only 15–20 genera are present in sufficiently large numbers (more than one million cells per ml of rumen liquor) to perform any important function in the rumen. But as soon as a new component in the diet is included, the bacteria required for its utilization/detoxification increase in numbers and shift the fermentation in that direction, where the new component is utilized or detoxified. When the diet of Holstein cows was shifted from hay- to grain-based diet, within 3 days, there was a 20- and tenfold decrease in the numbers of cellulose-degrading bacteria, e.g. *F. succinogenes* and *R. flavefaciens*, and those of starch-utilizing bacteria like *P. ruminicola* and *P. bryantii* increased seven- and 263-fold. The xylan-degrading bacteria *E. ruminantium* also decreased in numbers by 14-fold on the shift of diet from hay to grain based (Tajima et al. 2001).

As far as the comparative studies on the rumen microbiome of the domesticated animals (like buffalo, cattle, sheep and goat) are concerned, there are only a few studies in which the comparison has been made simultaneously in all species fed on the same diet and under similar management conditions. But there are some studies in which large ruminants (buffalo and cattle) and small ruminants (sheep and goat) have been compared between the two animals of almost similar size.

Between sheep and goat, the latter is reported to be a better utilizer of lignocellulosic and tannin-rich feeds and, therefore, is able to survive in much harsher conditions as compared to other domesticated ruminants such as sheep, cattle and buffaloes. One of the major reasons might be higher retention time of digesta in the gut and more efficient recycling of urea in the rumen (Devendra 1989).

Selenomonas ruminantium is capable of growing on tannic acid as a sole energy source and has been isolated from feral goats browsing on high tannin-containing acacia species (Skene and Brooker 1995). Trans-inoculation of these bacteria from feral goats to domestic goats and sheep fed with tannin-rich foliage (*Acacia aneura*) increased feed intake and nitrogen retention in inoculated animals as compared with uninoculated ones. Similarly, the tannin-degrading/tannin-tolerating bacteria isolated from the goat rumen fed on tannin-rich diet on a medium containing 1 % tannic acid were found to be Gram-positive and exhibited tannase activity with a maximum of 29.0 units and were identified as *Streptococcus gallolyticus* (Accession no. HM771331). The isolate was able to increase in vitro pakar leaves' degradability by 9.7 % by inclusion of live cultures of isolate 6 in the incubation medium (Singh et al. 2011).

In goats fed with either green oats (*Avena sativa*) or tannin-rich pakar (*Ficus infectoria*) leaves, the dry matter intake was significantly higher ($p<0.05$), and digestibilities of DM, OM, CP, EE, NDF and ADF were reduced in experimental (leaves-fed group) as compared with the control group (green oats fed) animals. The rumen microbial profile as obtained by MPN technique showed no change in total bacterial population, but total fungi and cellulolytic bacteria were reduced ($p<0.05$), whereas tannin-degrading/tannin-tolerant bacteria increased with feeding of pakar leaves. Real-time PCR data revealed a decrease in *Ruminococcus flavefaciens*, an increase in methanogens and no change in the *Fibrobacter succinogenes* population by feeding of pakar leaves (Singh et al. 2011).

In the rumen, mimosine-degrading bacteria are a specialized group of bacteria performing a specific function of mimosine degradation/modification. Mimosine, an anti-nutritional compound present in *Leucaena leucocephala*, is a good source of protein, but due to the presence of mimosine, it is toxic for the animals. A microbe capable of hydrolyzing mimosine was identified as *Synergistes jonesii* (Allison et al. 1992). *S. jonesii isolated* from the rumen hydrolyses mimosine into 2,3 DHP which is a nontoxic substance. The trans-inoculation of *Synergistes jonesii* from Hawaiian goats to Australian goats enabled them to degrade mimosine present in leucaena leaves. However, the research on microorganisms has demonstrated that there is inconsistency in the effectiveness of the *S. jonesii* inoculum in degrading leucaena toxins. This might be either due to loss of the effectiveness of original inoculum of *S. jonesii* or due to functional diversity of different strains of *S. jonesii* due to geographic variations.

Exhaustive work done on rumen manipulation suggests that the bacterial diversity in the rumen is not very much related to the species, but rather is more influenced by the type of diet taken by the animal or geographical variation. For example, the specific groups of bacteria like tannin-degrading or mimosine-degrading bacteria are present in all the rumens but in very low number, but as the animals are provided with the respective feeds, the population of these bacteria rises tremendously, and the ecosystem starts working efficiently in utilizing the newly entering feed into the rumen.

2.4.2 Protozoa

Ciliate protozoa play a vital role in rumen fermentation. The protozoa have two types of functions, i.e. general functions of feed fermentation where lignocellulosic feed ingredients are fermented to volatile fatty acids and microbial proteins, which are used by the animals as a source of energy and amino acids. The second and the more important function being performed by the ciliate protozoa is the protection of easily fermentable carbohydrates (sugars and starch) from sugar-/starch-utilizing bacteria (Kamra 2005) so that organic acids are not produced in plenty immediately after the feeding of animals; however, these sugars are released slowly during the day so that there is a constant supply of energy for the animals in the form of short-chain volatile fatty acids.

Rumen protozoa constitute only a small fraction (10^4–10^6/ml depending upon the diet) of the total number of microbes (10^{10}–10^{11}/ml), but in terms of protozoal mass, it is almost equal to that of bacteria present in the rumen (Table 2.1). Therefore, rumen ciliate protozoa play a very significant role in the fermentation of feed and making the energy and protein available to the animals. Even among the protozoa (which are as small as 4–5 μm and as large as 250–300 μm in size), the relative contributions of a specific group, small spirotrichs, present in larger numbers, contribute only a small fraction of the total biomass as compared to holotrichs, which are smaller in numbers, but contribute a larger portion of protozoal biomass (Kamra et al. 1991).

The number of ciliate protozoa in the rumen content of buffalo, cattle, sheep and goat varied between 11.35 and 28.13×10^4/ml, representing

Table 2.1 Rumen microbial ecosystem

	Number/g rumen content	Mass (% of microbial mass)
Bacteria	10^{10}–10^{11}	40–50
Protozoa	10^4–10^6	40–50
Archaea	10^7–10^8	2–3
Fungi	10^3–10^5	3–4
Bacteriophages	10^8–10^9	<0.1

Table 2.2 Total protozoa count and generic composition of rumen ciliates in buffaloes, cattle, sheep and goats

Protozoa (no.)	Buffaloes (35)	Cattle (48)	Sheep (32)	Goats (35)
Total protozoa count (×10⁴/ml)	16.02±3.41	11.35±2.53	28.13±4.13	13.38±2.26
Entodinium (%)	89.50±2.62	90.40±1.87	90.40±	93.00±2.35
Diplodinium (%)	3.70±0.14	2.30±1.32	2.80±1.40	2.00±0.07
Epidinium (%)	2.00±0.34	0.70±0.51	3.60±1.62	0.50±0.03
Holotricha (%)	2.80±0.41	5.90±1.34	4.20±1.94	3.50±1.72
Ophryoscolex (%)	2.00±1.80	0.80±0.14	0.00±0.00	1.00±0.91
Holotrichs				
Buetschlia parva	+	–	+	–
Isotricha prostoma	+	+	–	+
Isotricha intestinalis	+	+	–	+
Dasytricha ruminantium	+	+	+	+
Charonina ventricularis	–	+	+	–
Spirotrichs				
Ent. caudatum f. caudatum	+	+	+	+
Ent. caudatum f. dubardi	–	+	–	+
Ent. longinucleatum	+	+	+	–
Diplodinium monocanthum	+	+	–	–
Diplodinium tetracanthum	–	+	–	–
Eudiplodinium magii	+	+	–	–
Ostrachodinium gracile	+	–	–	–
Metadinium medium	–	+	–	–
Elytroplastron bubali	–	+	–	–
Epidinium caudatum	+	+	+	+
Epidinium ecaudatum	–	+	+	+

Adapted from Baraka (2012)

9, 12, 6 and 7 genera and 22, 38, 14 and 19 species of protozoa, respectively (Table 2.2) (Baraka 2012), but their numbers might change beyond these limits depending upon many factors like chemical composition of diet, frequency of feeding, time of sampling, method of sampling, transportation of rumen liquor from animal sheds to the laboratory, etc. As the study has been conducted on sufficient numbers of animals (32–48 per type of animals), there are limited chances that the diversity of protozoa might change considerably.

The species of ciliate protozoa commonly observed in all the four animals are *Dasytricha ruminantium*, *Entodinium caudatum*, *Ent. exiguum*, *Epidinium caudatum* and *Epid. bicaudatum*, while there are seven species exclusively present in buffalo, ten in cattle and one each in sheep and goat. The exclusive representatives of the ciliate protozoa in different species of animals represent a very small fraction of the protozoa population, and therefore, these might not be able to affect rumen fermentation very significantly. The common protozoa species constituting a major part of the protozoa population in different animals might play a significant role in feed fermentation and nutrient availability to the animals.

2.4.3 Rumen Methanogens

Methanogens are one of the major components of domain Archaea represented in the rumen microbial ecosystem. In the rumen, methanogen population ranges from 10^8 to 10^9/g of rumen content. Previously, methanogens were grouped with the domain bacteria, but later on, based on peculiar cell wall structure and 16S rRNA sequence, they were classified in a separate domain Archaea (Woese and Fox 1977) though both bacteria and archaea share the same ancestors. Methanogens establish in the gastrointestinal tract within 1–3 days of birth, and within 3 weeks of age, they reach to the maximum population in lambs (Skillman et al. 2004). Methanogens are a unique group of microorganisms generating methane as a stoichiometric end product of their metabolism.

Methanogens are the essential part of rumen microbiome because they maintain steady-state fermentation in the rumen. During fermentation of monosaccharides in the rumen, hydrogen is generated which is used by the methanogens for methane generation in the rumen. Singhal et al. (2005) calculated the contribution of crossbred cattle, indigenous cattle, buffalo, goat and sheep in methane emission through enteric fermentation as 4.6 %, 48.5 %, 39.0 %, 4.7 % and 1.8 %, respectively. Normally, cattle and buffalo produce about 200–250 l, while sheep and goat produce about 30–40 l of methane per day. Productive animals produce more methane than the nonproductive animals. But the extent of methane production depends on many factors like the type of feed taken by the animal, frequency of feeding, breed of the animal and production potential of the animal, and one of the most important is methanogen archaeal community structure present in the rumen.

Looking at the economic losses due to methanogenesis in the rumen and its contribution to environmental pollution, methanogenic archaea became an area of interest for researchers. The animal nutritionists, microbiologists and biotechnologists joined hands in this venture to minimize methane emission by the livestock. The biochemical reactions and enzyme profile involved in methanogenesis are well identified and described. The major groups identified are Methanobacteriales, Methanococcales, Methanomicrob-iales, Methanosarcinales and Methanopyrales. Methanobacteriales are the predominant group and include *Methanobrevibacter* spp., *Methanobacterium* spp. and *Methanosphaera* spp. and constitute the major portion of methanogen community in majority of the ruminants. Methanobacteriales (which include *Methanobrevi-bacter* spp., *Methanobacterium* spp. and *Methanosphaera* spp. constituting 30–99 % of the archaeal community) being the most prevalent, Methanomicrobiales (which include *Methanomicrobium* spp.) are the second most prevalent methanogen in the rumen environment. Methanosarcinales (Nicholson et al. 2007; Wright et al. 2007) and Methanococcales are the least populated methanogens in the rumen. A total of 28 genera and 113 species have been identified in the rumen (Garrity et al. 2007). According to Leahy et al. (2010), previously, only three genera were identified from the rumen, but now, the rumen has 22 genera of methanogenic archaea as worked out by using advanced biotechnological tools.

In buffalo (*Bubalus bubalis*) fed on wheat straw and concentrate mixed diet, the total archaeal community comprised of 1.94 %, 0.72 % and 0.47 % of Methanomicrobiales, Methanobacteriales and Methanococcales, respectively, as retrieved from 16S rRNA clone library (Singh et al. 2013). Janssen and Kirs (2008), after reviewing the global data, described rumen methanogen archaea (92.3 %) structure as *Methanobrevibacter* (61.6 %), *Methanomicrobium* (14.9 %) and a large group of uncultured rumen archaea affiliated with Thermoplasmatales (15.8 %) and named this uncultured archaea group as rumen cluster C (RCC) for the first time.

In the rumen, some of the methanogens are closely associated with protozoa, and some are free living, and the association is species specific. Tymensen et al. (2012) described the community structure of free-living and protozoa-associated methanogens by analysing 16S rRNA and methyl coenzyme M reductase (*mcrA*) gene clone libraries. Both the groups comprised of *Methanobrevibacter*, *Methanomicrobium* and rumen cluster C (RCC), which is distantly related to Thermoplasmata. *Methanobrevibacter* was predominant in protozoa-associated methanogens, whereas *Methanomicrobium* were predominant in free-living methanogens. Phylogenetic analysis of archaeal 16S rRNA gene clone libraries from defaunated and different protozoa-inoculated sheep revealed the presence of a large cluster of uncultured archaeal sequences not related to the known cultured methanogens, which was mainly associated with protozoa of the family Ophryoscolecidae (Ohene-Adjei et al. 2007). The authors claim that this is the first report of the presence of such organisms in the rumen of sheep and has not yet been reported in cattle.

Association of methanogens with anaerobic fungi has been established recently (Jin et al. 2014). The rumen fungus was collected from the goat rumen and subcultured to obtain uniform colonies. The methanogens found with fungal cultures were the members of novel RCC cluster and were confirmed by using *mcr*A gene. The methanogen-associated fungus was identified as *Candidatus Methanomethylophilus alvus* but represents a very small part of RCC cluster. The authors suggested that this is a new approach to identify the novel methanogens. This was first reported in the goat rumen. This small group of methanogens associated with protozoa and fungi in the rumen of sheep is distantly related to Thermoplasmatales; hence, a new order Methanoplasmatales was proposed (Paul et al. 2012). Poulsen et al. (2013) observed decreased methylotrophic methanogen population by feeding rapeseed oil-supplemented diet to lactating cows. These methylotrophic methanogens degrade methylamines as they grew with the inclusion of methylamine in in vitro incubations. The authors suggested that these methylotrophic methanogens can be targeted to reduce methane emission in the ruminants. Still, the Thermoplasmata archaea are very poorly characterized.

It is believed that the diversity in methanogen community mainly varies with the variation in diet and geographical variation, and this variation is only in the form of shifts among the various groups of methanogens. Wright et al. (2004) compared rumen methanogen diversity in sheep on three different feeding systems, viz., grazing, oaten hay-based diet and lucerne hay-based diet. The predominant methanogen was *Methanobrevibacter* spp. on all the three feeding systems. Two new sequences were encountered from the grazing sheep and the authors proposed that a new order should be created for this group, but their functional contribution in the rumen has yet to be established. The 16S rRNA clone libraries from ten Hereford-cross cattle from Ontario and ten from Prince Edward Island were compared and showed predominance of Methanobacteriales (51.5 % clones), and about 42 % of the clones revealed a high degree of sequence similarity to three methanogen species, *Mbb. smithii*, *Mbb. ruminantium* and *Mbb. thaueri*, but did not find any sequence showing close similarity to *Methanobacterium*, *Methanomicrobium* or *Methanosarcina* (Wright et al. 2007). Screening of 241 clones from both the 16S rRNA libraries revealed 28 phylotypes in which only 5 were common in both, whereas 11 phylotypes were unique to cattle from Ontario (38 clones) and 7 phylotypes (42 clones) were unique to cattle from Prince Edward Island. The authors identified nine new genera belonging to orders Methanobacteriales and Methanosarcinales. The authors could not reach the conclusion that the variation in the methanogenic archaeal community of the two groups of cattle was due to difference in the feeding or it was geographical variation. Comparative studies of clone libraries conducted on sheep from CSIRO Western Australia and Queensland, Australia, revealed that in the former case, the rumen methanogens were dominated by *Mbb. gottschalkii* (75.3 %) and *Mbb. ruminantium* (19.5 %), whereas, in the latter case, the uncultured group was dominated (80.8 %) followed by *Mbb. gottschalkii* clade (9.0 %), *Methanomicrobium* (7.7 %) and *Mbb. ruminantium* at non-detectable levels (Wright et al. 2006, 2007). In sheep from Venezuela also, genus *Methanobrevibacter* was the predominant group with the highest number of *Mbb. gottschalkii* (Wright et al. 2008). The results indicated that there were so many factors which influenced the microbial community structure in the rumen like diet, environment, animal genotype and age (McSweeney et al. 2007).

The distribution pattern of methanogen archaea among various domesticated ruminants is almost similar with the only variation in the population density of different methanogen groups. Whatever shifts among various methanogen populations reported so far are mainly due to the difference in the diet of the animal or sometimes related to geographical region, and geographical variation is also up to some extent related to the feed available in that region.

2.4.4 Fungi

Rumen fungi constitute 5–8 % of total biomass indicating a very small population in the rumen and establish within 8–10 days of birth. Its presence in the rumen was established very late as it was confused with flagellate protozoa and was categorized with them (Liebetanz 1910; Braune 1913). It was only in 1975 that rumen fungi were identified for the first time due to the presence of chitin in its cell walls (Orpin 1975). The first rumen fungi were identified as *Neocallimastix frontalis* in the rumen of sheep which had both motile (zoospores) and non-motile zoosporangium (Orpin 1975). The anaerobic gut fungi have been classified as *Chytridiomycetes*, based on their thallus morphology (Orpin 1975) and the presence of chitin in their cell walls (Orpin 1976, 1977). Bauchop (1979) worked out a mode of action of rumen fungi in cattle and sheep and established their role in fibre digestion. The fungus was located throughout the gastrointestinal tract, but about 90 % of the total fungal population is present in the reticulo-rumen, and 10 % is distributed in the rest of the GIT (Davies et al. 1993). But it was expected that fungi did not grow in any other part but only in rumen and the presence of fungi in other parts originated from the rumen. The work on rumen fungi started very late because of the following reasons: its least significance in the rumen function, very low number, slow growth rate and difficulty to grow in the laboratory and to preserve for a long period. But later on, when its mode of action and enzyme profile were studied in detail, it was realized that rumen fungi were a better fibre degrader as compared to bacteria, but because of its very low numbers, this group was not given any importance in the past.

Rumen fungi have a unique mode of action. The zoospores attack freshly ingested lignocellulosic feed particles due to chemotactic movement. The zoospores generally prefer stomata, damaged surfaces and cut ends of the plant fragments for attachment and encystment (Orpin and Bountiff 1978). They attach through flagella, encyst and develop a rhizoidal system which penetrates the substrate with the help of polysaccharide-degrading enzymes. The attachment of zoospores is very fast usually within 15–30 min of incubation of feed in the rumen. The fungi are able to completely hydrolyze the non-lignified tissues of plants (mesophyll, parenchyma and phloem), and the lignified tissues (xylem and sclerenchyma ring tissue in the stem) are partially degraded (Akin and Benner 1988), but sclerenchyma tissue of the leaf blade (which is softer than that of stem tissue) is extensively hydrolyzed as reported by Akin et al. (1989). Thus, fungi degrade the plant tissue and reduce its tensile strength. The stems of Bermuda grass and alfalfa incubated with fungi were weakened by 57 % and 55 %, while bacteria could weaken only to an extent of 29 % and 42 %, respectively (Akin et al. 1990). Hillaire and Jouany (1989) reported that in the rumen simulation technique, the dry matter degradation of wheat straw by rumen bacteria was 15 % higher and that of neutral-detergent fibre was 30 % higher in the presence of fungi (*Neocallimastix* sp.) as compared to bacteria alone. Rumen fungi secrete an array of enzymes including esterases (feruloyl esterase, p-coumaryl esterase and acetyl esterase) which break the ester bonds between hemicelluloses and lignin, thus releasing free celluloses and hemicelluloses for the other microbes to attack (Wubah and Kim 1996; Yue et al. 2009).

Only six genera of rumen anaerobic fungi have been identified so far, namely, *Neocallimastix*, *Piromyces* (previously known as *Piromonas*), *Caecomyces* (previously known as *Sphaeromonas*), *Orpinomyces*, *Anaeromyces* (previously known as *Ruminomyces*) and *Cyllamyces*. Identification of rumen fungi is a tedious job because a small subunit of rRNA cannot be used for phylogenetic identification of genera or species of anaerobic fungi due to high degree of conserved sequence. The polymorphic and homoplasious internal transcribed spacer 1 (ITS1) region has been identified as a molecular marker to compare fungal community structure. Kittelmann et al. (2012) described ten different groups of fungi, by using different sets of primers in density gradient gel electrophoresis fingerprinting of ITS 1 region of rumen fungi of sheep, cattle and deer fed on a variety of diets. The authors suggested ten groups instead of six already known with four new ones named as SK1, SK2, SK3 and SK4. The authors were of the opinion that either these sequences not yet have been retrieved in any of the studies or they have been wrongly placed in the already known six genera. The diversity in rumen fungal community was also diet based. Structurally, they are present in two forms; one is polycentric where the nucleus migrates throughout the hyphal mass, and the other one is monocentric where the nucleus concentrates in a zoosporangium. The monocentric forms include *Neocallimastix*, *Piromyces* and *Caecomyces*, and the polycentric forms include *Orpinomyces* and *Anaeromyces*. Other morphological characteristics, such as shape and size of sporangia and zoospores, have been used in the characterization of anaerobic *Chytridiomycetes* (Barr et al. 1989; Breton et al. 1989; Orpin 1975, 1976). The various species of anaerobic fungi isolated from the rumen of buffalo, goat, sheep and cattle are mentioned in Table 2.3.

Table 2.3 Rumen anaerobic fungi reported in different ruminants

Genus	Species	Source(s)	Reference(s)
Caecomyces	*communis*	Sheep	Gold et al. (1988)
		Cattle	Orpin (1976), Wubah and Fuller (1991)
	sympodialis	Yellow cow	Chen et al. (2007)
	equi	Sheep	
Piromyces	*communis*	Sheep	Gold et al. (1988)
		Cow	Julliand et al. (1998)
	spiralis	Goat	Ho et al. (1993)
	polycephalus	Water buffalo	Chen et al. (2002)
	mae	Sheep	Wong et al (1995)
Neocallimastix	*frontalis*	Sheep	Orpin (1975), Heath et al. (1983)
		Cow	Orpin (1977)
	patriciarum	Sheep	Orpin and Munn (1986)
	hurleyensis	Cow	Webb and Theodorou (1988)
		Sheep	Webb and Theodorou (1991)
	variabilis	Cow	Ho et al. (1993)
Anaeromyces	*elegans*	Cow	Ho et al. (1990)
	mucronatus	Sheep	Breton et al. (1990)
		Cow	Breton et al. (1990)
Orpinomyces	*joyonii*	Sheep	Breton et al. (1989)
		Cow	Sridhar et al. (2010)
		Buffalo	
	intercalaris	Cow	Ho et al (1994), Sridhar et al. (2010)
		Buffalo	Sridhar et al. (2010)
	bovis	Cow	Barr et al. (1989)
Cyllamyces	*aberensis*	Cow	Ozkose et al. (2001)
	icaris	Buffalo	Sridhar et al. (2014)
Piromyces	*communis*	Sheep	Gold et al. (1988)
		Cow	Julliand et al. (1998)
	spiralis	Goat	Ho et al. (1993)
	polycephalus	Water buffalo	Chen et al. (2002)
	mae	Sheep	Wong et al. (1995)

Being very active fibre-degrading organisms in the rumen, even better than bacteria, work has been conducted to use the fungal isolates as a microbial feed additive. In majority of the reports, the feeding of anaerobic fungal isolates was found beneficial in terms of improved fibre digestibility (Lee et al. 2000; Paul et al. 2004a, b). Looking at the significance of rumen anaerobic fungi, their superior enzyme profile, researchers are looking more towards this group of microbes and trying to explore them as target for rumen manipulation.

References

Aerts JV, De Boever JL, Cottyn BG et al (1984) Comparative digestibility of feedstuffs by sheep and cows. Anim Feed Sci Technol 12:47–56

Akin DE, Benner R (1988) Degradation of polysaccharides and lignin by ruminal bacteria and fungi. Appl Environ Microbiol 54:1117–1125

Akin DE, Lyon CE, Windham WR et al (1989) Physical degradation of lignified stem tissues by ruminal fungi. Appl Environ Microbiol 55:611–616

Akin DE, Borneman WS, Lyon CE (1990) Degradation of leaf blades and stems by monocentric and polycentric isolates of ruminal fungi. Anim Feed Sci Technol 31:205–221

Allison MJ, Maybery WR, Stahl DA (1992) *Synergistes Jonesii* Gen. Nov. Sp. Nov. A rumen bacterium that degrades toxic pyridinediols. Syst Appl Microbiol 15:522–529

Arman P, Hopcraft D (1975) Nutritional studies on East African herbivores. 1. Digestibilities of dry matter, crude fibre and crude protein in antelope, cattle and sheep. Br J Nutr 33:255–264

Baraka TA (2012) Comparative studies of rumen pH, total protozoa count, generic and species composition of ciliates in camel, buffalo, cattle, sheep and goat in Egypt. J Am Sci 8(2):448–462

Barr DJS, Kudo H, Jackober KD et al (1989) Morphology and development of rumen fungi: *Neocallimastix* sp., *Piromyces communis*, *Orpinomyces bovis*, gen. nov. sp. nov. Can J Bot 67:2815–2824

Bauchop T (1979) Rumen anaerobic fungi of cattle and sheep. Appl Environ Microbiol 38:148–158

Braune R (1913) Untersuchungenuber die Wiederkauemagenvorkommendenprotozoen. Arch Protistenkd 32:111–170

Breton A, Bernalier A, Bonnemoy F, Fonty G, Gaillard-Martinie B, Gouet P (1989) Morphological and metabolic characterization of a new species of strictly anaerobic rumen fungus: *Neocallimastix joyonii*. FEMS Microbiol Lett 58:309–314

Breton A, Bernalier A, Dusser M (1990) *Anaeromyces mucronatus* nov.gen, nov. sp. A new strictly anaerobic rumen fungus with polycentric thallus. FEMS Microbiol Lett 70:177–182

Chanthakhoun V, Wanapat M, Kongmun P et al (2012) Comparison of ruminal fermentation characteristics and microbial population in swamp buffalo and cattle. Livest Sci 143:172–176

Chaudhary LC, Verma DN, Husain KQ (1987) Studies on the utilization of nutrients in Jersey, Haryana and Murrah heifers. N D J Agric Res 2:126

Chen YC, Hseu RS, Chien CY (2002) Piromyces *polycephalus* (Neocallimastigaceae), a new rumen fungus. Nova Hedwigia 75:409–414

Chen YC, Tsai SD, Cheng HL (2007) *Caecomyces sympodialis* sp. nov., a new rumen fungus isolated from *Bos indicus*. Mycologia 99(1):125–130

Davies DR, Theodorou MK, Lawrence MI et al (1993) Distribution of anaerobic fungi in the digestive tract of cattle and their survival in faeces. J Gen Microbiol 139:1395–1400

Devendra C (1989) Comparative aspects of digestive physiology and nutrition in goats and sheep. In: Proceedings of the VII international symposium on ruminant physiology, Sendai, p 100

Garrity GM, Lilburn TG, Cole JR et al (2007) Taxonomic outline of the bacteria and archaea. Part 1. The archaea, phyla Crenarchaeota and Euryarchaeota. Release 7.7. Michigan State University, Lansing. www.taxonomicoutline.org

Gold JJ, Heath LB, Bauchop T (1988) Ultrastructural description of a new chytrid genus of caecum anaerobes, *Caecomyces equi* gen. nov., sp. nov., assigned to the Neocallimasticaceae. BioSystem 21:403–415

Heath LB, Bauchop T, Skipp RA (1983) Assignment of the rumen anaerobe *Neocallimastix frontalis* to the Spizellomycetales (Chytridiomycetes) on the basis of its polyflagellate zoospore ultrastructure. Can J Bot 61:295–307

Hillaire MC, Jouany JP (1989) Effects of anaerobic rumen fungi on the digestion of wheat straw and the end products of microbial metabolism studies on a semi continuous in vitro system. In: Nolan JV, Leng RA, Demeyer DI (eds) The roles of protozoa and fungi in ruminant digestion. Penumble Books, Armidale, pp 269–272

Ho YW, Bauchop T, Abdullah N et al (1990) *Ruminomyces elegans* gen. et sp. nov., a polycentric anaerobic rumen fungus from cattle. Mycotaxon 38:397–405

Ho Y, Barr DS, Abdullah N et al (1993) *Piromyces spiralis*, a new species of anaerobic fungus from the rumen of goat. Mycotaxon 48:59–68

Ho YW, Abdullah N, Jalaludin S (1994) *Orpinomyces intercalaris*, a new species of polycentric anaerobic rumen fungus from cattle. Mycotaxon 50:139–150

Holtenius K, Bjornhag G (1989) The significance of water absorption and fibre digestion in the omasum of sheep, goats and cattle. Comp Biochem Physiol A Physiol 94:105–109

Huston JE, Rector BS, Ellis WC et al (1986) Dynamics of digestion in cattle, sheep, goats and deer. J Anim Sci 62:208–215

Janssen PH, Kirs M (2008) Mini review: Structure of the archaeal community of the rumen. Appl Environ Microbiol 74:3619–3625

Jetana T, Tasripoo K, Vongpipatana C et al (2009) The comparative study digestion and metabolism of nitrogen and purine derivatives in male, Thai, Swamp buffalo and Thai, Brahman cattle. Anim Sci J 80:130–139

Jin W, Cheng YF, Mao SY, Zhu WY (2014) Discovery of a novel rumen methanogen in the anaerobic fungal culture and its distribution in the rumen as revealed by real-time PCR. BMC Microbiol 14:104

Julliand V, Riondet C, De Vaux A et al (1998) Comparison of metabolic activities between *Piromyces citronii*, an equine fungal species and *Piromyces communis*, a ruminal species. Anim Feed Sci Technol 70:161–168

Kamra DN (2005) Rumen microbial ecosystem. Curr Sci 89:124–135

Kamra DN, Sawal RK, Pathak NN, Kewalramani N, Agarwal N (1991) Diurnal variations in ciliate protozoa in the rumen of blackbuck *(Antilope cervicapra)* fed green forages. Lett Appl Microbiol 13:165–167

Kittelmann S, Naylor GE, Koolaard JP et al (2012) A proposed taxonomy of anaerobic fungi (class Neocallimastigomycetes) suitable for large-scale sequence-based community structure analysis. PLoS ONE. www.plosone.org

Leahy SC, Kelly WJ, Altermann E et al (2010) The genome sequence of the rumen methanogen *Methanobrevibacter ruminantium* reveals new possibilities for controlling ruminant methane emissions. PLoS One 5(1):e8926, 20126622 Cit:7

Lee SS, Ha JK, Cheng KJ (2000) Influence of an anaerobic fungal culture administration on in vivo ruminal fermentation and nutrient digestion. Anim Feed Sci Technol 88:201–217

Liebetanz E (1910) Die parasitischen Protozoen des Widenkauermagens. Arch Protistenkd 19:19–80

McSweeney CS, Denman SE, Wright DG et al (2007) Application of recent DNA/RNA-based techniques in rumen ecology. Asian Aust J Anim Sci 20:283–294

Nicholson MJ, Evans PN, Joblin KN (2007) Analysis of methanogen diversity in the rumen using temporal temperature gradient gel electrophoresis: identification of uncultured methanogens. Microb Ecol 54:141–150

Ohene-Adjei S, Teather RM, Ivan M et al (2007) Post inoculation protozoan establishment and association patterns of methanogenic archaea in the ovine rumen. Appl Environ Microbiol 73:4609–4618

Crpin CG (1975) Studies on the rumen flagellate *Neocallimastix frontalis*. J Gen Microbiol 91:249–262

Orpin GC (1976) The characterization of the rumen bacterium Eadi's oval *Magnoovum* gen. nov. *eadii* sp. nov. Arch Microbiol 111:155–159

Orpin CG (1977) The occurrence of chitin in the cell walls of the rumen organisms *Neocallimastix frontalis, Piromonas communis* and *Sphaeromonas communis*. J Gen Microbiol 99:215–218

Orpin CG, Bountiff L (1978) Zoospore chemotaxis in the rumen phycomycete *Neocallimastix frontalis*. J Gen Microbiol 104:113–122

Orpin CG, Munn E (1986) *Neocallimastix patriciarum* sp. nov., a new member of the Neocallimasticaceae inhibiting the rumen of sheep. Trans Br Mycol Soc 86:178–181

Ozkose E, Thomas BJ, Davies DR et al (2001) *Cyllamyces aberensis gen*. nov. sp. nov., a new anaerobic gut fungus with branched sporangiophores isolated from cattle. Can J Bot 79:666–673

Paul SS, Kamra DN, Sastry VRB et al (2004a) Effect of administration of an anaerobic gut fungus isolated from wild blue bull (*Boselaphus tragocamelus*) to buffaloes (*Bubalus bubalis*) on in vivo ruminal fermentation and digestion of nutrients. Anim Feed Sci Technol 115:143–157

Paul SS, Kamra DN, Sastry VRB et al (2004b) Effect of anaerobic fungal isolates from wild and domestic ruminants on in vitro digestibility, fermentation characteristics and fibrolytic enzyme feed by production on lignocellulosic mixed rumen microflora of buffalo. Nutr Reprod Dev 44:313–319

Paul K, Nonoh JO, Mikulski L, Brune A (2012) Methanoplasmatales, Thermoplasmatales-related archaea in termite guts and other environments, are the seventh order of methanogens. Appl Environ Microbiol 78:8245–8253

Playne MJ (1978) Differences between cattle and sheep in their digestion and relative intake of a mature tropical grass hay. Anim Feed Sci Technol 3:41–49

Poulsen M, Schwa BC, Jensen BB et al (2013) Methylotrophic methanogenic Thermoplasmata implicated in reduced methane emissions from bovine rumen. Nat Commun 4:1428. doi:10.1038/ncomms2432

Singh B, Chaudhary LC, Agarwal N et al (2011) Phenotypic and phylogenetic characterisation of tannin degrading/tolerating bacterial isolates from the rumen of goats fed on pakar (*Ficus infectoria*) leaves. J Appl Anim Res 39:120–124

Singh KM, Tripathi AK, Pandya PR et al (2013) Molecular genetic diversity and quantitation of methanogen in ruminal fluid of buffalo (*Bubalus bubalis*) fed ration (wheat straw and concentrate mixture diet). Genetics Research International, Article ID 980191. 10.1155/2013/980191

Singhal KK, Mohini M, Jha AK et al (2005) Methane emission estimates from enteric fermentation in Indian livestock: dry matter intake approach. Curr Sci 88:119

Skene IK, Brooker JD (1995) Characterization of tannin acylhydrolase activity in the ruminal bacterium *Selenomonas ruminantium*. Anaerobe 1:321–327

Skillman LC, Evans PN, Naylor GE et al (2004) 16S ribosomal DNA-directed PCR primers for ruminal metha-

nogens and identification of methanogens colonising young lambs. Anaerobe 10:277–285
Sridhar M, Kumar D, Anandan S et al (2010) Morphological and molecular characterization of rumen fungi belonging to the genus *Orpinomyces* isolated from Indian cattle and buffaloes. Res J Microbiol 5:581–594
Sridhar M, Kumar D, Anandan S (2014) *Cyllamyces icaris sp. nov.*, a new anaerobic gut fungus with nodular sporangiophores isolated from Indian water buffalo (*Bubalus bubalis*). Int J Curr Res Aca Rev 2:7–24
Tajima K, Aminov RI, Nagamine T et al (2001) Diet-dependent shifts in the bacterial population of the rumen revealed with real-time PCR. Appl Environ Microbiol 67:2766–2774
Tymensen LD, Beauchemin KA, McAllister TA (2012) Structures of free-living and protozoa-associated methanogen communities in the bovine rumen differ according to comparative analysis of 16S rRNA and mcrA genes. Microbiol 158:808–1817
Webb J, Theodorou MK (1988) A rumen anaerobic fungus of the genus *Neocallimastix*: ultrastructure of the polyflagellate zoospore and young thallus. Biosystems 21:393–401
Webb J, Theodorou MK (1991) *Neocallimastix hurleyensis* sp. nov., an anaerobic fungus from the ovine rumen. Can J Bot 69:1220–1224
Woese CR, Fox G (1977) Phylogenetic structure of the prokaryotic domain: the primary kingdoms. Proc Natl Acad Sci U S A 74:5088–5090
Wong MVL, Ho YW, Tan SG et al (1995) Isozyme and morphological characteristics of the anaerobic fungus *Piromyces mae* isolated from the duodenum, rumen and faeces of sheep. FEMS Microbiol Lett 134:9–14
Wright A-DG, Williams AJ, Winder B et al (2004) Molecular diversity of rumen methanogens from sheep in Western Australia. Appl Environ Microbiol 70:1263–1270
Wright A-DG, Toovey AF, Pimm CL (2006) Molecular identification of methanogenic archaea from sheep in Queensland, Australia, reveal more uncultured novel archaea. Anaerobe 12:134–139
Wright A-DG, Auckland CH, Lynn DH (2007) Molecular diversity of methanogens in feedlot cattle from Ontario and Prince Edward Island, Canada. Appl Environ Microbiol 73:4206–4210
Wright A-DG, Ma X, Obispo NE (2008) Methanobrevibacter phylotypes are the dominant methanogens in sheep from Venezuela. Microb Ecol 56:390–394
Wubah DA, Fuller MS (1991) Studies on *Caecomyces communis*: morphology and development. Mycologia 83:303–310
Wubah DA, Kim SH (1996) Chemoattraction of anaerobic ruminal fungi zoospores to selected phenolic acids. Microbiol Res 151:257–262
Yue Q, Yanga HJ, Cao YC et al (2009) Feruloyl and acetyl esterase production of an anaerobic rumen fungus *Neocallimastix* sp. YQ2 effected by glucose and soluble nitrogen supplementations and its potential in the hydrolysis of fibrous feedstuffs. Anim Feed Sci Technol 153:263–277

3 Domesticated Rare Animals (Yak, Mithun and Camel): Rumen Microbial Diversity

S.S. Paul and A. Dey

Abstract

The importance of rumen microbial ecology and diversity has gained significance in recent years as microbes have a profound effect on bioconversion of feed to end products which in turn affect productivity and also the environment. Recent research indicates that herbivores and their gut microbes have co-evolved while adapting to their climate and botanical environment. Rare domesticated ruminants like yak, mithun and camel are adapted to hostile environment and are of economic and social significance to more remote communities that rely heavily on the productivity of these rare ruminant livestock for their wellbeing. This review presents an account of the progress that has been made in rumen microbiology research and our current understanding of the rumen microbial ecosystem and its role in improving productivity of livestock and the environment with respect to yak, mithun and camel. The amount of information on rumen microbial diversity of these animals is limited. Only recently, high-throughput techniques are being employed to understand structural and functional diversity of rumen microbes of these animals. Limited data available so far indicate that there is a substantial host specificity in community structure of rumen microbes of these rare domesticated animals.

Yak (*Bos grunniens*) and mithun (*Bos frontalis*) are found at high altitude and have a grazing habit. They usually migrate to low altitude in search of pastures in the winter season. The scarcity of natural pasture at high altitude developed the animals to extract more nutrients from feed by increasing retention time and ability of microbes of rumen habitat for production of an array of enzymes for utilization of fibrous feeds. Camels (*Camelus dromedarius*) are adapted to extreme temperature and rely on scarce natural forages for their nutritional needs.

S.S. Paul (✉) • A. Dey
Rumen Microbiome Laboratory, Division of Animal Nutrition and Feed Technology, Central Institute for Research on Buffaloes, Hisar 125 001, India
e-mail: sspaulcirb@rediffmail.com

A.K. Puniya et al. (eds.), *Rumen Microbiology: From Evolution to Revolution*, DOI 10.1007/978-81-322-2401-3_3

Keywords
Camel • Mithun • Yak • Rumen • Diversity • Microbe • Nutrition

3.1 Yak

Yak (*Bos grunniens*) is an incredible bovine species of economic importance in high hill and snow-bound areas. Distinct from other bovine species, yak has been considered as multipurpose animal and provides milk, meat, fibre, hide and dung at places where arable farming, including other livestock, is unavailable. Animal power of yak is also exploited for transportation of goods in hilly topography. Nutritional management especially during winter season is extremely important when yak farmers come down to the mid-altitude along with their animals and face acute feed and fodder crisis.

3.1.1 Intake and Nutrition of Yak

The volume of the yak rumen contents by using polyethylene glycol (PEG) as a marker revealed that for a yak of 150 kg body weight, the rumen content varied from 32.3 to 35.8 l. The capacity of rumen is smaller than that of cattle; therefore, they consume less feed (Han et al. 1990). The animal has adapted itself to extract more nutrients from feeds by increasing the retention time of feed in the gastrointestinal tract. The average retention time of feed is about 72 h, which is about 20 % higher than that observed in domesticated cattle (Schaffer et al. 1978).

3.1.2 Microbial Ecology

Like other ruminants, bacteria, protozoa and fungi are the predominant microbes present in the yak rumen and are responsible for digestion of fibrous feeds. Among the protozoa, ciliates are the predominant group. Earlier study (Dogiel 1934) on Mongolian yak reported to have 8 genera including 8 species with 4 formae, viz. *Isotricha prostoma*, *Entodinium simplex*, *Diplodinium anisacanthum f. anisacanthum*, *Eudiplodinium maggii*, *Metadinium medium*, *Ostracodinium obtusum f. obtusum*, *Enoploplastron triloricatum*, *Epidinium ecaudatum f. ecaudatum* and *Epidinium ecaudatum f. caudatum*. However, the composition of ciliates depends upon the type of feed they are offered and the management of yak. Cattle kept together in proximity to yak influence the ciliate numbers and species (Guirong et al. 2000). *Entodinium* was reported as the most predominant (51.9–61.0 %) ciliate in the rumen of yak, and total ciliate densities were reported as $0.7–8.5 \times 10^5$/ml (Guirong et al. 2000), which is much lower than those reported in cattle and buffaloes (Han 1984; Ito et al. 1994; Chaudhary et al. 2000). Good protein and carbohydrate sources increase the number of *Entodinia* (Hungate 1966). Poor-quality feeds consumed by yak are reflected by the lower number of *Entodinia* in the rumen with increased number of *Eudiplodinium* and *Metadinium* having the ability to utilize plant fragments (Ogimoto and Imai 1981).

Surveys of the 16S rRNA gene diversity showed that microorganisms in the yak rumen were less diverse than those in cattle rumen; however, a greater proportion was uncultured in yak (An et al. 2005). Thus, the yak rumen may harbour a unique microbiome for efficient conversion of fibrous feeds. An obligate anaerobic, Gram-negative, mesophilic cellulolytic bacterium, *Cellulosilyticum ruminicola*, was isolated by Cai and Dong (2010), which was able to hydrolyse cellulose and xylan to produce acetate as one of the major end products. The researchers also reclassified *Clostridium lentocellum* as *Cellulosilyticum lentocellum* based on the similarity of 16S rRNA sequence data of *Clostridium* with *Cellulosilyticum* and reassigned to the family *Lachnospiraceae*. The enzymatic activities, viz. endoglucanases, cellobiohydrolases, xylanases, mannanase, pectinases and feruloyl esterases and acetyl esterases, break the inter-bridge cross-link, and the enzymes that degrade the glycosidic bonds were reported from *Cellulosilyticum ruminicola*. This bacterium appears to produce polymer hydrolases that act on both soluble and crystalline celluloses. Approximately half of the cellulolytic activities, including cellobiohydrolase (50 %) and feruloyl esterase (45 %), and one third of xylanase (31 %) and endoglucanase (36 %) activities were bound to cellulosic fibres. The bacterium seems to degrade the plant-derived polysaccharides by producing individual fibrolytic enzymes, whereas the majority of polysaccharide hydrolases contain carbohydrate-binding module (Cai et al. 2010). Dai et al. (2012) based on metagenomic and bacterial artificial chromosome-based functional approaches reported that the majority of glycosyl hydrolase proteins in yak rumen were from *Bacteroides* and *Firmicutes* and only a few from *Fibrobacteres*. This implies that the *Bacteroides* and *Firmicutes* play a major role in plant cell wall degradation in the rumen of yak which is similar to the situation in rumen of other herbivores.

Phylogenetic analysis of yak rumen bacteria by constructing a 16S rRNA gene clone library revealed the presence of a diverse group of fibrolytic bacteria such as *Butyrivibrio fibrisolvens*, *Pseudobutyrivibrio ruminis*, *Ruminococcus flavefaciens*, *Succiniclasticum ruminis*, *Selenomonas ruminantium* and *Prevotella ruminicola*. *Proteobacteria* have also been isolated from yak rumen; however, sequence analysis indicated that more than half of the species harboured in the yak rumen belonged to the not yet cultured group (An et al. 2005). It was thus reported that yak possesses a diverse range of bacteria; some are different from cattle and the other ruminants (Yang et al. 2010a). Moreover, the dominant bacteria of yak mainly belonged to phyla *Firmicutes* and *Bacteroidetes*, and the known dominant bacteria included *Ruminococcus flavefaciens*, *Butyrivibrio fibrisolvens*, *Pseudobutyrivibrio ruminis*, *Schwartzia succinivorans* and *Clostridiales*, most of which are common rumen bacteria (Leng et al. 2011).

Rumen fungus *Neocallimastix* sp. YAK11 isolated from faeces revealed to have higher activities of feruloyl esterase, acetyl esterase and xylanase (Cao and Yang 2011) to utilize low-grade roughages in the high altitude.

Yak produces less methane in comparison to cattle fed on similar diets (Ding et al. 2010). Based upon 16S rRNA gene sequence data, it was revealed that yak possesses methanogens, significantly different from cattle. Common species of methanogens belong to genera *Methanobrevibacter*, *Methanomicrobium*, *Methanobacterium* and *Methanosarcina* (Jarvis et al. 2000). Recent study of Huang et al. (2012) revealed that the majority of methanogen sequences were very distantly related to *Methanomassiliicoccus luminyensis* and were found to belong to the Thermoplasmatales-affiliated Lineage C, a group of uncultivated and uncharacterized rumen archaea that is a distantly related sister group to the order Thermoplasmatales. A lesser per cent of Methanobacteriales, Methanomicrobiales and Methanosarcinales were reported in yak than cattle. *Methanobrevibacter olleyae*, *Methanobrevibacter ruminantium*, *Methanobrevibacter woesei*, *Methanobrevibacter smithii*, *Methanobrevibacter millerae*, *Methanobrevibacter gottschalkii* and *Methanobrevibacter thaueri* were reported in the yak as unique methanogens. However, clones related to *Methanimicrococcus blatticola* and *Methanomicrobium mobile* were found in both yak and cattle. Therefore, yak has higher methanogen diversity and significantly different methanogen community structures than cattle, which could explain the reduced methane emission from yak. Significantly high levels of propionic acid (Long 2003), which led to efficient energy utilization and lower acetate-propionate ratio (Huang et al. 2012), further suggest the lower methane production in this species.

3.2 Mithun

Mithun (*Bos frontalis*) is a massive semi-domesticated rare ruminant species mainly reared for meat, besides milk and leather. This strongly built hill animal of Southeast Asia plays an important role in the socio-economic and cultural life of the local population (Simoons 1984;

Mondal et al. 2010). Mithun is believed to have originated more than 8,000 years ago and considered to be descendent from wild Indian gaur (Simoons 1984; Mondal and Pal 1999). It is mainly found in the north-eastern hilly states of Arunachal Pradesh, Nagaland, Manipur and Mizoram of India. Besides, it is also available in small numbers in Myanmar, China, Bangladesh and Bhutan.

Mithun thrives on jungle forages, tree fodders, shrubs, herbs and other natural vegetations (Das et al. 2011). Though the animals are owned by the farmers, they are kept under natural forests in a semiwild condition. Farmers usually do not provide any additional feeding except common salt, occasionally during restraining of animals for some purposes. Phylogenetic analysis of 16S rRNA gene clone library of rumen bacteria of mithun fed on either mixed tree leaves- and rice straw-based diet or bamboo leaves- and twigs-based diet revealed the presence of a diverse group of fibrolytic bacteria (Deng et al. 2007b), most of which (60 %) are under the group of uncultured rumen bacteria (Das et al. 2014). The sequence similarity of clones was observed for *Prevotella ruminicola*, as a major fibrolytic bacterium in the rumen of mithun. Other bacteria reported were *Butyrivibrio fibrisolvens*, *Pseudobutyrivibrio ruminis*, *Succinivibrio dextrinosolvens*, *Ruminococcus flavefaciens*, *Sporanaerobacter acetigenes*, *Clostridiales* and *Bacteroidetes*. Therefore, four phyla, namely *Bacteroidetes*, *Firmicutes*, *Proteobacteria* and *Tenericutes*, were described for the rumen bacteria of mithun (Das et al. 2014). However, the presence of *Ruminococcus albus*, *R. flavefaciens*, *R. bromii*, *R. gnavus*, *Eubacterium cellulosolvens*, *Clostridium fusiformis*, *Butyrivibrio fibrisolvens*, *Quinella ovalis*, *Clostridium symbosium*, *Succiniclasticum ruminis*, *Selenomonas ruminantium*, *Pseudobutyrivibrio ruminis* and *Allisonella histaminiformans* was reported by Deng et al. (2007b) in mithun fed on bamboo leaves- and twigs-based diet. Comparative evaluation of rumen microbial community of mithun and Yunnan Yellow cattle fed with pelleted lucerne (*Medicago sativum*) revealed more numbers of total viable bacteria as well as cellulolytic and amylolytic bacteria with no difference in proteolytic bacteria and protozoa in mithun than cattle (Deng et al. 2007a).

The analysis of denaturing gradient gel electrophoresis (DGGE) profiles, identification of dominant bands and phylogenetic analysis of 16S rDNA sequences in DGGE profiles revealed that mithun has a diverse microbial population with less intraspecies similarities; most sequences appeared to be unculturable bacteria, accounting for 85 % of the total sequences (Leng et al. 2011). According to phylogenetic analysis, the rumen dominant bacteria of mithun were mainly phylogenetically placed within phyla *Firmicutes* and *Bacteroidetes*, and the known bacteria mainly belonged to the family *Lachnospiraceae*, *Ruminococcus flavefaciens* and *Clostridium celerecrescens* (Leng et al. 2011). Animal-specific bacteria, *Allisonella histaminiformans*, were also reported from mithun (Yang et al. 2010b).

3.3 Camel

Camelids developed independently of the ruminants; ruminants developed more recently in the evolutionary process than camelids. Both have large forestomachs with extensive microbial fermentation. Unlike true ruminants, the forestomach of camels consists of three different sections. The feeding behaviour of camels showed considerable differences in the microbial activity in the forestomach as compared to other ruminants. Under desert conditions, camels browse on a range of forage plants that are of little nutritional value and high in tannins or are not palatable to cattle, sheep, goats and other herbivores (Kayouli et al. 1993). Camels have a complex gut microbiome that includes bacteria, archaea, protozoa and fungi to coordinate plant biomass breakdown.

Taxonomic analysis of the metagenomic reads indicated that camel faecal microbiome was dominated by the bacterial domain and phylum *Firmicutes* with major genera as *Clostridium.* In adult camels, the next major phylum was *Bacteroidetes* followed by *Verrucomicrobia* and

Proteobacteria (Sena and Patil 2013). Other bacteria identified from camel faecal metagenomes were *Bacteroides fragilis*, *Eubacterium eligens*, *Prevotella ruminicola*, *Pseudoflavonifractor* sp., *Ruminococcus albus*, *Butyrivibrio* sp., etc. Functional metagenomic studies revealed that genes associated with protein metabolism, RNA metabolism and carbohydrate metabolism were the most abundant in the systems (Sena and Patil 2013).

16S rRNA gene sequence of the foregut bacterial community in the dromedary camel (*Camelus dromedarius*) of central Australia revealed the major bacterium phylum as *Firmicutes* (67 %) that were related to the classes *Clostridia*, *Bacilli* and *Mollicutes* followed by *Bacteroidetes* (25 %) that were mostly represented by the family *Prevotellaceae*. The remaining phyla were represented by *Actinobacteria*, *Chloroflexi*, *Cyanophyta*, *Lentisphaerae*, *Planctomycetes*, *Proteobacteria* and *Spirochaetes*. Moreover, bacteria were identified as *Brevundimonas* sp., *Butyrivibrio fibrisolvens*, *Prevotella* sp. and *Ruminococcus flavefaciens* (Samsudin et al. 2012).

Taxonomic analysis of the metagenomic reads of gut microbes of Indian dromedary camel revealed *Bacteroidetes* (55.5 %), *Firmicutes* (22.7 %) and *Proteobacteria* (9.2 %) phyla as predominant camel rumen taxa. *Bacteroides* sp. dominated the camel rumen metagenome. Functional analysis exposed that clustered based subsystems and carbohydrate metabolism were the most abundant SEED subsystems of camel metagenome (Bhat et al. 2013).

Whereas in all true ruminants *Entodinium* spp. is known to be the major ruminal ciliate protozoa, in camels *Diplodinium* sp. was the dominant component of ciliates, and levels of *Epidinium*, *Holotricha* and *Ophryoscolex* spp. were higher than that in other ruminants (Baraka and Dehority 2003). Four camel-specific protozoan species were reported, namely, *Buetschlia neglectum*, *Buetschlia omnivore*, *Dasytricha kabbani* and *Caloscolex camelinus* (Baraka 2012).

16S rRNA gene-based high-throughput analysis of rumen microbiome indicated that a proportion of unclassified and uncultured microorganisms are much higher in domesticated rare ruminants especially in yak and mithun as compared to cattle, buffalo, sheep and goats. Ruminal protozoal community structure in domesticated rare ruminants is substantially different from that of cattle. In the case of yak, density of *Entodinium* genus was shown to be much lower than that of cattle. In the case of camel, *Diplodinium* sp. was shown to be predominant as compared to *Entodinium* sp. found in domesticated ruminants. Methanogen community in rumen of mithun was shown to be substantially different from those of cattle, thus contributing to lower methane and higher propionate production. Many host-specific novel organisms have also been reported from these animals. The substantial difference in nature of diet and physiology and anatomical difference of rumen (like rate of passage of digesta) might have contributed to the development of host-specific microbial structure in these rare domesticated ruminants. More research efforts need to be directed to characterize structural and functional diversity of gut microbes of these animals.

References

An D, Dong X, Dong Z (2005) Prokaryote diversity in the rumen of yak (*Bos grunniens*) and Jinnan cattle (*Bos taurus*) estimated by 16 s rDNA homology analyses. Anaerobe 11:207–215

Baraka TA (2012) Comparative studies of rumen PH, total protozoa count, generic and species composition of ciliates in camel, buffalo, cattle, sheep and goats in Egypt. J Am Sci 8(2):448–462

Baraka TA, Dehority BA (2003) Diagnostic value of forestomach ciliate protozoa of dromedary camels. J Egypt Vet Med Assoc 1:287–295

Bhat VB, Dande SS, Patil NV, Joshi CG (2013) Molecular analysis of the bacterial microbiome in the fore stomach fluid from the dromedary camel (*Camelus dromedarius*). Mol Biol Rep 40(4):3363–3371

Cai S, Dong X (2010) *Cellulosilyticum ruminicola* gen. nov., sp. nov., isolated from yak rumen contents, and reclassification of *Clostridium lentocellum* as *Cellulosilyticum lentocellum* comb. nov. Int J Syst Evol Microbiol 60:845–849

Cai S, Li J, Hu FZ et al (2010) *Cellulosilyticum ruminicola*, a newly described rumen bacterium that possesses redundant fibrolytic-pectin-encoding genes and degrades lignocellulose with multiple carbohydrate borne fibrolytic enzymes. Appl Environ Microbiol 76(12):3818–3824

Cao YC, Yang HJ (2011) Effect of roughage fibre content on fibrolytic activities and volatile fatty acid profiles of *Neocallimastix* sp. YAK11 isolated from rumen fluids of yak (*Bos grunniens*). Anim Feed Sci Technol 170:284–290

Chaudhary LC, Kamra DN, Singh R et al (2000) Effect of feeding bentonite on rumen fermentation, enzyme activity and protozoal population in adult buffaloes (*Bubalus bubalis*). Buffalo J 16:73–79

Dai X, Zhu Y, Luo Y et al (2012) Metagenomic insights into the fibrolytic microbiome in yak rumen. PLoS One 7(7):e40430

Das KC, Haque N, Baruah KK et al (2011) Comparative nutrient utilization, growth, and rumen enzyme profile of Mithun (*Bos frontalis*) and Tho-tho cattle (*Bos indicus*) fed on tree leaves based ration. Trop Anim Hlth Prod 43(1):209–214

Das KC, Paul SS, Sahoo L et al (2014) Bacterial diversity in the rumen of mithun (*Bos frontalis*) fed on mixed tree leaves and rice straw based diet. Afr J Microb Res 8(13):1426–1433

Deng W, Wang L, Ma S et al (2007a) Comparison of Gayal (*Bos frontalis*) and Yunnan yellow cattle (*Bos taurus*): rumen function, digestibilities and nitrogen balance during feeding of pelleted lucerne (*Medicago sativum*). Asian-Aust J Anim Sci 20(6):900–907

Deng W, Wanapat M, Ma S et al (2007b) Phylogenetic analysis of 16 s rDNA sequences manifest rumen bacterial diversity in Gayals (*Bos frontalis*) fed fresh bamboo leaves and twigs (*Sinarumdinaria*). Asian-Aust J Anim Sci 20(7):1057–1066

Ding X, Long R, Kreuzer M et al (2010) Methane emissions from yak (*Bos grunniens*) steers grazing or kept indoors and fed diets with varying forage: concentrate ratio during the cold season on the Qinghai-Tibetan Plateau. Anim Feed Sci Technol 162:91–98

Dogiel VA (1934) Angaben uber die Ophryoscolecidae des Wildschafes aus Kamtschatka des Elches und des Yaks, nebst deren Zoogeographischen Verwertung. Arch Protistenkd 82:290–297

Guirong SN, Hua Z, Se Z et al (2000) Rumen ciliated protozoan fauna of the yak (*Bos grunniens*) in China with the description of *Entodinium monuo n. sp.* J Eukaryotic Microbiol 47:178–182

Han SS (1984) Rumen ciliate protozoal fauna of the native cattle in Korea. Jpn J Zootech Sci 55:279–286

Han X, Xie A, Hu L (1990) Feed intake of growing yak. J Qinghai Anim Husb Vet Med 20(6):5–6

Huang XD, Tan HY, Long R et al (2012) Comparison of methanogen diversity of yak (*Bos grunniens*) and cattle (*Bos taurus*) from the Qinghai-Tibetan plateau, China. BMC Microbiol 12:237

Hungate RE (1966) The Rumen and its Microbes. Academic, New York

Ito A, Imai S, Ogimoto K (1994) Rumen ciliate composition and diversity of Japanese beef black cattle in comparison with those of Holstein- Friesian cattle. J Vet Med Sci 56:707–714

Jarvis GN, Strompl C, Burgess DM et al (2000) Isolation and identification of ruminal methanogens from grazing cattle. Curr Microbiol 40:327–332

Kayouli C, Jouany JP, Demeyer AA et al (1993) Comparative studies on the degradation and mean retention time of solid and liquid phases in the forestomachs of dromedaries and sheep fed on low quality roughages from Tunisia. Anim Feed Sci Technol 40:343–355

Leng J, Xie L, Zhu R et al (2011) Dominant bacterial communities in the rumen of Gayals (*Bos frontalis*), Yaks (*Bos grunniens*) and Yunnan Yellow Cattle (*Bos taurus*) revealed by denaturing gradient gel electrophoresis. Mol Biol Rep 38(8):4863–4872

Long R (2003) Yak nutrition- a scientific basis. In: Gerald WN, Han JL, Long R (eds) The Yak, 2nd edn. RAP Publication, Bangkok

Mondal SK, Pal DT (1999) Mithun: historical perspective. Asian Agri-History 3(4):245–260

Mondal M, Karunakaran M, Lee K et al (2010) Characterization of Mithun (*Bos frontalis*) ejaculate and fertility of cryopreserved sperm. Anim Reprod Sci 118:210–216

Ogimoto K, Imai S (1981) Atlas of rumen microbiology. Japan Science Society Press, Tokyo

Samsudin AA, Wright ADG, Jassim RA (2012) Cellulolytic bacteria in the foregut of the dromedary camel (*Camelus dromedarius*). Appl Environ Microbiol 78(24):8836–8839

Schaffer AL, Young BA, Chimwano AM (1978) Ration digestion and retention times of digesta in domestic cattle (*Bos taurus*), American bison (*Bison bison*) and Tibetan Yak (*Bos grunniens*). Can J Zool 56:2355

Sena DS, Patil NV (2013) Camel fecal metagenomics. In: ICAR sponsored short course on Metagenomics: role of next generation sequencing and bioinformatics. Anand Agricultural University, 15–24 Oct. pp 98–105

Simoons FJ (1984) Gayal or Mithun. In: Mason I, Longman L (eds) Evolution of domesticated animals. Longman, London

Yang LY, Chen J, Cheng XL (2010a) Phylogenetic analysis of 16 s rRNA gene sequences reveals rumen bacterial diversity in Yaks (*Bos grunniens*). Mol Biol Rep 37(1):553–562

Yang S, Ma S, Chen J et al (2010b) Bacterial diversity in the rumen of Gayals (*Bos frontalis*), Swamp buffaloes (*Bubalus bubalis*) and Holstein cow as revealed by cloned 16 s rRNA gene sequences. Mol Biol Rep 37(4):2063–2073

4 Wild Ruminants

Suzanne L. Ishaq and André-Denis G. Wright

Abstract

There are not many studies on the rumen microbial ecosystems of the wild ruminants. As these animals survive in natural conditions, which are sometimes very harsh, this might have a significant influence on the microbiome and the role played by it in fermentation of feed. This chapter reviews studies of the rumen microbiome of wild ruminants, including various species belonging to the families Antilocapridae, Bovidae, Cervidae, and Giraffidae.

Keywords

16S rRNAgene • 18S rRNA gene • Bacteria • Archaea • Protozoa • Fungi • Rumen • Gut microbiome

4.1 Introduction

There are relatively few studies on the microbiome of wild ruminants, despite what is currently known about their ability to digest toxic plant compounds, forage on low-quality feedstuffs, and survive in harsh environments. This is largely due to the difficulty in acquiring fresh rumen or even fecal samples from a large sample population. Fecal samples tend to be the easiest, cheapest, and least invasive collection method. However, fecal/colonic bacteria are not an efficient model for estimating rumen bacterial populations (Neumann and Dehority 2008; Ishaq and Wright 2012). The problem of sample viability has been somewhat ameliorated by the increasing sophistication and wider availability of culture-independent methods, although PCR-based methods generate many biases of their own. Many studies also report data in terms of operational taxonomic units (OTUs), which represent species-level clades, wherein all the sequences within an OTU are within 2–3 % genetic distance of each other, depending on the domain of the microorganism. A summary of the studies cited in this review can be found in Table 4.1.

S.L. Ishaq (✉) • A.-D.G. Wright
Department of Animal Science, College of Agriculture and Life Sciences, University of Vermont, 570 Main St, Burlington, VT 05401, USA
e-mail: slpelleg@uvm.edu

A.K. Puniya et al. (eds.), *Rumen Microbiology: From Evolution to Revolution*, DOI 10.1007/978-81-322-2401-3_4

Table 4.1 Summary of studies by wild ruminant species

Ruminant	Bacteria	Archaea	Protozoa	Fungi
Antilocapridae				
Pronghorn antelope		–	Dehority (1995)	Liggenstoffer et al. (2010)
Bovidae				
Antilope cervicapra Sable antelope	–	–	Kamra et al. (1991)	Liggenstoffer et al. (2010)
American bison	–	–	–	Liggenstoffer et al. (2010)
Bontebok	–	–	–	Liggenstoffer et al. (2010)
Eland	Nelson et al. (2003)	–	–	–
Grant's gazelle	Nelson et al. (2003)	–	–	Liggenstoffer et al. (2010)
Thompson's gazelle	Nelson et al. (2003)	–	–	–
Southern gerenuk	–	–	–	Liggenstoffer et al. (2010)
Goral	–	–	–	Liggenstoffer et al. (2010)
Greater kudu	–	–	–	Liggenstoffer et al. (2010)
Kafue lechwe		–	Imai et al. (1992)	–
Nile lechwe	–	–	–	Liggenstoffer et al. (2010)
Sassaby antelope	–	–	Ito et al. (2007)	–
Tsessebe	–	–	Van Hoven (1975)	–
Cervidae				
Sika deer	Li et al. (2013b)	Li et al. (2013a)	Ichimura et al. (2004)	Liggenstoffer et al. (2010)
Mule deer	–	–	Dehority (1995)	–
White-tailed deer	–	–	Dehority (1995)	–
Elk	–	–	Dehority (1995)	–
North American moose	Ishaq and Wright (2012)	–	Dehority (1974)	–
Norwegian reindeer	Sundset et al. (2007)	Sundset et al. (2009b)	Sundset et al. (2009b)	Sundset et al. (2009b)
Svalbard reindeer	Sundset et al. (2007), Pope et al. (2012)	Sundset et al. (2009a)	–	–
Giraffidae				
Giraffe	–	–	Kleyhans and Van Hoven (1976)	Liggenstoffer et al. (2010)
Okapi	–	–	–	Liggenstoffer et al. (2010)

4.2 Antilocapridae

The ruminant family Antilocapridae contains just one small (40–70 kg) living species in North America: the pronghorn (*Antilocapra americana)*. In one study on the rumen protozoa in pronghorn, Dehority (1995) found that several individuals contained either no visible protozoa or protozoa exclusively dominated by either *Entodinium dubardi* or *Entodinium ovibos*. In a separate study, the ruminal fungi were identified in the pronghorn as belonging to the genera *Piromyces* and *Anaeromyces* and from two novel groups NG3 and NG5, which may represent new genera (Liggenstoffer et al. 2010). To date, no studies on rumen bacteria or archaea exist for pronghorns. See Table 4.1.

4.3 Bovidae

The Bovidae family contains over 50 genera and more than 100 species of grazing and browsing ruminants around the world, with most species occurring in Africa and Asia. Common bovids include cattle, oxen, buffalo, sheep, goats, and antelope. Antelope is the largest category, which contains over two-thirds of all species of bovid. The family Bovidae is divided into three subfamilies, the Bovinae, the Cephalophinae, and the Antilopinae. Within the subfamily Bovinae, there are spiral-horned antelope (i.e., kudus, elands, nyalas, and bushbucks) and four-horned antelopes (i.e., nilgai). The subfamily Cephalophinae contains only duikers. The subfamily Antilopinae contains dwarf antelopes (i.e., dik-dik), gazelles, springboks, gerenuks, blackbucks, oryxes, addaxes, roan and sable antelopes, reedbucks, kobs, lechwes, waterbucks, hartebeests, wildebeests/gnu, and topis.

Wild bovids are known to forage on plants which contain large amounts of plant secondary compounds. These compound scans either reduce digestibility or be toxic to the animal. Diaminobutyric acid (DABA) and 4-*N*-acetyl-2,4-diaminobutyric acid (ADAB), an acetylated form of DABA, prevent collagen formation and are both found in acacia tree species. Several African ruminant species, such as bush duikers, Kenyan goats, hartebeest, impala, eland, and Grant's gazelle, have rumen bacteria, which are tolerant to one or both chemicals and can also degrade them (McKie et al. 2004). Adaptations such as this offer many ruminants a unique advantage in expanding their forage options.

4.3.1 Bacteria

The phylum Firmicutes was dominant in wild eland (*Taurotragusoryx*), Thompson's gazelle (*Gazella rufifrons*), and Grant's gazelle (*Gazella granti*) (Nelson et al. 2003). Of the 252 full-length 16S rRNA sequences, 48 % were identified as being related to the genus *Clostridium* (especially *Clostridium cellulovorans*), with a large proportion also identified as being related to the species *Ruminococcus gnavus*, *Ruminococcus torques*, *Ruminococcus flavefaciens*, *Ruminococcus bromii*, and *Eubacterium cellulolyticum* (Nelson et al. 2003). In addition to the Firmicutes, several other phyla were detected, including sequences distantly related to *Rikenella microfusus* and *Streptococcus sobrinus* from the phylum Bacteroidetes (Nelson et al. 2003). *Bilophila wadsworthia*, a common bacterium in clinical samples, was also identified (Nelson et al. 2003). Interestingly, *Verrucomicrobium spinosum* was identified as belonging to Verrucomicrobia, a phylum which is dominated by bacteria found in soil. Bacteria belonging to the phylum Proteobacteria (notably the subphyla delta- and gammaproteobacteria) were also identified, including the genus *Delftia* (subphylum betaproteobacteria) (Nelson et al. 2003).

4.3.2 Methanogens

To date, there are no published studies of the rumen methanogens of wild bovids. However, there are several studies on domestic yaks and cattle (An et al. 2005; Huang et al. 2012), water buffalo (Franzolin et al. 2012; Singh et al. 2012) and many other domesticated Bovidae. For a recent review, see St-Pierre and Wright (2013).

4.3.3 Protozoa

Ciliated protozoa from the African sassaby antelope (*Damaliscus lunatus lunatus*), one of the largest antelopes in the world, included the following genera: *Entodinium*, *Diplodinium*, *Eudiplodinium*, *Ostracodinium*, *Enoploplastron*, and *Opisthotrichum* (Ito et al. 2007). Another study comparing blue wildebeest (*Connochaetes taurinus*) and black wildebeest (*Connochaetes gnou*) identified rumen protozoa, which were distinct between the two species (Booyse and Dehority 2012). Sixteen ciliate protozoal species were identified in the blue wildebeest, with *Dasytricha ruminantium*, *Opisthotrichum janus*, and *Ostracodinium gracile occurring in all blue wildebeest samples, while 23 species were identified in the black wildebeest, with Diplodinium bubalidis* and *Ostracodinium damaliscus occurring in all black wildebeest samples* (Booyse and Dehority 2012).

In a separate study of the ciliate protozoa found in the tsessebe (*Damaliscus lunatus lunatus*), a grass grazing ruminant related to the topi, 18 species of ciliate protozoa were found and identified as belonging to the family Ophryoscolescidae (Van Hoven 1975). In that study, three new species were identified: *Entodinium fyferi*, *Enoploplastron garstangi*, and *Epidinium lunatus*, though no vestibuliferids (i.e., *Isotricha* and *Dasytricha*) (Van Hoven 1975). The Kafue lechwe antelope (*Kobus leche kafuensis*) from the Kafue flats in Zambia differs from other African herbivores because it subsists mainly on aquatic species from marshland and swamps, which often have a high salinity. In a study of their rumen protozoa, 24 species from 5 genera were identified, and 4 new species were described: *Diplodinium lochinvarense*, *Diplodinium leche*, *Diplodinium zambiense*, and *Metadinium ossiculi* (Imai et al. 1992).

In North America, one study of wild musk ox (*Ovibos moschatus) identified primarily Entodinium species, including Entodinium exiguum*, most of which had been previously identified in reindeer/caribou (Dehority 1974).

4.3.4 Fungi

A recent study identified rumen fungi (e.g., *Neocallimastix*, *Piromyces*, *Caecomyces*, *Anaeromyces, Orpinomyces*, and *Cyllamyces*) in the following wild bovids located at an American zoo: sable antelope (*Hippotragus niger*), American bison (*Bison bison*), bontebok (*Damaliscus pygargus*), Grant's gazelle (*Nanger granti*), Southern gerenuk (*Litocranius walleri walleri*), goral (*Naemorhedus goral*), greater kudu (*Tragelaphus strepsiceros*), and Nile lechwe (*Kobus megaceros*) (Liggenstoffer et al. 2010). There were also six novel groups (NG1–NG6) found across the various wild ruminants (Liggenstoffer et al. 2010).

4.4 Cervidae

There are over 20 genera and almost 50 species belonging to the family Cervidae, such as deer, elk, moose, caribou, and muntjacs. Many cervids, such as moose and reindeer, are known to consume a large proportion of lichens and mosses, especially in the winter, as a source of carbohydrates and water. This poses a problem as many contain high concentrations of phenolic compounds (i.e., usnic acid) and can accumulate environmental pollutants over time. Many ruminants are able to overcome these dietary constraints in large part due to the microorganisms they host, which are able to breakdown toxic compounds. Several bacteria have been isolated from arctic reindeer, which can withstand high levels of usnic acid, including the species *Eubacterium rangiferina* (Sundset et al. 2008) and some *Clostridia* isolates (Glad et al. 2009). *Selenomonas* spp., a common bacterium in the rumen, has also been shown to be resistant to tannic acid (Odenyo and Osuji 1998). Some browsing ruminants, such as deer and moose, are also able to produce a tannin-binding protein in their saliva, something not found in most grazing ruminants (i.e., cattle and sheep) (Austin et al. 1989).

Arctic reindeer (*Rangifer tarandus*), also known as caribou in North America, are a compelling species to study, not only because of their ability to digest toxic plant compounds (Glad et al. 2009; Sundset et al. 2008, 2009a, b, 2010), but because they annually lose up to 20 % of their body weight over the winter (Aagnes et al. 1995). In Scandinavia (i.e., Denmark, Norway, and Sweden) and Russia, there are two reindeer subspecies. *Rangifer tarandus* is found on the mainland and *Rangifer tarandus platyrhynchus* is isolated on the archipelago of Svalbard located between Norway and the North Pole (74–81 °N). This allows for a comparison of rumen microorganisms between the two subspecies, especially as mainland reindeer are often kept semi-wild for farming, and wild Svalbard reindeer endure severe weather and forage shortages.

4.4.1 Bacteria

The isolation of Svalbard reindeer has led to a 6–14X greater concentration of cellulolytic bacteria in the rumen of these reindeer over their mainland relatives in Norway (Sundset et al. 2007), which would seem to indicate that the Firmicutes would be the consistently dominant bacterial phylum found in Svalbard reindeer. Later that same year, bacteria belonging to the phylum Firmicutes were shown to be the most dominant prokaryotes in the rumen of Svalbard reindeer (Sundset et al. 2007). A clone library of 121 bacterial clones identified sequences as 55 % Firmicutes and 42 % Bacteroidetes (Sundset et al. 2007). In the same 2007 study, using semi-domesticated Norwegian reindeer grazing on wild pasture, 71 % of sequences were identified as belonging to the phylum Firmicutes and 29 % to the phylum Bacteroidetes (Sundset et al. 2007). Yet, in two other studies of the bacteria from the rumen of Svalbard reindeer, species belonging to the Bacteroidetes was shown to be the dominant bacteria based upon traditional culturing techniques (Orpin et al. 1985) and Roche 454 pyrosequencing (Pope et al. 2012).

Using culturing techniques, *Streptococcus bovis* (phylum Bacteroidetes) was found to increase during the summer, and the cellulolytic bacterium *Butyrivibrio fibrisolvens* (phylum Firmicutes) was found to increase during the winter (Orpin et al. 1985). Moreover, during digestion trials using rumen microbes found in fresh reindeer rumen fluid, starch utilization, proteolysis, and lactate utilization were all increased during the summer, and cellulolysis and xylanolysis were increased during the winter (Orpin et al. 1985). Recently, a study using high-throughput sequencing reported that the Bacteroidetes represented ~61 % of OTUs, while the Firmicutes represented ~27 % of OTUs in Svalbard reindeer (Pope et al. 2012). Interestingly, within the phylum Bacteroidetes, a novel Bacteroidales clade (SRM-1) was created (Pope et al. 2012). A small number of bacteria were identified as belonging to either the phyla Chloroflexi (5 % in 1 animal, 0 % in others), Actinobacteria, and Proteobacteria (<1 % of sequences each) or as "other" bacteria (4–19 %) (Pope et al. 2012).

Sundset and her colleagues (2009b) conducted a rumen microbiome study of Norwegian reindeer, using denaturing gradient gel electrophoresis (DGGE) to identify the following bacteria: *Peptostreptococcus anaerobius*, *Lachnospira multiparus*, *Butyrivibrio fibrisolvens*, *Eubacterium ruminantium*, *Selenomonas ruminantium*, *Fibrobacter succinogenes*, *Eubacteriumpyruvovorans*, and *Fusocillus* sp. (*C. proteoclasticum*-like).

More recently, the bacteria in the rumen and colon of moose in Vermont (*Alces alces americana*) were identified using the second-generation (G2) PhyloChip DNA microarray (Ishaq and Wright 2012). As in most other wild ruminants, bacteria belonging to the phylum Firmicutes were dominant in the rumen, with most OTUs belonging to the class Clostridia, followed by the phyla Proteobacteria and Bacteroidetes (Ishaq and Wright 2012). Rumen samples consistently clustered separately from colon samples, and there were 73 OTUs (21 families) exclusive to the rumen with an additional 71 OTUs (22 families) exclusive to the colon, indicating that bacteria from the colon were too dissimilar to be used to estimate rumen bacteria populations (Ishaq and Wright 2012). Bacterial families exclusive to the rumen included the following: Acetobacteraceae, Acidobacteriaceae, Anammoxales, Desulfoarculaceae, Desulfobacteraceae, Desulfovibrionaceae, Flavobacteriaceae, Flexibacteraceae, Helicobacteraceae, Lachnospiraceae, Peptococc/Acidaminococc, Peptostreptococcaceae, Pirellulae, Porphyromonadaceae, Prevotellaceae, Rikenellaceae, Streptococcaceae, and Syntrophobacteraceae, as well as Verrucomicrobia subdivisions 5 and 7, and unclassified bacteria (Ishaq and Wright 2012).

Sika deer are small grazers in Asia which select a wild variety of forage types. A study on wild sika deer reported that total ruminal bacterial concentration decreased during the winter, and, of those, Gram-negative bacteria increased in proportion over the winter (Ichimura et al. 2004). More recently, a study of domesticated sika deer compared rumen bacteria between deer eating a typical domestic diet (concentrate with high-fiber corn stalks) and deer eating a diet supplemented with forage (concentrate with oak leaves) (Li et al. 2013b). The phylum Bacteroidetes was dominant in both diet groups, the phylum Firmicutes was only found in the oak leaf diet group, and the corn stalk diet group contained a higher proportion of the phylum Proteobacteria (Li et al. 2013b). *Prevotella*-like bacteria were the dominant group in both diet groups. However, the two groups contained differential abundances of different *Prevotella* species (Li et al. 2013b).

There are also several studies which investigate the infectious potential of fecal bacteria in species such as red deer (*Cervus elaphus*), roe deer (*Capreolus capreolus*), or white-tailed deer (*Odocoileus virginianus*) (Aschfalk et al. 2003; Gimenez et al. 2009; Guan et al. 2002; Hristov 2011; Lillehaug et al. 2005; Lyautey et al. 2007).

4.4.2 Methanogens

An investigation of rumen methanogens from Svalbard reindeer used 97 clones from a 16S rRNA gene library to reveal 22 OTUs (Sundset et al. 2009a). The largest proportion of OTUs (11 OTUs) were found to be highly similar to the methanogenic orders Methanobacteriales, Methanomicrobiales, and Methanosarcinales (Sundset et al. 2009a). Norwegian reindeer are more often kept as semidomesticated herds for farming, and fewer studies exist on wild Norwegian reindeer on natural pasture. A microbiome study of the rumen of Norwegian reindeer used 16S rRNA clones to identify methanogens, including two common families Methanobacteriaceae and Methanosarcinaceae (Sundset et al. 2009b). Clones within the family Methanobacteriaceae were identified as *Methanobrevibacter gottschalkii*, *Mbr. smithii*, *Mbr. ruminantium*, and *Mbr. thaueri*, and there were several unclassified clones which were similar to sequences, which had never been cultured and classified (Sundset et al. 2009b).

An investigation of the rumen methanogens in domesticated sika deer found that the genus *Methanobrevibacter* was dominant in deer eating a typical domestic sika deer diet (concentrate plus corn stalks) or a supplemental forage diet (concentrate plus oak leaves) (Li et al. 2013a). Using a 16S rRNA gene clone library of 197 clones, it was found that *Methanobrevibacter millerae*-related clones were predominant in deer fed corn stalks, while *Mbr. smithii*- and *Mbr. ruminantium*-related clones were not found in deer fed the oak leaf diet at all (Li et al. 2013a).

4.4.3 Protozoa

Ciliated rumen protozoa in reindeer tend to be very similar across geographically isolated species, implying that they may have differentiated from the protozoa found in other ruminants and have now stabilized (Imai et al. 2003). Common rumen protozoal taxa include the families Isotrichidae and Ophryoscolecidae. Protozoa identified in reindeer from Inner Mongolia, China, included *Entodinium anteronucleatum*, *Entodinium bicornutum*, *Enoploplastron confluens*, and *Epidinium gigas* (Imai et al. 2003). These protozoal species had previously been detected in reindeer from Alaska (Dehority 1975). *Entodinium caudatum*, *Epidinium caudatum*, and *Isotricha intestinalis* were also detected in Norwegian reindeer using DGGE (Sundset et al. 2009b).

Entodinium alces was originally identified by light microscopy in a moose in Alaska (*Alces alces gigas*), a study in which all the protozoa found belonged to the genus *Entodinium* (Dehority 1974). One small study on rumen protozoa in mule deer (*Odocoileus hemionus*), white-tailed deer (*Odocoileus virginianus*), and elk (*Cervus canadensis*) in the United States showed that mule deer contained *Entodinium dubardi* and *Entodinium quadricuspis*, whereas white-tailed deer contained *Entodinium dubardi* and *Eudiplodinium impalae* (Dehority 1995). Elk contained a more diverse collection of ophryoscolecid protozoa, including a new species called *Entodinium wapiti* (Dehority 1995).

To date, most studies on Asian ruminants investigated domestic species in order to improve animal production. However, there have been several published studies on the rumen protozoa of wild Asian ruminants. Eight species of *Entodinium* (*E. simplex*, *E. ovinum*, *E. dubardi*, *E. longinucleatum*, *E. exiguum*, *E. parvum*, *E. nanellum*, and *E. dilobum*) and *Diplodinium anisacanthum* were isolated from wild sika deer

(*Cervus nipponyesoensis*) in Hokkaido Island, Japan (Ichimura et al. 2004). In that study, it was noted that protozoal diversity was reduced during the winter, as was ruminal bacterial concentration (Ichimura et al. 2004).

4.4.4 Fungi

The rumen fungi identified from the rumen of Norwegian reindeer targeted the Neocallimastigales population, which confirmed the presence of several different phylotypes that the study did not fully sequence (Sundset et al. 2009b). Rumen fungi identified from sika deer included the genera *Piromyces* and *Anaeromyces* and three novel groups (NG1, NG3, NG5) as likely new genera (Liggenstoffer et al. 2010).

4.5 Giraffidae

There are currently only two species in the family Giraffidae: giraffes (*Giraffa camelopardalis*) and okapi (*Okapia johnstoni*). Unfortunately, little gut microbial research has been done on either.

4.5.1 Protozoa

Twenty six species of ciliated protozoa in the rumen of the giraffe were found using microscopic identification (Kleyhans and Van Hoven 1976). Eleven species of *Entodinium* represented the largest proportion of rumen protozoa at 41 %, followed by the vestibuliferids (i.e., *Isotricha* and *Dasytricha*) at 19 %, *Enoploplastron* at 16 %, and *Eodinium* at 11 % (Kleyhans and Van Hoven 1976). Two novel species were also identified: *Entodinium longicorpum* and *Eodinium bispinosum* (Kleyhans and Van Hoven 1976).

4.5.2 Fungi

Recently, a large study identified the anaerobic rumen fungi of various wild and domestic ruminants (Liggenstoffer et al. 2010). This study identified the genera *Anaeromyces* and *Orpinomyces* and four novel groups (NG1, NG3, NG5, and NG6), which likely represent new genera, from Rothschild's giraffe and *Neocallimastix*, *Piromyces*, *Anaeromyces*, NG1, and NG6 from the okapi (Liggenstoffer et al. 2010).

4.6 Concluding Remarks

While many studies exist on the microorganisms of wild ruminants, few investigate the entire microbiome, or even multiple domains. Most have small sample sizes and are therefore unable to indicate trends between rumen microorganism populations and other factors, such as host species, geographic location, age, weight, gender, or reproductive status. There is also a general lack of dietary information in many studies of wild ruminants, which is a problem inherent to using wild-caught species. This chapter has summarized a limited body of work that exists on wild ruminant microbiology. It is clear that a large amount of work remains to be done on cataloging the microbiome of wild ruminants.

References

Aagnes TH, Sørmo W, Mathiesen SD (1995) Ruminal microbial digestion in free-living, in captive lichen-fed, and in starved reindeer (*Rangifer tarandus tarandus*) in winter. Appl Environ Microbiol 61(2):583–591

An D, Dong X, Dong Z (2005) Prokaryote diversity in the rumen of yak (*Bos grunniens*) and Jinnan cattle (*Bos taurus*) estimated by 16SrDNA homology analyses. Anaerobe 11:207–215

Aschfalk A, Hundertmark KJ, Bendiksen HR et al (2003) Serosurvey for antibodies against *Salmonella* species in free-ranging moose (*Alces alces*) from Norway. Berl Munch Tierarztl Wochenschr 116(9–10):417–420

Austin PJ, Suchar LA, Robbins CT et al (1989) Tannin-binding proteins in saliva of deer and their absence in saliva of sheep and cattle. J Chem Ecol 5(4):1335–1347

Booyse DG, Dehority BA (2012) Protozoa and digestive tract parameters in blue wildebeest (*Connochaetes taurinus*) and black wildebeest (*Connochaetes gnou*), with description of *Entodinium taurinus* n sp. Eur J Protistol 48(4):283–289

Dehority BA (1974) Rumen ciliate fauna of Alaskan moose (*Alces americana*), musk-ox (*Ovibos moschatus*) and Dall mountain sheep (*Ovis dalli*). J Eukaryot Microbiol 21(1):26–32

Dehority BA (1975) Characterization studies on rumen bacteria isolated from Alaskan reindeer (*Rangifer tarandus L*). In: Proceedings of the 1st internat reindeer and caribou symposium, University of Alaska, Fairbanks, pp 228–240

Dehority BA (1995) Rumen ciliates of the pronghorn antelope (*Antilocapra americana*), mule deer (*Odocoileus hemionus*), white-tailed deer (*Odocoileus virginianus*) and elk (*Cervus canadensis*) in the Northwestern United States. Arch Protistenkd 146(1):29–36

Franzolin R, St-Pierre B, Northwood KS et al (2012) Analysis of rumen methanogen diversity in water buffaloes (*Bubalus bubalis*) under three different diets. Microb Ecol 64:131–139

Gimenez C, Casado N, Criado-Fornelio A et al (2009) A molecular survey of Piroplasmida and Hepatozoon isolated from domestic and wild animals in Burgos (northern Spain). Vet Parasitol 162(1–2):147–150

Glad T, Falk A, Barboza PS et al (2009) Fate and effect of usnic acid in lichen on the bacterial population in the reindeer rumen. Microb Ecol 57:570–571

Guan S, Xu R, Chen S et al (2002) Development of a procedure for discriminating among *Escherichia coli* isolates from animal and human sources. Appl Environ Microbiol 68(6):2690–2698

Hristov AN (2011) Historic, pre-European settlement, and present-day contribution of wild ruminants to enteric methane emissions in the United States. J Anim Sci 90(4):1375–1379

Huang XD, Tan HY, Long R et al (2012) Comparison of methanogen diversity of yak (*Bos grunniens*) and cattle (*Bos taurus*) from the Qinghai-Tibetan plateau, China. BMC Microbiol 12:237

Ichimura Y, Yamano H, Takano T et al (2004) Rumen microbes and fermentation of wild sika deer on the Shiretoko peninsula of Hokkaido Island, Japan. Ecol Res 19(4):389–395

Imai S, Tsutsumi Y, Yumara S et al (1992) Ciliate protozoa in the rumen of Kafue Lechwe, *Kobus leche kafuensis*, in Zambia, with the description of four new species. J Protozool 39(5):564–572

Imai S, Oku Y, Morita T et al (2003) Rumen ciliate protozoal fauna of reindeer in Inner Mongolia, China. J Vet Med Sci 66(2):209–212

Ishaq SL, Wright A-DG (2012) Insight into the bacterial gut microbiome of the North American moose (*Alces alces*). BMC Microbiol 12:212

Ito A, Arai N, Tsutsumi Y et al (2007) Ciliate protozoa in the rumen of Sassaby antelope, *Damaliscus lunatus lunatus*, including the description of a new species and form. J Eukaryot Microbiol 44(6):586–591

Kamra DN, Sawal RK, Pathak NN et al (1991) Diurnal variations in ciliate protozoa in the rumen of blackbuck (*Antilope cervicapra*) fed green forages. Lett Appl Microbiol 13:165–167

Kleyhans CJ, Van Hoven W (1976) Rumen protozoa of the giraffe with a description of two new species. Afr J Ecol 14(3):203–214

Li ZP, Liu HL, Jin CA et al (2013a) Differences in the methanogen population exist in Sika deer (*Cervus nippon*) fed different diets in China. Microb Ecol 66(4):879–888

Li ZP, Liu HL, Li GY et al (2013b) Molecular diversity of rumen bacterial communities from tannin-rich and fiber-rich forage fed domestic Sika deer (*Cervus nippon*) in China. BMC Microbiol 13:151

Liggenstoffer AS, Youssef NH, Couger MB et al (2010) Phylogenetic diversity and community structure of anaerobic gut fungi (phylum Neocallimastigomycota) in ruminant and non-ruminant herbivores. ISME J 4(10):1225–1235

Lillehaug A, Bergsjø B, Schau J et al (2005) *Campylobacter* spp, *Salmonella* spp, verocytotoxic *Escherichia coli*, and antibiotic resistance in indicator organisms in wild cervids. Acta Vet Scand 46(1–2):23–32

Lyautey E, Hartmann A, Pagotto F et al (2007) Characteristics and frequency of detection of fecal *Listeria monocytogenes* shed by livestock, wildlife, and humans. Can J Microbiol 53(10):1158–1167

McKie MR, Brown DL, Melesse A et al (2004) Rumen microbes from African ruminants can degrade Acacia angustissima diamino acids. Anim Feed Sci Technol 117(3–4):179–195

Nelson KE, Zinder SH, Hance I et al (2003) Phylogenetic analysis of the microbial populations in the wild herbivore gastrointestinal tract: insights into an unexplored niche. Environ Microbiol 5(11):1212–1220

Neumann LM, Dehority BA (2008) An investigation of the relationship between fecal and rumen bacterial concentrations in sheep. Zoo Biol 27(2):100–108

Odenyo AA, Osuji PO (1998) Tannin-tolerant ruminal bacteria from East African ruminants. Can J Microbiol 44(9):905–909

Orpin CG, Mathiesen SD, Greenwood Y et al (1985) Seasonal changes in the ruminal mircoflora of the high-arctic Svalbard reindeer (*Rangifer tarandus platyrhynchus*). Appl Environ Microbiol 50(1):144–151

Pope PB, Mackenzie AK, Gregor I et al (2012) Metagenomics of the Svalbard reindeer rumen microbiome reveals abundance of polysaccharide utilization loci. PLoS One 7(6):e38571

Singh KM, Tripathi AK, Pandya PR et al (2012) Methanogen diversity in the rumen of Indian Surti buffalo (*Bubalus bubalis*), assessed by 16SrDNA analysis. Res Vet Sci 92:451–455

St-Pierre B, Wright A-DG (2013) Diversity of gut methanogens in herbivorous animals. Animal 7(s1):49–56

Sundset MA, Praesteng KE, Cann IKO et al (2007) Novel rumen bacterial diversity in two geographically separated sub-species of reindeer. Microb Ecol 54(3):424–438

Sundset MA, Kohn A, Mathiesen SD et al (2008) *Eubacterium rangiferina*, a novel usnic acid-resistant bacterium from the reindeer rumen. Naturwissenschaften 95(8):741–749

Sundset MA, Edwards JE, Cheng YF et al (2009a) Rumen microbial diversity in Svalbard reindeer, with particular emphasis on methanogenic archaea. FEMS Microbiol Ecol 70(3):553–562

Sundset MA, Edwards JE, Cheng YF et al (2009b) Molecular diversity of the rumen microbiome of Norwegian reindeer on natural summer pasture. Microb Ecol 57(2):335–348

Sundset MA, Barboza PS, Green TK et al (2010) Microbial degradation of usnic acid in the reindeer rumen. Naturwissenschaften 97(3):273–278

Van Hoven W (1975) Rumen ciliates of the Tsessebe (*Damaliscus lunatus lunatus*) in South Africa. J Eukaryot Microbiol 22(4):457–462

5 Structure and Function of a Nonruminant Gut: A Porcine Model

Kiyoshi Tajima and Rustam Aminov

Abstract

In many aspects, the anatomical, physiological, and microbial diversity features of the ruminant gut are different from that of the monogastric animals. Thus, the main aim of this chapter is to give a comparative overview of the structure and function of the gastrointestinal tract of a nonruminant monogastric animal, and here it is represented by a pig model. In this chapter, we describe and discuss (i) microbial diversity in different parts of the porcine gut; (ii) differences between the ruminant and nonruminant gut; (iii) main events during colonization and succession of microbiota in the porcine gut; (iv) effects of various feed additives including antibiotics, phages, probiotics, and prebiotics on pigs; and (v) the use of the porcine model in translational medicine.

Keywords

Porcine gut • Microbial diversity • Colonization • Succession • Feed additives • Translational medicine

5.1 Introduction

For a number of years, rumen microbiology has been on the forefront of gut microbiology research. This concerns both stages in the gut microbiology research history including cultivation-based and cultivation-independent periods. Straight from the commencement of the microbiological science, when the father of microbiology, Antonie van Leeuwenhoek, invented the microscope, it has been heavily dependent on the technological advances. In line with these requirements, the overwhelmingly anaerobic microbiota of gut ecosystems could be accessed and studied only with the use of sophisticated anaerobic techniques allowing to create oxygen-free conditions in order to the anaerobes to grow.

K. Tajima
National Institute of Livestock and Grassland Science, National Agriculture and Food Research Organization, Tsukuba, Ibaraki 305-0901, Japan

R. Aminov (✉)
National Veterinary Institute, Technical University of Denmark, Bülowsvej 27, 1870 Frederiksberg C, Denmark
e-mail: rusam@vet.dtu.dk

A.K. Puniya et al. (eds.), *Rumen Microbiology: From Evolution to Revolution*, DOI 10.1007/978-81-322-2401-3_5

Although a roll tube for isolation and quantification of microorganisms was invented long time ago (Esmarch 1886), it remained an obscure technique, rarely used in practice. A few modifications followed, with the most important contributions coming from a former zoologist turned into a rumen microbiologist, Robert Hungate (1979; Chung and Bryant 1997). Probably the most important of his pioneering inventions were the use of pre-reduced anaerobically sterilized medium in "roll tubes" and the development of strict anaerobic techniques (butyl rubber stoppers, resazurin as a redox indicator, a gas (N_2 or CO_2) to displace air, boiling the medium to remove dissolved oxygen, and other improvements allowing access to a greater diversity of gut anaerobes, in particular, in the rumen (Hungate 1966). For example, the initial estimates of bacterial numbers in the rumen were in the range of 10^6 colony-forming units (CFU) per mL of rumen liquid since the count was done using non-reduced medium (Johnson et al. 1944). The use of the strict anaerobic techniques developed by Hungate made the rumen microbiology one of the best-studied anaerobic ecosystems in the 1960s and 1970s and contributed to the fast progress in microbiology of other anaerobic ecosystems.

Despite the impressive progress in cultivating anaerobic microbiota, the majority of the rumen inhabitants resisted cultivation attempts, the so-called great plate count anomaly known for many microbial ecosystems (Amann et al. 1995). The explosive development of molecular microbial ecology that we are witnessing in a recent couple of decades has been largely driven by the development and application of various advanced molecular techniques (Nocker et al. 2007). At the heart of these technology developments were the fundamental and seminal works of Carl Woese, who discovered three domains of life on Earth and demonstrated how the evolution of every living being can be reconstructed on the basis of molecular markers (in particular of the SSU rRNA molecule) and how the phylogenetic positioning defines the taxonomic position and identity of an organism (Woese and Fox 1977; Fox et al. 1980; Olsen and Woese 1993). The new technologies revolutionized the field and brought a better understanding of how microbes contribute to every aspect of our life, from the global cycling of major components on our planet, such as carbon, nitrogen, methane, and others, down to every human being, who carry a sizeable amount of commensal gut microbiota that contribute to the nutrition, normal development, and health.

In rumen microbiology, the molecular microbial ecology methods used were 16S rRNA hybridization (Stahl et al. 1988; Lin et al. 1994; Forster et al. 1997), RFLP (Wood et al. 1998), competitive PCR (Reilly and Attwood 1998; Kobayashi et al. 2000; Koike and Kobayashi 2001), denaturing gradient gel electrophoresis (DGGE) (Kocherginskaya et al. 2001), 16S rDNA clone libraries (Whitford et al. 1998; Tajima et al. 1999, 2000, 2001a; Ramsak et al. 2000), real-time PCR (Tajima et al. 2001b), and metagenomic analysis (Hess et al. 2011; Ross et al. 2012). The current genomic studies of the rumen-specific microbiota include 21 species belonging to 14 different bacterial families, while the number of the currently sequenced archaeal genomes is limited to 13 (Leahy et al. 2013). In terms of diversity and numbers, the rumen archaea represent a much smaller population, but the disproportionate interest in this group is mainly dictated by the aspirations to find methane mitigation strategies that are based on genomic information. Given the much higher species richness of the bacterial community in the rumen, the current genome sequencing efforts seems insufficient. To address these issues, the Hungate1000 project has been launched (www.hungate1000.org.nz).

In general, the stages of gut microbiology development in other agricultural animals such as a pig have followed the same routes described above for the rumen. In the beginning, the pig gut microbiology field has been based and absolutely relied on cultivation, identification, and physiological and biochemical characterization of microorganisms. The limitations of this approach in pig gut microbiology, similarly to the rumen microbiology, could be demonstrated by the huge discrepancy between the cultivated count and direct microscopic enumeration that have been observed fairly early, even during the cultivation-based era (Russell 1979). Cultural counts of porcine gut bacteria averaged 56.2 % of the direct microscopic counts in luminal content and surface layer and 20.2 % in intestinal tissue. Then the adoption of molecular microbial ecology techniques in swine gut microbiology resulted in a much more accurate description of this ecosystem. The next part of the chapter briefly describes the principal differences between the gastrointestinal tract of the ruminants and a monogastric animal such as a pig.

5.2 Comparison of Ruminant and Pig Gut

The porcine gastrointestinal tract is markedly different from that of ruminants in many aspects including anatomical, morphological, physiological, and microbiological. First of all (and this is stated in the name of this animal group) is the presence of the rumen, a prominent characteristic of the ruminant animals. In monogastric animals, the stomach is a single compartment of the gastrointestinal tract but divided into four sections, the cardia, fundus, corpus, and pylorus. In the ruminants, the stomach is much more complex and consists of the four anatomically, morphologically, and physiologically different compartments: the rumen, reticulum, omasum, and abomasum. The primary function of the stomach in monogastric animals is the digestion of food, aided by the secretion of various endogenous proteases and hydrochloric acid, thus creating extremely acid environment of the stomach. The rumen, on the contrary, is characterized by neutral pH, and it is where the microbiological fermentation of plant material takes place. In addition to bacteria, the rumen harbors protozoa, archaea, and fungi, which form the microbial consortium that degrades plant polysaccharides, the nutritional value of which cannot be retrieved otherwise because the host lacks the necessary fibrolytic enzymes. The rumen fermentation in the first 3 compartments results in the production of short-chain fatty acids (SCFAs), ammonia, and methane. Thus, the production of SCFAs by microbiota and their consumption by host ruminants are happening in the rumen, while the corresponding processes in monogastric animals are shifted to the large intestine. The fourth rumen compartment, abomasum, is functionally very much analogous to the stomach of monogastric animals, the feed being digested in a similar way.

As portrayed in this book and elsewhere (White et al. 1998; Tajima et al. 1999, 2001a; Ramsak et al. 2000; Kocherginskaya et al. 2001; Wright et al. 2006; Mackie et al. 2013), microbial diversity of the rumen is extraordinary and includes the representatives of all three domains of life, *Archaea*, *Bacteria*, and *Eucarya*. The microbial composition of the stomach in monogastric animals such as pigs is studied much less because it has been thought that the harsh acidic environment of the stomach can be tolerated only by a limited number of bacterial taxonomic groups. This may be true for the luminal content of the stomach, but the mucosa-adherent microbial diversity in the pig stomach is substantial and, in fact, even the cultivation-based approach can recover dominant bacterial groups belonging to *Lactobacillus*, *Enterobacteriaceae*, *Enterococcus*, and *Propionibacterium* (Collado and Sanz 2007). Less abundant bacteria in the pig stomach lining include the *Bacteroides*, *Bifidobacterium*, *Clostridium*, and *Staphylococcus* genera, with a small number of yeast cells. Culture-independent approach can recover even greater diversity of bacteria associated with the stomach mucosa in pigs, with the groups belonging to the *Atopobium*, *Coriobacterium*, and *Eubacterium* genera as well as the sulfate-reducing bacteria, in addition to the groups recovered by the cultivation-based approach (Collado and Sanz 2007).

In contrast to ruminants, in the case of the porcine gastrointestinal tract, the numbers and proportion of obligate anaerobic microbiota increases from the proximal to distal parts, thus essentially following the general trend of anaerobic conditions and redox potential within the alimentary tract. Comparatively limited research has been conducted on the microbiota of the lower parts of the gastrointestinal system in ruminants compared to the rumen microbiota. There are indications that, similarly to the rumen, the fibrolytic bacteria can also be encountered in the lower parts of the gastrointestinal system of ruminants (Ozutsumi et al. 2005). In general, however, the bulk of degradation and fermentation of plant polysaccharides in ruminants is performed in the rumen, while in monogastric animals, the majority of bacteria capable of plant polysaccharide degradation and fermentation are located in the large intestine. Because of this, bacteria similar to rumen bacteria such as those belonging to the genera *Ruminococcus* and *Megasphaera* can be found in the lower gastrointestinal tract of pigs. The microbiota in the large intestine of pigs, however, is nowhere close to the high efficiency of plant material degradation achieved by the rumen microbiota. While the ruminants almost completely depend on microbial fermentation in the rumen for nutritional needs, the products of microbial fermentation in the large intestine in the form of SCFAs supply only 10–30 % of the total energy requirements in pigs (Collinder et al. 2003).

The porcine gut microbiota has been examined for the complex carbohydrate degradation capabilities by both cultivation-based and cultivation-free methods. Enumeration of xylanolytic and cellulolytic bacteria from the fecal samples of 8-month-old gilts fed either a control or a 40 % alfalfa meal (high-fiber) showed a substantial number of bacteria capable of degrading both types of substrates (Varel et al. 1987). The numbers of xylanolytic bacteria were 1.6×10^8 and 4.2×10^8 colony-forming units (cfu)/g (dry weight) of feces in control pigs and pigs fed the high-fiber diet, respectively. The overall numbers of cellulolytic bacteria were 0.36×10^8 and 4.1×10^8 cfu/g for the control and high-fiber fed pigs, respectively. The hydrolyzing activities were largely attributed to *Prevotella ruminicola*, *Ruminococcus flavefaciens*, and *Fibrobacter succinogenes* suggesting similarity to the main plant material degraders in the rumen.

A recent study used a functional metagenomic approach to characterize the cellulolytic gene pool in pigs (Wang et al. 2012). The high-fat diet of pigs was supplemented with 10 % of Solka-Floc®-powdered cellulose in order to increase the carboxymethylcellulose-degrading population. Fecal metagenomic DNA was cloned into a plasmid vector, and the recombinant clones were screened for carboxymethylcellulase activities. An overall hit rate was 0.14 %, resulting in isolation of 11 cellulases, 4 hemicellulases, 1 polygalacturonase, 1 glycoside hydrolase family, 26 mannanase-family 5 cellulase chimeric enzyme gene, and 1 cellobiose phosphorylase. It is not clear, however, whether the original gut microbiota might have already possessed the fiber-degrading enzymes, without the dietary intervention. Besides, the high-fat diet might have suppressed the carbohydrate-utilizing populations of the porcine gut.

On the other hand, metagenomic analysis of the cow rumen microbiota resulted in identification of 27,755 putative carbohydrate-active genes (Hess et al. 2011). Expression studies with 90 candidate proteins demonstrated that 57 % of them were enzymatically active against cellulosic substrates. Thus, one carbohydrate-active gene was encountered within an average of 9.7 Mb of rumen metagenomic DNA, while in the pig, a similar hit was encountered within an average of 23.9 Mb of porcine large intestinal metagenome (Wang et al. 2012). Thus, the prevalence of genes encoding for plant fiber degradation activities in the rumen is much higher than in the porcine large intestine, which is in general agreement with the main function of the rumen as a specialized organ for degradation and utilization of cellulosic plant material.

5.3 Porcine Gut Microbial Diversity

As discussed above, the initial works describing the composition of the pig gut microbiota have been based on cultivation methods, and only later the culture-independent methods have been introduced, the latter largely being implemented using the 16S rRNA gene sequence as a molecular marker. It is generally accepted that the characterization of the microbiota by cultivation methods results in a biased representation of the gut microbial diversity. Factors contributing to the bias include fastidious growth requirements, syntrophic relationships, and obligate anaerobic metabolism of gut microbiota, the conditions that are difficult to reproduce under laboratory conditions.

Nevertheless, the cultivation methods have been the only way to characterize the microbial diversity in the early days of gut microbiology, and these investigations resulted in important initial glimpses at the complex and rich diversity of microbiota inhabiting the gut. It has been established that, normally, the feces of adult mammals contain bacteria belonging to *Bacteroidaceae*, *Eubacterium*, *Peptococcaceae*, *Spirillaceae*, *Lactobacillus* and *Bifidobacterium*, *Enterobacteriaceae*, and *Streptococcus* (Mitsuoka and Kaneuchi 1977). The corresponding cultivation-based studies in pigs uncovered the presence of the *Streptococcus*, *Lactobacillus*, *Eubacterium*, *Fusobacterium*, *Bacteroides*, *Peptostreptococcus*, *Bifidobacterium*, *Selenomonas*, *Clostridium*, *Butyrivibrio*, and *Escherichia* genera in the gut (Moore et al. 1987). The microbiota composition in pigs is also dependent on the site sampled and cultivation media used (Allison et al. 1979; Robinson et al. 1981). Robinson and coworkers found that bacteria in the porcine cecum belong to the *Bacteroides*, *Selenomonas*, *Butyrivibrio*, *Lactobacillus*, *Peptostreptococcus*, and *Eubacterium* genera (Robinson et al. 1981). Gastrointestinal diseases in pigs drastically change the composition of the gut microbiota. For example, severe dysentery in pigs shifts the bacterial composition in the colon towards the predominance of Gram-negative bacteria (88 %), while in healthy pigs it is dominated by Gram-positive bacteria (71 %) (Robinson et al. 1984).

Introduction of molecular ecology tools into the porcine gut microbiology area broadened our knowledge on microbial diversity in this ecosystem. The first works published in the field included the use of denaturing gradient gel electrophoresis (DGGE) and sequencing of the 16S rRNA gene libraries (Pryde et al. 1999; Simpson et al. 1999). Unless sequenced, the bands on DGGE gels represent only the visual diversity of gut microbiota and allow visual monitoring of changes in the microbiota profiles, if any. Another drawback is that the diversity displayed is restricted to the most abundant phylotypes present and the rare phylotypes are not detected. For these reasons, DGGE is rarely used in contemporary gut microbiota analyses. Sequencing of the 16S rRNA gene libraries, on the other hand, becomes a method of choice in analysis of many microbial ecosystems, especially with the introduction of next-generation sequencing platforms. The first systematic and large-scale characterization of gut microbiota in pigs using a culture-independent approach was performed by Leser and coworkers (2002). They analyzed the bacterial composition in the contents of the ileum, cecum, and large intestine from 24 pigs reared in different environments using the 16S rRNA gene clone libraries. In total, 4,270 clones were analyzed that belonged to the *Firmicutes*, *Bacteroidetes*, *Actinobacteria*, and *Proteobacteria* phyla. Of these, the *Firmicutes* (also formerly called the low G + C Gram-positive bacteria) were dominant (81 %). At the phylotype level, however, only 17 % of sequences were sufficiently similar to that of the cultivated strains, thus allowing taxonomic affiliation at the species level. But the remaining 83 % of sequences had the level of similarity with reference species below the threshold suggesting that the vast majority of bacterial species and strains in the pig gut are unknown. These results suggested that the microbial diversity in the porcine intestine is much more rich and complex than previously anticipated on the basis of cultivation studies.

Further studies of the porcine gut microbial diversity employed the next-generation sequencing techniques that allow much better coverage of microbial diversity because of deep sequencing. Dowd and coworkers used a pyrosequencing approach to investigate the bacterial diversity in the ileum of newly weaned piglets (Dowd et al. 2008). Besides the ubiquitous sequences related to *Clostridium* spp., *Lactobacillus* spp., and *Helicobacter* spp., what they found was a plethora of other sequences, with the predicted diversity of as many as 821 different species associated with the ileum of piglets.

A total of 1,031,130 16S rRNA sequences were retrieved in a recent longitudinal study of pig fecal microbiomes of 20 pigs from two commercial farms in the USA (Kim et al. 2011). The majority of sequences were classified into five bacterial phyla, *Firmicutes*, *Bacteroidetes*, *Proteobacteria*, *Actinobacteria*, and *Spirochaetes*, with the first 2 contributing to the 90 % of the overall diversity. At the genus level, *Prevotella* was predominant representing up to 30 % of all classifiable sequences when the pigs were 10 weeks of age. At 22 weeks of age, the proportion of *Prevotella* decreased to 3.5–4.0 %, accompanied by the increase of *Anaerobacter*. Similarly, at the higher taxonomic level, the proportion of the *Firmicutes* increased, whereas the proportion of bacteria in the phylum *Bacteroidetes* decreased as the pigs aged (Kim et al. 2011). This investigation suggested that the porcine fecal microbiome is not static after reaching the adult configuration but continues to restructure during aging.

A recent large-scale assessment of the porcine gut mucosal microbiota retrieved 447,849 16S rRNA sequences, which were clustered into 997 operational taxonomic units (OTUs) (Mann et al. 2014). OTUs were assigned to 198 genera belonging to 14 different phyla. This study discovered a previously unknown high diversity and species richness at the *pars nonglandularis* of the stomach and revealed a distinct mucosa-associated bacterial community at different gut sites.

The limitation of the 16S rRNA gene-based approach in molecular ecology is that it provides mostly a community structure information but is limited in terms of the functionality and metabolic activities of microbiota. A more straightforward approach to assess the metabolic and other functional properties of various microbiota is the metagenomic approach (Hugenholtz and Tyson 2008). This approach, for example, in application to human gut metagenomes identified 237 gene families commonly enriched in adult-type and 136 families in infant-type microbiomes, with a little overlap, suggesting that these gene sets encode the core functions of the adult- and infant-type gut microbiota (Kurokawa et al. 2007).

A recent metagenomic study of the swine fecal microbiota revealed that this ecosystem is exceptionally enriched by genes encoding resistance to antibiotics and carbohydrate metabolism (Lamendella et al. 2011). Up to 13 % of sequences were related to carbohydrate metabolism; genes associated with stress, virulence, cell wall, and cell capsule were also abundant. The abundance of antibiotic-resistant genes in the porcine gut microbiota may be explained by the fact that the animal feed in this study was routinely supplemented with feed-grade antibiotics for the improvement of growth performance (Lamendella et al. 2011). These antibiotics consisted of chlortetracycline and penicillin at the concentration of 20 g per ton of feed. Interestingly, hierarchical clustering based on taxonomy and functional gene repertoire suggested a close relationship of the swine gut microbiome with the intestinal microbiomes of chickens and the cow rumen. This observation possibly reflects the dietary influence in these animals, which consume mostly plant-based diets.

The composition of the porcine gut has been mainly studied for the occurrence and diversity of the *Bacteria* representatives. But this ecosystem also harbors the representatives of the other two domains of life, the *Archaea* and *Eukarya*. Molecular diversity of the fecal *Archaea* has been recently assessed in Erhualian (obese type) and Landrace (lean type) pigs using a 16S rRNA gene library approach (Luo et al. 2012a). The majority of archaeal sequences (59.4–96.6 %) were associated with the *Methanobrevibacter* genus, with the rest belonging to the genus *Methanosphaera*. The lean breed harbored a higher diversity and density of fecal methanogens compared to the obese breed.

Diversity of the *Eukarya* in the fecal microbiota of unweaned and weaned piglets was accessed by cultivation as well as by the amplification of the D1 domain of the 26S rRNA gene followed by separation of the resulting amplicons by DGGE (Urubschurov et al. 2011). Among the cultured yeast isolates, *Kazachstania slooffiae* was the dominant species. Establishment of yeasts, especially *K. slooffiae*, in the porcine gastrointestinal tract essentially coincided with the consumption of grain-based feed suggesting the importance of diet in the maintenance of yeast populations. Interestingly, *K. slooffiae* was the only yeast species detected by PCR-DGGE, thus supporting our previous notion in the chapter that this technique is suitable only for the detection of the major microbiota representatives, while the minor components may be easily missed. In this regard, cultivation using selective media may offer a better detection of minor microbiota representatives.

Besides the viruses implicated in pig diseases such as porcine epidemic diarrhea virus, porcine kobuvirus, porcine bocavirus, porcine group A rotavirus, and transmissible gastroenteritis virus, there is also an endogenous viral diversity in the gut of apparently healthy pigs. Thus, the fecal virome of the healthy and diarrheic piglets on a high-density farm was recently examined by Shan and coworkers using a metagenomic approach (Shan et al. 2011). The vast majority, 99 %, of the viral sequences were related to the RNA virus families *Picornaviridae*, *Astroviridae*, *Coronaviridae*, and *Caliciviridae*, with the rest related to the small DNA virus families *Circoviridae* and *Parvoviridae*. Interestingly, the viral load was not markedly different in healthy and diarrheic piglets: the former shed an average of 4.2 different mammalian viruses, while the latter shed an average of 5.4 different mammalian viruses. An essentially similar diversity of the pig fecal virome encompassing the families *Astroviridae*, *Caliciviridae*, *Coronaviridae*, *Arteriviridae*, *Picornaviridae*, *Picobirnaviridae*, and *Parvoviridae* was recently reported in another study (Lager et al. 2012).

5.4 Microbial Diversity at Different Sites of the Porcine Gut

Most of our knowledge regarding the porcine gut microbiology has been obtained by analyzing fecal samples, which supposedly represent the luminal content of the large intestine (Stewart 1997). Relatively little information is available regarding the *in situ* situation in other parts of the porcine gastrointestinal tract. In pigs, the gastrointestinal tract consists of the esophagus, the stomach, the small intestine (duodenum, jejunum, and ileum), and the large intestine (cecum, colon, and rectum). There is a gradient of microbiota concentration in the gut that increases from the upper to the lower parts of the gastrointestinal tract. In the pig esophagus, lactobacilli and streptococci are closely associated with the squamous area known as the *pars esophagea* (Fuller et al. 1978). At this site, *Lactobacillus fermentum*, *L. salivarius*, and *Streptococcus salivarius* are more frequent in the suckling pigs, while in the weaned pigs, *L. acidophilus* and *S. bovis* are dominant. Besides the lactic acid bacteria such as *L. mucosae* the esophageal mucosa may also harbor *Enterococcus faecium* and *Weissella cibaria* (Dr. Tohno, personal communication). Interestingly, the total viable count of bacteria adhered to per gram of the *pars esophagea* tissue in pigs is not different from the other tissues along the gastrointestinal tract (Krause et al. 1995). This suggests that the density of the adherent bacteria is largely independent from the density of bacteria in the lumen.

The luminal content of the stomach and small intestine are thought to harbor low numbers of microorganisms that can persist under the conditions of extreme pH (acidic in the stomach and alkaline in the small intestine), pancreatic juices, bile, and high passage rates. Indeed, in the pig stomach, the numbers of dominant acid-tolerant lactobacilli and streptococci rarely exceeds 10^3–10^5 cfu/g of the luminal content (Canibe et al. 2005). In the small intestine and particularly in the ileum, the numbers may reach up to 10^8 cfu/g of the luminal content. In our estimates of microbial molecular diversity in the porcine ileum by the use of 16S rRNA gene clone libraries, the sequences belonged to *L. acidophilus*, *L. crispatus*, *L. delbrueckii*, *L. fermentum*, *L. johnsonii*, *L. reuteri*, and *L. vaginalis* (Tajima et al. 2010a).

The lower parts of the gut, the cecum and colon, are characterized by normal pH, slower passage rates, and, in general, higher nutrient availability. Thus, the distal parts of the porcine gut are inhabited by more stable, diverse, and dense microbiota, with the numbers reaching up to 10^{12} cfu/g of digesta (Pryde et al. 1999; Gaskins 2001). The majority of cultivated bacteria belong to the genera *Streptococcus*, *Lactobacillus*, *Peptostreptococcus*, *Clostridium*, *Prevotella*, and *Bacteroides*. Our analysis of the porcine cecum content by using a 16S rRNA clone library approach has established that the molecular diversity in this section of the gut is substantially higher (Tajima et al. 2010a). In addition to lactobacilli (*L. acidophilus*, *L. crispatus*, *L. fermentum*, *L. johnsonii*, *L. reuteri*, and *L. vaginalis*), the representatives of other taxonomic groups included *Dorea*, *Coprococcus*, *Roseburia*, *Lachnospiraceae*, *Ruminococcus*, *Faecalibacterium*, *Subdoligranulum*, and *Bacteroides*. The pig colon, analyzed by 16S rRNA gene clone libraries, demonstrated an overwhelming dominance of the *Firmicutes* (70–88 %), with much lower proportions of *Bacteroidetes* (4–25 %), *Spirochaetes* (1–3 %), *Actinobacteria* (0–2 %), *Proteobacteria* (0–1 %), *Verrucomicrobia* (0–1 %), *Acidobacteria* (0–1 %), and unclassified bacteria (0–6 %) (Tajima et al. 2010b). These findings have been confirmed by others as well (Isaacson and Kim 2012).

The use of quantitative PCR (qPCR) to assess the relative numbers of bacteria in the stomach, small intestine, and large intestine of weaned piglets confirmed earlier observations that the large intestine possesses a significantly higher total bacterial count compared to the stomach and ileum (Metzler-Zebeli et al. 2013). While the *Lactobacillus* group was dominant in the stomach and ileum, the anaerobic *Firmicutes* and *Bacteroidetes* were dominant in the large intestine. Thus, regarding the luminal microbiota content along the porcine gut, there is a general consensus that the majority (>90 %) of the bacteria in the pig large intestinal microbiome belong to the *Firmicutes* and *Bacteroidetes* phyla, whereas the ileum may have a larger proportion (up to 40 %) of the *Proteobacterium* representatives (Isaacson and Kim 2012).

As mentioned above, there are substantial differences between the mucosa-adherent and luminal bacterial populations in the porcine gut (Pryde et al. 1999). The microbial composition of the stomach in monogastric animals such as pigs has been studied much less in the past because of the prevailing assumption that the high passage rate and harsh acidic environment of the stomach can be tolerated only by a limited number of microorganisms. This may be true for the luminal content of the stomach, but the mucosa-adherent microbial diversity in the pig stomach is considerable and, in fact, even the cultivation-based approach can recover dominant bacterial groups belonging to *Lactobacillus*, *Enterobacteriaceae*, *Enterococcus*, and *Propionibacterium* (Collado and Sanz 2007). Less abundant bacteria in the pig stomach lining include the *Bacteroides*, *Bifidobacterium*, *Clostridium*, and *Staphylococcus* genera, with a small number of yeast cells. Culture-independent approach can recover even greater diversity of bacteria associated with the stomach mucosa in pigs, with the groups belonging to the *Atopobium*, *Coriobacterium*, and *Eubacterium* genera, as well as the sulfate-reducing bacteria, in addition to the groups recovered by the cultivation-based approach (Collado and Sanz 2007).

Assessment of the mucosal bacterial populations in the pig gut by the pyrosequencing approach revealed unexpectedly high diversity and species richness at the *pars nonglandularis* of the stomach in weaned pigs (Mann et al. 2014). Most of the sequences belonged to the *Firmicutes*, with four highly abundant OTUs matching *Lactobacillus* sequences (e.g., *L. johnsoni*, *L. amylovorus*, *L. delbrueckii*, and *L. mucosae*). They accounted for over 50 % of sequences at this site and were significantly increased compared to the lower gut sites. Surprisingly, *Prevotella*-related OTUs (exemplified by cultivated *P. copri*), which are usually functionally associated with complex carbohydrate degradation, can also be recovered from this site (Mann et al. 2014).

Besides the role in nutrition, the small intestine is also a focal point of mucosal immunity, which includes organized and diffused tissues (Neutra 1998). The former includes mesenteric lymph nodes and Peyer's patches. Multiple Peyer's patches are located in the terminal part of the small intestine, the ileum, and they actually assist as a discriminating anatomical feature of the ileum from the duodenum and jejunum. Peyer's patches are the main antigen sampling points in the gut, involved in acquisition of antigens from the lumen and delivering them to the antigen-presenting cells. In this process, luminal antigens are transported across the follicle-associated epithelium of the Peyer's patches via specialized microfold cells or by a paracellular route. In the Peyer's patches, antigens are sampled by dendritic cells, which then migrate to the mesenteric lymph nodes and present the antigens to naïve T cells. This results in T cells activation, differentiation, relocation, and initiation of immune responses. In our investigations, we established that the mucosal lining of the lower parts of the gastrointestinal tract of pigs such as the ileum is associated with bacterial diversity dominated by lactic acid bacteria belonging to *Lactobacillaceae*, *Leuconostocaceae*, *Streptococcaceae*, and, in lower abundance, *Enterococcaceae* and *Aerococcaceae* (Schmidt et al. 2011). Thus, the most abundant phylotypes belonged to the *Firmicutes* (67.5 % of all sequences), followed by the *Proteobacteria* (17.7 %), *Bacteroidetes* (13.5 %), and *Actinobacteria* (0.1 %).

Another approach, targeting metabolically active populations of the ileal mucosal surfaces of pigs, was implemented by us using the RNA-based methodology (Tajima et al. 2013). Mucosal samples of the terminal ileum were collected from 45 weaned piglets, and, after RNA extraction, an RT-PCR-based method was used to retrieve the 16S rRNA gene sequences. Clone libraries were constructed, cloned DNA inserts sequenced, and the resulting sequences subjected to similarity search and phylogenetic analysis. Surprisingly, the predominant bacterial sequences were identified as segmented filamentous bacteria (SFB), which are still uncultivated but functionally are well recognized as potent stimulators of Th17 cell differentiation and proliferation. SFB contribute to the increase in the number of αβ intraepithelial lymphocytes (IELs), the mitotic activity and ratio of number of columnar cells to goblet cells, and induce MHC class II in small intestinal epithelial cells (Umesaki et al. 1995). SFB also interact with intraepithelial mononuclear cells in Peyer's patches (Meyerholz et al. 2002) and stimulate SIgA production (Klaasen et al. 1993; Talham et al. 1999). Other sequences from the ileal mucosal clone libraries were associated with *Helicobacter rappini*, *Clostridium colinum*, *C. disporicum*, *L. johnsonii*, and *L. amylovorus* (Tajima et al. 2013). *Helicobacter-related* OTUs such as those similar to *H. rappini* were also recovered from the ileal mucosa of pigs in another study (Mann et al. 2014). The role of helicobacteria in human gastrointestinal diseases is well established, but the consequences of colonization of the pig ileal mucosa by this group of bacteria are presently unknown.

As mentioned above, the mucosa-adherent community of lactobacilli is very similar in terms of abundance and diversity along the porcine gastrointestinal tract (see also Krause et al. 1995). Microbial molecular diversity analyses of biopsy samples from the human colon performed by Zoetendal and coworkers (2002) established that the predominant bacterial communities from different locations of the colon are similar to each other, but are significantly different compared to fecal samples of the same subject. Mucosal surfaces of the pig gut are also associated with distinct populations that are present at different gut sites and include the *Lactobacillus*, *Prevotella*, and *Helicobacter* genera (Mann et al. 2014). At the same time, the luminal contents harbor a substantially different microbial diversity, especially in the distal gut (Leser et al. 2002; Tajima et al. 2010b; Isaacson and Kim 2012).

These observations suggest that mucosa-associated and luminal microbial populations are dissimilar and may play different functional roles in the gut. Given the probiotic potentials of lactic acid bacteria, their close association with the gut mucosa may confer health benefits to the host. This suggestion is supported by recent experiments that included the incubation of porcine intestinal epithelial cells with lactic acid bacteria (Hosoya et al. 2011; Shimazu et al. 2012). In particular, exposure of the porcine intestinal epitheliocyte cell line to *Lactobacillus casei* MEP221106 beneficially regulated the antiviral immune response triggered by poly(I:C) stimulation (Hosoya et al. 2011). Another *Lactobacillus* species, *L. jensenii*, elicited a potent anti-inflammatory activity in porcine intestinal epithelial cells (Shimazu et al. 2012). This effect is achieved by modulating the negative regulators of the Toll-like receptor-signaling pathway, namely, A20, Bcl-3, and MKP-1.

5.5 Colonization and Succession of Microbiota in the Porcine Gut

Our knowledge regarding the dynamics of microbiota composition of the gastrointestinal tract during the pig ontogeny is rather limited. In general, however, the dynamics of gut colonization is remarkably similar in humans, chicken, pigs, and cows (Mackie et al. 1999), thus complementing the insufficient knowledge on colonization and succession in the porcine gut by other models. The important factors governing colonization and succession of microbiota can be internal such as host genetics and immunity and external such as diet and environmental microbial load. In general, *in utero* conditions are considered to be sterile, although there are some observations contradicting this assumption and suggesting that gut colonization may start before birth (Jiménez et al. 2008; Satokari et al. 2009; Hong et al. 2010; Mshvildadze et al. 2010; Gosalbes et al. 2013). It needs to be emphasized here that the presence of bacteria in the latter cases has been detected by molecular methods; the cultivation-based approach is less sensitive and can give negative results on the samples that are found to be positive for bacteria by molecular tools. The bacteria consistently detected in the meconium of newborns are lactic acid bacteria and enterobacteria, with less frequently detected enterococci, streptococci, and staphylococci.

Previous cultivation-based studies (Smith 1965) demonstrated the presence of *Escherichia coli*, *Lactobacillus*, and *Streptococcus* in the stomach and small intestine 3 h after birth. In the large intestine, these bacteria are detectable 12 h after birth. The obligate anaerobic bacterium, *Bacteroides*, is not found in the stomach or small intestine, but it is found in the large intestine 2 days after birth. Thus, the initial colonization by aerobes and facultative anaerobes allows depleting oxygen in the pig gastrointestinal tract and creating the suitable conditions for the subsequent colonization by obligate anaerobes (Swords et al. 1993).

Inoue and coworkers analyzed the rectal microbiota of pigs from the first day after birth, at weaning, and post-weaning until the grower stage (49 days) using the PCR-TGGE technique (Inoue et al. 2005). The results demonstrated that the bacterial diversity as judged by the number of bands on TGGE gels continuously increases immediately after birth and becomes more diversified. The diversity is reduced temporarily at weaning, but later it is increasing again, and by day 49, the diversity values are three times higher compared to the first day after the birth. The analysis also showed that *Clostridium perfringens* disappears rapidly by day 7 following birth and coliform bacteria disappears by day 25 after birth. As discussed before, these changes are thought to be due to the establishment of the anaerobic conditions in the gut allowing the growth of obligate anaerobic bacteria such as *Bacteroides*.

Similarly to other mammals, one of the important factors governing the initial colonization stages in the porcine gut is the milk-based nutrition. During the suckling period, the dominant bacterial groups include lactobacilli and streptococci that are able to utilize milk substrates such as oligosaccharides. Although present at this stage, the numbers of clostridia, eubacteria, bifidobacteria, and *Bacteroides* are lower (Mackie et al. 1999). Another important factor in pig gut colonization is the host-microbiota interaction implemented via the signaling components of the microbiota and the immune system. Similarly to other newborn mammals, the pig immune system is Th2-biased and possesses very few immune cells. At this stage, immunoglobulins supplied by the mothers' milk provide the protection against invading pathogens and also affect the composition of the commensal gut microbiota.

The most crucial stage in the development of pigs is the immediate post-weaning period. It is characterized by the increased morbidity and mortality rates, which results in major economic losses to the pig industry. The estimates for the EU countries revealed that the mortality rate at weaning is about 17 % of the pig population, and a considerable part of it is due to the gastrointestinal infections (Lalles et al. 2007). The current intensive farming practices resulted in the change of weaning to the earlier time, thus exposing the infection-susceptible piglets with immature immune system to potential pathogens. Another dramatic impact in this period is the stress caused by separation from the mother and changing environment. This is topped with the introduction to solid feed that leads to the reduced feed intake, change in dietary antigens, removal of the immune protection by immunoglobulins in the mother's milk, and drastic changes in gut microbiota. The main factor governing microbiota succession is the shift in nutrition, with the introduction of solid feed and the switch from the lipid- and protein-rich milk to the carbohydrate-rich diet. This nutritional switch contributes to the increased diversity of gut microbiota (Inoue et al. 2005), and it becomes more closely resembling the adult microbiota, with the prevalence of the two major phyla, *Bacteroidetes* and *Firmicutes* (Swords et al. 1993; Mackie et al. 1999). The prevalence of these two phyla in feces of pigs after weaning has been confirmed also by independent cultivation-free approaches using the 16S rRNA gene libraries and metagenomics (Tajima et al. 2010b; Lamendella et al. 2011). In the metagenomic approach, it has been established that the carbohydrate-rich diet forms a distinct gene repertoire in the swine distal gut, which is highly enriched in genes encoding carbohydrate metabolism proteins as well as proteins with homology to putative carbohydrate membrane transporters (Lamendella et al. 2011).

It is well known that the exposure to certain bacteria drives the further development of the immune system. For example, the previously discussed clostridial group, SFB, largely drives the differentiation and expansion of Th17 cells, while *Bacteroides* contributes to the development of the regulatory T helper cells. The abundance of the SFB on the ileal mucosal surfaces of piglets has been found by us using an RNA-based approach (Tajima et al. 2013). Although yet to be cultivated, the role of SFB in the maturation of gut immune system in rodents as well as the molecular mechanisms of this process have been well established (Klaasen et al. 1993; Umesaki et al. 1995; Talham et al. 1999; Meyerholz et al. 2002; Schnupf et al. 2013). SFB genomes from the rat and mouse hosts have been sequenced and demonstrated a substantial degree of genome reduction, with highly auxotrophic needs and adaptations to the intestinal environment (Prakash et al. 2011; Sczesnak et al. 2011). Our phylogenetic analysis with the 16S rRNA genes of SFB from mice, rats, pigs, humans, and rainbow trout suggested that these bacteria are host-specific and cluster accordingly to the host from which they have been isolated (Tajima et al. 2013). Although the functions of SFB in the porcine gut may be similar to the better-studied rodent models, this assumption has to be independently verified by direct experiments in pigs.

As discussed above, even after reaching the adult configuration, the composition of the pig intestinal microbiota is changing, at least in the large intestine (Kim et al. 2011). From 10 to 22 weeks of age, the proportion of the *Bacteroidetes* is decreasing, with the concomitant increase of the *Firmicutes*. The genera that are mostly responsible for this shift belong to *Prevotella* (in the *Bacteroidetes* phylum) and *Anaerobacter* (in the *Firmicutes* phylum). Principal coordinate analysis of the OTUs shows a tight clustering of animals of the same age compared to animals of different ages. Since the animals in this study have been maintained on the same diet during the trial, the microbiota changes may reflect the changes in host metabolism and physiology associated with the aging process (Kim et al. 2011).

5.6 Antibiotic Effects on Pigs and Porcine Gut Microbiota

The first report suggesting that the low-dose antibiotics, streptomycin and sulfasuxidine, in feed cause the increased weight gain in food animals (in this case, chickens) appeared in 1946 (Moore et al. 1946). This was followed by the discovery of growth-promoting properties of another antibiotic, aureomycin, on poultry (Stokstad et al. 1949; Stokstad and Jukes 1950). Further investigations established that the addition of streptomycin in combination with vitamin B_{12} to the feed substantially increases weight gain performance in young pigs (Luecke et al. 1950). Since then the use of subtherapeutic antibiotics in food animal industry (and particularly in pigs) rapidly gained a worldwide acceptance aimed at the promotion of growth and enhancement of feed conversion efficiency (Jukes 1972; Kiser 1976). In fact, more than a half of antibiotics produced in the USA are used in agriculture (Lipsitch et al. 2002).

The metaphylactic use of antibiotics in pigs mainly tackles the high morbidity and mortality rates during the weaning period as well as possible infection spread during all stages of pig production. The intensive and highly industrialized pig production systems dictated earlier weaning ages, thus exposing vulnerable piglets with underdeveloped immune system to potential pathogens. Besides the viral infections caused by porcine epidemic diarrhea virus, porcine kobuvirus, porcine bocavirus, porcine group A rotavirus, and transmissible gastroenteritis virus, many other gastrointestinal diseases of pigs are of bacterial origin and include *E. coli*, *C. perfringens*, *Lawsonia intracellularis*, *Salmonella enterica*, and *Brachyspira* (*Serpulina*) spp. infections (Moxley and Duhamel 1999). The preventive use of antibiotics to control the diseases of the weaning period has been widespread (Anderson et al. 2000; Dritz et al. 2002), and in fact they are still in the use for this purpose in many countries.

At the same time, there have been many disputes concerning potential dangers of this practice such as a contributing factor to the emergence of antibiotic-resistant human and animal pathogens (Swann 1969; Levy 2002; Wegener 2003; Singer et al. 2003; Angulo et al. 2004). As a precautionary measure, Sweden banned all growth-promoting antibiotics for all food animals in 1988; this was followed by the ban on avoparcin and virginiamycin in Denmark in 1995 and 1998, correspondingly (Casewell et al. 2003). The European Union (EU) banned the use of avoparcin in 1997, and the four remaining antibiotics used for growth promotion (bacitracin, spiramycin, tylosin, and virginiamycin) were banned in 1999. The complete ban on all growth-promoting antibiotics in agricultural animals in the EU came into effect in 2006 (Anadon 2006). In the USA, the Food and Drug Administration released a draft guidance in 2012, which recommends the farmers a judicious use of medically important antibiotics in food animals but on the voluntary basis (http://www.fda.gov/downloads/animalveterinary/guidancecomplianceenforcement/guidanceforindustry/ucm216936.pdf).

Despite the long-term practice and a lot of research devoted to the clarification of action of growth-promoting subtherapeutic antibiotics, the exact mechanisms of their action still remain unclear. The mechanisms initially proposed by Wallace (1970) included metabolic effect, nutrient sparing effect, and disease control effect. Recently we hypothesized that there might be also the direct effect of antibiotics on the host (Aminov 2013).

As discussed earlier, antibiotics in the swine industry have been largely used for growth-promoting and metaphylactic purposes. Besides the obvious selective effects of antibiotics resulting in the amplification of the antibiotic resistance gene pool among the gut inhabitants, antibiotics also alter the structure of gut populations. From the time of the discovery of the growth-promoting properties of tetracycline antibiotics, they remained one the most popular growth promoter additives in the swine industry. Chlortetracycline fed at concentration of 50 mg/kg of feed to weaning piglets results in three major shifts of ileal microbial populations: a decrease in *Lactobacillus johnsonii*, an increase in *L. amylovorus*, and a decrease in *Turicibacter* phylotypes (Rettedal et al. 2009).

Another most popular antibiotic in the pig industry of the USA is a macrolide antibiotic tylosin, which has multipurpose applications including growth promotion (grower and finisher pigs), metaphylaxis, and infectious disease treatment, in particular caused by *L. intracellularis* (Dewey et al. 1999). The effect of continuous tylosin administration or the rotation of five other antibiotic combinations (chlortetracycline, sulfathiazole, and penicillin; bacitracin and roxarsone; lincomycin; carbadox; and virginiamycin) on the porcine ileal microbiota was studied in 5-week-old barrows using PCR-DGGE approach followed by sequencing of DGGE bands (Collier et al. 2003). All antibiotics decreased the total number of bacteria in the ileum, while tylosin treatment increased the percentage of lactobacilli. The authors hypothesized that suppression of bacteria by antibiotics may reduce host-related intestinal or immune responses, thus diverting more energy for growth, while the selection of lactobacilli by tylosin may offer health benefits associated with this group of bacteria.

Another study of the effect of tylosin (at the commonly used growth-promoting concentration of 40 g/t of feed) on pig fecal microbiota used the amplification of the V3 hypervariable region of bacterial 16S rRNA genes followed by pyrosequencing of the resulting amplicons (Kim et al. 2012). At the phylum level, there was no significant difference between the tylosin-treated and tylosin-free groups regarding the proportion of the major phyla represented by the *Firmicutes*, *Bacteroidetes*, *Spirochaetes*, *Actinobacteria*, *Proteobacteria*, TM7, and unclassified phyla. The differences began to be evident at the lower taxonomic level: a total of 12 differentially abundant genera were identified at the genus level. The six most abundant genera consisted of *Prevotella*, *Lactobacillus*, *Sporacetigenium*, *Megasphaera*, *Blautia*, and *Sarcina*, and the less abundant genera consisted of *Barnesiella*, *Mitsuokella*, *Acetanaerobacterium*, *Anaerosporobacter*, *Succinivibrio*, and *Eggerthella*. *Lactobacillus*, *Sporacetigenium*, *Acetanaerobacterium*, and *Eggerthella* were detected more frequently in the tylosin-fed group, while the others were more frequent in the tylosin-free animals. The increase of *Lactobacillus* in antibiotic-treated pigs has been noticed in other works as well (Collier et al. 2003; Rettedal et al. 2009; Looft et al. 2012).

One of the popular drug mixes routinely used in the US pig industry for growth-promoting purposes and for the treatment of bacterial enteritis is ASP250, which consists of chlortetracycline, sulfamethazine, and penicillin (Looft et al. 2012). The impact of ASP250 on the swine gut microbiota was investigated using the phylogenetic, metagenomic, and quantitative PCR-based approaches (Looft et al. 2012). The ASP250 treatment resulted in the decrease of bacteria belonging to the *Bacteroidetes* phylum. At the genus level, there was a significant increase in *Proteobacteria*, which was largely driven by the increase of *E. coli* populations. The negative responders to the drug mix were *Anaerobacter*, *Barnesiella*, *Papillibacter*, *Sporacetigenium*, and *Sarcina*. Metagenomic analyses revealed the enrichment by microbial functional genes relating to energy production and conversion in the antibiotic-fed pigs. As expected, the pool and diversity of antibiotic resistance genes in the gut microbiota of antibiotic-fed animals substantially increased. Interestingly, some antibiotic-resistant genes such as conferring resistance to aminoglycosides also increased despite the absence of a direct selection by the antibiotics in the mix, chlortetracycline, sulfamethazine, and penicillin. This is a good example of antibiotic-resistant gene co-selection due to the close location of antibiotic resistance genes on mobile genetic elements (Aminov and Mackie 2007; Aminov 2011, 2012).

Another study involving the use of ASP250 (and also carbadox) investigated their impact on swine intestinal bacteria and viruses (Allen et al. 2011). As discussed before, the swine gut harbors a substantial diversity of phages (Shan et al. 2011; Lager et al. 2012). Feeding ASP250 caused significant population shifts in both the phage and bacterial communities (Allen et al. 2011). In particular, this resulted in an increase of the phage populations suggesting the induction of prophages with antibiotic treatment. In parallel, this caused a significant decrease of bacteria in the *Firmicutes* phylum, specifically members of the *Streptococcus* genus. Regarding carbadox, it has been established earlier that the antibiotic induces the production of phage particles and contributes to the transmission of tylosin and chloramphenicol resistance determinants between the strains of the etiologic agent of swine dysentery, *Brachyspira hyodysenteriae* (Stanton et al. 2008).

As mentioned above, in the face of the growing problem of antibiotic resistance among the human and animal pathogens, certain practices of antibiotic use have been revisited. In particular, the controversial practice of growth-promoting use of antibiotics has been reconsidered in the light of its possible contribution to the dissemination of antibiotic resistance genes. Presently the use of antibiotics as growth promoters in animal industry in the EU countries is prohibited, but the use is still allowed for therapeutic purposes. Thus, despite the ban, the use of antibiotics in animal industry, especially in swine production, remains high. Another concern associated with the excessive use of antibiotics in agriculture is the residual antibiotics in food. It is hypothesized that the chronic exposure to low-residue antibiotics in food may cause intestinal dysbiosis resulting in body physiology changes (Riley et al. 2013). This exposure is probably similar to the growth promotion effect of antibiotics in food animals, and in humans it is expressed as the current obesity epidemic in many countries, especially in the USA. Substitutes to antibiotics hence are actively sought, and one of them is the use of phage therapy as a possible alternative for antibiotic treatment.

5.7 Phage Therapy and Prevention in Pigs

Historically, the first observation of the transmissible lytic agent that is specifically active against *Bacillus anthracis* has been made by a Russian microbiologist Nikolay Gamaleya in 1898 (Gamaleya 1898). The potential of bacteriophages for infection treatment has been recognized after the acknowledged discoveries by Frederick Twort and Felix d'Hérelle in 1915 and 1917, respectively (Twort 1915; D'Herelle 1917). The further phage therapy developments, however, have been overshadowed mainly by the success of antibiotics in infection control and treatment, and the phage therapy research and development persisted only in small scale in the former Soviet Union countries, Russia and Georgia, and also in Poland.

There is a renewed interest in the replacement of antibiotic therapy by phage therapy and prevention of bacterial infections (Thiel 2004), and this attention is dictated by a number of advantages that this approach offers. First, unlike a wide range of bacteria targeted by antibiotics, bacteriophages are very specific and do not affect other beneficial bacteria in the complex microbial ecosystem of the gut. Complications such as dysbiosis and diarrhea frequently occur in pigs treated with antibiotics (Tsukahara et al. 2003), but the intervention by phages targets only disease-causing bacteria, thus preserving the integrity of the gut microbial ecosystem. Second, the bacteriophages are multiplied at the sites where the target presents, thus amplifying the local antibacterial effects. Third, no serious side effects of phage therapy have been so far detected. And finally, the phage-resistant bacteria remain sensitive to other bacteriophages, and selection of new bacteriophages or bacteriophage cocktails is a much faster process compared to the development of new antibiotics. Besides, the use of sufficiently high phage concentrations helps to overcome the phage resistance barrier (Jia and Mann 2009), the phenomenon known from the 1940s as "lysis from without" (Delbrück 1940).

The use of phages and phage cocktails in agriculture is expanding fast ranging from agricultural animals to plants (Jones et al. 2007; Johnson et al. 2008). In pigs, however, the progress has been rather slow, and the currently available literature covers only a narrow range of intestinal bacterial infections and conditions that are targeted by phage therapy. These are focused on the carriage of, and the contamination of pork by, *Salmonella*, and diarrhea caused by enterotoxigenic *E. coli* (ETEC) infections. Together with eggs and poultry meat, the contamination of pork by *Salmonella* is a major cause for the corresponding salmonellosis in humans (Fosse et al. 2008). Difficulty in controlling *Salmonella* in pigs has to do with the fact that its carriage and shedding are usually asymptomatic and the infection can be exposed only in susceptible pigs as well as during the transportation and slaughter.

Under experimental conditions, administration of an anti-Salmonella phage cocktail at the time of challenge by *Salmonella enterica* serovar Typhimurium reduces the *Salmonella* count in the tonsils, ileum, and cecum by two to three orders of magnitude (Wall et al. 2010). Under production-like settings, the effect is less pronounced, with 95 % and 90 % reduction of *Salmonella* in the cecum and ileum, correspondingly (Wall et al. 2010). In another set of *S.* Typhimurium challenge experiments, the administration of a phage cocktail at 24 and 48 h post-challenge also resulted in the decrease of the *Salmonella* count, but the decrease was significant only for the rectal population (Callaway et al. 2011). Administration of a microencapsulated anti-*Salmonella* phage cocktail in the feed too resulted in a substantial decrease of *Salmonella* colonization and shedding in pigs (Saez et al. 2011).

Jamalludeen and others investigated the effect of six phages or the combinations thereof on the prevention and treatment of diarrhea due to the experimental enterotoxigenic *E. coli* O149 (ETEC) infection of weaned pigs (Jamalludeen et al. 2009). When given prophylactically or therapeutically, the phage cocktails substantially moderated diarrhea in these challenge experiments. In another study, prophylactic dosing of pigs with a lytic phage CJ12 1 week in advance made them less susceptible to diarrhea caused by the ETEC challenge (Cha et al. 2012). Moreover, the phage-treated animals showed a reduced ETEC shedding compared to the control group. Thus, the preliminary results of prophylaxis and treatment of *Salmonella* and ETEC infections in pigs are promising, and it is expected that this approach will be extended to cover other bacterial infections in pigs.

5.8 Probiotics in Pigs

Among other potential substitutions for antibiotics, the use of probiotics in swine industry has received much more attention and associated research efforts compared to other alternatives because of the long history and good safety record of probiotic use in humans. The initial idea of colonization of the gastrointestinal tract by beneficial microorganisms belongs to Élie Metchnikoff, who noticed, more than a century ago, that the largest proportion of centenarians live in Bulgaria. He attributed this longevity to the extensive consumption of milk fermentation products, in particular Bulgarian yogurt (кисело мляко in Bulgarian, sour milk). He developed a theory that aging is caused by toxic bacteria in the gut, and he encouraged the use of Bulgarian yogurt and its principal component, *Lactobacillus delbrueckii* subsp. *bulgaricus*, to prevent this toxicity (Metchnikoff 1907). As mentioned earlier in this chapter, lactobacilli represent one of the dominant groups along the mucosal surfaces of the pig gut, possibly contributing to the animal health. In the meantime, the advocated use of lactic acid bacteria as agents beneficial for health has grown tremendously, and in fact, the global probiotics demand was worth USD 27.9 billion in 2011 and is expected to reach USD 44.9 billion in 2018 (Transparency Market Research 2013).

The original definition of probiotics introduced by Daniel Lilly and Rosalie Stillwell (1965), however, had a different meaning than that of the Metchnikoff's, and it has been applied to the protozoa, in particular to the growth promotion of *Tetrahymena pyriformis* in response to a factor produced by *Colpidium campylum*. Only at the later stages, through several refinements, the definition has been converged with the original idea of Metchnikoff, and the current definition of probiotics is "live microbial feed supplement which beneficially affects the host animal by improving its intestinal microbial balance" (Fuller 1989).

Commonly used probiotics include mainly lactic acid bacteria, particularly lactobacilli, bifidobacteria, lactococci, and streptococci. Less commonly used are yeasts, bacilli, and nonpathogenic *E. coli* strains (Fuller 1989; Tannock 1995; de Vrese and Schrezenmeir 2008). In pigs, the most investigated probiotics are lactobacilli, enterococci, and yeasts. Feeding of *L. plantarum* Lq80 to weaning piglets, for example, stimulated the growth of indigenous lactobacilli (Takahashi et al. 2007). Feeding of a lactic acid-producing bacterium, *Pediococcus acidilactici*, positively influenced weight and post-weaning average daily gain of treated piglets (Di Giancamillo et al. 2008). The treated piglets also showed a larger number of proliferating enterocytes and an increase in villi height and crypts depth. In pigs fed *Saccharomyces boulardii*, an increase of villus length in the small intestine and a decrease in the number of goblet cells were observed (Baum et al. 2002). The use of live yeast dietary supplementation in weanling piglets improved nursery growth performance and promoted a "healthy" intestine (Bontempo et al. 2006). Mallo and coworkers reported about the improved intestinal microbiota and growth performance in piglets fed the diet with the addition of *Enterococcus faecium* (Mallo et al. 2010). There are many other reports claiming the health and growth benefits of probiotic treatment in pigs, and they are summarized in a recent review (Cho et al. 2011).

The majority of probiotics demonstrated in the examples above are facultative anaerobes, and it is thought that their main effects are mediated through the production of lactate and other SCFAs that may inhibit the pathogens and affect the composition of gut microbiota. Probiotic bacteria can also produce a number of antibacterial agents, thus protecting from colonization by pathogens. Besides of SCFAs mentioned, these may include the production of hydrogen peroxide and various bacteriocins. For example, the Abp 118 bacteriocin produced by *L. salivarius* displays a potent protective effect against the invasive foodborne pathogen *Listeria monocytogenes* (Corr et al. 2007).

Another presumed action of probiotics that protects the animal from invading pathogens is supposedly through the competitive exclusion when the ecological niches of the gut targeted by pathogens are occupied and protected by the probiotics. This can be implemented through the competition for specific physicochemical and nutritional biological niches in the gut such as pH, oxygen, redox potential, and nutrient availability. Spatial localization of probiotics in close proximity to the gut mucosa also plays an important protective role, and it is realized largely via the adhesion-like proteins (ALPs) of probiotic strains. For probiotic bacteria, for example, adhesion to the host mucosa is considered to be as a very desirable characteristic, enhancing the probiotic potential (Juge 2012). Besides, close proximity to the host cells allows bacteria to exert potent physiological effects, reversing metabolic disorders and controlling inflammation, gut barrier function, and gut peptide secretion (Everard et al. 2013). Another function of ALPs may be blocking and "masking" the sites of attachment for pathogens, thus contributing to colonization resistance. Moreover, colonization resistance may be enhanced by the production of surface glycans resembling those found in the gut mucosa, which may serve as the "false targets" for the pathogen attachment.

Finally, as mentioned before, the probiotic bacteria potentially modulate the systemic immune response through the mucosal immune system (Corthésy et al. 2007; Hosoya et al. 2011; Shimazu et al. 2012). The immunomodulatory properties of probiotics, however, are not displayed consistently and depend on the strain used, age of animals, and a number of other factors (Ohashi and Ushida 2009). It is advocated, therefore, to search for high-potential health-beneficial bacteria from the human or animal microbiota that can be applied in a probiotic-like manner for this purpose (van Baarlen et al. 2013).

As mentioned above, similarly to the mechanisms of action of probiotics in humans, the production of large amounts of lactic acid is thought to create an unfavorable environment for certain bacteria, including pathogenic strains of *E. coli*, *Salmonella*, or *Shigella*. Although some probiotic lactic acid bacteria in the form of liquid feed may display beneficial effects in pigs (Tajima et al. 2010a), in general, the consistency of probiotic use as feed additives in pigs is poor, and it is difficult to discern a sufficient confidence from the results of the current probiotic use studies that are not properly designed, very variable, and lack statistical significance (Kenny et al. 2011). Also, the economic return from the use of conventional probiotics in pig production presently cannot be justified. The same problems of probiotic application outcomes such as poor reproducibility and the lack of significance compared to antibiotics are also typical for other agricultural animals (Allen et al. 2013). Besides probiotics, a number of other alternatives to antibiotics in swine industry have been tried, but they are too producing inconsistent results and rarely equal antibiotics in their effectiveness (Thacker 2013).

In order to deliver consistent results, probiotic concentrations should be sufficiently high, and they should be delivered frequently and regularly since they are not efficient colonizers of the pig gut. One approach that satisfies these requirements is the delivery of a probiotic strain with the fermented liquid feed. Since a strain is initially used in feed fermentation, the concentration of cells in the feed may reach up to 10^{10} CFU/g of feed (Tajima et al. 2010a). Fed twice daily, the fermented liquid feed regularly delivered a high number of *L. plantarum* to the gut of weaned piglets, with significant effects on a number of parameters. In particular, bacterial diversity in the ileum and cecum was increased, especially that of the *Firmicutes* phylum represented by the *Dorea*, *Coprococcus*, *Roseburia*, and *Faecalibacterium* genera. Also, there was a statistically significant difference between the microbiota composition of the antibiotic-fed and fermented liquid feed-fed animals suggesting different mechanisms implemented by these two treatments on the composition of gut microbiota.

Another way for a consistent delivery of high-titer probiotics is the use of by-products of fermentation in swine nutrition. A recent study investigated spent cider yeast as a dietary probiotic supplement for modifying gut microbiota in weaning pigs (Upadrasta et al. 2013). The daily total ingestion of viable yeast cells reached up to $\sim 1 \times 10^{10}$ CFU per piglet, and the feeding continued for 21 consecutive days. The bacterial composition of fecal samples was evaluated by pyrosequencing of 16S rRNA tags (V4 region). At the end of the feeding trial, the composition of the fecal bacterial microbiota substantially changed: the relative abundance of the *Firmicutes* was significantly lower in the treatment group at 47.6 % compared to control (60.74 %), while the *Bacteroidetes* increased to 42.4 % compared to control (15.4 %). Interestingly, a significant decrease in the plate counts of total numbers of *Lactobacillus* sp. was observed in the cider yeast-supplemented group compared to the control group, while the opposite trend was observed for butyrate-producing organisms, such as *Faecalibacterium* spp. Most importantly, however, the relative abundance of enteric pathogens such as *S. enterica* and *Escherichia fergusonii* was reduced in the cider yeast supplemented group compared to the control group (Upadrasta et al. 2013). In general, however, the effect of probiotics depends on many factors, and the current understanding is that specific probiotic strains, which are suitable for each animal species and their life stage as well as each individual, should be found (Ohashi and Ushida 2009). Also, in terms of cost-efficiency, the "classical mode" of probiotic delivery using the tablet forms similar to human applications is unlikely to be implemented in the pig industry.

5.9 Prebiotics in Pigs

The concept of prebiotics was initially introduced in 1995 by Gibson and Roberfroid, with the next definition: "a nondigestible food ingredient that beneficially affects the host by selectively stimulating the growth and/or activity of one or a limited number of bacteria in the colon, and thus improves host health" (Gibson and Roberfroid 1995). They addressed the limitations associated with the maintenance of health-promoting effects of probiotics that we discussed earlier in the previous section of this chapter. In particular, they indicated that the beneficial changes in microbiota balance due to use of probiotics are transient because the establishment and survival of the exogenous bacteria is limited. Instead, they proposed the use of prebiotics that are nondigestible food ingredients, which beneficially affect the microbiota balance by selectively stimulating the growth and/or activity of one or a limited number of endogenous microbial species. They illustrated these suggestions by the examples of prebiotics such as nondigestible oligosaccharides (in particular fructooligosaccharides) that stimulate the growth of endogenous bifidobacteria, which become predominant after a short feeding period. Moreover, they also proposed the concept of synbiotics that combines the rationale of pro- and prebiotics (Gibson and Roberfroid 1995).

Since many food oligosaccharides and polysaccharides have been claimed as having prebiotic activity, clear criteria to define prebiotics are needed (Gibson et al. 2004). The proposed criteria are: (1) resistance to gastric acidity, to hydrolysis by mammalian enzymes, and to gastrointestinal absorption; (2) fermentation by intestinal microflora; and (3) selective stimulation of the growth and/or activity of those intestinal bacteria that contribute to health and well-being (Gibson et al. 2004). Further refinement of the prebiotic definition to include other areas that may benefit from selective targeting of particular components of microbiota arrived at "a selectively fermented ingredient that allows specific changes, both in the composition and/or activity in the gastrointestinal microflora that confers benefits upon host well-being and health" (Roberfroid 2007). According to the definition and the criteria proposed, only two compounds, inulin and oligofructose, could fulfill these requirements at that time (Roberfroid 2007). Further studies suggested that lactulose and resistant starch could also meet all aspects of the definition and criteria (Slavin 2013). Other chemically defined structures such as galacto-oligosaccharides, transgalactooligosaccharides, and polydextrose as well as foods such as acacia gum, psyllium, banana, whole grain wheat, and whole grain corn may also have prebiotic effects (Slavin 2013).

There is a problem, however, on how the prebiotic approach may address the stimulation by specific substrates of other taxa representatives, which are more dominant components of the gut microbiota. The metabolic capabilities of the presently recognized beneficial bacteria such as lactobacilli and bifidobacteria are well established, and these allow specific stimulation by using, for example, bifidogenic prebiotic substrates. One of the major phyla in the colon is the *Bacteroidetes*, with some species of the phylum demonstrating clear health-promoting properties. For example, beneficial effects of *Bacteroides fragilis* on the immune system are well established (Troy and Kasper 2010). It also displays other health benefits such as correcting gut permeability, altering microbial composition, and ameliorating defects in communicative, stereotypic, anxiety-like, and sensorimotor behaviors (Hsiao et al. 2013). In the face of complexity of the gut microbiota with a large diversity of *Bacteroides* phylotypes, presumably having similar nutritional requirements, it is not clear how the *B. Fragilis* populations can be targeted for specific stimulation.

In pigs, the well-defined and confirmed prebiotic inulin (Roberfroid 2007) and inulin-type fructans were investigated for health-promoting properties (Tako et al. 2008). Inulin improves iron and calcium absorption, while the products of fermentation in the large intestine, SCFAs, stimulate the growth of health-promoting bacteria such as bifidobacteria and lactobacilli (Tako et al. 2008). Butyrate also has a direct effect on the colonic health as a preferred substrate for colonocytes and via regulation of proliferation, differentiation, and apoptosis of epithelial cells (Wong et al. 2006; Tako et al. 2008). Besides, it inhibits the enzyme histone deacetylase and decreases the transformation of primary to secondary bile acids as a result of colonic acidification (Wong et al. 2006). At the same time, there are no indications that inulin may have any effect on the growth performance of pigs (Jolliff and Mahan 2012). Besides, inulin is too expensive to be used in pig production on a regular basis.

In pigs, pectins promote fermentation at the terminal ileum and in the first part of the colon (Drochner et al. 2004). Chicory (*Cichorium intybus* L.) forage has high pectin content (80–90 g/kg dry matter), which is composed of polymers of galacturonic acid, which is highly soluble in comparison with other pectin sources (Ivarsson et al. 2011). The gut microbiota modulation properties of chicory have been demonstrated in several animal trials. For example, the inclusion of chicory forage and roots in the pig diet affected the intestinal microenvironment and the gut microbiota (Liu et al. 2012). In particular, the chicory forage diets increased the relative proportion of lactic acid bacteria in the distal ileum as well as butyrate-producing bacteria and *Megasphaera elsdenii* in the colon. The bacterial shifts were accompanied by the changes in the gut microenvironment such as the colonic concentration of SCFAs. In another investigation, the inclusion of chicory in the growing pig diet modulated the composition of the fecal gut microbiota including *Prevotella* and *Lactobacillus* species (Ivarsson et al. 2012). The cereals substitution by chicory forage in the pig diet led to the significant decrease of dominant *Prevotella* sp. and to the increase of dominant *Roseburia* sp. in the large intestine (Liu et al. 2013). Thus, the inclusion of chicory in the pig diet has beneficial effects on animals. Its widespread use, however, is limited by cost considerations; at least in the Danish pig industry, the economic returns cannot be justified (Dr. Mette Boye, personal communication).

There are a number of other compounds that have been tried in the porcine model for the prebiotic properties. These experimental trials are rather of the translational medicine nature than research intended for the use in pig industry. For instance, the human milk oligosaccharides tested in piglets exert potent prebiotic and anti-infective activities (Hester et al. 2013), but they are virtually absent in infant formulae. One of the ways to restore the effect of the natural oligosaccharides is the fortification of infant formulae by other prebiotics. Thus, the enrichment of infant formula by polydextrose displayed certain characteristics consistent with prebiotic effects in suckling piglets (Herfel et al. 2011). In particular, the enrichment increased ileal lactobacilli and propionate and lactate concentrations but decreased pH, with the concomitant changes in ileal cytokine expression. Interestingly, an artificial sweetener SUCRAM®, consisting of saccharin and neohesperidin dihydrochalcone, displayed prebiotic-like effects in pigs by dramatically increasing the cecal population of *Lactobacillus* (Daly et al. 2014).

5.10 Other Feed Additives in Pigs

A number of other feed additives for the enhancement of the pig growth and health such as metals, antimicrobial peptides, clay minerals, egg yolk antibodies, essential oils, eucalyptus oil-medium chain fatty acids, rare earth elements, and recombinant enzymes have been tried (Thacker 2013). The two most popular feed additives used in the pig industry are zinc and copper. These are trace elements, but when used at high concentrations (copper at 100–250 ppm and zinc at 2,000–3,000 ppm), they increase growth performance of young pigs (Jacela et al. 2010). Copper is usually administered in the form of copper sulfate as well as tribasic copper chloride (Cromwell et al. 1998). High levels of dietary copper contribute to the increase in the growth rate of pigs, which is comparable with in-feed antimicrobials (Cromwell 2001; Pettigrew 2006). The growth-promoting effect of copper is additive when used together with antimicrobials (Cromwell 2001). Contrary to antibiotics, however, the copper effects seem limited to the earlier stages of the pig development, and it is less effective when used for prolonged periods of time (Hastad et al. 2001). The exact mechanism of the growth-promoting effect of copper is lacking. One of the explanations proposed pointed to its antimicrobial activity as a possible mechanism (Dupont et al. 1994). Care must be taken to dose copper correctly: overdosing problem resulting in jaundice may be encountered when copper is used in excess of 250 ppm for long periods (National Research Council 1998). Another negative consequence of copper feeding to pigs is that it increases the unsaturated fat content, which results in a softer pork fat (Pettigrew and Esnaola 2001).

The growth-promoting properties of zinc have been noticed long time ago (Hahn and Baker 1993). Zinc oxide is also used for prophylaxis of diarrhea in weaning piglets (Poulsen 1995). The action of zinc can be implemented via effects on the host biology by modulating the host gene expression as well as by affecting the composition of the gut microbiota. In a recent study by Vahjen and others, it has been established that feeding high concentrations of zinc oxide to piglets significantly increased the relative abundance of ileal *Weissella*, *Leuconostoc*, and *Streptococcus*, decreased the numbers of *Sarcina* and *Neisseria*, and increased the numbers of all Gram-negative facultative anaerobic genera (Vahjen et al. 2010). In another study, however, bifidobacteria, enterococci, streptococci, *Weissella*, *Leuconostoc*, and the *Bacteroides-Prevotella-Porphyromonas* group in the small intestine were not influenced by high dietary zinc in piglets (Starke et al. 2013). At the same time, the *Enterobacteriaceae* and the *Escherichia* groups as well as *Lactobacillus* were reduced. High dietary zinc oxide intake, therefore, has a major impact on the bacterial composition of the porcine small intestine, but the results are not consistent.

Besides the effects on the porcine gut microbiota, zinc strongly affects the host biology as well. There are some indications, for example, that the stimulation of ghrelin secretion from the stomach by zinc may promote food intake by, and the growth of, piglets (Yin et al. 2009). The exact mechanisms of the growth promotion by zinc, however, still remain unknown. Various forms of zinc have been tested, and it appeared to be that its growth-promoting properties are largely independent of its relative bioavailability (Mavromichalis et al. 2000). Organic zinc sources do not show any advantage in terms of weight gain, feed intake, or feed efficiency compared to zinc oxide (Hollis et al. 2005). When overdosed, the toxic effects of zinc may be indicated by depressed pigs, arthritis, gastritis, and death (Jacela et al. 2010). Extensive use of copper and zinc in animal production raises environmental concerns since both are excreted with feces and may accumulate in the soil after manure application (Jondreville et al. 2003; Carlson et al. 2004).

Other feed additives for pigs do not have a proven record of consistent growth-promoting properties. Clay minerals, for example, have been shown to prevent diarrhea in weaned pigs, but the results of trials to find growth performance-enhancing properties have been inconclusive (Thacker 2013). Thus, clay minerals are not feasible substitutes for growth-promoting antibiotics. Analysis of data on many other compounds by the same author resulted in the conclusion that these compounds produce inconsistent results and rarely equals antibiotics in their effectiveness (Thacker 2013). More research, therefore, is needed to find viable alternatives to growth-promoting antibiotics.

5.11 Pigs in Translational Research

Despite being neglected in the past, currently the gut microbiota is considered as a separate and important organ of the body that is involved in numerous functions to maintain health and protect from disease (O'Hara and Shanahan 2006). These functions include protection against invading pathogens (colonization resistance), nutrient partitioning, and lipid metabolism (Backhed et al. 2004; Ley et al. 2006; Brown et al. 2012). Gut microbiota also play a major role in shaping and maintaining mucosal immunity by promoting gut barrier function and stimulating development of GALT as well as the systemic immune system (Kelly et al. 2005; O'Hara and Shanahan 2006). Gut microbiota contributes to the development and homeostasis in the gut as well (Stappenbeck et al. 2002; Rakoff-Nahoum et al. 2004).

Many human diseases are associated with the aberrant gut microbiome. The classical examples that include the gut *per se* are: inflammatory bowel disease (Dupaul-Chicoine et al. 2013), irritable bowel syndrome (Ohman and Simrén 2013), recurrent *Clostridium difficile* infections (Theriot and Young 2013), and a number of other functional gastrointestinal disorders (Simrén et al. 2013). The involvement of gut microbiota in human disease, however, extends well beyond the gut boundaries, and the spectrum of the gut microbiota-affected diseases include endotoxemia, systemic inflammation, and subsequent obesity and nonalcoholic fatty liver disease (Duseja and Chawla 2014), cardiovascular disease (Brown and Hazen 2014), as well as a number of allergic, autoimmune, and neuropsychiatric diseases (Khanna and Tosh 2014).

Given the importance of gut microbiota in health and disease, two large-scale microbiome sequencing projects have been initiated (Peterson et al. 2009; Qin et al. 2010). Launched in 2008, the Human Microbiome Project (HMP), for example, is aimed at extensive genomic and metagenomic characterization of human oral, skin, vaginal, gut, and nasal/lung microbiota. Another initiative, MetaHIT, is a project financed by the European Commission under the 7th FP program and mainly aimed at establishment of associations between the genes of the human intestinal microbiota and health and disease. The need to coordinate these efforts brought together USA, Europe, Canada, Japan, Australia, Korea, and Gambia under the umbrella of the International Human Microbiome Consortium (http://www.human-microbiome.org).

While important, the sequence data alone cannot provide an outcome that can be immediately translated into the applications in the form of therapies or interventions. First, the proposed interventions have to be tested for safety in animal models before proceeding further to clinical trials in humans. Another need in animal models is dictated by ethical concerns regarding the involvement of humans in initial drug or treatment development. There is a pressing demand for an adequate animal model that allows the use of experimental treatment but also permits a detailed analysis of systems, tissues, and organs of the body. Besides, animal models allow performing experiments with superior design and controls than it is possible with human subjects.

For many years, the rodents such as gnotobiotic mice have been instrumental in human health research studies of gut microbiology and host-microbe interactions (Faith et al. 2010; Goodman et al. 2011). Although the rodent models offer many advantages such as the uniform genetic background, the availability of genetic manipulation tools, and rapid growth and development, there are certain drawbacks as well. First of all, there are substantial differences in the immune system of mice and humans (Mestas and Hughes 2004). Second, the marked differences in microbiota composition between mice and humans have been noted (Ley et al. 2005). In particular, the strains belonging to the genera *Bifidobacterium*, *Lactobacillus*, and *Clostridium* from a human donor did not persist in recipient mice (Raibaud et al. 1980). Third, many aspects of human physiology are different from that of the murine models. Probably because of the combined differences in immunology and physiology, the rodent models do not display the clinical manifestations of enteric diseases typical for humans. And finally, the human diet and living conditions are obviously different from that of mice.

In many aspects, the anatomical, physiological, immunological, and microbial features of the porcine gastrointestinal tract are close to that of humans. In particular, physiology of digestion, food transit times, and associated metabolic processes are very similar between humans and pigs (Patterson et al. 2008). The human and pig genomes share very extensive conserved homology, and the immune systems between the two are very similar. Besides, many aspects of gut microbiology in humans and pigs are comparable (Heinritz et al. 2013). A somewhat limiting factor could be the large size and expenses associated with the maintenance of these animals, but this should not be a major problem for translational medicine.

As noted before, the use of pigs may allow checking experimentally the hypotheses that have only epidemiological support, and where direct experimentations in humans are not possible. For example, to explain the rapidly rising number of diseases with a significant proinflammatory component such as asthma and atopic eczema, the "hygiene hypothesis" was proposed (Strachan 1989). The main idea of the hypothesis is that there used to be a substantial exposure of humans to the environmental bacteria before the advent of industrial food production resulting in essentially sterile food protected by preservatives, conservatives, and freezing and that there are generally higher hygienic standards of living restricting the exposure to the environmental microbiota. This exposure seems to have been an essential component of the immune system education, and the lack of it results in an unbalanced immune system development, with a prominent proinflammatory bias.

In order to test this hypothesis experimentally, we used pigs that were differentially exposed to the environmental microbiota (Mulder et al. 2009). The pigs were housed at either an indoor (intensive) or an outdoor (extensive) facility. In addition, the indoor pigs were receiving antibiotics. Gut microbiota analyses with 16S rRNA clone libraries exposed major differences in the microbiota associated with the ileal mucosa. The outdoor pigs showed a predominance of the *Firmicutes*, in particular of lactobacilli, while animals housed in a hygienic indoor environment had a reduced number of lactobacilli and a greater number of potentially pathogenic phylotypes. The latter group also showed an increased expression of type 1 interferon genes, major histocompatibility complex class I, and several chemokines. The human homologues of some of these chemokines have been implicated in the etiology and chronic inflammation leading to tissue damage in IBD patients (McCormack et al. 2001; Banks et al. 2003).

A review detailing the use of pigs as a potential model for research into dietary modulation of the human gut microbiota has been recently published (Heinritz et al. 2013). Transplantation of the fecal microbiota from a human donor into cesarean section-delivered germfree piglets was performed by Pang and others (Pang et al. 2007). The composition and dynamics of the transplanted microbiota were monitored using enterobacterial repetitive intergenic consensus sequence-PCR fingerprinting-based community DNA hybridization, group-specific PCR-temperature gradient gel electrophoresis, and qPCR. The transplanted microbiota displayed the community structure comparable with that of the donor, with minimal individual variations. Moreover, the dynamic of the transplanted microbiota succession during aging was also similar to the process in humans, thus validating this model for human health studies (Pang et al. 2007). On the other hand, fecal transplantation from the heterologous (human) and homologous (pig) hosts into the germfree piglets may result in differential effects on the gut microbiota and intestinal mucosal immunity of recipients (Che et al. 2009). In particular, the heterologous transplant resulted in a higher number of bifidobacteria and affected a number of intestinal mucosal immunity parameters. Another recent work also proposed a pig model of the human gastrointestinal tract, which is an attempt to create a "humanized" animal microbiota model similar to the rodent models developed earlier (Zhang et al. 2013). The number of animals used, however, was very small, and it is not possible to draw any definite conclusions whether or not the dynamic of the "humanized" microbiota in the pig intestine actually properly emulates the composition and dynamics of the gut microbiota in the original human host. More research, therefore, is necessary in order to validate this approach.

5.12 Conclusions

The porcine gastrointestinal system represents one of the best-studied models of the monogastric animal gut. Presently, with a generally less emphasis placed on the research that deals exclusively with the porcine health and disease as a food animal *per se*, the attention is turned on the use of this model in translational medicine. As demonstrated in this chapter, there are many similarities between humans and pigs that may facilitate translational research. Also, certain pig breeds (e.g., Ossabaw minipigs) display the features consistent with the clinical presentation of human disease such as the metabolic syndrome as they exhibit obesity, insulin resistance, and hypertension (Litten-Brown et al. 2010). Potential drawbacks of the pig model such as the limited availability of genetically modified and transgenic pigs are currently tackled by many research groups. With the availability of pig genomic sequences and the establishment of efficient and precise techniques for genetic modification of pigs, the problem is currently addressed in a number of ongoing projects that are aimed at the development of porcine genetic models of neurodegenerative diseases, cardiovascular diseases, eye diseases, bone diseases, cancers and epidermal skin diseases, diabetes mellitus, cystic fibrosis, and inherited metabolic diseases (Aigner et al. 2010; Luo et al. 2012b; Flisikowska et al. 2013). Thus, in addition to supplying humans with high-quality and affordable protein, the pig also becomes a favorite model in translational medicine.

References

Aigner B, Renner S, Kessler B et al (2010) Transgenic pigs as models for translational biomedical research. J Mol Med (Berl) 88:653–664

Allen HK, Looft T, Bayles DO et al (2011) Antibiotics in feed induce prophages in swine fecal microbiomes. MBio 2, e00260-11

Allen HK, Levine UY, Looft T et al (2013) Treatment, promotion, commotion: antibiotic alternatives in food-producing animals. Trends Microbiol 21:114–119

Allison MJ, Robinson IM, Bucklin JA et al (1979) Comparison of bacterial populations of the pig cecum and colon based upon enumeration with specific energy sources. Appl Environ Microbiol 37:1142–1151

Amann RI, Ludwig W, Schleifer KH (1995) Phylogenetic identification and in situ detection of individual microbial cells without cultivation. Microbiol Rev 59:143–169

Aminov RI (2011) Horizontal gene exchange in environmental microbiota. Front Microbiol 2:158

Aminov RI (2012) The extent and regulation of lateral gene transfer in natural microbial ecosystems. In: Francino MP (ed) Horizontal gene transfer in microorganisms. Horizon Scientific Press, Norwich

Aminov RI (2013) Biotic acts of antibiotics. Front Microbiol 4:241

Aminov RI, Mackie RI (2007) Evolution and ecology of antibiotic resistance genes. FEMS Microbiol Lett 271:147–161

Anadon A (2006) The EU ban of antibiotics as feed additives alternatives and consumer safety. J Vet Pharmacol Ther 29:41–44

Anderson DB, McCracken VJ, Aminov RI et al (2000) Gut microbiology and growth-promoting antibiotics, in swine. Nutr Abstr Rev 70:101–108

Angulo FJ, Baker NL, Olsen SJ et al (2004) Antimicrobial use in agriculture: controlling the transfer of antimicrobial resistance to humans. Semin Pediatr Infect Dis 15:78–85

Backhed F, Ding H, Wang T et al (2004) The gut microbiota as an environmental factor that regulates fat storage. Proc Natl Acad Sci U S A 101:15718–15723

Banks C, Bateman A, Payne R et al (2003) Chemokine expression in IBD. Mucosal chemokine expression is unselectively increased in both ulcerative colitis and Crohn's disease. J Pathol 199:28–35

Baum B, Liebler-Tenorio EM, Enss ML et al (2002) *Saccharomyces boulardii* and *Bacillus cereus* var. Toyoi influence the morphology and the mucins of the intestine of pigs. Z Gastroenterol 40:277–284

Bontempo V, Di Giancamillo A, Savoini G et al (2006) Live yeast dietary supplementation acts upon intestinal morpho-functional aspects and growth in weanling piglets. Anim Feed Sci Technol 129:224–236

Brown JM, Hazen SL (2014) Metaorganismal nutrient metabolism as a basis of cardiovascular disease. Curr Opin Lipidol 25:48–53

Brown K, DeCoffe D, Molcan E et al (2012) Diet-induced dysbiosis of the intestinal microbiota and the effects on immunity and disease. Nutrients 4:1095–1119

Callaway TR, Edrington TS, Brabban A et al (2011) Evaluation of phage treatment as a strategy to reduce *Salmonella* populations in growing swine. Foodborne Pathog Dis 8:261–266

Canibe N, Højberg O, Højsgaard S et al (2005) Feed physical form and formic acid addition to the feed affect the gastrointestinal ecology and growth performance of growing pigs. J Anim Sci 83:1287–1302

Carlson MS, Boren CA, Wu C et al (2004) Evaluation of various inclusion rates of organic zinc either as polysaccharide or proteinate complex on the growth performance, plasma, and excretion of nursery pigs. J Anim Sci 82:1359–1366

Casewell M, Friis C, Marco E et al (2003) The European ban on growth-promoting antibiotics and emerging consequences for human and animal health. J Antimicrob Chemother 52:159–161

Cha SB, Yoo AN, Lee WJ et al (2012) Effect of bacteriophage in enterotoxigenic *Escherichia coli* (ETEC) infected pigs. J Vet Med Sci 74:1037–1039

Che C, Pang X, Hua X et al (2009) Effects of human fecal flora on intestinal morphology and mucosal immunity in human flora-associated piglet. Scand J Immunol 69:223–233

Cho JH, Zhao PY, Kim IH (2011) Probiotics as a dietary additive for pigs: a review. J Anim Vet Adv 10:2127–2134

Chung K-T, Bryant MP (1997) Robert E. Hungate: pioneer of anaerobic microbial ecology. Anaerobe 3:213–217

Collado MC, Sanz Y (2007) Characterization of the gastrointestinal mucosa-associated microbiota of pigs and chickens using culture-based and molecular methodologies. J Food Prot 70:2799–2804

Collier CT, Smiricky-Tjardes MR, Albin DM et al (2003) Molecular ecological analysis of porcine ileal microbiota responses to antimicrobial growth promoters. J Anim Sci 81:3035–3045

Collinder E, Björnhag G, Cardona M et al (2003) Gastrointestinal host-microbial interactions in mammals and fish: comparative studies in man, mice, rats, pigs, horses, cows, elk, reindeer, salmon and cod. Microb Ecol Health Dis 15:66–78

Corr SC, Li Y, Riedel CU et al (2007) Bacteriocin production as a mechanism for the antiinfective activity of *Lactobacillus salivarius* UCC118. Proc Natl Acad Sci U S A 104:7617–7621

Corthésy B, Gaskins HR, Mercenier A (2007) Cross-talk between probiotic bacteria and the host immune system. J Nutr 137(3 Suppl 2):781S–790S

Cromwell GL (2001) Antimicrobial and promicrobial agents. In: Lewis AJ, Southern LL (eds) Swine nutrition, 2nd edn. CRC Press, Boca Raton

Cromwell GL, Lindemann MD, Monegue HJ et al (1998) Tribasic copper chloride and copper sulfate as copper sources for weanling pigs. J Anim Sci 76:118–123

Daly K, Darby AC, Hall N et al (2014) Dietary supplementation with lactose or artificial sweetener enhances swine gut *Lactobacillus* population abundance. Br J Nutr 2:1–6

Delbrück M (1940) The growth of bacteriophage and lysis of the host. J Gen Physiol 23:643–660

De Vrese M, Schrezenmeir J (2008) Probiotics, prebiotics, and synbiotics. Adv Biochem Eng Biotechnol 111:1–66

Dewey CE, Cox BD, Straw BE et al (1999) Use of antimicrobials in swine feeds in the United States. Swine Health Prod 7:19–25

D'Herelle FH (1917) Sur un microbe invisible antagoniste des bacilles dysenteriques. C R Acad Sci 165:373–375

Di Giancamillo A, Vitari F, Savoini G et al (2008) Effects of orally administered probiotic *Pediococcus acidilactici* on the small and large intestine of weaning piglets. A qualitative and quantitative micro-anatomical study. Histol Histopathol 23:651–664

Dowd SE, Sun Y, Wolcott RD et al (2008) Bacterial tag-encoded FLX amplicon pyrosequencing (bTEFAP) for microbiome studies: bacterial diversity in the ileum of newly weaned *Salmonella*-infected pigs. Foodborne Pathog Dis 5:459–472

Dritz SS, Tokach MD, Goodband RD et al (2002) Effects of administration of antimicrobials in feed on growth rate and feed efficiency of pigs in multisite production systems. J Am Vet Med Assoc 220:1690–1695

Drochner W, Kerler A, Zacharias B (2004) Pectin in pig nutrition, a comparative review. J Anim Physiol Anim Nutr (Berl) 88:367–380

Dupaul-Chicoine J, Dagenais M, Saleh M (2013) Crosstalk between the intestinal microbiota and the innate immune system in intestinal homeostasis and inflammatory bowel disease. Inflamm Bowel Dis 19:2227–2237

Dupont DP, Duhamel GE, Carlson MP et al (1994) Effect of divalent cations on hemolysin synthesis by *Serpulina* (*Treponema*) *hyodysenteriae*: inhibition induced by zinc and copper. Vet Microbiol 41:63–73

Duseja A, Chawla YK (2014) Obesity and NAFLD: the role of bacteria and microbiota. Clin Liver Dis 18:59–71

Esmarch E (1886) Ueber eine Modification des Koch'schen Plattenverfahrens zur Isolirung und zum quantitativen Nachweis von Mikroorganismen. Z Hyg 1:293–301

Everard A, Belzer C, Geurts L et al (2013) Cross-talk between *Akkermansia muciniphila* and intestinal epithelium controls diet-induced obesity. Proc Natl Acad Sci U S A 110:9066–9071

Faith JJ, Rey FE, O'Donnell D et al (2010) Creating and characterizing communities of human gut microbes in gnotobiotic mice. ISME J 4:1094–1098

Flisikowska T, Kind A, Schnieke A (2013) Genetically modified pigs to model human diseases. J Appl Genet 55:53–64

Forster RJ, Gong J, Teather RM (1997) Group-specific 16S rRNA hybridization probes for determinative and community structure studies of *Butyrivibrio fibrisolvens* in the rumen. Appl Environ Microbiol 63:1256–1260

Fosse J, Seegers H, Magras C (2008) Foodborne zoonoses due to meat: a quantitative approach for a comparative risk assessment applied to pig slaughtering in Europe. Vet Res 39:1

Fox G, Stackebrandt E, Hespell R et al (1980) The phylogeny of prokaryotes. Science 209:457–463

Fuller R (1989) Probiotics in man and animals. J Appl Bacteriol 66:365–378

Fuller R, Barrow PA, Brooker BE (1978) Bacteria associated with the gastric epithelium of neonatal pigs. Appl Environ Microbiol 35:582–591

Gamaleya NF (1898) Bacteriolysins-ferments destroying bacteria. Russ Arch Pathol Clin Med Bacteriol 6:607–613

Gaskins HR (2001) Intestinal bacteria and their influence on swine growth. In: Lewis AJ, Southern LL (eds) Swine nutrition, 2nd edn. CRC Press, Florida

Gibson GR, Roberfroid MB (1995) Dietary modulation of the human colonic microbiota: introducing the concept of prebiotics. J Nutr 125:1401–1412

Gibson GR, Probert HM, Van Loo JAE et al (2004) Dietary modulation of the human colonic microbiota: updating the concept of prebiotics. Nutr Res Rev 17:257–259

Goodman AL, Kallstrom G, Faith JJ et al (2011) Extensive personal human gut microbiota culture collections characterized and manipulated in gnotobiotic mice. Proc Natl Acad Sci U S A 108:6252–6257

Gosalbes MJ, Llop S, Vallès Y et al (2013) Meconium microbiota types dominated by lactic acid or enteric bacteria are differentially associated with maternal eczema and respiratory problems in infants. Clin Exp Allergy 43:198–211

Hahn JD, Baker DH (1993) Growth and plasma zinc responses of young pigs fed pharmacologic levels of zinc. J Anim Sci 71:3020–3024

Hastad CW, Nelssen JL, Tokach MD et al (2001) Evaluation of different copper sources as a growth promoter in swine finishing diets. Kansas Agric Exp Sta Prog Rep 880:111–117

Heinritz SN, Mosenthin R, Weiss E (2013) Use of pigs as a potential model for research into dietary modulation of the human gut microbiota. Nutr Res Rev 26:191–209

Herfel TM, Jacobi SK, Lin X et al (2011) Polydextrose enrichment of infant formula demonstrates prebiotic characteristics by altering intestinal microbiota, organic acid concentrations, and cytokine expression in suckling piglets. J Nutr 141:2139–2145

Hess M, Sczyrba A, Egan R et al (2011) Metagenomic discovery of biomass-degrading genes and genomes from cow rumen. Science 331:463–467

Hester SN, Chen X, Li M et al (2013) Human milk oligosaccharides inhibit rotavirus infectivity *in vitro* and in acutely infected piglets. Br J Nutr 26:1–10

Hollis GR, Carter SD, Cline TR et al (2005) Effects of replacing pharmacological levels of dietary zinc oxide

with lower dietary levels of various organic zinc sources for weanling pigs. J Anim Sci 83:2123–2129
Hong PY, Lee BW, Aw M et al (2010) Comparative analysis of fecal microbiota in infants with and without eczema. PLoS One 5:e9964
Hosoya S, Villena J, Shimazu T et al (2011) Immunobiotic lactic acid bacteria beneficially regulate immune response triggered by poly (I:C) in porcine intestinal epithelial cells. Vet Res 42:111
Hsiao EY, McBride SW, Hsien S et al (2013) Microbiota modulate behavioral and physiological abnormalities associated with neurodevelopmental disorders. Cell 155:1451–1463
Hugenholtz P, Tyson GW (2008) Microbiology metagenomics. Nature 455:481–483
Hungate RE (1966) The rumen and its microbes. Academic Press, New York
Hungate RE (1979) Evolution of a microbial ecologist. Annu Rev Microbiol 33:1–20
Inoue R, Tsukahara T, Nakanishi N et al (2005) Development of the intestinal microbiota in the piglet. J Gen Appl Microbiol 51:257–265
Isaacson R, Kim HB (2012) The intestinal microbiome of the pig. Anim Health Res Rev 13:100–109
Ivarsson E, Frankow-Lindberg BE, Andersson HK et al (2011) Growth performance, digestibility and faecal coliform bacteria in weaned piglets fed a cereal-based diet including either chicory (*Cichorium intybus* L) or ribwort (*Plantago lanceolata* L) forage. Animal 4:558–564
Ivarsson E, Liu HY, Dicksved J et al (2012) Impact of chicory inclusion in a cereal-based diet on digestibility, organ size and faecal microbiota in growing pigs. Animal 6:1077–1085
Jacela JY, DeRouchey JM, Tokach MD et al (2010) Feed additives for swine: fact sheets – high dietary levels of copper and zinc for young pigs, and phytase. J Swine Health Prod 18:87–91
Jamalludeen N, Johnson RP, Shewen PE et al (2009) Evaluation of bacteriophages for prevention and treatment of diarrhea due to experimental enterotoxigenic *Escherichia coli* O149 infection of pigs. Vet Microbiol 136:135–141
Jia Y, Mann NH (2009) Anti-bacteria compositions. Patent WO2009044163 A2
Jiménez E, Marín ML, Martín R et al (2008) Is meconium from healthy newborns actually sterile? Res Microbiol 159:187–193
Johnson BC, Hamilton TS, Robinson WB (1944) On the mechanism of non-protein nitrogen utilisation by ruminants. J Anim Sci 3:287–302
Johnson RP, Gyles CL, Huff WE et al (2008) Bacteriophages for prophylaxis and therapy in cattle, poultry and pigs. Anim Health Res Rev 9:201–215
Jolliff JS, Mahan DC (2012) Effect of dietary inulin and phytase on mineral digestibility and tissue retention in weanling and growing swine. J Anim Sci 90:3012–3022
Jondreville C, Revy PS, Dourmad JY (2003) Dietary means to better control the environmental impact of copper and zinc by pigs from weaning to slaughter. Livest Prod Sci 84:147–156
Jones JB, Jackson LE, Balogh B et al (2007) Bacteriophages for plant disease control. Annu Rev Phytopathol 45:245–262
Juge N (2012) Microbial adhesins to gastrointestinal mucus. Trends Microbiol 20:30–39
Jukes T (1972) Antibiotics in animal feeds and animal production. Bioscience 22:526–534
Kelly D, Conway S, Aminov RI (2005) Commensal gut bacteria: mechanisms of immune modulation. Trends Immunol 26:326–333
Kenny M, Smidt H, Mengheri E et al (2011) Probiotics – do they have a role in the pig industry? Animal 5:462–470
Khanna S, Tosh PK (2014) A clinician's primer on the role of the microbiome in human health and disease. Mayo Clin Proc 89:107–114
Kim HB, Borewicz K, White BA et al (2011) Longitudinal investigation of the age-related bacterial diversity in the feces of commercial pigs. Vet Microbiol 153:124–133
Kim HB, Borewicz K, White BA et al (2012) Microbial shifts in the swine distal gut in response to the treatment with antimicrobial growth promoter, tylosin. Proc Natl Acad Sci U S A 109:15485–15490
Kiser JS (1976) A perspective on the use of antibiotics in animal feeds. J Anim Sci 42:1058–1072
Klaasen HL, Van der Heijden PJ, Stok W et al (1993) Apathogenic, intestinal, segmented, filamentous bacteria stimulate the mucosal immune system of mice. Infect Immun 61:303–306
Kobayashi Y, Forster RJ, Teather RM (2000) Development of a competitive polymerase chain reaction assay for the ruminal bacterium *Butyrivibrio fibrisolvens* OB156 and its use for tracking an OB156-derived recombinant. FEMS Microbiol Lett 188:185–190
Kocherginskaya S, Aminov RI, White BA (2001) Analysis of the rumen bacterial diversity under two different diet conditions using denaturing gradient gel electrophoresis, random sequencing, and statistical ecology approaches. Anaerobe 7:119–134
Koike S, Kobayashi Y (2001) Development and use of competitive PCR assays for the rumen cellulolytic bacteria: *Fibrobacter succinogenes*, *Ruminococcus albus* and *Ruminococcus flavefaciens*. FEMS Microbiol Lett 204:361–366
Krause DO, Easter RA, White BA et al (1995) Effect of weaning diet on the ecology of adherent lactobacilli in the gastrointestinal tract of the pig. J Anim Sci 73:2347–2354
Kurokawa K, Itoh T, Kuwahara T et al (2007) Comparative metagenomics revealed commonly enriched gene sets in human gut microbiomes. DNA Res 14:169–181
Lager KM, Ng TF, Bayles DO et al (2012) Diversity of viruses detected by deep sequencing in pigs from a

common background. J Vet Diagn Invest 24:1177–1179
Lalles JP, Bosi P, Smidt H et al (2007) Nutritional management of gut health in pigs around weaning. Proc Nutr Soc 66:260–268
Lamendella R, Domingo JW, Ghosh S et al (2011) Comparative fecal metagenomics unveils unique functional capacity of the swine gut. BMC Microbiol 11:103
Leahy SC, Kelly WJ, Ronimus RS et al (2013) Genome sequencing of rumen bacteria and archaea and its application to methane mitigation strategies. Animal 7(Suppl 2):235–243
Leser TD, Amenuvor JZ, Jensen TK et al (2002) Culture-independent analysis of gut bacteria: the pig gastrointestinal tract microbiota revisited. Appl Environ Microbiol 68:673–690
Levy SB (2002) The 2000 Garrod lecture. Factors impacting on the problem of antibiotic resistance. J Antimicrob Chemother 49:25–30
Ley RE, Bäckhed F, Turnbaugh P et al (2005) Obesity alters gut microbial ecology. Proc Natl Acad Sci U S A 102:11070–11075
Ley RE, Turnbaugh PJ, Klein S et al (2006) Microbial ecology: human gut microbes associated with obesity. Nature 444:1022–1023
Lilly DM, Stillwell RH (1965) Probiotics: growth-promoting factors produced by microorganisms. Science 147:747–748
Lin CZ, Flesher B, Capman WC et al (1994) Taxon specific hybridization probes for fiber-digesting bacteria suggest novel gut-associated *Fibrobacter*. Syst Appl Microbiol 17:418–424
Lipsitch M, Singer RS, Levin BR (2002) Antibiotics in agriculture: when is it time to close the barn door? Proc Natl Acad Sci U S A 99:5752–5754
Litten-Brown JC, Corson AM, Clarke L (2010) Porcine models for the metabolic syndrome, digestive and bone disorders: a general overview. Animal 4:899–920
Liu H, Ivarsson E, Dicksved J et al (2012) Inclusion of chicory (*Cichorium intybus* L.) in pigs' diets affects the intestinal microenvironment and the gut microbiota. Appl Environ Microbiol 78:4102–4109
Liu H, Ivarsson E, Lundh T et al (2013) Chicory (*Cichorium intybus* L.) and cereals differently affect gut development in broiler chickens and young pigs. J Anim Sci Biotechnol 4:50
Looft T, Johnson TA, Allen HK et al (2012) In-feed antibiotic effects on the swine intestinal microbiome. Proc Natl Acad Sci U S A 109:1691–1696
Luecke RW, McMillen WN, Thorp F Jr (1950) The effect of vitamin B12, animal protein factor and streptomycin on the growth of young pigs. Arch Biochem 26:326–327
Luo YH, Su Y, Wright AD et al (2012a) Lean breed Landrace pigs harbor fecal methanogens at higher diversity and density than obese breed Erhualian pigs. Archaea 2012:605289
Luo Y, Lin L, Bolund L et al (2012b) Genetically modified pigs for biomedical research. J Inherit Metab Dis 35:695–713
Mackie RI, Sghir A, Gaskins HR (1999) Developmental microbial ecology of the neonatal gastrointestinal tract. Am J Clin Nutr 69:1035S–1045S
Mackie RI, McSweeney CS, Aminov RI (2013) Rumen. In: Battista J (ed) eLS. John Wiley & Sons Ltd, New York
Mallo JJ, Rioperez J, Honrubia P (2010) The addition of *Enterococcus faecium* to diet improves piglet's intestinal microbiota and performance. Livest Sci 133:176–178
Mann E, Schmitz-Esser S, Zebeli Q et al (2014) Mucosa-associated bacterial microbiome of the gastrointestinal tract of weaned pigs and dynamics linked to dietary calcium-phosphorus. PLoS One 9:e86950
Mavromichalis I, Peter CM, Parr TM et al (2000) Growth-promoting efficacy in young pigs of two sources of zinc oxide having either a high or a low bioavailability of zinc. J Anim Sci 78:2896–2902
McCormack G, Moriarty D, O'Donoghue DP et al (2001) Tissue cytokine and chemokine expression in inflammatory bowel disease. Inflamm Res 50:491–495
Mestas J, Hughes CCW (2004) Of mice and not men: differences between mouse and human immunology. J Immunol 172:2731–2738
Metchnikoff E (1907) Essais optimistes. In: Mitchell PC (ed) The prolongation of life: optimistic studies. Heinemann, London
Metzler-Zebeli BU, Mann E, Schmitz-Esser S et al (2013) Changing dietary calcium-phosphorus level and cereal source selectively alters abundance of bacteria and metabolites in the upper gastrointestinal tracts of weaned pigs. Appl Environ Microbiol 79:7264–7272
Meyerholz DK, Stabel TJ, Cheville NF (2002) Segmented filamentous bacteria interact with intraepithelial mononuclear cells. Infect Immun 70:3277–3280
Mitsuoka T, Kaneuchi C (1977) Ecology of the bifidobacteria. Am J Clin Nutr 30:1799–1810
Moore PR, Evenson A, Luckey TD et al (1946) Use of sulfasuxidine, streptothricin, and streptomycin in nutritional studies with the chick. J Biol Chem 165:437–441
Moore WE, Moore LV, Cato EP et al (1987) Effect of high-fiber and high-oil diets on the fecal flora of swine. Appl Environ Microbiol 53:1638–1644
Moxley RA, Duhamel GE (1999) Comparative pathology of bacterial enteric diseases of swine. Adv Exp Med Biol 473:83–101
Mshvildadze M, Neu J, Shuster J et al (2010) Intestinal microbial ecology in premature infants assessed with non-culture-based techniques. J Pediatr 156:20–25
Mulder IE, Schmidt B, Stokes CR et al (2009) Environmentally-acquired bacteria influence microbial diversity and natural innate immune responses at gut surfaces. BMC Biol 7:79
National Research Council (1998) Minerals. In: Cromwell GL (ed) Nutrient requirements of swine, 10th edn. National Academy Press, Washington, DC

Neutra MR (1998) Current concepts in mucosal immunity. V. Role of M cells in transepithelial transport of antigens and pathogens to the mucosal immune system. Am J Physiol Gastrointest Liver Physiol 274:G785–G791

Nocker A, Burr M, Camper AK (2007) Genotypic microbial community profiling: a critical technical review. Microb Ecol 54:276–289

O'Hara AM, Shanahan F (2006) The gut flora as a forgotten organ. EMBO Rep 7:688–693

Ohashi Y, Ushida K (2009) Health-beneficial effects of probiotics: its mode of action. Anim Sci J 80:361–371

Ohman L, Simrén M (2013) Intestinal microbiota and its role in irritable bowel syndrome (IBS). Curr Gastroenterol Rep 15:323

Olsen GJ, Woese CR (1993) Ribosomal RNA: a key to phylogeny. FASEB J 7:113–123

Ozutsumi Y, Hayashi H, Sakamoto M et al (2005) Culture-independent analysis of fecal microbiota in cattle. Biosci Biotechnol Biochem 69:1793–1797

Pang X, Hua X, Yang Q et al (2007) Inter-species transplantation of gut microbiota from human to pigs. ISME J 1:156–162

Patterson JK, Lei XG, Miller DD (2008) The pig as an experimental model for elucidating the mechanisms governing dietary influence on mineral absorption. Exp Biol Med (Maywood) 233:651–664

Peterson J, Garges S, Giovanni M et al (2009) The NIH Human Microbiome Project. Genome Res 19:2317–2323

Pettigrew JE (2006) Reduced use of antibiotic growth promoters in diets fed to weanling pigs: dietary tools, part 1. Anim Biotechnol 4:207–215

Pettigrew JE, Esnaola MA (2001) Swine nutrition and pork quality: a review. J Anim Sci 79 (E-suppl):E316–E342

Poulsen HD (1995) Zinc oxide for weanling piglets. Acta Agric Scand 45:159–167

Prakash T, Oshima K, Morita H et al (2011) Complete genome sequences of rat and mouse segmented filamentous bacteria, a potent inducer of Th17 cell differentiation. Cell Host Microbe 10:273–284

Pryde SE, Richardson AJ, Stewart CS et al (1999) Molecular analysis of the microbial diversity present in the colonic wall, colonic lumen, and cecal lumen of a pig. Appl Environ Microbiol 65:5372–5377

Qin J, Li R, Raes J et al (2010) A human gut microbial gene catalogue established by metagenomic sequencing. Nature 464:59–65

Raibaud P, Ducluzeau R, Dubos F et al (1980) Implantation of bacteria from the digestive tract of man and various animals into gnotobiotic mice. Am J Clin Nutr 33:2440–2447

Rakoff-Nahoum S, Paglino J, Eslami-Varzaneh F et al (2004) Recognition of commensal microflora by Toll-like receptors is required for intestinal homeostasis. Cell 118:229–241

Ramsak A, Peterka M, Tajima K et al (2000) Unravelling the genetic diversity of ruminal bacteria belonging to the CFB phylum. FEMS Microbiol Ecol 33:69–79

Reilly K, Attwood GT (1998) Detection of *Clostridium proteoclasticum* and closely related strains in the rumen by competitive PCR. Appl Environ Microbiol 64:907–913

Rettedal E, Vilain S, Lindblom S et al (2009) Alteration of the ileal microbiota of weanling piglets by the growth-promoting antibiotic chlortetracycline. Appl Environ Microbiol 75:5489–5495

Riley LW, Raphael E, Faerstein E (2013) Obesity in the United States-dysbiosis from exposure to low-dose antibiotics? Front Public Health 1:69

Roberfroid M (2007) Prebiotics: the concept revisited. J Nutr 137(3 Suppl 2):830S–837S

Robinson IM, Allison MJ, Bucklin JA (1981) Characterization of the cecal bacteria of normal pigs. Appl Environ Microbiol 41:950–955

Robinson IM, Whipp SC, Bucklin JA et al (1984) Characterization of predominant bacteria from the colons of normal and dysenteric pigs. Appl Environ Microbiol 48:964–969

Ross EM, Moate PJ, Bath CR et al (2012) High throughput whole rumen metagenome profiling using untargeted massively parallel sequencing. BMC Genet 13:53

Russell EG (1979) Types and distribution of anaerobic bacteria in the large intestine of pigs. Appl Environ Microbiol 37:187–193

Saez AC, Zhang J, Rostagno MH et al (2011) Direct feeding of microencapsulated bacteriophages to reduce *Salmonella* colonization in pigs. Foodborne Pathog Dis 8:1269–1274

Satokari R, Gronroos T, Laitinen K et al (2009) *Bifidobacterium* and *Lactobacillus* DNA in the human placenta. Lett Appl Microbiol 48:8–12

Schmidt B, Mulder IE, Musk CC et al (2011) Establishment of normal gut microbiota is compromised under excessive hygiene conditions. PLoS One 6:e28284

Schnupf P, Gaboriau-Routhiau V, Cerf-Bensussan N (2013) Host interactions with segmented filamentous bacteria: an unusual trade-off that drives the post-natal maturation of the gut immune system. Semin Immunol 25:342–351

Sczesnak A, Segata N, Qin X et al (2011) The genome of Th17 cell-inducing segmented filamentous bacteria reveals extensive auxotrophy and adaptations to the intestinal environment. Cell Host Microbe 10:260–272

Shan T, Li L, Simmonds P et al (2011) The fecal virome of pigs on a high-density farm. J Virol 85:11697–11708

Shimazu T, Villena J, Tohno M et al (2012) Immunobiotic *Lactobacillus jensenii* elicits anti-inflammatory activity in porcine intestinal epithelial cells by modulating negative regulators of the Toll-like receptor signaling pathway. Infect Immun 80:276–288

Simpson JM, McCracken VJ, White BA et al (1999) Application of denaturant gradient gel electrophoresis for the analysis of the porcine gastrointestinal microbiota. J Microbiol Methods 36:167–179

Simrén M, Barbara G, Flint HJ et al (2013) Intestinal microbiota in functional bowel disorders: a Rome foundation report. Gut 62:159–176

Singer RS, Finch R, Wegener HC et al (2003) Antibiotic resistance – the interplay between antibiotic use in animals and human beings. Lancet Infect Dis 3:47–51

Slavin J (2013) Fiber and prebiotics: mechanisms and health benefits. Nutrients 5:1417–1435

Smith HW (1965) The development of the flora of the alimentary tract in young animals. J Pathol Bacteriol 90:495–513

Stahl DA, Flesher B, Mansfield HR et al (1988) Use of phylogenetically based hybridization probes for studies of ruminal microbial ecology. Appl Environ Microbiol 54:1079–1084

Stanton TB, Humphrey SB, Sharma VK et al (2008) Collateral effects of antibiotics: carbadox and metronidazole induce VSH-1 and facilitate gene transfer among *Brachyspira hyodysenteriae* strains. Appl Environ Microbiol 74:2950–2956

Stappenbeck TS, Hooper LV, Gordon JI (2002) Developmental regulation of intestinal angiogenesis by indigenous microbes via Paneth cells. Proc Natl Acad Sci U S A 99:15451–15455

Starke IC, Zentek J, Vahjen W (2013) *Ex vivo* – growth response of porcine small intestinal bacterial communities to pharmacological doses of dietary zinc oxide. PLoS One 8:e56405

Stewart CS (1997) Microorganisms in hindgut fermentation. In: Mackie RI, White BA, Isaacson RE (eds) Gastrointestinal microbes and host interactions, vol 2. Chapman and Hall, New York

Stokstad ELR, Jukes TH (1950) Further observations of the "Animal Protein Factor". Exp Biol Med (Maywood) 72:523–528

Stokstad ELR, Jukes TH, Pierce J et al (1949) The multiple nature of the animal protein factor. J Biol Chem 180:647–654

Strachan DP (1989) Hay fever, hygiene, and household size. BMJ 299:1259–1260

Swann M (1969) Joint Committee on the use of antibiotics in animal husbandry and veterinary medicine. HMSO, London

Swords WE, Wu CC, Champlin FR et al (1993) Postnatal changes in selected bacterial groups of the pig colonic microflora. Biol Neonate 63:191–200

Tajima K, Aminov RI, Nagamine T et al (1999) Rumen bacterial diversity as determined by sequence analysis of 16S rDNA libraries. FEMS Microbiol Ecol 29:159–169

Tajima K, Arai S, Ogata K et al (2000) Rumen bacterial community transition during adaptation to high-grain diet. Anaerobe 6:273–284

Tajima K, Nagamine T, Matsui H et al (2001a) Phylogenetic analysis of archaeal 16S rRNA libraries from the rumen suggests the existence of a novel group of archaea not associated with known methanogens. FEMS Microbiol Lett 200:67–72

Tajima K, Aminov RI, Nagamine T et al (2001b) Diet-dependent shifts in the bacterial population of the rumen revealed with real-time PCR. Appl Environ Microbiol 67:2766–2774

Tajima K, Ohmori H, Aminov RI et al (2010a) Fermented liquid feed enhances bacterial diversity in piglet intestine. Anaerobe 16:6–11

Tajima K, Kobashi Y, Ohmori H et al (2010b) Bacterial community composition in faeces from pigs in an outdoor production system without prophylactic or growth-promoting antibiotics. Livest Sci 133:110–112

Tajima K, Ohmori H, Tohno M et al (2013) Segmented filamentous bacteria are a major group in terminal ileum of piglets. Anaerobe 23:109–111

Takahashi S, Egawa Y, Simojo N et al (2007) Oral administration of *Lactobacillus plantarum* strain Lq80 to weaning piglets stimulates the growth of indigenous lactobacilli to modify the lactobacilli population. J Gen Appl Microbiol 53:325–332

Tako E, Glahn RP, Welch RM et al (2008) Dietary inulin affects the expression of intestinal enterocyte iron transporters, receptors and storage protein and alters the microbiota in the pig intestine. Br J Nutr 99:472–480

Talham GL, Jiang HQ, Bos NA et al (1999) Segmented filamentous bacteria are potent stimuli of a physiologically normal state of the murine gut mucosal immune system. Infect Immun 67:1992–2000

Tannock GW (1995) Role of probiotics. In: Gibson GR, Macfarlane GT (eds) Human colonic bacteria: role in nutrition, physiology and pathology. CRC Press, London

Thacker PA (2013) Alternatives to antibiotics as growth promoters for use in swine production: a review. J Anim Sci Biotechnol 4:35

Theriot CM, Young VB (2013) Microbial and metabolic interactions between the gastrointestinal tract and *Clostridium difficile* infection. Gut Microbes 5(1):86–95

Thiel K (2004) Old dogma, new tricks – 21st century phage therapy. Nat Biotechnol 22:31–36

Transparency Market Research (2013) Probiotics market (Dietary supplements, animal feed, foods & beverages) – global industry size, share, trends, analysis, growth and forecast 2012–2018. http://www.transparencymarketresearch.com/probiotics-market.html

Troy EB, Kasper DL (2010) Beneficial effects of *Bacteroides fragilis* polysaccharides on the immune system. Front Biosci (Landmark Ed) 15:25–34

Tsukahara T, Iwasaki Y, Nakayama K et al (2003) Microscopic structure of the large intestinal mucosa in piglets during an antibiotic-associated diarrhea. J Vet Med Sci 65:301–306

Twort FW (1915) An investigation on the nature of ultramicroscopic viruses. Lancet 186:1241–1243

Umesaki Y, Okada Y, Matsumoto S et al (1995) Segmented filamentous bacteria are indigenous intestinal bacteria that activate intraepithelial lymphocytes and induce MHC class II molecules and fucosyl asialo GM1 glycolipids on the small intestinal epithelial cells in the ex-germ-free mouse. Microbiol Immunol 39:555–562

Upadrasta A, O'Sullivan L, O'Sullivan O et al (2013) The effect of dietary supplementation with spent cider yeast on the swine distal gut microbiome. PLoS One 8:e75714

Urubschurov V, Janczyk P, Souffrant WB et al (2011) Establishment of intestinal microbiota with focus on yeasts of unweaned and weaned piglets kept under different farm conditions. FEMS Microbiol Ecol 77:493–502

Vahjen W, Pieper R, Zentek J (2010) Bar-coded pyrosequencing of 16S rRNA gene amplicons reveals changes in ileal porcine bacterial communities due to high dietary zinc intake. Appl Environ Microbiol 76:6689–6691

Van Baarlen P, Wells JM, Kleerebezem M (2013) Regulation of intestinal homeostasis and immunity with probiotic lactobacilli. Trends Immunol 34:208–215

Varel VH, Robinson IM, Jung HJ (1987) Influence of dietary fiber on xylanolytic and cellulolytic bacteria of adult pigs. Appl Environ Microbiol 53:22–26

Wall SK, Zhang J, Rostagno MH et al (2010) Phage therapy to reduce preprocessing *Salmonella* infections in market-weight swine. Appl Environ Microbiol 76:48–53

Wallace H (1970) Biological responses to antibacterial feed additives in diets of meat producing animals. J Anim Sci 31:1118–1126

Wang W, Archbold T, Kimber MS et al (2012) The porcine gut microbial metagenomic library for mining novel cellulases established from growing pigs fed cellulose-supplemented high-fat diets. J Anim Sci 90(Suppl 4):400–402

Wegener HC (2003) Antibiotics in animal feed and their role in resistance development. Curr Opin Microbiol 6:439–445

White BA, Cann IKO, Kocherginskaya SA et al (1998) Molecular analysis of *Archaea*, *Bacteria* and *Eucarya* communities in the rumen – review. Asian-Aust J Anim Sci 12:129–138

Whitford MF, Forster RJ, Beard CE et al (1998) Phylogenetic analysis of rumen bacteria by comparative sequence analysis of cloned 16S rRNA genes. Anaerobe 4:153–163

Woese CR, Fox GE (1977) Phylogenetic structure of the prokaryotic domain: the primary kingdoms. Proc Natl Acad Sci U S A 74:5088–5090

Wong JM, de Souza R, Kendall CW et al (2006) Colonic health: fermentation and short chain fatty acids. J Clin Gastroenterol 40:235–243

Wood J, Scott KP, Avgustin G et al (1998) Estimation of the relative abundance of different *Bacteroides* and *Prevotella* ribotypes in gut samples by restriction enzyme profiling of PCR-amplified 16S rRNA gene sequences. Appl Environ Microbiol 64:3683–3689

Wright A, Tajima K, Aminov RI (2006) 16S/18S ribosomal DNA clone library analysis of rumen microbial diversity. In: Makkar RSH, McSweeney CS (eds) Methods in gut microbial ecology for ruminants. Springer, New York

Yin J, Li X, Li D et al (2009) Dietary supplementation with zinc oxide stimulates ghrelin secretion from the stomach of young pigs. J Nutr Biochem 20:783–790

Zhang Q, Widmer G, Tzipori S (2013) A pig model of the human gastrointestinal tract. Gut Microbes 4:193–200

Zoetendal EG, Wright A, Vilpponen-Salmela T et al (2002) Mucosa-associated bacteria in the human gastrointestinal tract are uniformly distributed along the colon and differ from the community recovered from feces. Appl Environ Microbiol 68:3401–3407

Part II

Rumen Microbial Diversity

Rumen Bacteria

6

M. Zhou, Y. Chen, and L.L. Guan

Abstract

The rumen microbial ecosystem comprises a diverse population of microorganisms, and the microbial diversity can be affected by many factors. Rumen bacteria are the predominant components of the rumen microbiota, which account for more than 95 % of the population of the entire rumen microbial community. This chapter will mainly focus on rumen bacteria and recent developments in understanding their roles and function using the molecular-based approaches as well as the potential implications in applying the knowledge about rumen bacteria in improving animal production.

Keywords

Rumen • Bacteria

6.1 Rumen Bacteria

Ruminants and ruminal microorganisms have a symbiotic relationship, in which microbial fermentation of ingested feed plays a major role in providing energy for metabolic functions in cattle. In the rumen, microbes metabolize feedstuffs into volatile fatty acids (VFAs), microbial biomass, vitamins, and other substances for the host's nutritional requirements. The rumen microbial ecosystem comprises a diverse symbiotic population of anaerobic bacteria, archaea, ciliated protozoa, and fungi (Krause and Russell 1996), of which only 5–15 % have been isolated (Forster et al. 1998). Among microorganisms, bacteria are known to colonize the rumen soon after birth and contribute to carbohydrate and nitrogen metabolism through fermentation.

Bacteria are the predominant microbes residing in the rumen. The bacterial density in the rumen can be as high as 10^{11} cells per gram of rumen content measured by direct counts (Mackie et al. 2000), and the bacterial community comprises of more than 200 species (McSweeney et al. 2005). The rumen bacterial population can be subdivided into four major subpopulations:

M. Zhou • Y. Chen • L.L. Guan (✉)
Department of Agricultural, Food, and Nutritional Science, University of Alberta, Edmonton, AB T6G 2P5, Canada
e-mail: leluo.guan@ualberta.ca

A.K. Puniya et al. (eds.), *Rumen Microbiology: From Evolution to Revolution*, DOI 10.1007/978-81-322-2401-3_6

(1) liquid-associated population, which is composed of the planktonic bacteria in the rumen liquid and is either the population detached from the feed particles or the ones consuming soluble feed components from the rumen liquid (McAllister et al. 1994); (2) solid-associated population, which includes the bacteria that are loosely or tightly adhered to the feed particles and are fundamental in digesting the ingesta (McAllister et al. 1994) that accounts for up to 75 % of the total rumen bacterial population (Koike et al. 2003); (3) epithelium-associated population, which attaches to the rumen epithelium and counts for only 1 % of the rumen population (Cheng et al. 1979a; Czerkawski 1986), is more diverse than the other subpopulations (Malmuthuge et al. 2012), and is considered to be more closely related to the host metabolic activities than the other subpopulations (Wallace et al. 1979); and (4) eukaryote-associated population, which is represented by the bacteria attached to the surface of protozoa or fungal sporangia (Miron et al. 2001).

6.1.1 Rumen Digesta Associated Bacteria and Their Functions

To date, most of the researches have focused on the bacteria associated with rumen digesta, including those inhabiting in the rumen fluid and adhering to the feed particles. Early studies using culture-based and microscopic methods showed that these bacteria were rods, cocci, or ovals, under the light microscopy (Bryant 1959), and most of them being Gram-positive and Gram-negative bacteria (Stewart and Bryant 1988). The bacteria have specialized physiological and biochemical traits, with which they can be classified into different species and strains. Among these species and strains, the ones which are transiently observed and do not proliferate with similar conditions as the rumen may be the bypass species introduced by the feed and are not considered "real" rumen species. According to Bryant (1959), the species with their metabolism occur in the compatible environment with the rumen, being able to attack the substrates present in the feed (cellulose, hemicellulose, pectin, starch, protein, etc.) or to utilize the degraded products of these compounds, and be repeatedly observed from different hosts over a period of time from various locations, are generally considered to be the true rumen species. As the major roles of rumen bacteria are to ferment feed compounds and supply host with usable substrates, the following section will discuss the representative species according to their substrate preference.

Cellulose Digesters Cellulose digesters refer to the species that majorly digest the cellulose present in the feed. The three major cellulolytic species in the rumen are *Fibrobacter succinogenes*, *Ruminococcus flavefaciens*, and *Ruminococcus albus*. All of these three species are found in a large variety of ruminants such as cattle, cows, sheep, water buffalo, musk oxen, and reindeer (Dehority and Scott 1967; Bryant 1986; Dehority 1986; Van Gylswyk and Van der Toorn 1986; Varel and Dehority 1989; Jalaludin et al. 1992). *Fibrobacter succinogenes* (formerly *Bacteroides succinogenes*) Hungate cells are Gram-negative; predominant cells are rod shaped (can be in different shapes in young culture) which mainly occur singly. Cell sizes range from 0.3 to 0.5 μm in diameter and 1–2 μm in length. *F. succinogenes* can only grow on cellulose, cellobiose, and glucose and mainly produces acetic and succinic acids. This species is anaerobic and nonsporeforming (Hungate 1950). *Ruminococcus flavefaciens* Sijpesteijn cells are usually shown to be Gram-positive staining, the almost spherical cells generally occurring in chains, with cell diameter ranged between 0.7 and 1.6 μm. When grown on cellulose, a characteristic yellow pigment is produced. The *R. flavefaciens* can utilize cellulose and cellobiose, with some strains also able to use glucose, while other carbon sources such as maltose, lactose, xylose, and starch are not used. The main end products include formic and succinic acids, together with traces of hydrogen and lactate. This species is anaerobic and mesophilic (Kaars-Sijpesteijn 1951). *Ruminococcus albus* cells are shown to be Gram-negative to Gram-variable, with the cells usually single or diplococcic, 0.8–2.0 μm in diameter. *R. albus* can ferment cellulose and cellobiose but not glucose and other sugars and produces hydrogen, CO_2, ethanol, acetic, formic, and lactic acids with different combinations and proportions as the major fermentation products. Succinic acid is not produced, making it distinct from *R. flavefaciens* (Hungate 1957). Overall, these cellulose digesters share some features in common: Firstly, they are all strictly anaerobic and are not able to survive when exposed to oxygen. Secondly, the optimal pH range for cell growth is narrow (6–7). Besides, the cells are coated with an extracellular glycocalyx, which helps them to attach to cellulose. Lastly, these microbes are highly specialized in nutritional requirement, majorly restricted to cellulose and its hydrolytic products.

Although all of these three species mentioned above are the primary cellulolytic bacteria in the rumen, their digestibility of different form of cellulose is extremely different. Halliwell and Bryant (1963) compared the cellulose utilization among a *F. succinogenes* strain, two *R. flavefaciens* strains, and two *R. albus* strains and found that all strains can effectively use purified cellulose but not forage celluloses. A similar study conducted by Dehority and Scott (1967) also found that the digestibility of cellulose derived from various forages varied vigorously among the eight strains belonging to these three species, ranging from 12.3 % to 80.3 %.

Hemicellulose Digesters Hemicellulose is another important compound, which occurs in diet and represents a significant amount of the total forage polysaccharides (Dehority 1973). Bacteria species specifically fermented isolated hemicelluloses were firstly observed by Hungate (1950), followed by Howard et al. (1960), Dehority (1966), Clarke et al. (1969), and Hespell et al. (1987). Dehority (2004b) has concluded that four bacteria species appeared to be the predominant hemicellulose digesters in the rumen: *Butyrivibrio fibrisolvens*, *Prevotella ruminicola*, *Ruminococcus flavefaciens*, and *Ruminococcus albus*. *Butyrivibrio fibrisolvens* was described first by Hungate (1950). This species is an anaerobic, nonspore-forming, Gram-negative, motile, and slightly curved rod (Dehority 2004a). The size of the cells ranges between 0.4 and 0.8 μm in width and 1.0–3.0 μm in length. The fermentation substrates of this species include but not restricted to hemicelluloses. *Prevotella ruminicola* was first documented by Bryant et al. (1958a) and classified as *Bacteroides ruminicola*. While in the later study, due to its distinct characteristics from other *Bacteroides* species being observed, Shah and Collins (1990) have reclassified this species into a new genus *Prevotella* and thus named this species as *P. ruminicola* onwards. The cells of this species are Gram-negative, nonmotile, rod shaped, 0.8–1.0 μm wide, and 0.8–8.0 μm long (Bryant et al. 1958a). *P. ruminicola* is a strict anaerobe, fermenting many soluble sugars and producing succinic, formic, and acetic acid as end products. *R. flavefaciens* and *R. albus* are already described above. Besides these four major species, *Eubacterium ruminantium* was also found to extensively ferment glucose, cellobiose, and fructose (Bryant 1959).

Pectin Fermenters Pectin is the third most important carbohydrate present in forage and occurs in lower percentage compared to cellulose and hemicelluloses. Pectin is a compound often found in the cell walls, with D-galacturonic acid as its sugar unit, weighed between 3 and 25 % of the dry matter depends on the type of diet (Waite and Gorrod 1959; Aspinall 1970; Van Soest 1983). In the rumen, pectin is fermented at a faster rate and greater extent compared to other cell wall material breakdown (Chesson and Monro 1982). The degradation products of pectin are high in acetate and low in butyrate and lactate (Leedle and Hespell 1983; Marounek et al. 1985). The major pectin fermenters in the rumen include *Butyrivibrio fibrisolvens*, *Prevotella ruminicola*, *Lachnospira multiparus*, *Succinivibrio dextrinosolvens*, and *Fibrobacter succinogenes*. *Lachnospira multiparus* was established by Bryant and Small (1956). The cells of this species are motile curved rods with 0.4–0.6 μm in width and 2–4 μm in length. This species is very active in degrading purified pectin, but its ability to ferment pectin from the intact forage is very limited (Osborne and Dehority 1989). Cocultured with *Eubacterium limosum*, the growth rate of

L. multiparus is more rapid, and the optical density measurement is almost doubled (Rode et al. 1981), indicating that a better utilization of pectin of this species is achieved by interacting with other microbial species. *Succinivibrio dextrinosolvens* is a Gram-negative, anaerobic, nonspore-forming, and curved rod, documented by Bryant and Small (1956). The dimension of this cell is 0.3–0.5 μm in width and 1–5 μm in length. This species is of only a small portion of the rumen bacteriome. It can not only ferment pectin but also arabinose, xylose, glucose, fructose, galactose, maltose, sucrose, and dextrin (Dehority 2004c). Although *B. fibrisolvens* and *P. ruminicola* are commonly recognized as hemicellulolytic species, these two species also play important roles in fermenting ingested pectin in the rumen. Details of their pectin metabolism have been described by Marounek and Dušková (1999). Briefly, when supplying the pure culture of these two species, pectin was effectively fermented, producing acetate as the predominant products (Marounek and Dušková 1999). Other end products of fermenting pectin and L-arabinose and D-glucose were also compared in these two species: *B. fibrosolvens* cells grown on pectin yielded significantly less butyrate compared to the cells grown on L-arabinose and D-glucose, while *P. ruminicola* culture grown on pectin produced significantly high propionate and less lactate, succinate, and fumarate than the culture grown on L-arabinose and D-glucose (Marounek and Dušková 1999). *F. succinogenes* is also an important cellulose digester and has been described above. Its capacity of fermenting pectin has been studied by Sun et al. (2008) that the pectin in chicory leaves was degraded effectively and released large amount of uronic acids as the fermentation product.

Starch Digesters Starch is rich in grain-based diet, which also requires amylolytic bacteria to break its glycosidic bonds. Most strains belong to *Butyrivibrio fibrisolvens* and *Prevotella ruminicola*, and some strains within *Fibrobacter succinogenes* and *Clostridium* species are able to digest starch. Besides to the cellulolytic and hemicellulolytic species, *Streptococcus bovis*, *Ruminobacter amylophilus*, *Succinimonas amylolytica*, and *Selenomonas ruminantium* are also the species frequently found to digest starch in the rumen. *Streptococcus bovis* are Gram-positive, anaerobic streptococci. This is one of the species that can be commonly isolated from the rumen owing to its non-strict requirement for low oxidation-reduction potential (Hungate 1966). *S. bovis* can rapidly ferment starch and produce lactic acid as the end product of fermentation (Dehority 2004d). *Ruminobacter amylophilus* was formally recognized as *Bacteroides amylophilus* (Hamlin and Hungate 1956). The cells are Gram-negative, anaerobic, nonspore-forming, and nonmotile rods, with sizes varied between 0.9 and 1.6 μm in width and 1.6–4.0 μm in length. The fermentation substrates of this species are limited to starch and maltose, producing acetic, formic, and succinic acids as end products. *Succinimonas amylolytica* was first reported by Bryant et al. (1958a), which is a Gram-negative, anaerobic, nonspore-forming, and straight rod. The cells are 1.0–1.5 μm in width and 1.2–3 μm in length. *S. amylolytica* counts for a small number in the rumen when fed a variety of rations and only indicates a larger proportion of total rumen bacteria when starch is present in the diet (Bryant et al. 1958b). This species decomposes starch by hydrolysis. *Selenomonas ruminantium* was firstly observed by Certes in 1889, and cell arrangement and habitat (Hungate 1966) and physiological traits of this species (Bryant and Small 1956) were later reported. Cells of this species are Gram-negative, anaerobic, and motile rods with 0.8–1.0 μm in width and 2–7 μm in length. Most strains of this species hydrolyze starch, producing primarily the lactic, acetic, and propionic acids as end products (Dehority 2004d).

Other Fermenters Amino acids can also serve as substrates for ATP formation and bacterial growth in the rumen. Only certain rumen bacteria (e.g., *Prevotella* species) can ferment amino acids, but the ATP yield is low (Rychlik et al. 2003; Walker et al. 2005). They need to ferment more than 20 amino acids to gain enough energy to polymerize a single amino acid into protein (Russell 2002). The specialized bacteria in the rumen synthesize all eight vitamin B compounds, as well as vitamin K, inside their cells (Strobel 1992; Nagaraja et al. 1997), to serve as coenzymes for different bioprocesses. For example, vitamin B1 serves as coenzyme for enzymes involved in decarboxylation reactions; vitamin B_5 (panthothenic acid) is part of the coenzyme A. Several studies have confirmed that vitamin B_{12} is an important growth factor for some ruminal microorganisms (Tanner and Wolfe 1988; Strobel 1992). Other microorganisms can synthesize it in pathways that produce propionate (Chen and Wolin 1981), finally supplying these vitamins to the protozoa when the rumen bacteria are passed to the abomasum.

6.1.2 The Bacteria Attached to the Ruminal Epithelium (Epimural Bacteria) and Their Functions

The internal surface of the rumen is covered with small flattened nipple or fingerlike projections (papillae) of the ruminal epithelium surface. The important function of this extended surface area is to facilitate the absorption of fermentation products, especially VFAs. An early study using microscopy showed the bacterial population attached to this surface (Cheng et al. 1979b). However, the understanding of the ecology and the function of such population are very limited.

The existence of the bacteria associated with the rumen epithelial surface of sheep was demonstrated by Bauchop et al. (1975) using scanning electron microscopy. They found that most of the bacteria were on the dorsal, caudal, and lateral surfaces of the rumen wall and that the densest populations were on the top of the dorsal rumen and on the bottom of the caudodorsal blind sac. Mueller et al. (1984) identified the diversity of the epimural community succession in young lambs, and the species *Lactobacillus ruminus, Clostridium ramosum, Butyrivibrio fibrisolvens, Ruminococcus albus, Streptococcus* sp., *Bacteroides* sp., *Succinivibrio dextrinosolvens, Acidaminococcus* sp., *Streptococcus bovis*, and *Ruminococcus flavefaciens* were present at different ages. The first study on bovine epithelial wall associated bacteria was carried out by Tamate et al. (1971), and it revealed similar colonization patterns in the rumen of cattle to those of sheep. Following these studies and still using culture-based methods and microscopy, McCowan et al. (1978) and Cheng et al. (1979a, b) provided further knowledge when they found that the bacteria attached to the rumen wall of cattle were taxonomically distinct from those in the rumen fluid and/or rumen solid particles. They reported that the bacteria attached to the rumen wall of cattle included *Micrococcus, Staphylococcus, Streptococcus, Corynebacterium, Lactobacillus, Fusobacterium, Propionibacterium*, and other unidentified anaerobic species.

Recently developed molecular-based techniques (molecular identification), such as PCR-DGGE (Sadet et al. 2007; Sadet-Bourgeteau et al. 2010; Chen et al. 2011) and 16S rRNA sequence analysis (Mitsumori et al. 2002; Cho et al. 2006; Pei et al. 2010; Li et al. 2012), have been applied to identify the rumen epithelial tissue microbial community and have confirmed that the profiles of the microbes attached to the ruminal tissue are different from those found in the ruminal digesta (Chen et al. 2011; Li et al. 2012). Although bacterial diversity and density under different diets have been compared, the population of bovine epimural bacteria has not yet been accurately defined. Early study showed that the counts of epimural bacteria in hay-fed sheep ranged from 4.4×10^7 to 2.2×10^8/g of wet tissue (Wallace et al. 1979). However, the count was found lower at $1.4 \sim 1.8 \times 10^7$ CFU/cm^2 of rumen epithelial tissue surface in young lambs (Mueller et al. 1984). In cattle, it is estimated that approximately 1–2 % of the total bacterial population resides in the rumen epithelium (Russell et al. 2002), while the density of such bacteria can be up to 2.0×10^{10} /g of ruminal tissue in beef heifers (Chen et al. 2011).

It is believed that the epimural bacteria are involved in oxygen scavenging (Cheng et al. 1979b), hydrolysis of urea entering the rumen across the wall (Fay et al. 1979; Wallace et al. 1979), and tissue recycling (McCowan et al. 1978). Moreover, Mitsumori et al. (2002) detected *Nitrosomonas* from the rumen epithelium and suggested the possibility that the bacterium oxidizes ammonia and methane on the rumen surface. These functions are distinct from those of the bacteria in the rumen content, suggesting that the epimural bacteria may play a different role during rumen fermentation. Due to the intimate contact between animal tissue and epimural bacteria, they also play some role in host-microbial interactions since they also have barrier functions, such as biofilm formation (Macfarlane and Dillon 2007). Recent studies on the association between epimural bacterial community and fermentation characteristics (Chen et al. 2011) as well as rumen tissue gene expression (Chen et al. 2012) have indicated a variety of different types of relationships with the host.

6.2 Rumen Bacterial Identification and Enumeration

Revealing this microbial community in the rumen has been of research interest for a long period, since it is one of the key factors determining the digestion of ingested plant materials, being important in supplying host animals with absorbable and ready-to-use substrates for host biological processes and energy to maintain normal functions. Rumen bacteria are highly diverse in sizes, shapes, organizations, substrate utilization, and so on. Firstly documented by von Tappeiner (1884), attempts have been made on revealing the composition and functions of bacteria which are living in the rumen. Little was achieved until the late 1940s, when Baker and Harriss (1947) successfully observed the bacteria through microscopic method for the first time to illustrate that cellulose degradation in the rumen was primarily conducted by the cocci. Around the same time, independent studies conducted by Hungate (1947), Sijpesteijn (1948), and Marston (1948) successfully isolated bacteria from the rumen sample and cultured them in vitro. Since then, significant improvements in the culturing methods have been achieved, and many more species have been isolated and cultured in the labs.

Although the morphological, physiological, and biochemical traits of the isolated species have been studied on the culturing media, it is believed that a large amount of bacteria living in the rumen are still missing from culture-based methods. This can be caused by the low recovery rate of isolation and the strict requirements for culturing cells in vitro. Culture-independent methods have been invented later to open a door for studying the non-culturable species. With these methods, the overall structure of the rumen bacterial communities has been identified, and the genomic features recovered through culture-independent methods may in turn improve the success rate for isolating and culturing the cells for studying their physiological and metabolic characteristics.

6.2.1 Cultivation-Based Methods

The rumen is a strict anaerobic environment, with a pH that normally ranges between 5.5 and 6.5 and a temperature between 39 and 40 °C. As such, the anaerobic techniques introduced by Hungate (1969) and modified by Bryant (1972) have been widely adopted in most of the microbial labs studying the rumen bacteria. The early attempts for detecting ruminal bacteria have been fulfilled by isolating cells and maintaining cultures in vitro. Culturing medium usually contains both nonselective compartment and selective compartment to be specific for particular species (McSweeney et al. 2005). The medium is also applied for cell isolation, and the majority of the culturable species can be stored in anaerobic diluent at −70 °C.

Roll-tube method was firstly described by Hungate (1969) and is still being extensively used today for isolating bacteria from the rumen. Olson (1992) has invented a modified agar bottle plate for cultivation and isolation of the strict anaerobes while allowing easier streaking as well as picking of colonies. The development of flexible plastic anaerobic glove box chamber makes the colony isolation even easier, where standard microbiological techniques such as agar spread plates, replica plating, and dispensing of medium can be achieved within the anaerobic environment.

Besides isolating different bacteria species, enumeration of each bacteria species as well as the whole microbial population was another key indicator for studying the bacterial communities. Direct counting through microscope was good for studying the rumen liquid-associated bacteria, but not suitable for the counting of solid phase. Therefore, direct counting did not reflect the real abundance of the rumen bacteria. The most probable number technique (MPN) was applied in the late 1980s and has been widely used afterwards (Dehority et al. 1989). This method enumerated the examined species by observing the microbial growth among a serial dilution of inoculum. However, this method can only be applied to the culturable species; thus, methods for quantifying the non-culturable species were required.

6.2.2 Molecular Identification

Owing to the strict requirement for growth and maintenance, only a very small fraction (<1 %) of the microbial community has been successfully identified by culturing methods, and the majority of the rumen bacteria have never been isolated and cultured (Amann et al. 1995). Besides, when the culturable bacteria are in the "viable but non-culturable (VBNC) state" they also fail to grow on the traditional media that they can normally grow and develop colonies (Oliver et al. 2006). As such, the bacteria which are either in a very low active stage or are with low abundance will be missed from the identification and the description of the rumen bacterial communities can be very misleading. The advent of molecular identification has been warranted a more precise and complete description of the entire bacterial community within the rumen.

The small subunit of ribosomal gene, 16S rRNA gene in bacteria, has been proposed to be the best indicator for the microbial diversity (Amann et al. 1995). The advantages of targeting 16S rRNA gene for describing the bacterial communities in the rumen are as follows: This gene is expressed among all species with similar sequence length (Zoetendal et al. 2008); it contains both conserved regions, variable regions, and hypervariable regions (V regions) with discriminatory potentials (Juste et al. 2008); and the databases (e.g., Ribosomal Database Project, http://www.cme.msu.edu) for this particular gene have been well established and enable accurate comparisons of obtained sequences and existing sequences (Chakravorty et al. 2007). Conventional molecular identification methods targeting this gene include terminal restriction fragment length polymorphism (T-RFLP), fluorescence in situ hybridization (FISH), denaturing gradient gel electrophoresis (DGGE), temperature gradient gel electrophoresis (TGGE), clone library, and so on. Quantification of bacteria can also be achieved by measuring the copy number of 16S rRNA gene of each species (Nadkarni et al. 2001), which coped with the shortages of culture-based microbial enumeration.

Although the conventional molecular identification has illustrated a more complex rumen bacterial community compared to the cultivation methods, the detection limits of these methods have prevented the discovery of bacteria of very low abundance. The invention of next-generation sequencing might reveal the entire rumen bacteriome by targeting either all of the amplicons of 16S rRNA gene or by targeting the whole metagenome. The high coverage of the sequences has assured a more in-depth analysis of the phylogenies and functional potentials of the bacterial communities.

6.2.3 Structure of the Rumen Bacterial Community

The key species of rumen bacteria have been described in the previous section, but in the rumen, the bacteria are not present as single colonies; instead, they interact with other microbes and form a complex community. Revealing the entire composition of the bacteria in the rumen may supply overview of all digestive potentials of the microbes and help to elucidate the metabolic potentials based on the characteristics of each component species.

The early description of the composition of the rumen bacteriome achieved by molecular identification was based on cloning analyses of the 16S rRNA genes in the rumen fluid of dairy cattle (Whitford et al. 1998). In their study, 20 novel Gram-positive bacteria and 6 previously uncharacterized Gram-negative bacteria were observed, forming a community predominantly containing *Prevotella*, *Bacteroides*, and *Clostridia*. Tajima et al. (1999) also applied clone library analysis to reveal the bacterial communities in the rumen of Holstein dry cows, finding that low G+C Gram-positive bacteria, *Cytophaga-Flexibacter-Bacteroides*, *Proteobacteria*, and *Spirochaetes* are the predominant phylotypes in the rumen liquid, whereas low G+C Gram-positive bacteria, *Cytophaga-Flexibacter-Bacteroides*, and *Spirochaetes* are the predominant phylotypes in the rumen content. Proportion of each bacterial group varied depending on its origin (liquid phase vs. content phase). Edwards et al. (2004) applied 16S rDNA library-based method to study the rumen bacterial community and reported low G+C Gram-positive bacteria (54 %) as the most predominant group followed by *Cytophaga-Flexibacter-Bacteroides* (40 %).

The rumen bacterial communities can be separated into four fractions: planktonic, loosely attached to rumen digesta particles, tightly attached to rumen digesta particles, and epithelial attached. Bacterial composition of each fraction has been examined and compared, and noticeable differences among the four fractions have been reported. *Cytophaga-Flexibacter-Bacteroides* group was the predominant bacteria in rumen fluid followed by low G+C Gram-positive bacteria and *Proteobacteria*; low G+C Gram-positive bacteria was the predominant bacteria in rumen content followed by *Cytophaga-Flexibacter-Bacteroides*, *Proteobacteria*, high G+C Gram-positive bacteria, and *Spirochaetes*; while in the epithelial-associated communities, only *Cytophaga-Flexibacter-Bacteroides* and low G+C Gram-positive bacteria have been identified (Cho et al. 2006). In a latter study by Kong et al. (2010), bacterial populations tightly attached to the particulates were found to be more diverse than the planktonic and loosely attached ones.

Traditional molecular identification has revealed a much higher diverse bacteriome; however, owing to the detection limit, the phylotypes presenting at very low abundance may escape detection. High-throughput sequencing method allows a deeper description of the rumen bacteriome. Pyrosequencing of the 16S rRNA gene amplified with indexed primers has been widely used to examine the complex rumen bacteriome. Jami and Mizrahi (2012) have examined the rumen microbiome of 16 Holstein cows by amplifying the V2 and V3 regions of 16S rRNA gene and reported that *Bacteroidetes* to be the most abundant phylum in the rumen (51 %) followed by *Firmicutes* (42 %). The bacterial communities in dairy cows fed with different diets were also assessed by sequencing the V3 to V4 regions of the 16S rRNA genes (Zened et al. 2013), and *Bacteroidetes* and *Firmicutes* appeared to be the top two phyla in all animals regardless of the dietary differences. Kittlemann et al. (2013) used universal primers amplifying bacteria and archaea simultaneously and have revealed a similar bacterial community that phylotypes belonging to *Bacteroidetes* and *Firmicutes* were the main phyla of the rumen bacterial community.

However, although the main phyla of different studies appeared to be identical when the rumen bacteriome was assessed by the amplicon pyrosequencing, description of the microbiome may be biased by the selected primers (Cai et al. 2013). Therefore, analyzing the microbial structure using metagenomic methods may be more precise since the microbial profiling is not achieved by amplification. Brulc et al. (2009) have examined the rumen microbial metagenome using pyrosequencing method, and the phylogenic data retrieved from their study confirmed with amplicon-based pyrosequencing analyses that the predominant bacterial phylotypes were *Firmicutes* and *Bacteroidetes*.

Moreover, functional potentials of rumen bacteriome can also be identified with metagenomic analysis. In a gene-centric analysis of rumen metagenome, fundamental differences in digestive potentials were observed between bovine rumen and termite hindgut, and the rumen microbiome has been found to develop specialized digestive ability attacking cellulose by the forage-specific glycoside hydrolases (Brulc et al. 2009). In a latter study, more than 27,000 putative carbohydrate-active genes were identified from cattle metagenome, displaying a more complete description of the genes participating in cellulose decomposition (Hess et al. 2011). Each bacteria species has its unique features, and the optimal growth conditions for the cells are thus different. Therefore, alteration in the rumen conditions can undoubtedly affect the symbiotic bacteria, and each bacterial species might respond to these changes in varied manners.

6.2.4 Whole Genome Sequencing of Key Rumen Bacteria Species

Having the whole rumen bacterial communities being characterized, the next stage of studying this microbiota is to understand the functional potentials of the communities. Although metagenomic analysis can identify the functional genes from the whole rumen microbiota, it has been widely noticed that particular species instead of the entire microbiome contribute to host variance (Zhou et al. 2009; Hook et al. 2010; Hernandez-Sanabria et al. 2012). In this context, getting insights of the species-specific features of each bacterial species might lead to a more effective redirection of the rumen bacterial community. A detailed summary of the available genome sequences of rumen bacteria can be found in Leahy et al. (2013). By comparing the genome sequences of the key rumen species, specialized targets for microbial manipulation may be identified. For example, Purushe et al. (2010) compared the genome sequences of *P. ruminicola* and *P. bryantii* and have found that although genome sizes and average gene lengths between the two species were similar, the genome of each species has embedded genes encoding more than 1,300 unique proteins compared to the other species. The exclusive proteins identified from each bacterial species may have contributed to its uniqueness compared to other species and make it suitable for particular niche within the rumen environment.

6.3 Factors Affecting Rumen Bacterial Community

Many factors have been reported to have an effect on the rumen microbial community, among which are changes in diet (Tajima et al. 2001; Bevans et al. 2005), age of the animal (Fonty et al. 1987, 2007), antibiotic usage (Kleen et al. 2003), health of the host animal (Kleen et al. 2003; Rustomo et al. 2006a), geographic location (Sundset et al. 2007), season (Orpin et al. 1985; Crater et al. 2007), photoperiod (McEwan et al. 2005), stress level (Uyeno et al. 2010), environment (Romero-Pérez et al. 2011), and feeding regimen (Rustomo et al. 2006b). In addition, the microbial species and activities have also been shown to be affected by feed intake levels (Crater et al. 2007) and frequency of feeding (Pulido et al. 2009), suggesting that it is possible to manipulate the composition of ruminal bacteria by diet and management. Recent studies revealed that the host may also have an effect on selecting rumen bacteria (Li et al. 2009; Hernandez-Sanabria et al. 2013), indicating that it may be possible to breed selected rumen bacteria animals for better performance.

6.3.1 Diet

Among the factors that can impact the rumen bacteria, the effects of diet on the changes of rumen microbial community have been the most studied. To date, the rumen bacterial diversity and population changes under various dietary conditions in many species have been widely reported. As stated in previous section, each microbial species has evolved specialized substrate preference. When dietary changes occur to ruminants, the symbiotic bacterial communities shifted to a structure that is more suitable for digesting the compounds within the diets. Therefore, the primary factor inducing the compositional variation in the rumen bacterial communities is the diet transition.

Using cattle as one of the examples, the diets used to feed dairy/beef cattle usually contain the following components: hay/forage (grass or legume) and/or grain (corn, barley, wheat, oats, and sorghum). The animals are usually fed with high-forage diet and then switched to the high-grain diet to achieve a higher level of productivity, since feeding animals with highly fermentable (high-grain) diets can increase VFAs production in the rumen (Penner et al. 2010), thereby increasing the total metabolizable energy for the animal. Through such dietary transition, the changes of fibrolytic species and starch fermenters can change significantly. For example, the quantity of the fibrolytic bacterium, *F. succinogenes* fell 20-fold, *R. flavefaciens* declined to 10 % of its initial level, *P. ruminicola* decreased threefold, and *E. ruminantium* and *T. bryantii* dropped 14-fold and sevenfold, respectively, while *P. bryantii* and *S. ruminantium-M. multiacida* increased tenfold and twofold, respectively, on the third day and 57-fold on day 28 after the switching from hay to grain (Tajima et al. 2001). A recent study using high-throughput sequencing analysis revealed that the bacterial diversity OTUs (operational taxonomic units) in the rumen digesta were highest with forage with 38 unique OTUs identified as compared to only 11 with the high-grain diet in beef heifers (Petri et al. 2013a). In addition, the dietary effect on rumen bacterial community has been also reported in other ruminants. The rumen bacterial communities of domestic sika deer fed with tannin-rich and fiber-rich forage also appeared to be different, where *P. shahii-like*, *P. veroralis*-like, *P. albensis*-like, and *P. salivae*-like sequences were more abundant in oak leaf-based diets, while *S. dextrinosolvens-like* and *P. ruminicola*-like sequences were more prevalent in corn stalk-based diet (Li et al. 2013).

Furthermore, the changes of bacterial community in response to dietary change have also been observed in different populations. Romero-Perez et al. (2011) examined the bacterial communities in rumen liquid of 12 steers and found that the bacterial population clustering was affected by the two diets (alfalfa silage vs. sainfoin silage). Hernandez-Sanabria et al. (2012) reported that the rumen bacterial PCR-DGGE profiles of rumen fluid of beef steers clustered majorly according to the two diets (low energy vs. high energy). Recent studies on ruminal epimural bacteria also revealed that this community can be impacted by diet both during a high-forage diet to a high-grain diet transition in beef heifers (Chen et al. 2011; Petri et al. 2013b), between high- and low-energy diet in beef steers (Hernandez-Sanabria et al. 2013), and between high-forage and high-grain diet in sheep (Sadet et al. 2007; Sadet-Bourgeteau et al. 2010).

6.3.2 Environment

Rumen Microenvironment One of the above changes in response to diet may be due to the ruminal pH changes in addition to the changes in fermentation substrate. It is known that changing the proportion of forage and concentration in diets affects ruminal fermentation characteristics such as volatile fatty acid (VFA) and ruminal pH (Schwartzkopf-Genswein et al. 2003). For example, low pH (<6.0) has negative effects on fibrolytic bacteria in the rumen, and the population of amylolytic bacteria decreases as pH continues to decline (Martin et al. 2002; Nagaraja and Titgemeyer 2007). Therefore, rumen pH is another factor that can also impact the bacterial growth (Russell and Dombrowski 1980; Hoover 1986; Grant and Mertens 1992; Allen et al. 2006). When bacteria are exposed to a low rumen pH, their ability to bind cellulose, or catabolism of hydrolytic products, is reduced or inhibited, and they are unable to survive.

Macroenvironment Although other environmental factors such as geographic location, management, and season may also be considered to impact rumen bacterial community, there are limited studies to date. Using culture-based methods, no difference was detected on counts of bacteria from rumen contents of deer feeding in their natural habitat that could be related to location, sex, or age of the animals (Pearson 1969). Research by Orpin et al. (1985) indicated that rumen bacterial concentrations in the rumen of Svalbard reindeer could be affected by the change of season. During the summer months, the estimated rumen bacterial concentration was 2.1×10^{10}/g, while in the winter season, it decreased by more than 20 % (Orpin et al. 1985). Using molecular-based techniques, the enhanced digestibility of poor-quality feeds reduced the rate of feed passage within the digestive tract and increased recycling of nitrogen observed in North American buffalo when compared to Asian water buffalo, suggesting that geographic location might have an impact on the rumen microbiota (McAllister 2009).

6.3.3 Host

In many occasions, providing identical feed to the animals in the same herd did not necessarily establish identical microbial composition among individuals (Hernandez-Sanabria et al. 2010; Welkie et al. 2010), suggesting that host-specific conditions may be another important factor that influences the rumen microbiota. For example, different rumen methanogen profiles were observed among deer, sheep, and cows when they were grazed on the same pasture, with the ecology of ruminal methanogens that was more similar within the same host species than across different host species (Jeyanathan et al. 2011). Distinguishable methanogenic communities were even observed from five alpaca when they were raised under the same conditions (St-Pierre and Wright 2012). Although these studies have been mainly focused on rumen methanogens, the rumen bacteria have also shown to be impacted by the host animals. Breed of animals has shown to influence rumen bacteria diversity in beef cattle (Guan et al. 2008). In addition, the sire breed of steers had impact on the rumen microbial phylotypes of their offspring (Hernandez-Sanabria et al. 2013), and each animal responded to dietary changes differently (Zhou et al. 2013). Dairy cows with same breed host a more related rumen microbiome in terms of diversity than those with different breeds (King et al. 2011), while there is further evidence that different individuals within the same breed displayed varied ruminal microbial ecology (St-Pierre and Wright 2012). In a preliminary study by Weimer et al. (2010), rumen bacteria of dairy cattle displayed host specificity, and each microbiome reestablished itself with varying success when they received a microbial community from a different host. To date, we could not address this discrepancy with respect to host genetic difference or simply individual variance since most of the studies about rumen microbial community did not include host genetic information such as parentage test as well as genotypes. Therefore, future studies identifying the linkage between host genetic markers and rumen microbial ecology may provide more accurate estimation of how each individual animal responds to exotic microbiota.

6.4 Rumen Bacteria and Their Implications in Animal Production

Improvement of feed efficiency continues to be an issue for dairy/beef industry due to higher inputs of feed costs, increased demand for feed grains from ethanol production, and increased demand for animal protein to meet a growing human population and is set against the reduction of agricultural land area and other resources (such as energy and water). As described above, the rumen microbial fermentation with a complex bacterial ecosystem is responsible for the breakdown of feed components, enabling ruminants to derive about 70 % of their metabolic energy from this process (Bergman 1990). The rumen bacterial community intensively involve in the digestion process in the rumen. Therefore, characteristics of the bacteriome such as the composition of the community, the physiology of each particular species, and the interactions between the bacteria and other microbes and/or host animals, are inevitable to be the factors impacting the rumen fermentation processes, the digestive end products, and the energy supply to host animals. Therefore, host performance and health are largely dependent on its symbiotic bacteriome.

Presently, the understanding of the biological process of microbes in the rumen has been very limited, and the linkage between the microbial functions and host performance, for example, feed efficiency, has not been well understood. Recent studies using a small population (58 steers with different feed efficiency) revealed that particular microbes may be associated with cattle feed efficiency including the measurements such as average daily gain, dry matter intake, feed conversion ratio, and residual feed intake (RFI) under growing and finishing diets (Guan et al. 2008; Hernandez-Sanabria et al. 2010). From the same population, the effect of diet on association between rumen function (presence or absence of particular microbes, fermentation measurements), RFI (Hernandez-Sanabria et al. 2012), and methane emission (Zhou et al. 2010) has also been detected. From these studies, populations of three bacterial species (*Succinivibrio* sp., *Eubacterium* sp., and *Robinsoniella* sp.) were identified to be correlated with feed efficiency measurements; their predicted metabolic mechanisms (propionate synthesis, formate production, and cross-feeding interaction with methanogens) may influence feed efficiency (Hernandez-Sanabria et al. 2012). Furthermore, particular microbial phylotypes in cattle differing sire breed can influence rumen microbial metabolic processes and, ultimately, in productive performance (Hernandez-Sanabria et al. 2013). Further analysis of rumen bacterial profiles showed that *Prevotella* sp. was abundant in Angus rumen liquid and contents, while *Clostridium* sp. was present in contents and tissue from Charolais steers, suggesting a strong association between host and the colonization of rumen bacteria. Based on these preliminary findings, it strongly suggests that rumen function (presence or absence of particular microbes) can be regulated by the interaction between "host (genotypes)" and "environment (diet, management)", which subsequently impact on the rumen fermentation efficiency. Future studies are necessary on host genetics (DNA marker) for rumen function and the particular host mechanisms responsible for the variations in the microbial populations and their interactions with diet and impact on host feed efficiency.

6.5 Conclusions

The rumen contains a diverse microbiota producing many bio-products, which are fundamental components for cattle's daily nutrition source. Bacteria are the predominant microbes in the rumen, which play essential roles in fiber degradation and carbohydrate fermentation for those predominating in the rumen digesta (liquid and feed particles associated). In the meantime, more evidences have appeared to reveal that ruminal epithelial-attached bacterial community may also play a role in VFA metabolism and/or host barrier function. A better understanding of basic physiological functions, such as nutrient transport and absorption mechanisms, as well as factors that affect these functions may help to explain the roles of such microbial population in the rumen. Many evidences have shown that various factors, including those discussed above, can alter or disrupt the normal bacterial diversity and density in the rumen. Therefore, to improve animal production by altering the rumen function, it is essential to first understand what species are there, what they are doing, and how they can respond to various factors. To date, the understanding of the biological process of microbes in the rumen is insufficient, and the linkage between the microbial functions with host's performance is largely under discovered. Current function-directed metagenomic-/metatranscriptomic-based approaches may prove to be useful in shedding additional information on the impact of specific phylotypes/taxa and highlighted metabolic pathways on rumen functions.

References

Allen AE, Vardi A, Bowler C (2006) An ecological and evolutionary context for integrated nitrogen metabolism and related signaling pathways in marine diatoms. Curr Opin Plant Biol 9:264–273

Amann RI, Ludwig W, Schleifer KH (1995) Phylogenetic identification and in-situ detection of individual microbial-cells without cultivation. Microbiol Rev 59:143–169

Aspinall GO (1970) Pectins, plant gums, and other plant polysaccharides. In: Pigman R, Heorton D (eds) The carbohydrates, vol 2B, 2nd edn. Academic, New York, pp 515–536

Baker F, Harris ST (1947) The role of the microflora of the alimentary tract of herbivora with special reference to ruminants. 2. Microbial digestion in the rumen (and caecum) with special reference to the decomposition of structural cellulose. Nutr Abstr Rev 18:3–12

Bauchop T, Clark RTJ, Newhook JC (1975) Scanning electron microscope study of bacteria associated with the rumen epithelium of sheep. Appl Microbiol 30:668–675

Bergman EN (1990) Energy contributions of volatile fatty acids from the gastrointestinal tract in various species. Physiol Rev 70(2):567–590

Bevans DW, Beauchemin KA, Shwartzkopf-Genswein KS et al (2005) Effect of rapid or gradual grain adaptation on subacute acidosis and feed intake by feedlot cattle. J Anim Sci 83:1116–1132

Brulc JM, Antonopoulos DA, Miller ME et al (2009) Gene-centric metagenomics of the fiber-adherent bovine rumen microbiome reveals forage specific glycoside hydrolases. Proc Natl Acad Sci U S A 106:1948–1953

Bryant MP (1959) Bacterial species of the rumen. Bact Rev 23:125–153

Bryant MP (1972) Commentary on the Hungate technique for culture of anaerobic bacteria. Am J Clini Nutr 25:1324–1328

Bryant MP (1986) Genus *Ruminococcus*. In: Sneath PHA, Mair NS, Sharpe ME et al (eds) Bergey's manual of systematic bacteriology, vol 2. Williams & Wilkins, Baltimore, pp 1093–1097

Bryant MP, Small N (1956) Characteristics of two new genera of anaerobic curved rods isolated from the rumen of cattle. J Bacteriol 72:22–26

Bryant MP, Small N, Bouma C et al (1958a) *Bacteroides ruminicola* n.sp. and *Succinimonas* amylolytica the new genus and species. Species of succinic acid-producing anaerobic bacteria of the bovine rumen. J Bacteriol 76:15–23

Bryant MP, Small N, Bouma C et al (1958b) Studies on the composition of the ruminal flora and fauna of young calves. J Dairy Sci 41:1747–1767

Cai L, Ye L, Tong AHY et al (2013) Biased diversity metrics revealed by bacterial 16s pyrotags derived from different primer sets. Plos One 8:e53649

Chakravorty S, Helb D, Burday M et al (2007) A detailed analysis of 16S ribosomal RNA gene segments for the diagnosis of pathogenic bacteria. J Microbiol Methods 69:330–339

Chen M, Wolin MJ (1981) Influence of heme and vitamin B_{12} on growth and fermentations of *Bacteroides* species. J Bacteriol 145:466–471

Chen Y, Penner GB, Li M et al (2011) The epithelial tissue associated bacterial diversity changes in the rumen of beef cattle during dietary transition to high grain diets. Appl Environ Microbiol 77(16):5770–5781

Chen Y, Oba M, Guan LL (2012) Variation of bacterial communities and expression of Toll-like receptor genes in the rumen of steers differing in susceptibility to subacute ruminal acidosis. Vet Microbiol 159(3–4):451–459

Cheng KJ, Bailey CBM, Hironaka R et al (1979a) A technique for depletion of bacteria adherent to the epithelium of the bovine rumen. Can J Anim Sci 59:207–209

Cheng KJ, McCowan RP, Costerton JW (1979b) Adherent epithelial bacteria in ruminants and their roles in digestive tract function. Am J Clin Nutr 32:139–148

Chesson A, Monro JA (1982) Legume pectic substances and their degradation in the ovine rumen. J Sci Food Agric 33:852–859

Cho SJ, Cho KM, Shin EC et al (2006) 16S rDNA analysis of bacterial diversity in three fractions of cow rumen. J Microbiol Biotechnol 16:92–101

Clarke RTJ, Bailey RW, Gaillard BDE (1969) Growth of rumen bacteria on plant cell wall polysaccharides. J Gen Microbiol 56:79–86

Crater AR, Barboza PS, Forster RJ (2007) Regulation of rumen fermentation during seasonal fluctuations in food intake of muskoxen. Comp Biochem Physiol Part A 146:233–241

Czerkawski JW (1986) An introduction to rumen studies. Pergamon Press, Oxford/New York

Dehority BA (1966) Characterization of several bovine rumen bacteria isolated with a xylan medium. J Bacteriol 91:1724–1729

Dehority BA (1973) Hemicellulose degradation by rumen bacteria. Fed Proc 32:1819–1825

Dehority BA (1986) Protozoa of the digestive tract of herbivorous mammals. Insect Sci Appl 7:279–296

Dehority BA (2004a) Cellulose digesting rumen bacteria. In: Dehority BA (ed) Rumen microbiology. Nottingham University Press, Nottingham, pp 177–208

Dehority BA (2004b) Species of rumen bacteria active in the fermentation of hemicellulose. In: Dehority BA (ed) Rumen microbiology. Nottingham University Press, Nottingham, pp 209–228

Dehority BA (2004c) Pectin-fermenting species of rumen bacteria. In: Dehority BA (ed) Rumen microbiology. Nottingham University Press, Nottingham, pp 229–242

Dehority BA (2004d) Starch digester, other less numerous species, and facultative anaerobes in the rumen. In: Dehority BA (ed) Rumen microbiology. Nottingham University Press, Nottingham, pp 243–264

Dehority BA, Scott HW (1967) Characterization of the predominant bacteria occurring in the rumen of goats (*Capra hircus*). Appl Environ Microbiol 33:1030–1036

Dehority BA, Tirabasso PA, Grifo JRAP (1989) Most-Probable-Number procedures for enumerating ruminal bacteria, including the simultaneous estimation of total and cellulolytic numbers in one medium. Appl Environ Microbiol 55:2789–2792

Edwards JE, McEwan NR, Travis AJ et al (2004) 16S rDNA library-based analysis of ruminal bacterial diversity. AVL Int J Gen Mol Microbiol 86(3):263–281

Fay JP, Cheng KJ, Costerton JW (1979) Production of alkaline phosphatase by epithelial cells and adherent bacteria of the bovine rumen and abomasum. Can J Microbiol 25:932–936

Fonty G, Gouet PH, Jouany JP et al (1987) Establishment of the microflora and anaerobic fungi in the rumen of lambs. J Gen Microbiol 133:1835–1943

Fonty G, Joblin K, Chavarot M et al (2007) Establishment and development of ruminal hydrogenotrophs in methanogen-free lambs. Appl Environ Microbiol 73(20):6391–6403

Forster RJ, Whitford MF, Teather RM et al (1998) Investigation into rumen microbial diversity using molecular cloning and probing techniques. In: Ohmiya K, Hayashi K, Sakka K et al (eds) Genetics, biochemistry, and ecology of cellulose degradation. Sukuya, Japan, pp 571–581

Grant RJ, Mertens DR (1992) Influence of buffer pH and raw corn starch addition on in vitro fiber digestion kinetics. J Dairy Sci 75:2762–2768

Guan LL, Nkrumah JD, Basarab JA et al (2008) Linkage of microbial ecology to phenotype: correlation of rumen microbial ecology to cattle's feed efficiency. FEMS Microbiol Lett 288:85–91

Halliwell G, Bryant MP (1963) The cellulolytic activity of pure strains of bacteria from the rumen of cattle. J Gem Microbiol 32:441

Hamlin LJ, Hungate RE (1956) Culture and physiology of a starch-digesting bacterium (*Bacteroides amylophilus* n. sp.) from the bovine rumen. J Bacteriol 72:548–554

Hernandez-Sanabria E, Guan LL, Goonewardene LA et al (2010) Correlation of particular bacterial PCR-denaturing gradient gel electrophoresis patterns with bovine ruminal fermentation parameters and feed efficiency traits. Appl Environ Microbiol 76:6338–6350

Hernandez-Sanabria E, Goonewardene LA, Wang ZQ et al (2012) Impact of feed efficiency and diet on adaptive variations in the bacterial community in the rumen fluid of cattle. Appl Environ Microbiol 78:1203–1214

Hernandez-Sanabria E, Goonewardene LA, Wang ZQ et al (2013) Influence of sire breed on the interplay among rumen microbial populations inhabiting the rumen liquid of the progeny in beef cattle. PLoS One 8:e58461

Hespell RB, Wolf R, Bothast RJ (1987) Fermentation of xylans by *Butyrivibrio fibrisolvens* and other ruminal bacteria. Appl Environ Microbiol 53:2849–2853

Hess M, Sczyrba A, Egan R et al (2011) Metagenomic discovery of biomass-degrading genes and genomes from cow rumen. Science 331:463–467

Hook SE, Wright ADG, McBride BW (2010) Methanogens: methane producers of the rumen and mitigation strategies. Archaea 2010:945785

Hoover WH (1986) Chemical factors involved in ruminal fiber digestion. J Dairy Sci 69:2755–2766

Howard BH, Jones G, Purdon MR (1960) The pentosanases of some rumen bacteria. Biochem J 74:173–180

Hungate RE (1947) Studies on cellulose fermentation.III. the culture and isolation of cellulose-decomposing bacteria from the rumen of cattle. J Bacteriol 53:631–645

Hungate RE (1950) The anaerobic mesophilic cellulolytic bacteria. Bacteriol Rev 14:1–49

Hungate RE (1957) Microorganisms in the rumen of cattle fed a constant ration. Can J Microbiol 3(2):289–311

Hungate RE (1966) The rumen and its microbes. Academic, New York

Hungate RE (1969) A roll tube method for cultivation of strict anaerobes. In: Norris JR, Ribbons DW (eds) Methods in Microbiology. Academic, New York, pp 117–132

Jalaludin S, Ho YW, Abdullah N et al (1992) Rumen microorganisms of the water-buffalo. Buffalo J 3:211–220

Jami E, Mizrahi I (2012) Similarity of the ruminal bacteria across individual lactating cows. Anaerobe 18(3):338–343

Jeyanathan J, Kirs M, Ronimus RS et al (2011) Methanogen community structure in the rumens of farmed sheep, cattle and red deer fed different diets. FEMS Microbiol Ecol 76:311–326

Juste A, Thomma BPHJ, Lievens B (2008) Recent advances in molecular techniques to study microbial communities in food-associated matrices and processes. Food Microbiol 25:745–761

Kaars-Sijpesteijn A (1951) *Ruminococcus flavefaciens*, a cellulose-decomposing bacterium from the rumen of sheep and cattle. J Gen Microb 5:869

King EE, Smith RP, St-Pierre B et al (2011) Differences in the rumen methanogen populations of lactating Jersey and Holstein dairy cows under the same diet regimen. Appl Environ Microbiol 77:5682–5687

Kittlemann S, Seedorf H, Walters WA et al (2013) Simultaneous amplicon sequencing to explore co-occurrence patterns of bacterial, archaeal and eukaryotic microorganisms in rumen microbial communities. Plos One 8:e47879

Kleen JL, Hooijer GA, Rehage J et al (2003) Subacute ruminal acidosis (SASA): a review. J Vet Med A Physiol Pathol Clinical Med 50:406–414

Koike S, Yoshitani S, Kobayashi Y et al (2003) Phylogenetic analysis of fiber-associated rumen bacterial community and PCR detection of uncultured bacteria. FEMS Microbiol Lett 229:23–30

Kong YH, Teather R, Forster R (2010) Composition, spatial distribution, and diversity of the bacterial communities in the rumen of cows fed different forages. FEMS Microbiol Ecol 74(3):612–622

Krause DO, Russell JB (1996) An rRNA approach for assessing the role of obligate amino acid-fermenting bacteria in ruminal amino acid deamination. Appl Environ Microb 62(3):815–821

Leahy SC, Kelly WJ, Ronimus RS et al (2013) Genome sequencing of rumen bacteria and archaea and its application to methane mitigation strategies. Animal 7:235–243

Leedle JAZ, Hespell RB (1983) Brief incubations of mixed ruminal bacteria: effects of anaerobiosis and sources of nitrogen and carbon. J Dairy Sci 66:1003

Li M, Penner GB, Hernandez-Sanabria E et al (2009) Effects of sampling location and time, and host animal on assessment of bacterial diversity and fermentation parameters in the bovine rumen. J Appl Microbiol 107(6):1924–1934

Li M, Zhou M, Adamowicz E et al (2012) Characterization of bovine ruminal epithelial bacterial communities using 16S rRNA sequencing, PCR-DGGE, and qRT-PCR analysis. Vet Microbiol 155(1):72–80

Li ZP, Liu HL, Li GY et al (2013) Molecular diversity of rumen bacterial communities from tannin-rich and fiber-rich forage fed domestic Sika deer (Cervus nippon) in China. BMC Microbiol 13:151

Macfarlane S, Dillon JF (2007) Microbial biofilms in the human gastrointestinal tract. J Appl Microbiol 102:1187–1196

Mackie RI, Aminov RI, White BA et al (2000) Molecular ecology and diversity in gut microbial ecosystems. In: Cronjé PB (ed) Ruminant physiology: digestion, metabolism, growth and reproduction. CAB International, London, pp 61–77

Malmuthuge N, Li MJ, Chen YH et al (2012) Distinct commensal bacteria associated with ingesta and mucosal epithelium in the gastrointestinal tracts of calves and chickens. FEMS Microbiol Ecol 79:337–347

Marounek M, Dušková D (1999) Metabolism of pectin in rumen bacteria *Butyrivibrio fibrisolvens* and *Prevotella ruminicola*. Lett Appl Microbiol 29:429–433

Marounek M, Bartos S, Brezina P (1985) Factors influencing the production of volatile fatty acids from hemicellulose, pectin and starch by mixed culture of rumen microorganisms. Z Tierphysiol Tierernahr Futtermittelkd 53:50–58

Marston HB (1948) The fermentation of cellulose in vitro by organisms from the rumen of sheep. Biochem J 42:564–574

Martin C, Fonty G, Michalet-Doreau B (2002) Factors affecting the fibrolytic activity of the digestive microbial ecosystems in ruminants. In: Martin SA (ed) Gastrointestinal microbiology in animals. Research Signpost, Trivandrum, pp 1–17

McAllister TA (2009) Learning more about rumen bugs: genetic and environmental factors affecting rumen bugs. Government of Alberta Agriculture and Rural Development. 2, Edmonton

McAllister TA, Bae HD, Jones GA et al (1994) Microbial attachment and feed digestion in the rumen. J Anim Sci 72:3004–3018

McCowan RP, Cheng KJ, Bailey CBM et al (1978) Adhesion of bacteria to epithelial cell surfaces within the reticulo-rumen of cattle. Appl Environ Microbiol 35:149–155

McEwan NR, Abecia L, Regensbogenova M et al (2005) Rumen microbial population dynamics in response to photoperiod. Lett Appl Microbiol 41(1):97–101

McSweeney CS, Blackall LL, Collins E et al (2005) Enrichment, isolation and characterisation of ruminal bacteria that degrade non-protein amino acids from the tropical legume Acacia angustissima. Anim Feed Sci Technol 121:191–204

Miron J, Ben-Ghedalia D, Morrison M (2001) Adhesion mechanisms of rumen cellulolytic bacteria. J Dairy Sci 84:1294–1309

Mitsumori M, Ajisaka N, Tajima K et al (2002) Detection of Proteobacteria from the rumen by PCR using methanotroph-specific primers. Lett Appl Microb 35:251–255

Mueller RE, Asplund JM, Iannotti EL (1984) Isolation and identification of adherent epimural bacteria during succession in young lambs. Appl Environ Microbiol 47:724–730

Nadkarni MA, Marin FE, Jacques NA et al (2001) Determination of bacterial load by real-time PCR using a broad-range (universal) probe and primers set. Microbiology 148(1):257–266

Nagaraja TG, Titgemeyer EC (2007) Ruminal acidosis in beef cattle: the current microbiological and nutritional outlook. J Dairy Sci 90:E17–E38

Nagaraja R, MacMillan S, Kere J et al (1997) X chromosome map at 75-kb STS resolution, revealing extremes of recombination and GC content. Genome Res 7:210–222

Oliver CM, Schaefer A, Greenberg EP et al (2006) Novel phenacylhomoserine lactones: microwave synthesis and structure activity evaluation in bacteria and cancer. Acta Pharmacol Sin 27:54–55

Olson KD (1992) Modified bottle plate for the cultivation of strict anaerobes. J Microbiol Methods 14:267–269

Orpin CG, Mathiesen SD, Greenwood Y et al (1985) Seasonal changes in the ruminal microflora of the high-arctic Svalbard reindeer (*Rangifer tarandus platyrhynchus*). Appl Environ Microbiol 50:144–151

Osborne JM, Dehority BA (1989) Synergism in degradation and utilization of intact forage cellulose, hemicellulose, and pectin by three pure cultures of ruminal bacteria. Appl Environ Microbiol 55:2247–2250

Pearson H (1969) Rumen microbial ecology in mule deer. Appl Environ Microbiol 17:819-824

Pei CX, Mao SY, Cheng YF et al (2010) Diversity, abundance and novel 16 s rRNA gene sequences of methanogens in rumen liquid, solid and epithelium fractions of Jinnan cattle. Animal 4:20–29

Penner GB, Oba M, Gabel G et al (2010) A single mild episode of subacute ruminal acidosis does not affect ruminal barrier function in the short term. J Dairy Sci 93:4838–4845

Petri RM, Schwaiger T, Penner GB et al (2013a) Characterization of the core rumen microbiome in cattle during transition from forage to concentrate as well as during and after an acidotic challenge. PLoS One 8(12):e83424

Petri RM, Schwaiger T, Penner GB et al (2013b) Changes in the rumen epimural bacterial diversity of beef cattle

as affected by diet and induced ruminal acidosis. Appl Environ Microbiol 79(12):3744–3755
Pulido RG, Muñoz R, Lemarie P et al (2009) Impact of increasing grain feeding frequency on production of dairy cows grazing pasture. Livest Sci 125:109–114
Purushe J, Fouts DE, Morrison M et al (2010) Comparative genome analysis of *Prevotella ruminicola* and *Prevotella bryantii*: insights into their environmental niche. Microb Ecol 60(4):721–729
Rode LM, Genthner BRS, Bryant MP (1981) Syntrophic association by cocultures of the methanol- and –CO2-H2-utilizing species *Eubacterium limosum* and pectin-fermenting *Lachnospira multiparus* during growth in a pectin medium. Appl Environ Microbiol 42:20–22
Romero-Perez GA, Ominski KH, McAllister TA et al (2011) Effect of environmental factors and influence of rumen and hindgut biogeography on bacterial communities in steers. Appl Environ Microbiol 77(1):258–268
Russell JB (2002) Rumen microbiology and its role in ruminant nutrition. In: Rumen microbiology and its role in ruminant nutrition, Ithaca, pp 1–121
Russell JB, Dombrowski DB (1980) Effect of pH on the efficiency of growth by pure cultures of rumen bacteria in continuous culture. Appl Environ Microbiol 39:604
Russell JB, Garner MR, Flint JP (2002) *Allisonella histiformans*, sp. nov., a novel bacterium that produces histamine, utilizes histidine as its sole energy source, and could play a role in bovine and equine laminitis. Syst Appl Microbiol 25:498–506
Rustomo B, Cant JP, Fan MP et al (2006a) Acidogenic value of feeds. I. the relationship between the acidogenic value of feeds and in vitro ruminal pH changes. J Anim Sci 86:109–117
Rustomo B, Alzahal O, Odongo NE et al (2006b) Effects of rumen acid load from feed and forage particle size on ruminal pH and dry matter intake in the lactating dairy cow. J Dairy Sci 89:4758–4768
Rychlik R, Siedentop H, Pfeil T et al (2003) Cost-effectiveness of adjuvant treatment with acamprosate in maintaining abstinence in alcohol dependent patients. Eur Addict Res 9:59–64
Sadet S, Martin C, Meunier B et al (2007) PCR-DGGE analysis reveals a distinct diversity in the bacterial population attached to the rumen epithelium. Animal 1:939–944
Sadet-Bourgeteau S, Martin C, Morgavi DP (2010) Bacterial diversity dynamics in rumen epithelium of wethers fed forage and mixed concentrate forage diets. Vet Microbiol 146:98–104
Schwartzkopf-Genswein KS, Beauchemin KA, Gibb DJ et al (2003) Impact of bunk management on feeding behavior, ruminal acidosis and performance of feedlot cattle: a review. J Anim Sci 81(E. Suppl. 2): E149–E158
Shah HN, Collins DM (1990) *Prevotella*, new genus to include *Bacteroides melaninogenicus* and related species formerly classified in the genus *Bacteroides*. Int J Syst Bacteriol 40:205–208
Sijpesteijn AK (1948) Cellulose decomposing bacteria from the rumen of cattle. PhD thesis, Rijksuniversiteit te Leiden
Stewart CS, Bryant MP (1988) The rumen bacteria. In: Hobson PN, Stewart CS (eds) The rumen microbial ecosystem. Elsevier, Barking, Essex, p 21
St-Pierre B, Wright A-DG (2012) Molecular analysis of methanogenic archaea in the forestomach of the alpaca (*Vicugna pacos*) reveals differences in population structure between individual hosts. BMC Microb 12:1
Strobel HJ (1992) Vitamin B_{12}-dependent propionate production by the ruminal bacterium *Prevotella ruminicola* 23. Appl Environ Microbiol 58:2331–2333
Sun XZ, Joblin KN, Andrew IG et al (2008) Degradation of forage chicory by ruminal fibrolytic bacteria. J Appl Microbiol 105(5):1286–1297
Sundset MA, Præsteng KE, Cann IKO et al (2007) Novel rumen bacterial diversity in two geographically separated sub-species of reindeer. Microbial Ecol 54:424–438
Tajima K, Aminov RI, Nagamine T et al (1999) Rumen bacterial diversity as determined by sequence analysis of 16S rDNA libraries. FEMS Microbiol Ecol 29:159–169
Tajima K, Aminov RI, Nagamine T et al (2001) Diet-dependent shifts in the bacterial population of the rumen revealed with real-time PCR. Appl Environ Microbiol 67:2766–2774
Tamate H, Kikuchi T, Onodera A et al (1971) Scanning electron microscopic observation on the surface of the bovine rumen mucosa. Arch Histol Jpn 33:273
Tanner RS, Wolfe RS (1988) Nutritional requirements of *Methanomicrobium mobile*. Appl Environ Microbiol 54:625–628
Uyeno Y, Sekiguchi Y, Kamagata Y (2010) rRNA-based analysis to monitor succession of faecal bacterial communities in Holstein calves. Lett Appl Microbiol 51:570–577
Van Gylswky NO, Van der Toorn JJTK (1986) Enumeration of *Bacteroides succinogenes* in the rumen of sheep fed maize-straw diets. FEMS Microbiol Ecol 38:205–209
Van Soest PJ (1983) Nutritional ecology of the ruminant. Cornell University/O & B Books, Ithaca
Varel VH, Dehority BA (1989) Ruminal cellulolytic bacteria and protozoa from bison, cattle-bison hybrids, and cattle fed three alfalfa-corn diets. Appl Environ Microbiol 55(1):148–153
Von Tappeiner H (1884) Untersuchungen über die gärung der cellulose insbesondere über deren lösung in darmkanale. Z Biol 20:52–134
Waite R, Gorrod ARN (1959) The comprehensive analysis of grasses. J Sci Food Agri 10(6):317–326
Walker ND, Newbold CJ, Wallace RJ (2005) Nitrogen metabolism in the rumen. In: Pfeffer E, Hristov A (eds) Nitrogen and phosphorus nutrition of cattle. CABI Publishing, Cambridge, pp 71–115
Wallace RJ, Cheng K-J, Dinsdale D et al (1979) An independent microbial flora of the epithelium and its role in the ecomicrobiology of the rumen. Nature 279:424–426

Weimer PJ, Stevenson DM, Mertens DR (2010) Shifts in bacterial community composition in the rumen of lactating dairy cows under milk fat-depressing conditions. J Dairy Sci 93:265–278

Welkie DG, Stevenson DM, Weimer PJ (2010) ARISA analysis of ruminal bacterial community dynamics in lactating dairy cows during a feeding cycle. Anaerobe 16:94–100

Whitford MF, Forster RJ, Beard CE (1998) Phylogenetic analysis of rumen bacteria by comparative sequence analysis of cloned 16S rRNA genes. Anaerobe 4:153–163

Zened A, Combes S, Cauquil L et al (2013) Microbial ecology of the rumen evaluated by 454 GS FLX pyrosequencing is affected by starch and oil supplementation of diets. FEMS Microbiol Ecol 83:504–514

Zhou M, Hernandez-Sanabria E, Guan LL (2009) Assessment of the microbial ecology of ruminal methanogens in cattle with different feed efficiencies. Appl Environ Microbiol 75(20):6524–6533

Zhou M, Hernandez-Sanabria E, Guan LL (2010) Characterization of variation in rumen methanogenic communities under different dietary and host feed efficiency conditions, as determined by PCR-denaturing gradient gel electrophoresis analysis. Appl Environ Microbiol 76(12):3776–3786

Zhou M, Hünerberg M, Beauchemin KA et al (2013) Individuality of Ruminal methanogen/protozoa populations in beef cattle fed diets containing dried distillers' grain with solubles. Acta Agri Scand Sect A Anim Sci 62(4):273–288

Zoetendal EG, Smith AH, Sundset MA (2008) The BaeSR two-component regulatory system mediates resistance to condensed tannins in *Escherichia coli*. Appl Environ Microbiol 74:535–539

7 Rumen Fungi

Katerina Fliegerova, Kerstin Kaerger, Paul Kirk, and Kerstin Voigt

Abstract

Rumen fungi inhabiting the gastrointestinal tract of ruminants have been stepwise renamed into "gut fungi" after the discovery of their occurrence in intestines/hindgut of large herbivores. These only anaerobic fungi are phylogenetically unique and comprise a separate clade, the *Neocallimastigomycetes*, among the basal fungi. Six genera have been described, namely, the monocentric *Neocallimastix*, *Caecomyces* and *Piromyces* and the polycentric *Anaeromyces*, *Orpinomyces* and *Cyllamyces*. Recent research indicates, however, the existence of several new taxa in ruminant and nonruminant animals. The major traits of the gut fungi are both singular and exceptional. Physiologically, the gut fungi are the only representatives of *Fungi* obligately anaerobic and possess, therefore, hydrogenosomes instead of mitochondria. Genetically, gut fungi represent organisms with the lowest genome G+C content in the entirety of life. Enzymologically, gut fungi produce a broad range of excellent hydrolases, some of which are organized in cellulosomes, the enzyme factories otherwise known only in some bacteria, and include the most active (hemi) cellulases. This chapter brings the insight into the recent state of knowl-

K. Fliegerova (✉)
Institute of Animal Physiology and Genetics, Czech Academy of Sciences, v.v.i.,
Videnska 1083, Prague 14220, Czech Republic
e-mail: fliegerova@iapg.cas.cz

K. Kaerger
National Reference Centre for Invasive Mycoses, Leibniz Institute for Natural Product Research and Infection Biology – Hans Knöll Institute (HKI),
Adolf Reichwein Str. 23, 07745 Jena, Germany

P. Kirk
Royal Botanic Gardens, Kew,
Richmond, Surrey TW9 3DS, UK

K. Voigt
Jena Microbial Resource Collection, Leibniz Institute for Natural Product Research and Infection Biology – Hans Knöll Institute (HKI),
Adolf Reichwein Str. 23, 07745 Jena, Germany

Department of Microbiology and Molecular Biology, Institute for Microbiology, Friedrich Schiller University, Jena, Germany

A.K. Puniya et al. (eds.), *Rumen Microbiology: From Evolution to Revolution*, DOI 10.1007/978-81-322-2401-3_7

edge on anaerobic fungi with special attention to their diversity, new taxa proposition, the effect of diet and host phylogeny on anaerobic fungal community and the composition of their lignocellulolytic machinery.

Keywords
Rumen fungi • Herbivore • Anaerobic • Hydrolases

7.1 Introduction

Rumen fungi, an unusual group of symbiotic zoosporic fungi discovered only in 1975 by British scientist Colin Orpin, occupy a unique ecological niche, the anaerobic environment of the rumen and gastrointestinal tract of large herbivores. The original notion of "rumen fungi" has been stepwise replaced by the term "gut fungi" after the discovery of their occurrence in intestines of hindgut animals like the horse, rhinoceros or elephant. In the digestive tract, the gut fungi contribute, in association with bacteria and protozoa, to hydrolysis of diet fibre resulting in the production of fermentation end products (volatile fatty acids) that can be utilized by the host animal as sources of nutrition. Anaerobic gut fungi might account for 5 % up to 20 % of the total microbial biomass (depending on the quantification method) in the rumen (Rezaeian et al. 2004); however, their contribution to plant cell wall degradation has not yet been exactly determined although some studies suggest that they may greatly exceed the contribution by bacteria to fermentation (Lee et al. 2000). They produce all the enzymes necessary for plant biomass decomposition including cellulases, xylanases, mannases, esterases, glucosidases and glucanases. These enzymes (collectively hydrolases) enable the gut fungi to penetrate plant cell walls, access fermentable substrates not available to surface-acting bacteria, colonize the sturdy plant structures and weaken and degrade plant tissue, thus reducing the size of plant particles. Released zoospores attach to the feed fragments where they encyst and subsequently germinate to produce a thallus composed of rhizoids and the sporangium. The rhizoid system can be highly branched and is able to penetrate deeply into plant stems and thus possibly aid substrate decomposition mechanically. Due to the capability of colonizing the recalcitrant plant components such as sclerenchyma and the vascular system, gut fungi are known to decompose lignin-containing plant walls, even if they do not utilize phenolics or lignin. Not only do the effective hydrolases make the gut fungi so attractive for scientific research, they are very interesting also from the evolutionary point of view. Their unique traits, especially anaerobiosis, possession of hydrogenosomes instead of mitochondria, multiple flagella and extremely low genomic GC content, make these microorganisms extraordinary in the fungal kingdom. Despite intensive research, there are still many questions regarding morphology, genetics, enzymology, taxonomy, diversity and the importance of gut anaerobic fungi to the host animals. This contribution aims to highlight some recent advances achieved in this field.

7.2 Taxonomy

The rumen fungi, due to their extraordinary morphological features, have occupied different taxonomic positions and thus have been placed in different taxonomic groups since the time of their discovery. These flagellate microorganisms, observed for the first time in rumen fluid more than 100 years ago, were originally considered as protozoans. Only in 1975 did Orpin recognized these cells as a motile stage in the life cycle of a fungus and classified them as *Phycomycetes*. Description of chitin in their cell walls (Orpin 1977a) proved, conclusively, that these organisms were true fungi, and in 1980 they were assigned to the *Chytridiomycetes* (Barr 1980,

1988) following the abandonment of the excessively polyphyletic *Phycomycetes*. Further advance in the systematics of gut fungi has been driven by modern methods of molecular biology. Multi-gene dataset phylogenies of DNA sequences of ribosomal RNA operon (18S rRNA, 28S rRNA, ITS) combined with protein-coding genes (EF1α, RNA polymerase II subunits RPB1and RPB2) subsequently led to the separation of the anaerobic gut fungi from *Chytridiomycetes* and the formation of the new phylum, *Neocallimastigomycota*, including one class, *Neocallimastigomycetes*, one order, *Neocallimastigales*, and a single family *Neocallimastigaceae* (Hibbett et al. 2007; Gruninger et al. 2014). Within the family there are currently six genera: *Neocallimastix*, *Piromyces*, *Caecomyces*, *Orpinomyces*, *Anaeromyces* and *Cyllamyces*. The very recently the new genus Buwchfawromyces eastonii has been described (Callaghan et al. 2015) enlarging the number of genera up to seven.The genera are defined on the basis of a monocentric or polycentric thallus, filamentous or bulbous rhizoids, the shape of sporangia and mono- or polyflagellated zoospores, whereas species are delimited mainly on zoospore ultrastructure (Munn 1994). Table 7.1 shows all known cultivable species although recent research analysing the diversity of gut fungi indicates that the currently described genera and species are significantly incomplete (see Sect. 7.5). Recent research, however, indicates that the revision of the classification of the gut fungi has not yet finished. Phylogenomic analyses based on 46 slowly evolving and 107 moderately evolving, orthologous, protein-coding genes (Ebersberger et al. 2012) as well as analyses of Schoch et al. (2012) based on ITS, LSU (Dagar et al. 2011), SSU and RPB1sequences of 1,022 fungal species suggest a monophyletic origin of the *Chytridiomycota* and *Neocallimastigomycota*. The position of gut fungi thus can be reassessed in the future, and they can be again united with the chytrids in the *Chytridiomycota* (authors, unpublished).

7.3 Life Cycle and Zoosporogenesis

The life cycle of the anaerobic gut fungi alternates between a motile zoospore stage and a non-motile vegetative stage consisting of a digesta-associated thallus bearing a fruiting body (a sporangium) which is well known in the monocentric anaerobic fungi. The zoospores actively move by means of flagella, which differ in number for different genera. Zoospores are attracted to freshly ingested plant tissues presumably by chemotactic response to soluble sugars and/or phenolic acids (Wubah and Kim 1996). Zoospore liberation is influenced by the diet and induced by water-soluble hemes (Orpin and Greenwood 1986; Orpin 1994). Released zoospores attach to the feed particles and encyst before germinating to produce the fungal thallus comprising the rhizoids and the sporangium. Diagrammatic representation of the life cycle of a species of *Neocallimastix* is depicted well in Orpin and Joblin (1997), and a descriptive animated model created by Karin Christensen (2004) is available on the web at http://www.goatbiology.com/animations/funguslc.html.

The rhizoids can be highly branched as in species of *Orpinomyces*, *Anaeromyces*, *Neocallimastix* and *Piromyces* or may consist of spherical or ovoid bodies (holdfast or haustoria) as in *Caecomyces* and *Cyllamyces*. The rhizoidal system penetrates the plant tissue, and sporangia develop on the exterior surface of the plant fragment or within hollow stems. Two types of zoosporangial development can occur in the anaerobic gut fungi: endogenous development, where the nucleus is retained in the encysted zoospore which enlarges into a zoosporangium, or exogenous development, where the nucleus migrates out of the zoospore and the zoosporangium is formed in the germ tube or sporangiophore (Barr et al. 1989; Ho et al. 1993a). Concentration of nuclear material in a zoosporangium is typical for monocentric species (*Neocallimastix*, *Piromyces*, *Caecomyces*), while

Table 7.1 Survey of known cultivable strains of anaerobic gut fungi

Genus	Species	Thallus, rhizoids, flagella	References
Neocallimastix	*frontalis*	Monocentric, filamentous, polyflagellate	Heath et al. (1983)
Neocallimastix	*hurleyensis*	Monocentric, filamentous, polyflagellate	Webb and Theodorou (1991)
Neocallimastix	*patriciarum*	Monocentric, filamentous, polyflagellate	Orpin and Munn (1986)
Neocallimastix	*variabilis*	Monocentric, filamentous, polyflagellate	Ho et al. (1993a)
Anaeromyces	*elegans*	Polycentric, filamentous, uniflagellate	Ho et al. (1993b)
Anaeromyces	*mucronatus*	Polycentric, filamentous, uniflagellate	Breton et al. (1990)
Caecomyces	*communis*[a]	Monocentric, bulbous, uniflagellate	Orpin (1976)
Caecomyces	*equi*	Monocentric, bulbous, uniflagellate	Gold et al. (1988)
Caecomyces	*sympodialis*	Monocentric, bulbous, uniflagellate	Chen et al. (2007)
Cyllamyces	*aberensis*	Polycentric, bulbous, uniflagellate	Ozkose et al. (2001)
Orpinomyces	*bovis*	Polycentric, filamentous, polyflagellate	Barr et al. (1989)
Orpinomyces	*intercalaris*	Polycentric, filamentous, polyflagellate	Ho et al. (1994)
Orpinomyces	*joyonii*[b]	Polycentric, filamentous, polyflagellate	Breton et al. (1989)
Piromyces	*communis*[c]	Monocentric, filamentous, uniflagellate	Barr et al. 1989
Piromyces	*citronii*	Monocentric, filamentous, uniflagellate	Gaillard-Martinie et al. (1995)
Piromyces	*dumbonica*	Monocentric, filamentous, uniflagellate	Li et al. (1990)
Piromyces	*mae*	Monocentric, filamentous, uniflagellate	Li et al. (1990)
Piromyces	*minutus*	Monocentric, filamentous, uniflagellate	Ho et al. (1993d)
Piromyces	*polycephalus*	Monocentric, filamentous, uniflagellate	Chen et al. (2002)
Piromyces	*rhizinflata*	Monocentric, filamentous, uniflagellate	Breton et al. (1991)
Piromyces	*spiralis*	Monocentric, filamentous, uniflagellate	Ho et al. (1993c)

[a]Originally described as *Sphaeromonas communis*
[b]Originally described as *Neocallimastix joyonii*
[c]Originally described as *Piromonas communis*

nuclear migration throughout the hyphal mass is typical for polycentric species (*Anaeromyces*, *Orpinomyces*, *Cyllamyces*) (Gaillard et al. 1989). Mature zoosporangia release zoospores, and the life cycle is repeated. Zoosporogenesis can occur as early as 8 h after germination under appropriate conditions (Orpin 1977b; France et al. 1990; Theodorou et al. 1993), but normally the life cycle varies between 24 and 32 h (Lowe et al. 1987b; Wubah et al. 1991a). Knowledge of the life cycle was acquired mostly from monocentric species; that of polycentric species has not yet been fully determined. A common characteristic of axenic cultures of the polycentric gut fungi is the paucity or absence of zoosporogenesis. Although some polycentric isolates form sporangia readily and abundantly in certain media, the majority of the sporangia do not differentiate and release zoospores. It is assumed that polycentric anaerobic fungi are able to grow vegetatively and thus their survival is less dependent upon the formation of zoospores (Phillips and Gordon 1989; Ho and Bauchop 1991). However, it is not known whether vegetative growth occurs in vivo, in the rumen. Ho and Bauchop (1991) accomplished successful germination and development into mature thalli of polycentric fungi, when their mycelium was, at the appropriate stage of development, flooded with or transferred into fresh liquid media or clarified rumen fluid. These results can indicate that the life cycle of polycentric fungi under favourable conditions of their natural environment need not be subjected to propagation problems observed in in vitro cultures.

A predicted, but up to now not understood, phase of life cycle of gut fungi is a resting stage, providing their survival for long periods of desiccation and exposure to oxygen. The first indication of this ability was the presence of anaerobic fungi out of the rumen, in saliva and in faeces (Lowe et al. 1987a) and the survival of gut fungi in dried faeces (Milne et al. 1989; Davies et al.

1993). Studies seeking an explanation of these facts revealed the melanized resistant sporangia in *Neocallimastix* sp. (Wubah et al. 1991b) and chambered spore-like structures in *Anaeromyces* sp. (Brookman et al. 2000). These resistant bodies can represent either an alternative path in the life cycle or a missing part of the normal life cycle, similar to a common stage in, for example, zoosporic fungi in the *Blastocladiales* (Robertson and Emerson 1982). This oxygen-resistant dormant stage is presumed to be important in inter-animal transfer via accidental or intentional coprophagy in addition to transfer through salivary contact (Lowe et al. 1987a).

7.4 Quantification

The two-stage life cycle of gut fungi has always complicated their monitoring and quantification. Estimation of fungal population from the measurement of concentration of free zoospores (10^3–10^5 fungal zoospores per ml of the rumen fluid) is inaccurate because this approach does not cover the vegetative stage attached by the rhizoidal system to plant particles (Joblin 1981; France et al. 1990; Obispo and Dehority 1992). Direct counts of fungal colonies on plant fragments (Windham and Akin 1984) or agar strips (Ushida et al. 1989) incubated in the rumen can be misleading because the majority of anaerobic fungi are uncultivable. The most probable number (MPN) method for estimation of both free zoospores and thalli adherent to plant particles ($>10^6$ thallus-forming units per g of dry matter) from serially diluted whole rumen digesta is complicated by problems with dilution of digesta from fibrous diets and correlation between population and biomass (Theodorou et al. 1990; France and Theodorou 1994). The biomass of anaerobic fungi cannot be directly related to numbers, because of the huge variation in thallus size demonstrated between species.

Denman and McSweeney (2006) first described the estimation of fungal biomass by using real-time PCR and suggested this assay as more rapid and accurate than culture-dependent method. Unfortunately, this approach also meets with difficulties arising from different growth characteristics of monocentric and polycentric species. However, the positive correlation between both ITS1 copy number and fungal dry weight on the one hand and ITS1 copy number and zoospore counts on the other hand determined by Lwin et al. (2011) offers a way to overcome these difficulties and provides a powerful tool for estimating the population density of anaerobic gut fungi.

7.5 Diversity

Our current understanding of gut fungi populations in the gastrointestinal tract of herbivores is mostly derived from classical microbiological approaches based on isolation and cultivation in addition to morphological and enzymological characterization of cultivable strains. However, most of the recent environmental studies indicate that cultured taxa represent only a minor part of the microbial ecosystems and more than 90 % of *Fungi* are apparently awaiting discovery (Blackwell 2011; Mora et al. 2011). Modern DNA-based techniques offer assured methods for increasing our knowledge of diversity of gut fungi and open up the opportunity for the assessment of stability or change in fungal community structure and its dependence on diet, host animal species or geographical locality. Especially, the recent development of next-generation sequencing (NGS) makes this approach very powerful, by allowing the direct characterization of dozens of samples with several thousand sequences per PCR product. However, the number of studies addressing fungal diversity is still limited.

Analysis of diversity of gut fungi in cow manure using ITS1 fragments in a classical cloning library approach resulted in the phylogenetic study showing the dominancy of the *Cyllamyces/ Caecomyces* group (67 %), followed by the *Piromyces* group (24 %). This first insight into the diversity of anaerobic fungi in cowhouse manure supported the opinion of Nicholson et al. (2010) relating to the prevalence of fungi with bulbous morphology in faeces of bovids. However, the results of Sirohi et al. (2013)

indicate the prevalence of the polycentric genus *Orpinomyces* (48 %) in rumen fluid of cattle fed high-fibre diet. This discrepancy can be explained by considerable differences in the abundance of different fungal taxa between rumen and faeces. In the study of Griffith et al. (2009), the anaerobic fungi with a bulbous morphotype (*Caecomyces* and *Cyllamyces*) were the most abundant in fresh faeces but represented a lower proportion of the fungal population in rumen samples. Conversely, those with a polycentric morphotype (*Orpinomyces* and *Anaeromyces*) were less frequent in fresh samples but represented the majority of fungi in rumen digesta. Monocentric morphological types (*Neocallimastix* and *Piromyces*) were surprisingly the least abundant fungi in cow rumen. Liggenstoffer et al. (2010), who performed an extensive NGS study of gut fungi in 30 different animals, also described *Caecomyces* as the most numerous genus in faeces of domestic cattle. Nevertheless, in all studied samples, the sequences identified as *Caecomyces* and *Cyllamyces* were the least distributed and found only in low numbers, while *Piromyces* was the most abundant genus representing 36 % of all obtained sequences. Among domestic ruminants, however, the composition of faecal anaerobic fungal communities differed considerably. In cattle samples with 60 % of *Caecomyces* and 25 % of *Piromyces*, sequences of *Neocallimastix* (5 %), *Anaeromyces* (4 %) and especially *Orpinomyces* (2 %) represented only the minority groups. Domestic sheep with 44 % of *Neocallimastix*, 22 % of *Caecomyces*, and 18 % of *Piromyces* also exhibited very low occurrence of *Anaeromyces* (4 %) and *Orpinomyces* (3.6 %). The prevalence of polycentric *Anaeromyces* (48 %) was detected only in domestic goat followed by 33 % of *Piromyces*. The incidence of other fungal genera in goat samples was negligible or none; however, 20 % of sequences were affiliated to the novel group (NG5).

Kittelmann et al. (2012, 2013) studied the diversity of anaerobic fungi in New Zealand ruminants (sheep, cows and deers), but rumen content was sampled instead of faeces. Barcoded pyrosequencing of ITS1 fragment of 11 animals analysed together discovered monocentric *Neocallimastix* (28 %) and *Piromyces* (20 %) as prevalent genera, while bulbous *Caecomyces* (8 %) and *Cyllamyces* (5 %) as the least numerous genera. Polycentric *Orpinomyces* represented 12 % of sequences, and 24 % of sequences belonged to novel clades SK1, SK3 and SK4 and BlackRhino group. However, differences among individual animals were as high as 88 %, which indicates that each individual ruminant has its own specific anaerobic fungal community influenced by the host animal as well as by the diet. For example, *Caecomyces* prevailing (more than 60 %) in beef cow (Friesian cross) fed hay was not present in dairy cow (Friesian-Jersey) fed silage (Kittelmann et al. 2013).

7.5.1 New Groups of Gut Fungi

The independent studies using different methodological approaches recently provided evidence for a much greater fungal diversity in the digestive tract of wild and domesticated ruminant as well as hindgut-fermenting herbivores than has been detected by classical microbiology techniques. At least three novel groups of gut fungi were predicted by Nicholson et al. (2010), who studied diversity in animal faeces by gel-based methods, namely, denaturating gradient gel electrophoresis (DGGE). Gut fungal-specific primers MN100 and MNGM2, amplifying the ITS1 region, were used to monitor fungal population in seventeen domestic and ten wild ruminants as well as nonruminant herbivores from Australia and Africa. Faecal samples contained a broad spectrum of gut fungi sequences covering five of six known genera (the exception being *Caecomyces*), but more than half of the sequences were assigned to three novel groups of gut fungi. Novel group 2 sequences were identified only in African buffalos, group 4 sequences were identified only in one Australian cattle sample and group 3 were found in two different Australian locations. Two new groups of gut fungi have been predicted by Fliegerova et al. (2010) in cow manure based on the detection of novel *Piromyces*-like and *Anaeromyces*-like ITS1 sequences. The opportunity provided by the revised rules of fungal nomenclature permitted the rapid naming, for the first time, of a species known only from environmental nucleic acid sequence (ENAS) data as *Piromyces cryptodigmaticus* (Fliegerova et al. 2012, GenBank no. GQ850318, GQ850355, GQ850368).

Liggenstoffer et al. (2010) identified in faeces of 30 ruminant and nonruminant herbivores a significant proportion (38 %) of sequences that could not be assigned to any of the six existing genera of anaerobic fungi. Phylogenetic analysis divided these new sequences into eight different novel groups NG1–NG8. Some of these phylogenetically distinct novel lineages were widely distributed (namely, NG1 and NG3), whereas others were specifically related only to particular animals (NG4, NG6, NG7, NG8). Groups NG1 and NG3 were present in 23 hindgut and foregut samples in significant numbers creating, after *Piromyces*, the second (20 %) and third (12 %) most abundant lineages. Their prevalence was striking, especially in all hindgut Equidae samples. The sequences of the group NG1constituted the absolute majority (100 %) in faecal samples from Grant's and Grevy's zebra; the group NG3 was dominant in miniature donkey samples. Both groups were detected in significant proportion also in wild ruminants (3–40 % in deer, 49 % in gazelle, 48 % in goral) but missing in samples of domestic cattle, sheep and goat. Regarding other new groups, NG5 is noticeable in hippopotamus (10 %), sika and western tufted deer (13–27 %), sable antelope (11 %) and domestic sheep and goat (8 % and 20 %). NG6 represented all of the sequences in samples from greater kudu (99 %), okapi (15 %) and Rothschild's giraffe (34 %).

Kittelmann et al. (2012) during the study of 53 rumen samples from sheep, cattle and deer fed different diets detected another four novel clades of gut fungi. The novel clusters referred to as SK1–SK4 comprised 23 % of retrieved sequences. Moreover, 380 GenBank ITS1 sequences were re-evaluated, and the comparison of topology of six different phylogenetic trees resulted in the proposal of a revised phylogeny and pragmatic taxonomy of anaerobic gut fungi (Fig. 7.1) according to Kittelmann et al. (2012). ITS1 sequences grouped into 37 reproducible clusters. 18 clusters grouped with previously named genera: *Neocallimastix* 1 (AF170192), *Piromyces* 1 (FJ501270), *Piromyces* 2 (GQ850316), *Piromyces* 3 (GQ850318, *Piromyces cryptodigmaticus* Fliegerová, K. Voigt & P.M. Kirk), *Piromyces* 4 (JF423776), *Piromyces* 5 (HQ585904), *Piromyces* 6 (GQ850360), *Piromyces* 7 (GQ850304), *Caecomyces* 1 (JF423681), *Cyllamyces* 1 (FJ483845) and *Cyllamyces* 2 (FJ501277), *Orpinomyces* 1 (AY429671), *Orpinomyces* 2 (AM690076), *Orpinomyces* 3 (AM690067), *Orpinomyces* 4 (FJ951430), *Orpinomyces* 5 (JF423690), *Orpinomyces* 6 (JF423655) and *Anaeromyces* 1(FJ501280). Another 14 clusters represented new groups of gut fungi: eight novel groups described by Liggenstoffer et al. (2010, NG1–NG8) were renamed (according to initials of the first author) to AL1 (GQ815916), AL2 (GQ826454), AL3 (GQ829000), AL4 (GQ705077), AL5 (GQ772882), AL6 (GQ600856), AL7 (GQ686301) and AL8 (GQ630940); two novel groups described by Nicholson et al. (2010, NG3 and NG4) were renamed to MN3 (AM690041) and MN4 (AM690050); and four novel groups SK1 (JF423640), SK2 (JF423666), SK3 (JF423720) and SK4 (JF423490) were described by Kittelmann et al. (2012). The remaining five clusters contained sequences from published studies: BlackRhino (GQ738584, Liggenstoffer et al. 2010), DT1 (AM690051, Tuckwell et al. 2005), KF1 (GQ850301, Fliegerova et al. 2010), JH1 (GU911240, Herrera et al. 2011) and UC1 (FJ951427, unpublished, deposited by Chaudhary et al. 2009). Eighteen clusters contained at least one reference sequence from an isolated gut fungus, but 17 clusters contained only environmental clone sequences (ENAS). Kittelmann et al. (2012) in their study declared that more than 29 % of ITS1 sequences of anaerobic fungal isolates or clones in the GenBank database are misnamed at the genus level, which corresponds with findings of Fliegerová et al. (2010).

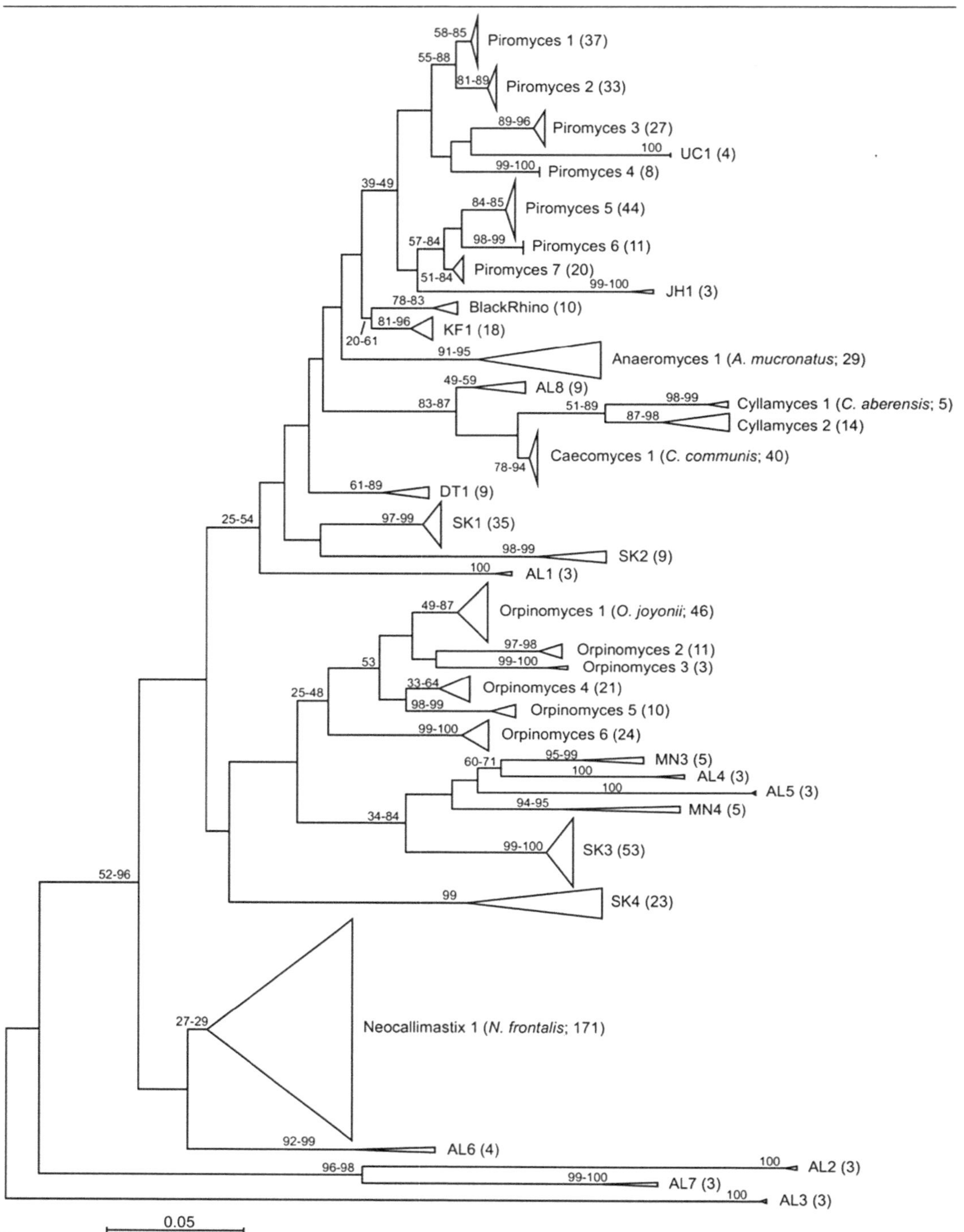

Fig. 7.1 Proposed taxonomy of the anaerobic gut fungi according to Kittelmann et al. (2012). Phylogenetic tree was constructed from 759 ITS gene sequences using the neighbour-joining algorithm with pairwise deletion and Jukes-Cantor correction. Coherent groups are clumped into triangles with a number of sequences shown in parentheses

7.5.2 Effect of Diet and Host Species

Knowledge of the effect of diet and host phylogeny on anaerobic gut fungal community is very important for a better understanding of the role of gut fungi in digestive processes and their influence on fermentation parameters, feed efficiency, animal growth rate and methane production. Liggenstoffer et al. (2010) concluded that feed type appeared to be the least relevant factor in shaping community structure, while the type of gut fermentation provided a better explanation of community relatedness and animal host phylogeny appeared to provide the best explanation for distribution of the genera of anaerobic fungi. On the other hand, several studies indicated greater fungal diversity (Denman et al. 2008) and higher fungal counts (Kumar et al. 2013) in animals fed by high-fibre diets compared with high-grain diet. DGGE study of Kittlemann et al. (2012) of gut fungal communities of cattle, sheep and deer fed three different diets (summer pasture, winter pasture and silage) clearly showed the effect of both diets and ruminant species on the diversity of anaerobic fungi and interaction of ruminant species and diet. Cows showed highly similar rumen fungal community profiles, independent of the feed (mean similarity 89 %). However, animal-to-animal variation among sheep and deer complicated the interpretation of the influence of diet on rumen fungi, which was, however, statistically significant.

No significant differences in anaerobic fungal diversity (assessed by ARISA fingerprinting method) were found in cow rumen content between diets differing in concentrate/forage ratio (Boots et al. 2012). However, the rumen solid and liquid phases reflected the diet changes dissimilarly. The *Neocallimastigales* assemblage was significantly affected in the liquid phase, but not in the solid phase. In the solid phase, the fungal number and species richness were decreased in high-concentrate diet, but the opposite trend was observed in the liquid phase.

A significant reduction in fragment number as well as species richness was, however, apparent when diets were supplemented with soya oil (6 %). The influence of soya oil addition was found much stronger than the influence of concentrate/forage ratio. Unfortunately, the interesting fingerprinting data of Boots et al. (2012) did not reveal any species identities. Therefore, it is not clear if soya oil influenced the *Neocallimastigales* community directly through the inhibition of specific species or indirectly by changing rumen fermentation features.

The host animal definitely affects the gut fungi community structure. The current state of our knowledge suggests a very high variability of gut fungal community not only among different animal species but also among individuals of the same breed. Diet composition also influences the gut fungal population. Recent research indicates that low-fibre/high-grain diets reduce the diversity and numbers of gut fungi although a deeper study is advisable to elucidate the mechanism of the feed effect.

7.6 Unique Features

The gut fungi differ considerably, in many characteristics, from all known members of the kingdom *Fungi*. This deviation from aerobic fungi is very probably caused by the specialized nature of their biotope. The most obvious exceptional feature is the anaerobiosis. Gut fungi are the only fungi for which no oxygen is required for their life cycle and, on the contrary, the presence of oxygen is toxic. Anaerobiosis indicates that gut fungi do not breathe and do not, therefore, require mitochondria, including the whole machinery of the respiratory chain; energy generation organelles are represented by hydrogenosomes. These organelles, under anoxic conditions, decarboxylate malate into acetate, CO_2 and H_2 with concomitant production of energy in the form of ATP (Yarlett and Hackstein 2005; van der Giezen 2009). The hydrogenosomes of gut fungi are unique, differ partially from protozoan hydrogenosomes and most likely followed an alternative pathway to adapt to anaerobic conditions. The available functional and phylogenetic evidences offer two different views on the evolutionary history of these organelles. One hypothesis (Yarlett and Hackstein 2005) suggests that hydrogenosomes of gut fungi, trichomonads and ciliates are substantially different and evolved independently. The other hypothesis (Benchimol 2009) demonstrates a similarity between hydrogenosomes of gut fungi and trichomonad protozoa, considers hydrogenosomes as homologous organelles in unrelated species, weakens the hypothesis of their polyphyletic origin and reinforces the hypothesis that fungal and trichomonad hydrogenosomes are derived from an ancestral endosymbiont preceded by a singular event of endosymbiosis during the course of evolution.

Polyflagellate zoospores are another exceptional morphological feature of gut fungi. Flagellate zoospores are formed during asexual reproduction in *Chytridiomycota* and *Blastocladiomycota*, but the vast majority of species in these two phyla produce exclusively uniflagellate zoospores. Only two genera of gut fungi, the polycentric *Orpinomyces* and the monocentric *Neocallimastix*, produce polyflagellate motile cells. Their zoospores have 7–20 flagella inserted in two rows. Zoospores of species in other gut fungi are uniflagellate, although sometimes biflagellate and tri-flagellate zoospores can be observed. Flagellate fungi are assumed to form early-diverging clades within the *Fungi*, because the simple aquatic forms with flagellate spores are considered to represent the ancestral state (James et al. 2006). The aerobic chytrids forming a single opisthokont flagellum at the end of their zoospores appear to be the closest phylogenetic ancestors of the anaerobic gut fungi. Thus, the presence of a flagellum appears to be an important synapomorphic character for the group of zoosporic fungi as a whole taxon.

Another unique feature of gut fungi is concealed in their DNA composition. The percentage representation of guanine and cytosine bases ranging from 5 % to 20 % in the genomic DNA exhibits the lowest GC content ever reported in any organism. This extreme nucleotide bias is reflected in both the coding and non-coding regions of the genome, with codon usage tending towards more AT-rich codons (Garcia-Vallvé et al. 2000). The non-coding regions are even more AT-rich with many sections expected to be near or above 95 mol% AT content (Nicholson et al. 2005). The codon bias resulting from the low G+C content leads to a depletion in amino acids coded by GC-rich codons (e.g. arginine, proline, glycine) and an overabundance of amino acids coded by AT-rich codons (e.g. lysine, phenylalanine, tyrosine) influencing thus the amino acid composition of encoded proteins.

The main attractive feature of anaerobic gut fungi is, however, their excellent enzymatic system, which is connected with their unique ability to organize hydrolases in large multienzyme complexes called cellulosomes. The gut fungi are the only known members of the kingdom *Fungi* possessing cellulosomes, and this can explain their cellulolytic superiority over aerobic cellulolytic fungi. These exocellular enzyme complexes are extremely active and can degrade both amorphous and crystalline celluloses (Bayer et al. 2004). Research indicates that cellulosomes of anaerobic gut fungi are similar to the most investigated cellulosome, found in the bacterium *Clostridium thermocellum*, although some important differences have been already identified. The composition of the fungal cellulosome is similar to the bacterial one, including the non-enzymatic scaffolding protein(s) with enzyme binding sites called cohesions and a variety of cellulosomal enzymes with dockerins, which interact with the cohesins in the scaffolding protein (Doi 2008; Fontes and Gilbert 2010). The fungal non-catalytic dockerin domain has a function similar to the bacterial dockerins (Ljungdahl 2008), but no sequence homology to clostridial dockerins has been found (Qiu et al. 2000; Aylward et al. 1999; Li et al. 1997). Moreover, thus far, no scaffolding polypeptide has been isolated from a cellulosome of any anaerobic fungus, and there is no detailed knowledge regarding cohesins in scaffolding peptides from gut fungi. Fungal cellulosomes have been described for species of *Neocallimastix*, *Piromyces* and *Orpinomyces*, and molecular biological studies indicate that enzymes associated with the fungal cellulosomes from species in these three genera are modular like those of the anaerobic bacteria.

7.7 Enzymes

The anaerobic gut fungi synthetize a variety of hydrolytic enzymes, including cellulases, xylanases, mannases, esterases, glucosidases and glucanases, which effectively hydrolyze plant biomass, consisting mainly of cellulose and hemicellulose. All genera of gut fungi degrade these polysaccharides, the most abundant in the biosphere, but the range of their hydrolysis depends on the degree of crystallinity. Moreover, gut fungi exhibit genus-dependent differences in their activity towards (hemi) celluloses and recalcitrant substrates. Species of *Neocallimastix* and *Piromyces* are presumed to be the most effective (Tripathi et al. 2007; Paul et al. 2010; Nagpal et al. 2011), while species of *Caecomyces* the least effective (Nielsen et al. 2002). Table 7.2 shows the set of enzymes produced by gut fungi for degradation of main and side chains of structural and storage polysaccharides.

The complicated structure of the cellulose chain is decomposed by several types of enzymes, including endoglucanase (cellulase), splitting β-1,4-glycosidic bonds randomly within the chain; exoglucanases (cellobiosidase or cellobiohydrolases), cleaving cellobiose from the ends of the chain; and β-glucosidases (cellobiases), which convert cellobiose and other low molecular mass cellodextrins into glucose. Cellulolytic hydrolases of gut fungi belong, according to amino acid sequences, hydrophobic clusters and stereochemistry, to several different glycoside hydrolase (GH) families. The β-glucosidases represent usually GH1 or sometimes GH3; exoglucanases are in GH6. Endoglucanases are generally of GH5, but some are classified in GH6. Unusual cellulases of GH9 (Steenbakkers et al. 2002a),

Table 7.2 Spectrum of enzymes produced by gut fungi

Enzyme[a]	EC number	Systematic name
Amylase	EC 3.2.1.1	4-α-D-glucan glucanohydrolase
Cellulase	EC 3.2.1.4	4-(1,3;1,4)-β-D-glucan 4-glucanohydrolase
Endoglucanase	EC 3.2.1.6	3-(1 → 3;1 → 4)-β-D-glucan 3(4)-glucanohydrolase
Xylanase	EC 3.2.1.8	Endo-1,4-β-xylanase
Endochitinase	EC 3.2.1.14	(1 → 4)-2-acetamido-2-deoxy-β-D-glucan glycanohydrolase
Pectinhydrolase	EC 3.2.1.15	(1 → 4)-α-D-galacturonan glycanohydrolase
α-Glucosidase	EC 3.2.1.20	α-D-glucoside glucohydrolase
β-Glucosidase	EC 3.2.1.21	β-D-glucoside glucohydrolase
β-Galactosidase	EC 3.2.1.23	β-D-galactoside galactohydrolase
α-Mannosidase	EC 3.2.1.24	α-D-mannoside mannohydrolase
Mannanase	EC 3.2.1.25	β-D-mannoside mannohydrolase
β-Glucuronidase	EC 3.2.1.31	β-D-glucuronoside glucuronosohydrolase
β-Xylosidase	EC 3.2.1.37	Xylan 1,4-β-xylosidase
Laminarinase	EC 3.2.1.39	3-β-D-glucan glucanohydrolase
N-acetylglucosaminidase	EC 3.2.1.52	β-L-*N*-acetylhexosaminidase
Licheninase	EC 3.2.1.73	(1 → 3)-(1 → 4)-β-D-glucan 4-glucanohydrolase
Cellobiohydrolase	EC 3.2.1.91	Cellulose 1,4-β-cellobiosidase
α-Xylosidase	EC 3.2.1.-	α-D-xylosidase
Acetylxylan esterase	EC 3.1.1.72	Acetylxylan esterase
Ferulic acid esterase	EC 3.1.1.73	4-Hydroxy-3-methoxycinnamoyl-sugar hydrolase
Deacetylase	EC 3.5.1.41	Chitin amidohydrolase
Pectate lyase	EC 4.2.2.2	(1 → 4)-α-D-galacturonan lyase
Arabinase	EC 5.3.1.3	D-arabinose aldose-ketose-isomerase
Exochitinase	(no EC)	N-acetyl-β-D-hexosaminidase

[a]Enzymes are arranged according to EC numbers

GH45 (Eberhardt et al. 2000) and GH48 (Steenbakkers et al. 2002b) have been discovered in species of *Piromyces*. Cellulases of GH5 are intron-less, and their catalytic domains show homology with those of GH from several anaerobic bacteria. It has been suggested that the gene for GH5 cellulases was originally transferred to the gut fungi from a ruminal bacterium and subsequently underwent gene duplication (Garcia-Vallvé et al. 2000). Cellulases of GH6 from gut fungi are often intron-containing, and this finding is considered to be good evidence for a fungal (eukaryotic) origin of this gene.

Enzymatic conversion of xylan, the principal type of hemicellulose, to its monomeric components requires the participation of several enzymes, which are effectively secreted by gut fungi. The main hydrolase is xylanase (endoxylanase), cleaving the β-1,4-xylosidic linkage randomly within the main xylan chain, with the addition of xylosidase, which removes successive D-xylose residues from the nonreducing termini. The ester bonds on xylan and the side chains are hydrolyzed by feruloyl acid esterases, acetylxylan esterases, arabinases or β-glucuronidases depending on the type of branching group. Xylanases of gut fungi belong usually to GH11, rarely to GH10. Other types of hemicellulases like licheninase and mannanase are from GH16 and GH26, respectively. Esterases of gut fungi are also very variable, belonging to carbohydrate esterase (CE) families 1, 2, 3 and 6. The fate of pentoses derived from the hydrolysis of hemicellulose has been elucidated only recently from studies using a species of *Piromyces* (Harhangi et al. 2003). Enzymatic and molecular analysis provided evidence that xylose is converted to xylulose by xylose isomerase and xylulose is phosphorylated by D-xylulokinase to xylulose-5-phosphate, which is a key intermediate in pentose metabolism. This metabolic route resembles that found in bacteria, because in the majority of yeast-forming and filamentous fungi, xylose is converted via xylose reductase and xylitol dehydrogenase to xylulose, which is subsequently phosphorylated (Jin and Jeffries 2003). Xylose isomerase from several strains of gut fungi thus has been intensively studied with respect to its biotechnological exploitation in ethanol production. This effort up to now resulted in a genetically modified strain of *Saccharomyces cerevisiae* able to convert xylose into xylulose using the isomerase from a species of *Piromyces* and/or *Orpinomyces* (van Maris et al. 2007; Madhavan et al. 2009).

7.8 Conclusions

Recent insights into anaerobic gut fungi research indicate significant progress, mainly in the domains of taxonomy, phylogeny and diversity, of these exceptional fungi. Many other investigation efforts have, however, not been mentioned in this chapter, but have strong impact on enlarging our knowledge of gut fungi. Of significance are certainly the sequencing project of the Joint Genome Initiative group on *Piromyces* E2 (http://genome.jgi.doe.gov), high-throughput expressed sequence tag (EST) analysis of complementary DNA of *Neocallimastix frontalis* (Kwon et al. 2009) and the latest outstanding analysis of the *Orpinomyces* genome by Youssef et al. (2013). Comparative genomic analysis of this polycentric rumen fungus identified multiple genes and pathways that are absent in the genomes of *Dikarya* but present in early-branching fungal lineages and/or non-fungal opisthokonts. These included genes for posttranslational fucosylation, the production of specific intramembrane proteases and extracellular protease inhibitors, the formation of a complete axoneme and intraflagellar trafficking machinery and a near-complete focal adhesion machinery. Analysis of the lignocellulolytic machinery in the *Orpinomyces* genome revealed an extremely rich repertoire consisting of 357 glycoside hydrolase (GH) genes, 24 polysaccharide lyases (PL) and 92 carbohydrate esterases (CEs), with evidence of horizontal gene acquisition from multiple bacterial lineages. The glycohydrolytic system is moreover markedly different from that of aerobic lignocellulolytic fungi. Among the surprisingly large number of 16,347 protein-coding genes identified by Youssef et al. (2013), about 42 % genes are unique and have not been previously encountered within the *Fungi*. The broad range of potent polysaccharide-degrading enzymes produced by gut anaerobic fungi thus still represents a big challenge for the scientific and biotechnological community.

References

Aylward JH, Gobius AS, Xue GP et al (1999) The *Neocallimastix patriciarum* cellulase, CelD, contains three almost identical domains with high specific activities on Avicel. Enzyme Microb Technol 24:609–614

Barr DJS (1980) An outline for the classification of the Chytridiales, and for a new order, the Spizellomycetales. Can J Bot 58:2380–2394

Barr DJS (1988) How modern systematics relates to the rumen fungi. BioSystem 21:351–356

Barr DJS, Kudo H, Jakober KD et al (1989) Morphology and development of rumen fungi: *Neocallimastix* sp., *Piromyces communis*, and *Orpinomyces bovis* gen. nov., sp. nov. Can J Bot 67:2815–2824

Bayer EA, Belaich JP, Shoham Y et al (2004) The cellulosomes: multienzyme machines for degradation of plant cell wall polysaccharides. Annu Rev Microbiol 58:521–554

Benchimol M (2009) Hydrogenosomes under microscopy. Tissue Cell 41:151–168

Blackwell M (2011) The Fungi: 1, 2, 3 … 5.1 million species? Am J Bot 98:426–438

Boots B, Lillis L, Clipson N, Kenny DA, Boland TM, Doyle E (2012) Responses of anaerobic rumen fungal diversity (phylum Neocallimastigomycota) to changes in bovine diet. J Appl Microbiol 114:626–635

Breton A, Bernalier A, Bonnemoy F et al (1989) Morphological and metabolic characterization of a new species of strictly anaerobic rumen fungus: *Neocallimastix joyonii*. FEMS Microbiol Lett 58:309–314

Breton A, Bernalier A, Dusser M et al (1990) *Anaeromyces mucronatus* nov. gen. nov. sp. A strictly anaerobic rumen fungus with polycentric thallus. FEMS Microbiol Lett 70:177–182

Breton A, Dusser M, Gaillard-Martinie B et al (1991) *Piromyces rhizinflata*, a species of strictly anaerobic fungus from feces of the Saharan ass – a morphological, metabolic and ultrastructural study. FEMS Microbiol Lett 82:1–8

Brookman JL, Ozkose E, Rogers S et al (2000) Identification of spores in the polycentric anaerobic gut fungi which enhance their ability to survive. FEMS Microbiol Ecol 31:261–267

Callaghan TM, Podmirseg SM, Hohlweck D, Edwards JE, Puniya AK, Dagar SS, Griffith GW (2015) *Buwchfawromyces eastonii* gen. nov., sp. nov.: a new anaerobic fungus (Neocallimastigomycota) isolated from buffalo faeces. MycoKeys 9:11–29

Chen YC, Hseu RS, Chien CY (2002) *Piromyces polycephalus* (Neocallimastigaceae), a new rumen fungus. Nova Hedwigia 75:409–414

Chen YC, Tsai SD, Cheng HL et al (2007) *Caecomyces sympodialis* sp. nov., a new rumen fungus isolated from *Bos indicus*. Mycologia 99:125–130

Christensen K (2004) Life cycle of *Neocallimastix* species. http://www.goatbiology.com/animations/funguslc.html

Dagar SS, Kumar S, Mudgil P, Singh R et al (2011) D1/D2 domain of large subunit rDNA for differentiation of Orpinomyces spp. Appl Environ Microbiol 77(18):6722–6725

Davies DR, Theodorou MK, Brooks AE et al (1993) Influence of drying on the survival of anaerobic fungi in rumen digesta and faeces of cattle. FEMS Microbiol Lett 106:59–64

Denman SE, McSweeney CS (2006) Development of a real-time PCR assay for monitoring anaerobic fungal and cellulolytic bacterial populations within the rumen. FEMS Microbiol Ecol 58:572–582

Denman SE, Nicholson MJ, Brookman JL et al (2008) Detection and monitoring of anaerobic rumen fungi using an ARISA method. Lett Appl Microbiol 47:492–499

Doi RH (2008) Cellulases of mesophilic microorganisms – cellulosome and noncellulosome producers. Ann NY Acad Sci 1125:267–279

Eberhardt RY, Gilbert HJ, Hazlewood GP (2000) Primary sequence and enzymatic properties of two modular endoglucanases, Cel5A and Cel45A, from the anaerobic fungus *Piromyces equi*. Microbiology 146:1999–2008

Ebersberger I, deMatosSimoes R, Kupczok A et al (2012) A consistent phylogenetic backbone for the fungi. Mol Biol Evol 29:1319–1334

Fliegerová K, Mrázek J, Hoffman K et al (2010) Diversity of anaerobic fungi within cow manure determined by ITS1 analysis. Folia Microbiol 55:319–325

Fliegerová K, Voigt K, Kirk PM (2012) *Piromyces cryptodigmaticus*. Index Fungorum 1:1

Fontes CMGA, Gilbert HJ (2010) Cellulosomes: highly efficient nanomachines designed to deconstruct plant cell wall complex carbohydrates. Annu Rev Biochem 79:655–681

France J, Theodorou MK (1994) Differential and integral equation and their application in quantifying the fungal population in the rumen. In: Orpin CG, Mountfort DO (eds) The anaerobic fungi. Marcel Dekker, New York, pp 257–270

France J, Theodorou MK, Davies D (1990) Use of zoospore concentration and life cycle parameters in determining the population of anaerobic fungi in the rumen ecosystem. J Theor Biol 147:413–422

Gaillard B, Breton A, Bernalier A (1989) Study of the nuclear cycle of four species of strictly anaerobic rumen fungi by fluorescence microscopy. Curr Microbiol 19:103–107

Gaillard-Martinie B, Breton A, Dusser M et al (1995) *Piromyces citronii* sp. nov., a strictly anaerobic fungus from the equine cecum- a morphological, metabolic, and ultrastructural study. FEMS Microbiol Lett 130:321–326

Garcia-Vallvé S, Romeu A, Palau J (2000) Horizontal gene transfer of glycosyl hydrolases of the rumen fungi. Mol Biol Evol 17:352–360

Gold JJ, Heath IB, Bauchop T (1988) Ultrastructural description of a new chytrid genus of caecum anaerobe, *Caecomyces equi* gen nov. BioSystems 125:311–415

Griffith GW, Oskoze E, Theodorou MK et al (2009) Diversity of anaerobic fungal population in cattle revealed by selective enrichment culture using different carbon sources. Fungal Ecol 2:87–97

Gruninger RJ, Puniya AK, Callaghan TM et al (2014) Anaerobic Fungi (Phylum Neocallimastigomycota): advances in understanding of their taxonomy, life cycle, ecology, role, and biotechnological potential. FEMS Microbiol Ecol. doi:10.1111/1574-6941.12383

Harhangi HR, Akhmanova AS, Emmens R et al (2003) Xylose metabolism in the anaerobic fungus Piromyces sp. strain E2 follows the bacterial pathway. Arch Microbiol 180:134–141

Heath IB, Bauchop T, Skipp RA (1983) Assignment of the rumen anaerobe *Neocallimastix frontalis* to the Spizellomycetales (Chytridiomycetes) on the basis of its polyflagellate zoospore ultrastructure. Can J Bot 61:295–307

Herrera J, Poudel R, Khidir HH (2011) Molecular characterization of coprophilous fungal communities reveals sequences related to root-associated fungal endophytes. Microb Ecol 61:239–244

Hibbett DS, Binder M, Bischhoff JF et al (2007) A higher-level phylogenetic classification of the Fungi. Mycol Res 111:509–547

Ho YW, Bauchop T (1991) Morphology of three polycentric rumen fungi and description of a procedure for the induction of zoosporogenesis and release of zoospores in cultures. J Gen Microbiol 137:213–217

Ho YW, Barr DJS, Abdullah N et al (1993a) *Neocallimastix variabilis*, a new species of anaerobic fungus from the rumen of cattle. Mycotaxon 46:241–258

Ho YW, Barr DJS, Abbdulah N et al (1993b) *Anaeromyces*, an earlier name for *Ruminomyces*. Mycotaxon 47:283–284

Ho YW, Barr DJS, Abdullah N et al (1993c) A new species of *Piromyces* from the rumen of deer in Malaysia. Mycotaxon 47:285–293

Ho YW, Barr DJS, Abdullah N et al (1993d) *Piromyces spiralis*, a new species of anaerobic fungus from the rumen of goat. Mycotaxon 48:59–68

Ho YW, Abdullah N, Jalaludin S (1994) *Orpinomyces intercalaris*, a new species of polycentric anaerobic fungus from cattle. Mycotaxon 50:139–150

James TY, Kauff F, Schoch CL et al (2006) Reconstructing the early evolution of fungi using a six-gene phylogeny. Nature 443:818–822

Jin YS, Jeffries TW (2003) Changing flux of xylose metabolites by altering expression of xylose reductase and xylitol dehydrogenase in recombinant *Saccharomyces cerevisiae*. Appl Biochem Biotechnol 105:277–285

Joblin KN (1981) Isolation, enumeration, and maintenance of rumen anaerobic fungi in roll tubes. Appl Environ Microbiol 42:1119–1122

Kittelmann S, Naylor GE, Koolard JP et al (2012) A proposed taxonomy of anaerobic fungi (class *Neocallimastigomycetes*) suitable for large-scale sequence-based community structure analysis. PlosOne 7:e36866

Kittelmann S, Seedorf H, Walters WA (2013) Simultaneous amplicon sequencing to explore co-occurrence patterns of bacterial, archaeal and eukaryotic microorganisms in rumen microbial communities. Plos One 8:e47879

Kwon M, Song J, Ha JK et al (2009) Analysis of functional genes in carbohydrate metabolic pathway of anaerobic rumen fungus *Neocallimastix frontalis* PMA02. Asian Aust J Anim Sci 22:1555–1565

Kumar S, Dagar SS, Sirohi SK et al (2013) Microbial profiles, in vitro gas production and dry matter digestibility based on various ratios of roughage to concentrate. Ann Microbiol 63:541–545

Lee SS, Ha JK, Cheng KJ (2000) Relative contributions of bacteria, protozoa, and fungi to in vitro degradation of orchard grass cell walls and their interaction. Appl Environ Microbiol 66:3807–3813

Li J, Heath JB, Bauchop T (1990) *Piromyces mae* and *Piromyces dumbonica*, two new species of uniflagellate anaerobic chytridiomycete fungi from the hindgut of the horse and elephant. Can J Bot 68:1021–1033

Li XL, Chen H, Ljungdahl LG (1997) Two cellulases, CelA and CelC, from the polycentric anaerobic fungus Orpinomyces strain PC-2 contain N-terminal docking domains for a cellulase-hemicellulase complex. Appl Environ Microbiol 63:4721–4728

Liggenstoffer AS, Youseff NH, Couger MB (2010) Phylogenetic diversity and community structure of anaerobic gut fungi (phylum Neocallimastigomycota) in ruminant and non-ruminant herbivores. ISME J 4:1225–1235

Ljungdahl LG (2008) The cellulase/hemicellulase system of the anaerobic fungus Orpinomyces PC-2 and aspects of its applied use. Ann NY Acad Sci 1125:308–321

Lowe SE, Grifith GG, Milne A (1987a) The life cycle and growth kinetics of an anaerobic rumen fungus. J Gen Microbiol 133:1815–1827

Lowe SE, Theodorou MK, Trinci APJ (1987b) Isolation of anaerobic fungi from saliva and faeces of sheep. J Gen Microbiol 133:1829–1834

Lwin KO, Hayakawa M, Ban-Tokuda T (2011) Real-time PCR assays for monitoring anaerobic fungal biomass and population size in the rumen. Curr Microbiol 62:1147–1151

Madhavan A, Tamalampudi S, Ushida K (2009) Xylose isomerase from polycentric fungus *Orpinomyces*: gene sequencing, cloning, and expression in *Saccharomyces cerevisiae* for bioconversion of xylose to ethanol. Appl Microbiol Biotechnol 82:1067–1078

Milne A, Theodorou MK, Jordan MGC (1989) Survival of anaerobic fungi in feces, in saliva and in pure cultures. Exp Mycol 13:27–37

Mora C, Tittensor DP, Adl S, Simpson AGB et al (2011) How many species are there on Earth and in the ocean? PLoS Biol 9:e1001127

Munn EA (1994) The ultrastructure of anaerobic fungi. In: Mountfort DO, Orpin CG (eds) The anaerobic fungi: biology, ecology, and function. Marcel Dekker, New York, pp 47–105

Nagpal R, Puniya AK, Sehgal JP et al (2011) *In vitro* fibrolytic potential of anaerobic rumen fungi from ruminants and non-ruminant herbivores. Mycoscience 52(1):31–38

Nicholson MJ, Theodorou MK, Brookman JL (2005) Molecular analysis of the anaerobic rumen fungus Orpinomyces – insight into an AT-rich genome. Microbiol-SGM 151:121–133

Nicholson MJ, McSweeney CS, Mackie RI et al (2010) Diversity of anaerobic gut fungal populations analyzed using ribosomal ITS1 sequences in faeces of wild and domesticated herbivores. Anaerobe 16:66–73

Nielsen BB, Zhu WY, Dhanoa MS (2002) Comparison of the growth kinetics of anaerobic gut fungi on wheat straw in batch culture. Anaerobe 8:216–222

Obispo NE, Dehority BA (1992) A most probable number method for enumeration of rumen fungi with studies on factors affecting their concentration in the rumen. J Microbiol Methods 16:259–270

Orpin CG (1975) Studies on the rumen flagellate *Neocallimastix frontalis*. J Gen Microbiol 91:249–262

Orpin CG (1976) Studies on the rumen flagellate *Sphaeromonas communis*. J Gen Microbiol 94:270–280

Orpin CG (1977a) The occurrence of chitin in the cell walls of the rumen organisms *Neocallimastix frontalis*, *Piromonas communis* and *Sphaeromonas communis*. J Gen Microbiol 99:215–218

Orpin CG (1977b) On the induction of zoosporogenesis in the rumen phycomycetes *Neocallimastix frontalis*, *Piromonas communis* and *Sphaeromonas communis*. J Gen Microbiol 101:181–189

Orpin CG (1994) Anaerobic fungi: taxonomy, biology and distribution in nature. In: Mountfort DO, Orpin CG (eds) The anaerobic fungi: biology, ecology, and function. Marcel Dekker, New York, pp 1–46

Orpin CG, Greenwood Y (1986) Effects of haems and related compounds on growth and zoosporogenesis of the rumen phycomycete *Neocallimastix frontalis* H8. J Gen Microbiol 132:2179–2185

Orpin GC, Joblin KN (1997) In: Hobson PN, Stewart CS (eds) The rumen anaerobic fungi. Elsevier, London

Orpin CG, Munn EA (1986) *Neocallimastix patriciarum*: new member of the Neocallimasticaceae inhabiting the sheep rumen. Trans Br Mycol Soc 86:178–181

Ozkose M, Thomas BJ, Davies DR et al (2001) *Cyllamyces aberensis* gen. nov., a new anaerobic gut fungus with branched sporangiophores isolated from cattle. Can J Bot 79:666–673

Paul SS, Kamra DN, Sastry VRB (2010) Fermentative characteristics and fibrolytic activities of anaerobic gut fungi isolated from wild and domestic ruminants. Arch Anim Nutr 64:279–292

Phillips MW, Gordon GLR (1989) Growth characteristics on cellobiose of three different anaerobic fungi isolated from the ovine rumen. Appl Environ Microbiol 55:1695–1702

Qiu X, Selinger B, Yanke LJ (2000) Isolation and analysis of two cellulase cDNAs from *Orpinomyces joyonii*. Gene 245:119–126

Rezaeian M, Beakes GW, Parker DS (2004) Distribution and estimation of anaerobic zoosporic fungi along the digestive tracts of sheep. Mycol Res 108:1227–1233

Robertson JA, Emerson R (1982) Two new members of the Blastocladiaceae II. Morphogenetic responses to O_2 and CO_2. Am J Bot 69:812–817

Schoch CL, Seifert KA, Huhndorf S et al (2012) Nuclear ribosomal internal transcribed spacer (ITS) region as a universal DNA barcode marker for Fungi. Proc Natl Acad Sci U S A 109:6241–6246

Sirohi SK, Choudhury PK, Puniya AK et al (2013) Ribosomal ITS1 sequence-based diversity analysis of anaerobic rumen fungi in cattle fed on high fiber diet. Ann Microbiol 63:1571–1577

Steenbakkers PJM, Freelove A, van Cranenbroek B et al (2002a) The major component of the cellulosomes of anaerobic fungi from the genus Piromyces is a family 48 glycoside hydrolase. DNA Seq 13:313–320

Steenbakkers PJM, Ubhayasekera W, Goossen HJ et al (2002b) An intron-containing glycoside hydrolase family 9 cellulase gene encodes the dominant 90 kDa component of the cellulosome of the anaerobic fungus *Piromyces* sp. strain E2. Biochem J 365:193–204

Theodorou MK, Gill M, King-Spooner C et al (1990) Enumeration of anaerobic chytridiomycetes as thallus-forming units: novel method for quantification of fibrolytic fungal populations from the digestive tract ecosystem. Appl Environ Microbiol 56:1073–1078

Theodorou MK, Davies DR, Jordan MGC (1993) Comparison of anaerobic fungi in faeces and rumen digesta of newly-born and adult ruminants. Mycol Res 97:1245–1252

Tripathi VK, Sehgal JP, Puniya AK et al (2007) Hydrolytic activities of anaerobic fungi from wild blue bull (*Boselaphus tragocamelus*). Anaerobe 13:36–39

Tuckwell DS, Nicholson MJ, McSweeney CS et al (2005) The rapid assignment of ruminal fungi to presumptive genera using ITS1and ITS2 RNA secondary structures to produce group-specific fingerprints. Microbiology 151:1557–1567

Ushida K, Tanaka H, Kojima Y (1989) A simple *in situ* method for estimating fungal population size in the rumen. Lett Appl Microbiol 9:109–111

Van der Giezen M (2009) Hydrogenosomes and mitosomes: conservation and evolution of functions. J Eukaryot Microbiol 56:221–231

Van Maris AJ, Winkler AA, Kuyper M et al (2007) Development of efficient xylose fermentation in *Saccharomyces cerevisiae*: xylose isomerase as a key component. Adv Biochem Eng Biotechnol 108:179–204

Webb J, Theodorou MK (1991) *Neocallimastix hurleyensis* sp. nov., an anaerobic fungus from the ovine rumen. Can J Bot 69:1220–1224

Windham WR, Akin DE (1984) Rumen fungi and forage fiber degradation. Appl Environ Microbiol 48:473–476

Wubah DA, Kim DSH (1996) Chemoattraction of anaerobic ruminal fungi zoospores to selected phenolic acids. Microbiol Res 151:257–262

Wubah DA, Fuller MS, Akin DE (1991a) *Neocallimastix*: a comparative morphological study. Can J Bot 69:835–843

Wubah DA, Fuller MS, Akin DE (1991b) Resistant body formation in *Neocallimastix* sp., an anaerobic fungus from the rumen of cow. Mycologia 83:40–47

Yarlett N, Hackstein JHP (2005) Hydrogenosomes: one organelle, multiple origins. Bioscience 55: 657–668

Youssef NH, Couger MB, Struchtemeyer CG et al (2013) The genome of the anaerobic fungus *Orpinomyces* sp. strain C1A reveals the unique evolutionary history of a remarkable plant biomass degrader. Appl Environ Microbiol 79:4620–4634

Rumen Protozoa

8

André-Denis G. Wright

Abstract

The interrelationships between the rumen ciliate protozoa and the livestock animal are discussed here. These include the distribution of protozoa, their classification and the effect of protozoa on and co-evolution along with livestock. The purpose of this chapter is to emphasise those aspects which are of importance to the rumen metabolism.

Keywords

Ciliate • *Dasytricha* • Rumen protozoa • *Isotricha* • *Entodinium* • *Epidinium* • *Ophryoscolex* • *Polyplastron*

8.1 Introduction

Discovered over 160 years ago (Gruby and Delafond 1843), ciliates are the most abundant protozoa found in the rumen of both domesticated and wild ruminants (Williams and Coleman 1992), hence their collective name, the rumen ciliates. They are characterised by generally having (1) a somatic kinetid that is typically made up of one kinetosome with usually two transverse microtubular ribbons, (2) microtubular bundles (nematodesmata) that extend into the cytoplasm from the bases of the kinetids that surround the cytostome (Lynn and Corliss 1991) and (3) unspecialised oral ciliature. Other ciliates, closely related to the rumen ciliates, also inhabit the forestomach and large intestine of a variety of vertebrate animals (camels, elephants, fishes, gorillas, hippopotamuses, horses, humans, kangaroos, rhinoceroses, rodents and warthogs) and are sometimes incorrectly referred to as rumen ciliates. They are morphologically dissimilar to the rumen ciliates, and very little is known about the phylogenetic relationships within and among the gut endosymbiotic ciliates. Although only ciliates from ruminant animals should be called rumen ciliates, it should be noted that a few

A.-D.G. Wright (✉)
Department of Animal Science, College of Agriculture and Life Sciences, University of Vermont, 570 Main Street, Burlington, VT 05401, USA
e-mail: agwright@uvm.edu

A.K. Puniya et al. (eds.), *Rumen Microbiology: From Evolution to Revolution*, DOI 10.1007/978-81-322-2401-3_8

exceptions have been observed. Rumen ciliates have been found in the hindgut of the capybara (Dehority 1987) and in camels (Selim et al. 1996). Interestingly, the so-called hindgut ciliates and nonruminant foregut ciliates have also been sometimes observed in the rumen (Dehority 1986).

8.1.1 Sibling Species

Rumen ciliates range in size from 15 to 250 μm in length and 10 to 200 μm in width (Dehority 2003). *Entodinium* is the smallest rumen ciliate and the most troublesome to classify into species because of the large number of similar forms (Fig. 8.1a). There are over 100 presumed species of *Entodinium*, and undoubtedly some, or most of these species, are the same (Williams and Coleman 1992) or are sibling species. For example, the *Entodinium dubardi* species complex is made up of about 12 presumed species (*E. bimastus*, *E. bovis*, *E. caudatum*, *E. convexum*, *E. dubardi*, *E. exiguum*, *E. longinucleatum*, *E. nanellum*, *E. ovinum*, *E. ovoideum*, *E. parvum* and *E. simplex*), whose identification is difficult and sometimes arbitrary (Dehority 1994). Although Dehority (1994) suggested using 18S rRNA gene sequences from various *E. dubardi* types to differentiate these species, the more variable internally transcribed spacer regions 1 and 2 (ITS-1 and ITS-2) may be more suitable.

8.1.2 Genetic Divergence of Species

Rumen ciliate species like *Isotricha prostoma* have been observed worldwide in almost all domestic and wild ruminants, plus pseudoruminants and nonruminants. Even though the distribution of *I. prostoma* is global, there has been only one study on the extent of genetic divergence among isolates. In that study, eight isolates of *I. prostoma* from three countries, representing two continents, had remarkably conserved ITS-1 and ITS-2 regions, as well as conserved 5.8S rDNA gene sequences with no nucleotide differences between isolates from cattle and sheep hosts (Wright 1999). This suggests that populations of *I. prostoma* on two continents diverged very recently, consistent with human migration and their domestic animals to these continents in the eighteenth and nineteenth centuries. *Isotricha-like* organisms have also been reported from marsupials (e.g. quokkas, kangaroos, wallabies) (Obendorf 1984; Dellow et al. 1988; Dehority 1996) and the hoatzin (Domínguez-Bello et al. 1993), a unique foregut-fermenting bird of South America.

8.2 Importance in Livestock

The rumen ciliates are an agriculturally important group that are involved in host metabolism and digestion of plant material. The concentration of ciliates in the rumen contents of cattle and sheep ranges between 6×10^4 and 4×10^6 per ml (Hungate 1966), with higher concentrations reported sometimes in wild ruminants, 3.39×10^7 per ml (Dehority 1994). Young ruminants isolated at birth do not contain rumen protozoa (Eadie 1962; Dehority 1978; Fonty et al. 1988), but become faunated as a result of adults regurgitating food and rumen contents back into the mouth during rumination and salivating on feed, which is then consumed by the young animal, or the protozoa are passed directly by the mother to its offspring during grooming (Dehority 1993).

8.2.1 Ciliate Biomass and Microbial Protein

Ruminal protozoa can account for half of the microbial biomass in the rumen and contribute up to one-third of fibre digestion (Hungate 1966; Williams and Coleman 1997). *Isotricha* and *Dasytricha* are very important in utilising soluble sugars and controlling the rate of carbohydrate fermentation, especially when large quantities of soluble carbohydrates are present in the diet,

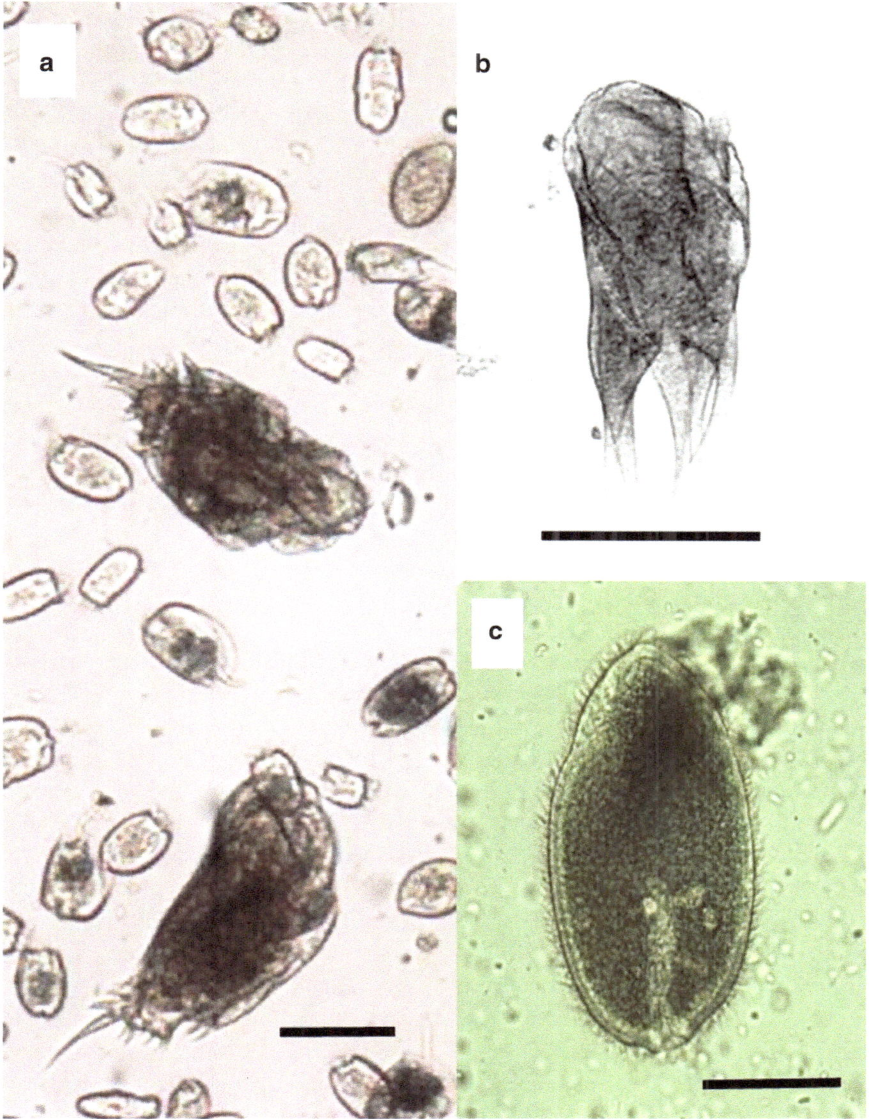

Fig. 8.1 (**a**) Micrograph of *Ophryoscolex caudatum* with a variety of *Entodinium* species. (**b**) Micrograph of *Epidinium cattanei*. (**c**) Micrograph of *Isotricha prostoma*. Bars indicate 50 μm length

whereas some entodiniomorphid ciliates are responsible for controlling starch digestion by engulfing whole starch granules. However, some rumen ciliates can also have a negative impact on ruminant protein metabolism. Microbial protein accounts for as much as 90 % of the amino acids reaching the small intestine of the animal.

The rumen ciliates, also a source of microbial protein, do not appear to pass as rapidly to the small intestine. Rumen ciliates also consume bacterial protein that could otherwise be used by the animal and are only able to convert about 50 % of bacterial nitrogen to protozoal protein (Coleman 1975). In addition, the larger ciliates consume smaller ciliates. As a result, the total microbial protein flow to the small intestine is generally reduced, and excess ruminal ammonia is also increased in faunated animals (Bird and Leng 1978). Also, there is a complex relationship between rumen protozoa and methane-producing archaea, also called methanogens. Methanogens living on and within the rumen ciliates may generate up to 37 % of the total methane emissions from ruminant animals (Finlay et al. 1994; Hegarty 1999).

8.3 Rumen Ciliate Classification

In the older literature (Kudo 1947), the rumen ciliates were divided into two groups: the holotrichs (order Holotricha) and the oligotrichs (order Spirotricha). In 1980, Levine et al. (1980) eliminated the subclass Holotricha and placed the rumen ciliates into two new orders, the Trichostomatida and the Prostomatida, within the class Kinetogragminophorea. Although the term holotrich is no longer strictly appropriate, it is still used in the current literature to refer the non-entodiniomorphid ciliates. Later, Lynn and coworkers (Small and Lynn 1981, 1985; Lynn and Corliss 1991; Lynn and Small 2002), using primarily ultrastructural features of the ciliate cortex, recognised two orders of rumen ciliates, the Entodiniomorphida and the Vestibuliferida within the subclass Trichostomatia, which is the sister group to the free-living haptorians (subclass Haptoria) within the class Litostomatea. In contrast, Grain (1994) also using ultrastructural features of the ciliate cortex recognised three orders (Trichostomatida, Entodiniomorphida and Blepharocorythida) of rumen ciliates within his class Vestibuliferea, the sister group to the class Litostomatea. Despite the differences in rank for the rumen ciliates, both schemes were basically quite similar.

With the introduction of molecular data, phylogenetic analyses of the 18S rRNA gene from the rumen protozoa and the free-living haptorians supported the entodiniomorphids (*Entodinium*, *Epidinium*) as the sister group to the vestibuliferids (*Isotricha*, *Dasytricha*), consistent with the classification of Lynn and Small (2002) (Wright et al. 1997; Wright and Lynn 1997a, b, c; Cameron and O'Donoghue 2004). Interestingly, the rumen ciliates, like all members of the class Litostomatea, share a similar secondary structure of their 18S rRNA gene sequences, which may serve as a diagnostic feature for this class. The secondary structure of the 18S rRNA gene from these ciliates is a missing helix E23-5 and has decreased length of other helices within variable region 4, making these 18S rRNA gene sequences approximately 120 bases shorter than the average length of the ciliate 18S rRNA gene (Leipe et al. 1994; Wright et al. 1997; Wright and Lynn 1997a, b).

8.4 The Entodiniomorphids

The entodiniomorphids are abundant and easily distinguishable from other rumen ciliates. The entodiniomorphids are generally characterised by the presence of a rigid pellicle, spines, skeletal plates (except for *Entodinium* and *Diplodinium*) and tufts of ciliature confined to the adoral and dorsal surfaces of the cell. Species of *Entodinium* only have an adoral zone of syncilia (AZS), whereas all other ophryoscolecids have an AZS and an additional dorsal zone of syncilia (DZS) (Fig. 8.2) that can be located either on the same transverse plane as the AZS, between one-quarter and one-third of the way down the cell, or at the equator of the cell.

More than half of the genera of rumen ciliates belong to the family Ophryoscolecidae, making it the largest family of rumen ciliates. Examples of common genera within this family include *Diplodinium*, *Elytroplastron*, *Entodinium*, *Enoploplastron*, *Epidinium*, *Eudiplodinium*, *Metadinium*, *Ophryoscolex*, *Ostracodinium* and *Polyplastron* (Fig. 8.1). There are 12 additional families of gastrointestinal tract ciliates

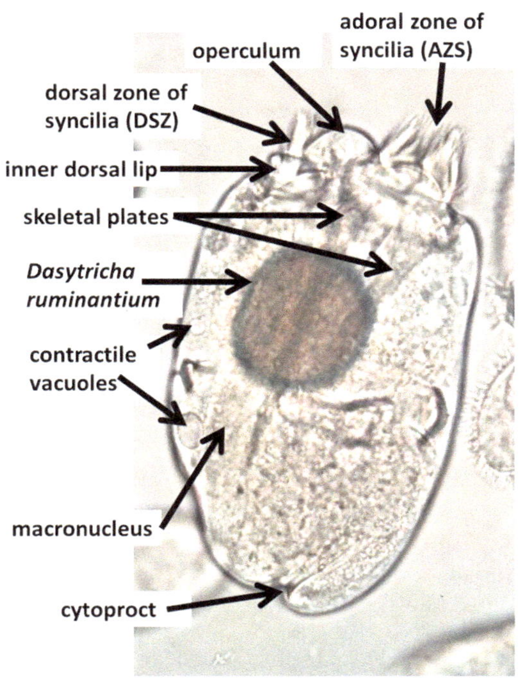

Fig. 8.2 Micrograph of *Polyplastron multivesiculatum* with an ingested *Dasytricha ruminantium*. Length about 160 μm

(Amylovoracidae, Buetschliidae, Blepharocorythidae, Cycloposthiidae, Ditoxidae, Macropodiniidae, Polycostidae, Polydiniellidae, Rhinozetidae, Spirodiniidae, Telamodiniidae and Troglodytellidae) within the order Entodiniomorphida that inhabit the forestomach and large intestine of herbivorous animals. The taxonomy of these 12 families of entodiniomorphids is inconsistent (Small and Lynn 1985; Grain 1994; Lynn and Small 2002).

Recent phylogenetic trees for the trichostomes depicted the vertebrate endosymbionts as a monophyletic group (Wright et al. 1997; Wright and Lynn 1997a, b; Cameron et al. 2001; Cameron and O'Donoghue 2004). However, the marsupial ciliates *Amylovorax*, *Bandia*, *Macropodinium* and *Polycosta*, representing three families (Amylovoracidae, Macropodidae and Polycostidae), grouped on a separate branch and were distantly related to the entodiniomorphids. This would suggest that the marsupial ciliates could represent a separate radiation (Cameron and O'Donoghue 2004; Cameron et al. 2001, 2003). Because the macropod marsupials are more primitive than placental mammals, it is possible that they diverged before the radiation of the endosymbionts of placental mammals. A more comprehensive review of the origin of the marsupial ciliates is presented in a paper by Cameron and O'Donoghue (2004). Interestingly, *Cycloposthium* grouped with the ophryoscolecids to form a clade, consistent with its placement within the order Entodiniomorphida. *Cycloposthium* inhabits herbivorous animals (hippopotamus, rhinoceros) that have diverged before the appearance of the ruminants, and similarly, it is the earliest branching entodiniomorphid before a clade consisting of rumen ciliates.

8.4.1 Evolution Within the Ophryoscolecidae

With more than half of the genera of rumen ciliates belonging to the family Ophryoscolecidae, there has been much speculation on evolution within this family (Crawley 1923; Dogiel 1925, 1947; Lubinsky 1957a, b, c). It was believed that *Entodinium* species is the most ancestral of these ciliates because it had only one AZS, had one contractile vacuole and lacked skeletal plates. It was also concluded that *Entodinium*-like species were probably the first to colonise the rumen. Lubinsky (1957c) observed that the right side of *Diplodinium* is remarkably similar to the left side of *Entodinium* and imagined that, with acquisition of skeletal plates, a torsional displacement of structures occurred across the left side of the cell repositioning the contractile vacuole, lateral groove and nuclei, giving rise to the more derived diplodiniines (Fig. 8.2), such as *Diplodinium*, *Eudiplodinium* and *Polyplastron*. Finally, the DZS was imagined to have migrated posteriorly from the transverse plane of the AZS, leading to the derivation of the evolutionarily advanced ophryoscolecines (*Epidinium*, *Ophryoscolex*; Fig. 8.1b) (Lubinsky 1957c). Interestingly, *Epidinium* has a tube-like cytopharynx and shows the greatest development of cytoalimentary organisation over *Polyplastron* (an evolutionarily intermediate form) and *Entodinium* (ancestral representative). This allows *Epidinium* to ingest large plant fragments in the ruminal contents not available to the intermediate or smaller forms, which ingest smaller plant fibres and bacteria, respectively.

Molecular phylogenetic trees depicted the ophryoscolecids as a monophyletic group with *Entodinium* (subfamily Entodiniinae) as the earliest branching ciliate before a dichotomy containing *Epidinium* and *Ophryoscolex* on one branch (subfamily Ophryoscolecinae) and *Polyplastron*, *Diplodinium* and *Eudiplodinium* (subfamily Diplodiniinae) on the other branch. These groupings corresponded to Lubinsky's (1957c) subfamilial division of the Ophryoscolecidae based upon morphological characters. Further, *Entodinium*'s basal position to the other ophryoscolecids also supported the idea that *Entodinium* was a representative of the ancestral ophryoscolecids (Crawley 1923; Dogiel 1925, 1947; Lubinsky 1957b, c).

8.5 The Vestibuliferids

The vestibuliferids are characterised by having more flexible pellicles entirely covered in cilia (Fig. 8.1c). Three families are recognised within the order Vestibuliferida: Balantidiidae, Isotrichidae and Paraisotrichidae (Lynn and Small 2002). The balantidiids are typically found in the intestine of vertebrate animals and include *Balantidium coli*, the only ciliate that is known to be harmful to humans. Species belonging to the Paraisotrichidae are typically found in horses, whereas species belonging to the Isotrichidae, *Isotricha*, *Dasytricha* and *Oligoisotricha*, are more widespread than the other vestibuliferids, being found in a wide variety of ruminants. Storage polysaccharides are abundant within the endoplasm of *Dasytricha* and *Isotricha*, and species belonging to these two genera are normally found together and feed on starch grains.

8.5.1 Evolution Within the Vestibuliferida

For more than half a century, researchers have speculated on the evolution of the vestibuliferids (Dogiel 1947; Grain 1966; Corliss 1979). Molecular phylogenetic trees revealed that the vestibuliferids, *Dasytricha*, *Isotricha* and *Balantidium*, consistently formed a monophyletic group on a branch as the sister group to the entodiniomorphids and not with the free-living haptorians. These findings are consistent with the view that the vestibuliferid ciliates evolved from a haptorian-like ancestor (Wright and Lynn 1997b). Wright and Lynn (1997b) also proposed that an ancestral holotrichous vestibuliferid, like *Balantidium*, evolved first, establishing symbiotic relationships with fishes (Grim 1985, 1989). As the vertebrate animals diversified, so did the ancestral vestibuliferids, giving rise to the other vestibuliferid genera and to the entodiniomorphids.

8.6 Maximum Ages of the Rumen Ciliates

Because *Entodinium* is found in pseudoruminants like camels, llamas, alpacas, vicunas and guanacos (Williams and Coleman 1992), but not in nonruminants (hippopotamus), *Entodinium* must have arisen after the nonruminants diverged from the order Artiodactyla 50 million years ago and before the separation of the pseudoruminants 40 million years ago (Williams and Coleman 1992). Based on this information, a maximum and a minimum age (40–50 million years) was established for the appearance of *Entodinium* (Wright and Lynn 1997c). Based upon this benchmark, the divergence rate for the rumen ciliates was determined to be 1 % per 8–11 million years. This rate of nucleotide substitution for rumen ciliates is almost a magnitude faster than that for free-living ciliates at 1 % per 72–80 million years (Wright and Lynn 1997c). It was speculated that a faster clock might be explained by intense selection on survivability as the ciliated protozoa invaded the rumen (Dykhuizen 1990) or by the relatively high ambient temperature (39 °C) of the rumen environment, as such high temperatures are known to decrease the efficiency of DNA repair mechanisms leading to higher mutation rates (Sancar and Sancar 1988). However, no further work has been done in this area. So, it is still unclear which explanation is more appropriate and must await future experiments.

8.7 Concluding Remarks

The rumen environment holds the key to developing new techniques capable of raising the level of production of food in an ecologically sustainable way. Methanogens and rumen ciliates are the major candidates for rumen manipulation. It is perhaps not surprising that a great deal of effort has been devoted to investigating methods for manipulation of the rumen ciliates.

References

Bird SH, Leng RA (1978) The effects of defaunation of the rumen on the growth rates of cattle on low protein high energy diets. Brit J Nutr 40:163–167

Cameron SL, O'Donoghue PJ (2004) Phylogeny and biogeography of the "Australian" trichostomes (Ciliophora: Litostomatea). Protist 156:215–235

Cameron SL, Adlard RO, Donoghue PJ (2001) Evidence for an independent radiation of endosymbiotic litostome ciliates within Australian marsupial herbivores. Mol Phylogen Evol 20:302–310

Cameron SL, Wright ADG, O'Donoghue PJ (2003) An expanded phylogeny of the Entodiniomorphida (Ciliophora: Litostomatea). Acta Protozool 42:1–6

Coleman GS (1975) The interrelationship between rumen ciliate protozoa and bacteria. In: McDonald IW, Warner ACI (eds) Digestion and metabolism in the ruminant. University of New England Publishing Unit, Armidale, pp 149–164

Corliss JO (1979) The ciliated protozoa: characterization classification and guide to the literature. Pergamon Press, Oxford, p 455

Crawley H (1923) Evolution in the family Ophryoscolecidae. Proc Natl Acad Nat Sci Phil 75:393–412

Dehority BA (1978) Specificity of rumen ciliate protozoa in cattle and sheep. J Protozool 25:509–513

Dehority BA (1986) Rumen ciliate fauna of some Brazilian cattle: occurrence of several ciliates new to the rumen including the Cycloposthid *Parentodinium africanum*. J Protozool 33:416–421

Dehority BA (1987) Rumen ophryoscolecid protozoa in the hindgut of the capybara *Hydrochoerus hydrochaeris*. J Protozool 34:143–145

Dehority BA (1993) The rumen protozoa. In: Kreier JP, Baker JR (eds) Parasitic protozoa, vol 3. Academic, Toronto, p 333

Dehority BA (1994) Rumen ciliate protozoa of the blue duiker (*Cephalophus monticola*) with observations on morphological variation lines within the species *Entodinium dubardi*. J Euk Microbiol 41:103–111

Dehority BA (1996) A new family of entodiniomorph protozoa from the marsupial forestomach with descriptions of a new genus and five new species. J Euk Microbiol 43:285–295

Dehority BA (2003) Rumen microbiology. Nottingham University Press, Nottingham

Dellow DW, Hume ID, Clarke RTJ et al (1988) Microbial activity in the forestomach of free-living macropodid marsupials: comparisons with laboratory studies. Aust J Zool 36:383–395

Dogiel VA (1925) New parasitic infusoria from the stomach of reindeer (*Rangifer tarandus*). Arch Russ Protistol 4:43–65

Dogiel VA (1947) The phylogeny of the stomach infusorians of ruminants in the light of palaeontological and parasitological data. Quart J Microscop Sci 88:337–343

Domínguez-Bello MG, Lovera M, Suarez P et al (1993) Microbial digestive symbionts of the crop of the Hoatzin (*Opisthocomus hoazin*): an avian foregut fermenter. Physiol Zool 66:374–383

Dykhuizen DE (1990) Experimental studies of natural selection in bacteria. Ann Rev Ecol Syst 21:373–398

Eadie JM (1962) The development of rumen microbial populations in lambs and calves under various conditions of management. J Gen Microbiol 29:563–578

Finlay BJ, Esteban G, Clarke KG et al (1994) Some rumen ciliates have endosymbiotic methanogens. FEMS Microbiol Lett 117:157–162

Fonty G, Senaud J, Jouany JP et al (1988) Establishment of ciliate protozoa in the rumen of conventional and conventionalised lambs: influence of diet and management conditions. Can J Microbiol 34:235–241

Grain J (1966) Étudecytologique de quelquescilliésholotrichesendocommensaux des ruminants et des équidés Microscopieélectroniqueetconsidérationssur la phylogénie des holotrichesgymnostomes et trichostomesendocommensaux. Protistologica:25–51

Grain J (1994) Classe Vestibuliferea de Puytorac et al 1974. In: de Puytorac P (ed) Traité de zoologie:anatomie systématique biologie infusoires ciliés tome II. Masson, Paris, pp 311–379

Grim JN (1985) *Balantidium prionurium* n sp. symbiont in the intestines of the surgeon fish *Prionurus punctatus*. J Protozool 32:587–588

Grim JN (1989) The vestibuliferan ciliate *Balantidium acanthurin* sp. from two species of the surgeonfish genus *Acanthurus*. Arch Protistenkd 137:157–160

Gruby D, Delafond HMO (1843) Recherchessur des animalcules se devéloppant en grand nombredansl'estomacetdan les intestins pendant la digestion des animaux herbivores et carnivores. C R Acad Sci Paris 17:1304–1308

Hegarty RS (1999) Reducing rumen methane emissions through elimination of rumen protozoa. Aust J Agric Res 50:1321–1327

Hungate RE (1966) The rumen and its microbes. Academic, New York, p 533

Kudo RR (1947) Protozoology. Charles C Thomas Publication, Springfield, 778 pp

Leipe DD, Bernhard D, Schlegel M, Sogin ML (1994) Evolution of 16S-like ribosomal RNA genes in the ciliophoran taxa Litostomatea and Phyllopharyngea. Eur J Protistol 30:354–361

Levine ND, Corliss JO, Cox FEG et al (1980) A newly revised classification of the protozoa. J Protozool 27:37–58

Lubinsky G (1957a) Studies on the evolution of the Ophryoscolecidae (Ciliata: Oligotricha) I A new species of *Entodinium* with *'caudatum' 'loboso-spinosum'* and *'dubardi'* forms and some evolutionary trends in the genus *Entodinium*. Can J Zool 35:111–133

Lubinsky G (1957b) Studies of the evolution of the Ophryoscolecidae (Ciliata: Oligotricha) II On the origin of the higher Ophryoscolecidae. Can J Zool 35:135–140

Lubinsky G (1957c) Studies of the evolution of the Ophryoscolecidae (Ciliata: Oligotricha) III Phylogeny of the Ophryoscolecidae based on their comparative morphology. Can J Zool 35:141–159

Lynn DH, Corliss JO (1991) Ciliophora. In: Harrison FR, Corliss JO (eds) Microscopic anatomy of invertebrates protozoa. Wiley, New York, pp 333–467

Lynn DH, Small EB (2002) Phylum Ciliophora. In: Lee JJ, Leedale GF, Bradbury P (eds) An illustrated guide to the protozoa. Society of Protozoologists Special Publication Allen Press, Lawrence, pp 371–656

Obendorf DL (1984) The macropodid oesophagus IV observations on the protozoan fauna of the macropodid stomach and oesophagus. Aust J Biol Sci 37:117–122

Sancar A, Sancar GB (1988) DNA repair enzymes. Ann Rev Biochem 57:29–67

Selim HM, Imai S, Yamato O et al (1996) Ciliate protozoa in the forestomach of the dromedary camel (*Camelus dromedarius*) in Egypt with description of a new species. J Vet Med Sci 58(9):833–837

Small EB, Lynn DH (1981) A new macrosystem for the phylum Ciliophora Doflein 1901. BioSystems 14:387–401

Small EB, Lynn DH (1985) Phylum Ciliophora. In: Lee J, Hutner SH, Bovee EC (eds) An illustrated guide to the Protozoa. Society of Protozoologists Special Publication Allen Press, Lawrence, pp 393–575

Williams AG, Coleman GS (1992) The rumen protozoa. Springer, New York, p 441

Williams AG, Coleman GS (1997) The rumen protozoa. In: Hobson PN, Stewart CS (eds) The rumen microbial ecosystem. Blackie Academic and Professional, London, UK, pp 73–139

Wright ADG (1999) Analysis of intraspecific sequence variation among eight isolates of *Isotricha prostoma* from two continents. J Euk Microbiol 46:434–435

Wright ADG, Lynn DH (1997a) Phylogenetic analysis of the rumen ciliate family Ophryoscolecidae based on 18S ribosomal RNA sequences with new sequences from *Diplodinium Eudiplodinium* and *Ophryoscolex*. Can J Zool 75:963–970

Wright ADG, Lynn DH (1997b) Monophyly of the trichostome ciliates (phylum Ciliophora: class Litostomatea) tested using new 18SrRNA sequences from the vestibuliferids *Isotricha intestinalis* and *Dasytricha ruminantium* and the haptorian *Didinium nasutum*. Eur J Protist 33:305–315

Wright ADG, Lynn DH (1997c) Maximum ages of ciliate lineages estimated using a small subunit rRNA molecular clock: crown eukaryotes date back to the Paleoproterozoic. Arch Protistenkd 148:329–341

Wright ADG, Dehority BA, Lynn DH (1997) Phylogeny of the Rumen Ciliates *Entodinium Epidininum* and *Polyplastron* (Litostomatea: Entodiniomorphida) inferred from small subunit ribosomal RNA sequences. J Euk Microbiol 44:61–67

Ruminal Viruses (Bacteriophages, Archaeaphages)

9

Rosalind A. Gilbert and Athol V. Klieve

Abstract

Viruses of prokaryotes (phages) are ubiquitous to the gastrointestinal tracts of all animals, and particularly dense and diverse populations occur in the rumen of herbivores. Although knowledge of their existence dates back to the 1960s, very few studies were undertaken until the late 1980s and 1990s, when a number of investigators examined rumen phages at both the individual and ecosystem level. Despite the fact that these viruses have characteristics that can be both detrimental (reduce feed efficiency, transfer toxin genes) and advantageous (bacterial population balance, lateral gene transfer, phage therapy, novel enzymes), very little is known about their biological properties or genetic make-up. With recent technical advances in molecular biology, particularly developments in high-throughput sequencing, the field of rumen phage research is predicted to rapidly change, with individual phage isolates and the entire virus fraction (viral metagenome or virome) being characterised in ways never previously possible. The overall importance of phage-host interactions in relation to the functioning of the rumen microbial ecosystem and the nutrition of the animal remains elusive, but these viruses are likely to impact on the bacterial population balance and the flow of genetic material between microorganisms within the ecosystem.

R.A. Gilbert (✉)
Department of Agriculture and Fisheries, EcoSciences Precinct, 41 Boggo Road, Dutton Park, QLD 4102, Australia

Centre for Animal Science, Queensland Alliance for Agriculture and Food Innovation, University of Queensland, St Lucia, QLD 4072, Australia

A.V. Klieve
Department of Agriculture and Fisheries, EcoSciences Precinct, 41 Boggo Road, Dutton Park, QLD 4102, Australia

Centre for Animal Science, Queensland Alliance for Agriculture and Food Innovation, University of Queensland, St Lucia, QLD 4072, Australia

School of Agriculture and Food Sciences, University of Queensland, Gatton Campus, Gatton, QLD 4343, Australia
e-mail: athol.klieve@daf.qld.gov.au; a.klieve@uq.edu.au

A.K. Puniya et al. (eds.), *Rumen Microbiology: From Evolution to Revolution*, DOI 10.1007/978-81-322-2401-3_9

Keywords
Phages • Lytic • Temperate • Ruminant • Microbial ecosystems • Viral metagenomes

9.1 Introduction

The most abundant and prolific viruses found in the rumen are those infecting bacteria, termed bacteriophages or phages. By definition, the general field of phage research covers all viruses of prokaryotes, including those of eubacteria, archaea, cyanobacteria and mycoplasmas (Ackermann and DuBow 1987b); however, the majority of viruses isolated from the ruminal ecosystem have been phages infecting bacteria. Despite occurring at very high densities (Klieve and Swain 1993), phages are the least studied inhabitants of the rumen. Attention is increasingly turning to the importance of studying rumen phages in order to further understand their genetics, interrelationships with bacteria and other organisms, possible impact on rumen function and efficiency and potential applications as biocontrol agents to control specific rumen microbial populations.

This chapter will review our current knowledge of the phage communities known to interact with the rumen biota, summarising early developments in rumen phage biology as well as more recent developments facilitated by rapid advances in molecular biology and high-throughput sequencing technologies.

9.2 Historical Overview

Phage research began shortly after 1914, after bacteriophages had been discovered independently by Twort in England and d'Herelle in France (Twort 1915; d'Herelle 1918; Ackermann and DuBow 1987a). The term bacteriophage was coined by d'Herelle, meaning 'eaters of bacteria' due to the altered appearance of infected colonies. The identity of phages as viruses had to wait until the arrival of the electron microscope, some considerable time later. The large majority of publications up to 1940 were orientated towards identifying a therapeutic use of phages against infectious disease. The discovery of antibiotics such as penicillin and sulphonamides brought this type of research to a halt. Max Delbruck and Salvador Luria, however, have often been credited with laying the foundations of modern phage research (Duckworth 1987). They were the major instigators and proponents of the idea that phages could be used to understand the basic coding properties of the gene, which revolutionised the study of molecular biology and genetics (Kellenberger 1995).

The first reported isolation of phages from the bovine rumen was made in the mid-1960s (Adams et al. 1966). Soon afterwards it was suggested that rather than being transient viruses ingested into the rumen with plant feed material, phages were common inhabitants of the rumen (Hoogenraad et al. 1967) using the rumen microbes as hosts to actively proliferate, and it is now accepted that phages are endemic to the rumen.

9.2.1 Traditional Morphology and Isolation Studies

The advent of electron microscopes with their high resolving power, combined with the development of negative staining in the late 1950s (Brenner and Horne 1959), led scientists to examine many different biological niches for the presence of virus particles. Microbial ecosystems examined in this way included aquatic microbial ecosystems and soil-associated and gut microbial communities (Barnet 1972; Flewett et al. 1974; Torrella and Morita 1979). These investigations quickly established that wherever large microbial populations were found, large numbers of viruses were also present. In this regard, the rumen proved to be no exception, with early electron microscopy surveys of rumen fluid establishing that the rumen contained a dense, morphologically diverse population of viruses, predominated by tailed phages of the viral order Caudovirales (Hoogenraad et al. 1967; Paynter et al. 1969; Klieve and Bauchop 1988).

Although there were several reports of phages in crude rumen fluid obtained from several different species of ruminants, including cattle, sheep and reindeer (Adams et al. 1966; Tarakanov 1974; Orpin and Munn 1974), only a few morphological surveys were conducted to describe the extent of rumen phage population diversity (Paynter et al. 1969; Ritchie et al. 1970; Klieve and Bauchop 1988). In one of the first morphological surveys of bovine rumen fluid, six morphologically distinct types were noted, with little diversity occurring amongst phage types (Paynter et al. 1969). By comparison in a later study (Ritchie et al. 1970), more than 40 distinct phage types representing tailed phages from the families of *Myoviridae*, *Siphoviridae* and *Podoviridae* were observed in both ovine and bovine rumen fluid samples. Interestingly, this study also observed intracellular phages in more than 20 morphologically different bacteria, providing the first evidence that phages may infect a wide variety of rumen bacterial hosts. Further studies of ovine and bovine rumen fluid (Klieve and Bauchop 1988) verified the findings of Ritchie et al. (1970), identifying a wide variety of phage morphotypes, with up to 26 distinct types of predominantly tailed phage observed. The only exception to this dominant morphotype was an icosahedral particle approximately 17×17 nm in size (Fig. 9.1). The early morphological surveys could not always distinguish between phages with filamentous or bacilliform-shaped particles of the viral family *Inoviridae* and broken phage tail fragments. Difficulties also arose when differentiating between the short-tailed *Podoviridae* and the tail-less, icosahedral particles belonging to the *Tectiviridae* family. Phages of other morphotypes representing additional viral families may also have been present in the rumen fluid samples examined but were not numerous nor distinct enough to be identified at this time.

Electron microscopy was also used to estimate the size of the rumen phage population; however, these early estimates were often inconsistent, with greater than 10^9 particles ml^{-1} rumen fluid (Ritchie et al. 1970) being initially reported, whilst another study estimated between 2×10^7–1×10^8 particles ml^{-1} rumen fluid (Klieve and Bauchop 1988). Further developments in molecular biology-based methods to examine rumen phage populations (detailed in Sect. 9.3.2) have since established that studies based solely on electron microscopy may have underestimated both the concentration and true extent of phage population diversity. Despite these technological limitations, these early studies played an important role in pioneering the field of rumen phage research. They firmly established that dense and highly diverse populations of phage were ubiquitous to the rumen and provided significant insights into the roles phages may have in modulating rumen bacterial populations.

9.2.1.1 Lytic Phages

The lytic cycle of phage reproduction is often thought of as the classical means of viral propagation. It is also the basis of most traditional phage detection and isolation methods, for example, the soft agar overlay technique (Klieve 2005). In the lytic cycle phage nucleic acid enters the host bacterium, phage genes are transcribed and replicated, the protein components of the particles produced and complete phage particles assembled. Finally, the host bacterium lyses, releasing the assembled progeny particles into the environment (Ackermann and DuBow 1987a).

The rumen has been a source of lytic phages that infect several rumen bacterial species. These host bacterial species and their respective phages, classified on the basis of particle morphology, are presented in Table 9.1. In addition to phages isolated for well-characterised and predominant rumen bacteria, such as *Prevotella (Bacteroides) ruminicola* and *Streptococcus bovis* (Iverson and Millis 1976a; Klieve and Bauchop 1991; Tarakanov 1974), lytic phages have also been isolated for rumen bacteria which are relatively uncommon, for example, a study which used methanotrophic bacteria of rumen and non-rumen

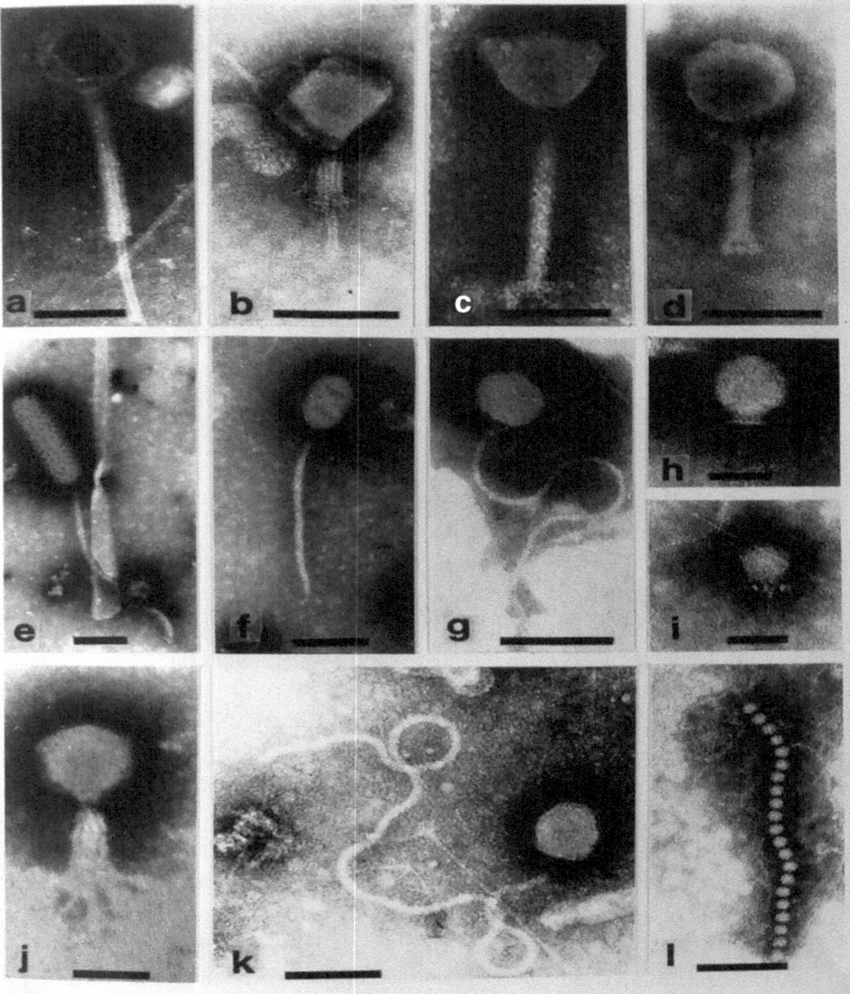

Fig. 9.1 Representative ruminal phage morphologies, determined using transmission electron microscopy (Klieve 1988). Images (**a–d**), *Myoviridae*; images (**e**, **f**, **g**, **k**), *Siphoviridae*; images (**h**, **i**), *Podoviridae*; image (**j**), Caudovirales but family not determinable; image (**l**), a chain of viral icosahedra

origin as hosts to isolate phages from cattle rumen fluid (Tyutikov et al. 1980). Poorly characterised bacteria were also employed as hosts for phage isolation, including novel, uncharacterized strains of acetogenic rumen bacteria (Jiang et al. 1995).

The lytic phages infecting rumen bacteria have mainly been small, tailed phage of the family Siphoviridae, similar to each other except in minor morphological details. Exceptions include the contractile tailed phages of the *Myoviridae* family infecting *Eubacterium* sp. W416 (Orpin and Munn 1974), *Lactobacillus plantarum* (Nemcova et al. 1993), *Selenomonas ruminantium* and *Fusobacterium* sp. (Hazlewood et al. 1983). There have also been reports of two small, short-tailed Podoviruses infecting *S. bovis* (Iverson and Millis 1976a) and a further two Podoviruses infecting *Ruminococcus albus* (Klieve et al. 2004). The only phages isolated to date without a typical tailed phage or Caudovirales morphology include an inducible (lysogenic) filamentous particle from *Butyrivibrio fibrisolvens* (Klieve et al. 1989) and two lytic, filamentous

Table 9.1 Rumen bacterial species from which phages (or phage-like particles) have been isolated and viral classification according to morphological description

Bacterial species	Morphotype (number of phage isolates)	Reference
Bifidobacterium ruminale RU27	*Siphoviridae* (1)	Matteuzzi and Sozzi (1971)
Butyrivibrio fibrisolvens AR14	*Inoviridae* (1)[a]	Klieve et al. (1989)
Eubacterium sp. W416	*Myoviridae* (1)[a]	Orpin and Munn (1974)
Eubacterium ruminantium AR35	*Siphoviridae* (1)[a]	Klieve et al. (1989)
Fusobacterium necrophorum	*Myoviridae* (1)	Tamada et al. (1985)
Fusobacterium sp.	*Myoviridae* (1)[a]	Hazlewood et al. (1983)
Lactobacillus plantarum	*Myoviridae* (3)	Nemcova et al. (1993)
Magnoovum eadii	*Siphoviridae* (1)[a]	Orpin and Munn (1974)
Methanobrevibacter spp.	Caudovirales (1)	Baresi and Bertani (1984)
Prevotella brevis	*Siphoviridae* (1)	Ambrozic et al. (2001)
Prevotella bryantii	nd. (>2)	
Prevotella ruminicola	*Siphoviridae* (1)	Klieve et al. (1991), Klieve et al. (1989)
	Siphoviridae (2)	
Ruminococcus albus	*Podoviridae* (2)	Klieve et al. (2004), Gilbert et al. (unpublished)
	Inoviridae (2)	
	Siphoviridae (1)	
Ruminococcus flavefaciens	*Siphoviridae* (1)[a]	Klieve et al. (1989)
Selenomonas ruminantium	*Myoviridae* (2)[a]	Hazlewood et al. (1983), Lockington et al. (1988), Cheong and Brooker (1998)
	Siphoviridae (1)[a]	
	Siphoviridae (1)[a]	
Serratia spp.	nd.	Adams et al. (1966)
Streptococcus bovis	nd.	Adams et al. (1966), Iverson and Millis (1976a), Klieve and Bauchop (1991), Klieve et al. (1999), Styriak et al. (1989, 1994b), Tarakanov (1974, 1976, 1994, 1996)
	Caudovirales (25)[b]	
	Siphoviridae (1)	
	nd. (3)	
	Siphoviridae (2)	
	Siphoviridae (1)	
	Caudovirales (>20)[b]	
Streptococcus durans	*Siphoviridae* (4)	Brailsford and Hartman (1968)
	Siphoviridae (1)[a]	
Streptococcus intermedius AR36	*Siphoviridae* (1)[a]	Klieve et al. (1989)
Unclassified rumen acetogen H3HH	*Siphoviridae* (1)[a]	Jiang et al. (1995)

[a]Lysogenic phage forming phage particles
[b]Including lytic and lysogenic phages; and not described (nd.)

phages of *R. albus* belonging to the *Inoviridae* family (Klieve et al. 2004). The latter are also the only phages isolated to date that do not contain dsDNA, both having circular ssDNA genomes.

The most detailed studies made on any group of rumen phages, however, have focused on the Caudovirales phages infecting *S. bovis*. These phages have been characterised in terms of their morphology, host range, stability, growth characteristics, serological properties and DNA restriction endonuclease digestion patterns (Tarakanov 1974, 1996; Iverson and Millis 1976a, b, 1977; Styriak et al. 1989; Klieve and Bauchop 1991; Styriak et al. 1991, 1994a; Styriak et al. 1998; Klieve et al. 1999). The majority of cultured *S. bovis* phages have belonged to either the *Siphoviridae* or *Podoviridae* families, with most isolates having long, non-contractile tails characteristic of the *Siphoviridae* family. Host ranges tend to be narrow with individual phages able to infect only one or two strains of *S. bovis* (Klieve et al. 1999) with only one report of a phage, designated F4 (Styriak et al. 1994b) having a wider host range, being able to infect 5 out of 10 *S. bovis* strains examined.

The focus on phages of *S. bovis* can be attributed to the importance of this amylolytic bacterial species in the rumen. *S. bovis* is present in relatively high numbers in rumen contents and can adhere to rumen epithelial cells (Styriak et al. 1994a). This bacterial species is also of economic importance, having been implicated in the disease syndrome, lactic acidosis, which affects cattle on high grain diets (Nagaraja and Titgemeyer 2007). *S. bovis* was also the first bacterial species for which phage therapy was considered as a possible means of population control in the rumen (Klieve et al. 1999; Tarakanov 2006).

9.2.1.2 Temperate Phages and Carrier States

Temperate or lysogenic phages are able to exist within their host bacterium in a relatively stable, dormant state without causing lysis. Infection proceeds as in the lytic cycle, but after entry of the phage nucleic acid into the bacterial cell, either the phage genome may be replicated and new particles produced or the phage genome enters an inactive or dormant state. In this state, the phage genome is often described as a prophage. Prophages may be integrated into the host genome either at a specific site or at several sites, with some phages integrating at or near palindromic structures, such as transfer RNA genes (Campbell 1992; Bobay et al. 2013), whilst other prophages can persist in the cytoplasm as a plasmid (Ackermann and DuBow 1987a). In addition, individual bacteria may play host to several integrated prophages (Refardt 2011) with many bacterial and archaeal genomes containing fragments of integrated prophages (Krupovic et al. 2011). Once a prophage becomes fragmented, it may lose the ability to produce intact, viable phage particles or may be completely unable to emerge from the lysogenic state. Intact prophages however are usually able to emerge from the lysogenic state in a process referred to as phage induction and enter the lytic cycle of phage production culminating in the lysis of the host cell and release of progeny phage particles.

Historically, the presence of temperate phages within rumen bacteria was verified by electron microscopy of pure cultures, for example, a large Siphovirus was detected in the culture medium of *Magnoovum eadii* (Orpin and Munn 1974), or through the application of an inducing agent to pure cultures of bacteria followed by electron microscopy (Iverson and Millis 1976b; Tarakanov 1974). Inducing agents may be physical, chemical or biological agents with mitomycin C being the most commonly employed chemical agent for inducing pure cultures of rumen bacteria, particularly those requiring the maintenance of strictly anaerobic conditions (Klieve 2005). An early study by a Russian researcher found that 50 % of rumen *S. bovis* isolates were hosts to temperate phage and suggested that the lysogenic lifecycle played a significant role in sustaining *S. bovis* phage numbers in the rumen (Tarakanov 1974). A further study detected lysogeny in only four percent of their rumen-derived *S. bovis* isolates (Iverson and Millis 1976b) and, in contrast to Tarakanov's claims, concluded that lysogenic strains of *S. bovis* were not a significant source of phages in vivo. This conclusion was later supported in an investigation that found lysogeny to be uncommon in isolates of *S. ruminantium* (Hazlewood et al. 1983).

The extent of lysogeny amongst rumen bacteria was later clarified in a study which examined a diverse range of rumen bacterial isolates for inducible phage particles (Klieve et al. 1989). Of the 38 ruminal bacteria studied, nine organisms (23.7 %), representing five genera, produced phage-like particles following induction with mitomycin C. It was therefore concluded that viral genetic material may be a normal genetic constituent of the genome of appreciable numbers of ruminal bacteria. The phages so far isolated from the rumen or induced from ruminal bacteria (Table 9.1), however, represent a small proportion of the morphological types observed in rumen fluid using electron microscopy (Fig. 9.1).

As an alternative to lysogeny, many phages cause persistent infections (pseudolysogeny or carrier states) characterised by the simultaneous presence of phages and bacterial cells in the same culture, with phages multiplying in only a fraction of the bacterial population (Weinbauer 2004). Carrier states have been shown to frequently occur in nature (Clokie et al. 2011) and have been observed in the microbial fraction of soil and aquatic environments (Ackermann and DuBow 1987b). It has also been suggested that carrier states may also occur in the rumen (Klieve et al. 1989); however, only a single study has demonstrated this interaction in vitro with the ruminal bacterium *P. ruminicola* AR20 and the phage ϕBrb01 (Klieve et al. 1991). In this study, bacterial cells surviving infection developed a thick extracellular capsule, presumably preventing the phage from reaching receptors on the cell wall. In the absence of the phage, the host strain had very few capsulated cells. It could be speculated that in the rumen microbial ecosystem, sustained growth of both capsulated and non-capsulated cells would maintain populations of both the phage and bacterial host therefore perpetuating the carrier state balance.

The remaining identified state by which a phage may be maintained and coexist in balance with its prokaryote host is referred to as chronic infection. In chronic infection, a host is infected by a phage which then produces progeny phage particles which may be constantly released from the host cells via budding or extrusion without causing cell death (Weinbauer 2004). Chronic infection occurs for some *Inoviridae* (filamentous phages), plasmaviruses (Clokie et al. 2011), archaeaphages (Porter et al. 2007) and *Siphoviridae* phages (Lood and Collin 2011). Although this state has not yet been demonstrated in vitro with rumen phage isolates, it is not unrealistic to expect that chronic phage infections may occur for some rumen microbes and also play a role in maintaining rumen phage populations.

9.2.2 Molecular Genetics of Rumen Phage Isolates (Pre-2000)

With the development of modern molecular biology techniques in the 1980s, phage biology shifted from using methods based on culturing (determination of plaque size, host range and burst size) and morphology (electron microscopy) to the use of nucleic acid-based methods for the characterisation of both phage isolates and environmental populations of phages. Attributes such as the nature of the nucleic acid and the size of the phage genome became important for phage classification.

Prior to the year 2000, the molecular genetics of several rumen phages were investigated, including phages of *P. ruminicola* (Klieve et al. 1991), *S. ruminantium* (Lockington et al. 1988) and *S. bovis* (Styriak et al. 1998). The techniques employed were similar to those used in molecular virology laboratories of the era, including (a) genome mapping using the relative location of phage-specific restriction enzyme sites, sometimes referred to as cut maps; (b) use of radiolabelled dNTPs to investigate the mechanisms of phage genome packaging and the occurrence of terminal redundancy and circular permutation of phage genomes; (c) southern blotting and hybridisation to identify phage genes of interest and determine interactions with host genomes; and (d) early Sanger sequencing of individual phage genes.

9.2.2.1 Restriction Enzyme Genome Mapping

Restriction enzyme mapping involves digesting purified phage genomic DNA with various restriction enzymes to create a fingerprint-like pattern of DNA lengths (cut map) based on the location of restriction enzyme sites within the phage genome. This method also allows the phage genome size to be determined (Lockington et al. 1988) and can assist in differentiating morphologically similar phages. For example, two phages isolated from the rumen infecting *S. bovis* designated as phages F1 and F3 and both classified by electron microscopy as *Siphoviridae* (Styriak et al. 1989) could be differentiated following digestion of phage genomic DNA with the restriction enzymes *Pst*I, *Bgl*I and *Eco*RI.

Conversely, the inability of specific phages to be digested with certain restriction enzymes can be used to characterise phages, for example, the lytic *S. bovis* phage F4 could not be digested with the enzymes which recognise GATC sequences including *Bam*HI and *Pvu*I (Styriak et al. 2000). Similarly, the genome of *S. bovis* phage φSb01 could not be digested with *Bam*HI (Klieve and Bauchop 1991). Investigating the susceptibility of phage genomes to restriction enzyme digestion also provided early insights into the existence of phage resistance mechanisms within rumen bacterial isolates, for example, the presence of a restriction-modification system was noted for the *S. bovis* isolate II/1 (Styriak et al. 1998).

9.2.2.2 Determination of Circular and Terminal Redundancy

In addition to determining the size of phage genomes, early rumen phage researchers sought to investigate the mechanism specific phages used for packaging the phage DNA genome into the proteinaceous phage particle. Whilst cut maps provided an initial indication as to whether the phage DNA was packaged into genomes of consistent length or not (described as terminal redundancy), additional molecular biology methods and restriction enzyme mapping, were needed to examine the exact nature of the phage genome terminal ends. In this way, the genome length of the *P. ruminicola* phage φBrb01 was shown to be actually closer to 33 kb, although a genome overlap occurring during phage particle packaging (terminal redundancy) of approximately 2 kb resulted in genome lengths of approximately 35 kb being obtained from most phage particle preparations (Klieve et al. 1991). Similarly, the terminal ends of a temperate phage of *S. ruminantium* strain M-7 were found to be cohesive in nature (Lockington et al. 1988), although this bacterial strain and the associated lysogenic phage have since been lost from culture. A further study of a temperate phage infecting *S. ruminantium* strain ML12 (Cheong and Brooker 1998) identified that the genome of this phage also had cohesive ends with no evidence of terminal redundancy, indicating that progeny phage particles should always contain genomes of the same length.

9.2.2.3 Characterisation of Phage Integrases and Excisionases

The replication strategies of phages isolated from the rumen have rarely been characterised on the molecular level except in studies investigating the use of phages and phage genes as novel cloning vectors specifically tailored for the genetic transformation of rumen bacteria (Cheong and Brooker 1998; Gregg et al. 1994). These studies sought to identify the genes used by temperate rumen phages to integrate into the host bacterial chromosomes. In most characterised phages, site-specific recombination functions, involving the integrase (*Int*) and exisionase (*Xis*) genes, are tightly clustered in a relatively small stretch of the phage genome and are characterised by short sequence repeats surrounding a core sequence required for integration (Mumm et al. 2006).

Genes for site-specific recombination functions were identified for only two temperate rumen phages (Cheong and Brooker 1998; Gregg et al. 1994). Examination of the temperate phage ϕAR29 of *P. ruminicola* strain AR29 (Klieve et al. 1989) located the point of integrative recombination (*att*P locus) (Gregg et al. 1994). Further sequencing indicated the presence of several direct and inverted repeats around the *attP* site and two open reading frames encoding the *Int* and *Xis* genes, both being required for the stable integration of the phage genome into the host chromosome. Similarly, the *attP* region on the temperate *S. ruminantium* phage M1 genome was identified (Cheong and Brooker 1998), and further sequence analysis of the surrounding region enabled detection of the *Int* and *Xis* genes for this phage. The rumen phage gene sequences derived from both of these studies were lodged in publically accessible sequence databases and are still available for further reference (GenBank Accession numbers AF034575 and S75733.1).

9.3 Recent Advances

9.3.1 Genome Sequencing of Rumen Bacteria, Prophage Identification and CRISPR

Despite the lack of published rumen phage sequences, the genetic knowledge of rumen phages is rapidly expanding due to the advent of bacterial whole genome sequencing. Phage-related sequences and DNA secondary structures are often encountered during the sequence analysis and annotation procedures of complete or near-complete genome sequences, for example, using the web-based genome annotation software pipelines of RAST (Aziz et al. 2008) and IMG (Markowitz et al. 2012). This approach was undertaken in a study which identified prophage-related genes such as integrases, recombinases and other phage-related proteins within 14 rumen bacterial reference genome sequences (Berg Miller et al. 2012).

Alternatively, once raw bacterial genome sequence data is closed into either long contigs or a complete, circular genome, the sequences may be further analyzed using custom in-house software pipelines or web-based prophage prediction tools such as PHAST (Zhou et al. 2011), Prophage Finder (Bose and Barber 2006) and Prophinder which utilises the ACLAME database (Lima-Mendez et al. 2008). These tools not only detect phage-related genes by checking for sequence similarity against specific, virus-based sequence databases but also take into consideration additional factors such as the relative GC content, sequence repeats, alignment with expected flanking regions (e.g., tRNA genes and integration fragments) and the orientation, number and arrangement (clustering) of open reading frames (ORFs) to more accurately predict entire regions pertaining to integrated prophages.

A survey of several publically available, closed circular genomes of rumen bacteria and their associated plasmids is detailed in Table 9.2, with only 2 of the 11 rumen bacteria examined not containing any prophage-related regions within their complete genetic complement of chromosomes and cell-associated plasmids. Of the detected prophage regions, six regions of approximately 30 kb or more in length were predicted to represent intact prophages. Differences in the parameters and algorithms used in respective phage detection programs can result in differing prophage-related regions being detected; therefore, further analysis is usually required in order to fully annotate prophage genes and verify the length of the predicted prophage regions. Presently, the PHAST tool has the most frequently updated reference database, can predict the closest-known phage relative and tends to detect longer, more intact prophage regions than Prophage Finder. As an example, the 37.3 kb predicted prophage region within the genome of *Basfia (Mannheimia) succiniciproducens* strain MBEL55E (Hong et al. 2004) incorporates a possibly intact prophage genome encoding for tail, capsid and a range of phage functional genes with 28 ORFs (representing 51 % of the 55 prophage ORFs identified by PHAST) related to the *Burkholderia* phage phiE255, a lysogenic Myovirus spontaneously produced in cultures of the soil saprophyte *Burkholderia thailandensis* strain E255 (Ronning et al. 2010).

The process of prophage detection in closed genome sequences has also been undertaken for the rumen-derived methanogenic archaea *Methanobrevibacter ruminantium* strain M1 (ATCC35063) (Attwood et al. 2008; Leahy et al. 2010) which was found to contain a long prophage element, designated φmru. This prophage region encoded both structural (capsid and tail proteins) and functional phage genes (enzymes for phage integration, DNA replication and packaging and host cell lysis) (Attwood et al. 2008); however, phage particles proved to be difficult to induce or cultivate in vitro.

It is anticipated that as more rumen bacteria are genome sequenced, the identification of integrated phage genetic material will rapidly progress. In some instances, integrated prophages may have undergone deletions or modifications during the replication of the host bacteria over many generations; therefore, only a subset of phage genes will be present and expression of phage genes will not result in the creation of viable phage particles. However, if a complete, viable phage genome is present within a bacterial host chromosome or plasmid, intact phage particles may be produced following treatment of host cells with an inducing agent such as mitomycin C. Phage particles may then be fractionated and concentrated from the culture medium and phage genomic material extracted and sequenced. To date, this procedure has only been undertaken with φAR29, the temperate phage of *P. ruminicola* strain AR29 (Seet 2005). This study obtained full length (35,558 bp) sequence for φAR29,

Table 9.2 Assessment of publically available, closed rumen bacterial genome sequences and associated plasmids for predicted prophage regions using the web-based tools Prophage Finder (Bose and Barber 2006) and PHAST (Zhou et al. 2011)

Rumen bacteria		PHAST		Prophage finder		Accession number
		Prophage region/s	Size (kb)	Prophage region/s	Size (kb)	
Butyrivibrio proteoclasticus B316 Chromosome 1		0		0		NC 014387.1
Chromosome 2		0		2	10.2, 16.4	NC 014388.1
Plasmids	pCY360	0		3	24.1, 9.6, 8.2	NC 014389.1
pCY186		0		2	14.1, 13.1	NC 014390.1
Ruminococcus albus 7 Chromosome		0		5	27.7, 19,5, 18.5, 15.1 11.0	NC 014833.1
Plasmids	pRUMAL01	0		3	17.7, 15.8, 7.9	NC 014824.1
pRUMAL02		1	28.7[a]	4	30.0, 28.9, 10.3, 6.0	NC 014825.1
pRUMAL03		0		0		NC 014826.1
pRUMAL04		0		0		NC 014827.1
Actinobacillus succinogenes 130Z		2	4.1, 39.1[a]	4	43.6, 16.5, 13.0, 11.3	NC 009655.1[b]
Desulfovibrio desulfuricans subsp. *desulfuricans* ATCC 27774		1	36.6[a]	3	10.9, 20.3, 29.3	NC 011883.1[b]
Basfia (Mannheimia) succiniciproducens MBEL55E		1	37.3[a]	5	33.9, 22.6, 17.0, 16.0, 17.0	NC 006300.1
Prevotella ruminicola 23		0		4	15.8, 14.2, 13.9, 10.8	NC 014033.1
Fibrobacter succinogenes S85		0		0		NC 017448.1
Slackia heliotrinireducens DSM 20476		2	19.5, 34.3[a]	9	39.1, 19.8, 18.3, 17.8, 16.1, 16.9, 12.0, 13.1, 6.6	NC 013165.1
Wolinella succinogenes DSM 1740		0		0		NC 005090.1
Megasphaera elsdenii DSM 20460		2	9.8, 24.0	6	39.3, 18.8, 16.5, 13.7, 9.6, 5.8	HE576794
Desulfotomaculum ruminis DSM 2154		6	54.0, 35.8[a], 25.4, 15.4, 13.0, 11.2	10	37.3, 27.2, 26.0, 16.3, 14.8, 14.4, 12.3, 10.9, 8.8, 4.5	CP002780.1

[a]Prophage region predicted to be intact by PHAST
[b]Joint Genome Institute, unpublished genome

identifying 53 open reading frames and enabling the identification of several functional phage genes involved in phage particle packaging and genome replication such as the large terminase subunit, helicase, amidase and transcription regulator.

As more bacterial genomes are sequenced and deposited in publically available sequence databases, it is anticipated that more commonalities between phage genomes will be identified, which will (a) provide insights into the evolutionary history and selective pressures experienced by phages in the rumen, (b) enable the roles phages play in the rumen ecosystem to be further understood and (c) facilitate the detection of previously unidentified rumen phage genes within both whole genome and metagenome sequence datasets.

The relatively recent identification of clustered regularly interspaced short palindromic repeat sequences (CRISPR), forming arrays in complete genomes of bacteria, has changed our understanding of how phages interact with their prokaryote hosts and have been identified in approximately 50 % and 90 % of sequenced bacterial and archaeal genomes, respectively (Sorek et al. 2008, 2013). These arrays, together with groups of associated (*Cas*) proteins, confer resistance to phages and extrachromosomal elements, using small adaptive RNAs in an RNA-interference-like mechanism (Sorek et al. 2013). They may also more broadly contribute to the regulation of endogenous bacterial genes including those associated with virulence (Sampson et al. 2013).

Short sequences called spacers are stored in between the repeated elements, and it is these spacer regions that are homologous to invasive nucleic acids such as those introduced into the cell by infecting phages. The spacer regions are transcribed and processed into small non-coding interfering CRISPR RNAs (crRNAs) that guide the Cas proteins towards the target nucleic acids for specific cleavage of the homologous sequence (Barrangou 2013). This process was demonstrated experimentally in a study which demonstrated that new spacer regions were acquired following phage challenge and conveyed resistance against further phage infection (Barrangou et al. 2007). Sequence analysis of spacer regions can therefore provide a history of the phage infections encountered by bacteria and archaea, although the spacer regions tend to be considerably shorter (26–72 bp) (Grissa et al. 2007a) than integrated prophage regions. Several bioinformatics programs and online resources have been developed to facilitate the detection of CRISPRs in genomic sequence data, for example, the CRISPR Recognition Tool (Bland et al. 2007), CRISPRFinder and a database of CRISPR arrays that allows the alignment and comparison of repeats and spacers against public sequence databases CRISPRdb (Grissa et al. 2007b).

These tools have been applied to the genomes of rumen bacteria, with CRISPR sequences identified in the genome sequences of several rumen bacteria including *P. ruminicola* strain 23 and *P. bryantii* B_14 (Purushe et al. 2010) and the rumen archaean *Methanobrevibacter ruminantium* M1 (Attwood et al. 2008; Leahy et al. 2010). A more comprehensive survey for the presence of CRISPR sequences (Berg Miller et al. 2012) in 13 rumen bacterial genomes, including the three aforementioned genomes and several representatives of the genus *Ruminococcus*, found CRISPR loci in the majority of genomes, with CRISPR absent from two *R. flavefaciens* isolates. An additional 15 draft genome sequences were examined, and eight were found to contain CRISPR-like sequences, although all contained CRISPR sequences deemed 'questionable' by the CRISPRFinder program. This study also found that the CRISPR spacer regions identified within rumen bacterial genomes often did not match with sequences found within current genomic sequence databases (e.g., the GenBank NR nucleotide database and the NR_Viral_DB), although a relatively small number of spacer regions (39 unique spacer regions (6.7 %) out of a total of 581 spacer sequences) were related to known viral or prophage sequences. Interestingly, the identified rumen bacterial spacer regions in the CRISPR loci of the *F. succinogenes* S85 and *P. ruminicola* 23 genome sequences had greater homology to the three rumen virome datasets generated in the Berg Miller et al. (2012) study. The absence of rumen phage sequences within the current genomic sequence databases may therefore be limiting the ability to identify homology and classify the newly discovered CRISPR spacer regions of rumen bacteria.

The detection of CRISPR arrays within the rumen virome sequence datasets (Berg Miller et al. 2012) provided the evidence to suggest that a long history of phage infection within the rumen has contributed to the development of a genetic memory or immunity within many species of rumen bacteria and archaea. It also suggests that both the rumen microbes and associated phages are constantly evolving to sustain a viable community balance, and this constant selective pressure is likely to have resulted in the wide diversity of phages found to be associated with the rumen microbial community.

9.3.2 Molecular Ecology of Rumen Phage Populations

9.3.2.1 Fingerprinting Techniques (PFGE)

Before sequencing technologies became more rapid and affordable, molecular biology-based methods for studying microbial ecology often involved extracting the genetic material from a sample and developing a population-specific profile or fingerprint pattern. For example, PCR amplification of phylogenetic marker genes such as specific regions of the highly conserved ribosomal RNA genes (16S rRNA genes), followed by denaturing gradient gel electrophoresis (DGGE), can be employed to determine differences in rumen bacterial and archaeal populations (Gantner et al. 2011; Watanabe et al. 2001). Phages do not encode nor require complete rRNA genes for replication, therefore, PCR amplification of these genes cannot be employed to characterise phage populations. An alternative molecular method was developed for enumerating and differentiating rumen phage populations, utilising pulsed field gel electrophoresis (PFGE) to create a genetic fingerprint of phage populations based on phage genome length (Klieve and Swain 1993). The pulsed field gels generally distinguished two major components: (1) a broad region (30–200 kb) including the genome lengths expected for most tailed phages and comprising DNA genomes from many different phages and (2) discrete bands of DNA arising from a single or several phage genotypes, presumably representing blooms of lytic phage activity. When combined with a DNA quantification step, a relative measure of overall phage concentration within a rumen sample could also be determined (Klieve and Gilbert 2005).

The PFGE method was employed to determine differences between (a) species of ruminants (sheep, cattle and goats), (b) animals at pasture and in pens, (c) animals penned together and on the same diet and (d) the same animal sampled at the same time daily over consecutive days (Klieve and Swain 1993; Swain et al. 1996).

These studies first suggested that diet and dietary components may affect phage activity after it was found that sheep grazing pasture had total phage populations that were significantly higher than their counterparts fed a diet of chopped oaten and lucerne hays (70:30) (Swain et al. 1996). In a further investigation (Klieve et al. 1998), the size of the rumen phage population was surveyed in sheep and cattle fed a variety of fresh and dry forages and grain with or without supplementation. This study confirmed that animals feeding on green pasture tended to have higher phage numbers (two- or threefold) than those on dry feed and that feedlot animals had even fewer phage (up to fivefold) than those on dry feed, with this trend being unaffected by animal species or the feeding of supplements.

PFGE was also used to demonstrate that the phage population present in the rumen could be both diverse in composition and highly dynamic. In an investigation where sheep were fed once daily, a distinct diurnal variation in the phage population was observed (Swain et al. 1996). Two hours after feeding, total phage DNA decreased to its lowest concentration, the phage DNA concentration then increased, reaching a maximum concentration 8–10 h post-feeding before declining over the next 4 h to reach a stable concentration for the rest of the day. Interestingly, the general trend in phage DNA concentration appeared to be similar to previously recorded diurnal fluctuations in ruminal bacterial populations in cattle fed once daily (Leedle et al. 1982) and this was thought to be a consequence of the intimate relationship existing between phages and their bacterial hosts. The reasons behind the cyclic fluctuations observed in the phage population may also be linked with ingestion of feed by the host (Klieve and Swain 1993; Swain et al. 1996), introducing factors such as non-specific adsorption of phage particles circulating within the rumen liquor to feed particles. It was also suggested that the rise in phage numbers observed several hours after feeding could be attributed to either feed components inducing temperate phages to enter the lytic cycle, or an increase in host numbers becoming available for lytic infection, or both (Swain et al. 1996).

The PFGE fingerprinting method enabled the analysis of large numbers of rumen fluid samples and contributed significantly to the understanding of the factors affecting rumen phage populations. Like many DNA fingerprinting methods, however, it only provides a measure of total phage concentration and a population-specific phage genome length profile. PFGE cannot accurately determine phage taxonomy nor indicate the closest phage relatives; this information can only be obtained through nucleic acid sequence-based approaches.

9.3.2.2 Virome Analysis

With the development of new sequencing technologies, the ability to characterise microbial populations using sequence-based data has escalated to the point where genetic material can be prepared from an environmental sample and sequence information obtained for essentially all the DNA contained in that sample. This approach generates a sequence dataset often referred to as a metagenome, and the rumen metagenome has been characterised in several studies (Brulc et al. 2009; Hess et al. 2011; Ross et al. 2012). As these studies seek to incorporate all of the genetic material contained within a rumen fluid sample, viral genetic material is also encountered, arising from either (a) phage DNA associated with microbial genomes such as CRISPR spacers or integrated prophage regions or (b) from phage or virus particles. Sequences pertaining to viruses however tend to make up a very small proportion of the overall number of classified metagenome sequences, for example, in one rumen metagenome study (Brulc et al. 2009), approximately 0.1 % of coding sequences described as environmental gene tags (EGTs) were of viral origin, with most of these sequences related to dsDNA viruses.

The relatively low number of virus or phage-associated sequences encountered during metagenome analysis can in part be attributed to the relatively small size of phage genomes, which are generally less than 300 kb (Hatfull 2008), compared to archaeal and bacterial genomes which can range in size from approximately 1.5 to 9 Mb, respectively (Bentley and Parkhill 2004). Rumen metagenome studies also incorporate eukaryote genetic material from rumen populations of protozoa and fungi; therefore, depending on the sequence depth (coverage) of the metagenome study, the relatively small virus or phage-related genomes may be under-represented.

In addition, most virus-related sequences found within metagenome datasets are often classified as bacterial in origin. Whilst not strictly inaccurate, this classification arises when integrated prophage regions and phage-associated genes are only incorporated into genome sequence databases as being part of a larger bacterial genome sequence and are not annotated independently and submitted as a virus (prophage) for inclusion into, for example, the National Center for Biotechnology Information (NCBI) reference sequence database, Entrez Genomes. As a consequence, lysogenic phage genes are often described with the host bacterial or archaeal taxonomy descriptors and the associated classification numbers (e.g., GenBank accession numbers). As a large proportion of the phage particles found within rumen fluid are thought to have arisen from integrated prophages, the corresponding proportion of phage-related genes found within metagenomic datasets may actually be given bacterial or archaeal taxonomical classification. Similarly host-associated taxonomy can be given to phage-related functional genes recognised during the sequence annotation process as being associated with phage DNA replication and packaging or host cell lysis. Interestingly, when rumen microbial metagenome datasets (Brulc et al. 2009; Hess et al. 2011) were assessed for the presence of CRISPR sequences using specific algorithms and a CRISPR-specific sequence database (Berg Miller et al. 2012), a larger number of CRISPR-like sequences (approximately 300 with over 30,000 designated as 'questionable') were detected within one of the rumen metagenome datasets (Hess et al. 2011). This finding emphasises not only the importance of CRISPR-associated sequences in determining phage dynamics but also the importance of including specific phage or virus-related sequence databases in virus sequence analysis protocols.

To address the issue of under-representation and coverage, high-throughput sequencing studies of virus populations have adopted sample preparation methodologies that physically exclude contaminating microbial matter through steps such as ultrafiltration through 0.45 or 0.22 μm filters and selective fractionation and concentration of the virus particles present in a sample. Genetic material for sequencing is then prepared from these particles, with the inclusion of additional steps required for the detection of RNA viruses (Thurber et al. 2009). In this way, all of the genetic material sent for sequencing arises only from virus or phage particles. Sequencing studies prepared using virus-specific methodology are often referred to as viral metagenome or virome studies and have been undertaken for many different environmental sample types including aquatic (freshwater, marine and extreme), soil and sediments, and human associated (lung, gut, skin) (Rosario and Breitbart 2011).

At the time of writing, there was only one published rumen virome study (Berg Miller et al. 2012). This study examined the virus fraction of rumen fluid prepared in a manner similar to that employed in PFGE studies (Klieve and Swain 1993) and therefore focused on assessing populations of DNA viruses, which predominate in the rumen. The rumen fluid was obtained from three Holstein cows at various stages of lactation, maintained on a concentrate diet. Virome sequences were compared to reference sequence databases and viral taxonomic groupings determined. In this way, the rumen was found to be predominated by dsDNA tailed phages belonging to the viral order Caudovirales, with phages belonging to the *Siphoviridae* family (long tails) being found in relatively higher concentrations (36 ± 3 % of the total viral families) than phages belonging to the shorter-tailed *Myoviridae* (28 ± 4 %) and *Podoviridae* (14 ± 2 %) families. A considerable proportion of identified sequences were also classified as bacterial in origin, related to bacterial prophages and CRISPR elements; thereby reinforcing the findings of previous studies suggesting that a large proportion of phage particles found circulating within the rumen were produced following episodes of lysogenic phage induction and that lysogenic phages played a major role in maintaining phage numbers within the rumen (Klieve et al. 1989).

Although a large proportion of the rumen virome sequences were not related to virus reference sequences, taxonomy-independent analysis using programs such as PHACCS (Angly et al. 2005) indicated that a wide diversity of phage types were encompassed, with up to 28,000 different viral types estimated to be present (Berg Miller et al. 2012). This diversity was considerably higher than indicated by earlier electron-microscopy-based surveys based solely on morphology reflecting the advantage of using a sequence-based approach to differentiate between morphologically similar phages and to detect previously uncharacterised phages.

When the three published rumen virome sequence datasets are compared to viromes from other environments (Fig. 9.2) on the basis of sequence similarity, the rumen-derived viromes have little similarity with aquatic environments including freshwater and marine-derived viromes, and although they are closer to the human faecal and gut-associated viromes, they cluster separately as a rumen-associated group. From this comparison, it can be speculated that the rumen contains a large proportion of phages unique to this microbial ecosystem. As further virome studies are undertaken for different environments and on a wider range of domesticated and native ruminants, this distinction can be further investigated.

9.4 Roles of Phages Within the Rumen Microbial Ecosystem

Owing to the relative paucity of information regarding the mechanisms of rumen phage-host interactions and the environmental factors affecting the relative proportions and dynamics of the phage population in the rumen, it is not possible to definitively determine whether the presence of phage in the rumen is disadvantageous or advantageous. Theories proposed to date regarding the roles phages play within the rumen microbial ecosystem have revolved around (a) the negative nutritional consequences of phage-induced bacterial lysis resulting in recycling of nutrients within the rumen, (b) the positive effects of maintaining bacterial population diversity and facilitating gene transfer and (c) the negative consequences of phage-mediated gene transfer.

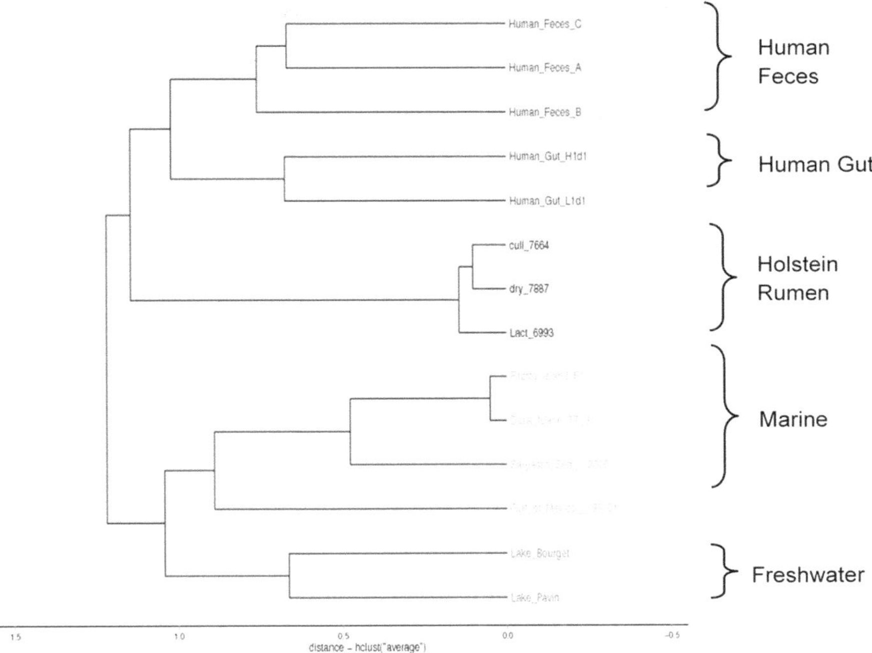

Fig. 9.2 Phylogenetic comparison of three rumen viromes from Holstein cows designated as cull_7664, dry_7887 and Lact_6993 (Berg Miller et al. 2012) and viromes from marine ecosystems including Fitzroy Island, Dunk Island (Hurwitz and Sullivan 2013), Sargasso Sea and the Gulf of Mexico (Angly et al. 2006), freshwater lake ecosystems including Lake Bourget and Lake Pavin (Roux et al. 2012) and human faecal samples and human gut samples (Kim et al. 2011). Viromes were compared using the Metavir server (Roux et al. 2011) to perform a tBLASTx search with a score matrix to cluster viromes using the pvclust R package (average linking distance, correlation distance) and 100 bootstraps

9.4.1 Nutritional Effects

In addition to the role rumen phages play in maintaining bacterial population diversity and balance, it has also been suggested that the high numbers and diurnal fluctuations in the phage population, associated with phage lytic activity, may play an important role in the turnover of microbial cells (Klieve and Swain 1993; Nolan and Leng 1972) by contributing to the inefficient process of intra-ruminal recycling of nitrogen and fermentation of microbial cellular materials.

Rumen microorganisms are able to utilise simple nitrogenous substances for the synthesis of cell proteins. These microorganisms, if not lysed in the rumen, ultimately pass from the rumen to the abomasum and small intestine, where microbial protein is digested and absorbed as in non-ruminant animals (Annison and Lewis 1959). The supply of nitrogenous matter within the feed promotes microbial growth up to the limit of the microbial nitrogen requirement. This requirement is set by the available fermentable carbohydrate, the ATP yield and the efficiency of conversion of metabolic substrates to microbial cells (Van Soest 1994). However, the nitrogen metabolism of rumen microorganisms especially with extensive grazing production systems is usually regarded as being inefficient. Dietary protein is broken down too rapidly relative to the breakdown of energy-containing plant fibre, excessive ammonia is produced and the biological value of the dietary protein is reduced (Wallace and Cotta 1988).

Research aimed at increasing the efficiency of rumen nitrogen utilisation has been generally focused on two objectives. The first capitalises on the capacity of rumen microorganisms to form protein from ammonia or other nonprotein nitrogen (NPN) sources, such as urea (Leng 1990; Wallace and Cotta 1988). The second is to minimise protein breakdown in the rumen so that more of the dietary protein escapes to the small intestine where the amino acids become available for adsorption by the host (Nagaraja et al. 1997; Wallace and Cotta 1988).

It has been suggested that intra-ruminal recycling and fermentation of microbial cellular materials will have the effect of reducing the net efficiency of microbial synthesis and the flow of microbial nitrogen to the small intestine (Dixon and Nolan 1986). There have been few quantitative estimates of intra-ruminal nitrogen recycling, but a widely cited study (Nolan and Leng 1972) estimated that 30–50 % of the nitrogen incorporated into microbial protein may later recycle through the rumen ammonia-N pool. Further studies of nitrogen kinetics in the rumen (Firkins et al. 1992; Wells and Russell 1996) indicated that many of the microbes in the rumen either secrete nitrogen during their metabolism or lyse and release materials that are subsequently fermented in the rumen. In addition, some actively growing rumen bacteria may synthesise amino acids in excess of their immediate requirements and excrete these into the medium.

Many investigations have indicated that all protozoa engulf and digest rumen bacteria, releasing soluble products contributing to intra-ruminal recycling of bacterial cellular materials (Bird et al. 1994). It has also been found that in the rumen of animals given starch-based diets, protozoa may at times also increase bacterial lysis, by starving the bacteria for substrates (Williams and Coleman 1997). Much of the protein component of the protozoal biomass, however, does not leave the rumen and is also recycled (Morrison and Mackie 1996).

In addition to protozoa-mediated lysis, bacteria may also be lysed in the rumen for no apparent reason, described as non-specific lysis (Wells and Russell 1996), or following the activity of bacteriophages (Hoogenraad et al. 1967; Klieve et al. 1989) or mycoplasma (Stewart et al. 1997). No studies have been undertaken to determine the extent to which phages contribute to bacterial lysis in the rumen. However, the diversity of phage types and large numbers observed in rumen fluid indicate that phages may play a major role in the lysis of rumen bacteria.

A novel study (Tarakanov 1994) attempted to use phages as a biological method for controlling *S. bovis* numbers in the rumen of dairy cows in an attempt to identify the long-term nutritional effects of phage addition to the rumen. It was found that addition of *S. bovis* phages resulted in an increase in milk fat content of 0.1–0.3 %. However, there are no further published investigations of this kind to verify that the addition of phages to the rumen can result in positive nutritional effects.

9.4.2 Phage-Mediated Gene Transfer

It has been suggested that phages may play a positive role in maintaining bacterial population diversity and balance within the rumen, allowing the ecosystem to rapidly adapt to changing conditions such as dietary changes (Swain et al. 1996) and contributing to normal, regular changes in the bacterial population by periodically reducing the numbers and relative density of a bacterial species from the rumen, through episodes of phage lytic activity. Genetic diversity of the bacterial population may also be catalyzed by phages in the rumen with the development of phage resistance within the bacterial population resulting in modified bacterial genotypes and phenotypes, thus increasing genetic variability amongst closely related rumen bacteria (Klieve and Bauchop 1991).

Phages may also act as a mechanism whereby genetic material can be transferred between bacteria, a process often described as horizontal or lateral gene transfer. Horizontal gene transfer can occur in three ways, either by transformation, conjugation or transduction. Transduction specifically involves the transfer of DNA from one bacterium to another through the activity of phages (Hendrickson 2012). This process may contribute to the high extent of genetic diversity found within the rumen microbial community and maintain the capacity for naturally occurring, heritable, genetic modification (Weinbauer 2004). This may have positive consequences for bacterial survival and persistence, for example, phage may transfer genes allowing for alternative or increased substrate utilisation ability, although this has not been demonstrated experimentally with rumen bacteria. Alternatively, phage transduction may have negative effects, for example, they may facilitate the transfer of genes for antibiotic resistance or toxin production, for example, the Shiga toxins of *Shigella dysenteriae* and *Escherichia coli* and the C, D and E neurotoxins of *Clostridium botulinum* are encoded by phages (Kuhl et al. 2012). In this way, phage-mediated gene transfer may contribute to bacterial virulence and lead to the evolution of new, potentially pathogenic bacteria. To date, no studies have been undertaken to experimentally verify whether phage-mediated transduction occurs between species of rumen bacteria, particularly as so few studies have determined biological factors such as phage-host range and the numbers of phage particles released upon phage-host cell lysis (burst size). It would also be particularly interesting to understand whether physical factors such as the rate of normal microbial turnover and outflow affect the extent and rate at which phage-mediated gene transfer can occur within the rumen.

9.5 Future Directions

The nature of rumen phage ecology research is currently changing, with this change being accelerated by the development of high-throughput sequencing of microbial ecosystems and microbial genomes. The analysis of the large amount of sequence data generated by high-throughput sequencing however is being hampered by a lack of relevant, rumen-derived phage viral sequences to facilitate accurate sequence comparison and taxonomic classification and a scarcity of general biological information for rumen phages such as host range, replication rates and burst size information.

As the agricultural production sector shifts away from the use of antibiotics, phage research is becoming increasingly focused on the development of phage therapy-based approaches for the biological control of bacterial populations. To date phage therapies have been developed for the control of pathogens in the control of biofilms (human clinical), plant, poultry, aquaculture and ruminant livestock industries (Donlan 2009; Monk et al. 2010). These therapies may employ either a cocktail of lytic phages which specifically infect the bacterial (or archaeal) species of interest (Callaway et al. 2008, 2011; Gill and Hyman 2010) or utilise preparations of phage-encoded proteins (O'Flaherty et al. 2009; O'Mahony et al. 2011). There has also been recent interest in the development of phage therapy approaches for the control of rumen methanogen (archaeal) populations in order to reduce ruminant enteric methane production (Klieve and Hegarty 1999; Cottle et al. 2011).

Phage therapy based on lytic phages has the advantage over conventional antibiotics and chemical antimicrobials in that phage (a) are naturally occurring biological agents which do not contribute to the development of therapeutic antibiotic resistance in the environment; (b) can be applied without the negative health side effects associated with some antibiotics; (c) can be administered directly to living tissue without causing harm; and (d) can attach, self-propagate and penetrate microbial biofilms which may be otherwise unaffected by chemical antimicrobials (Abedon 2012; Chan et al. 2013). Phages chosen for use in phage therapy also tend to have a very narrow spectrum of activity directed against a target organism, are free of toxins and transducing virulence factor genes and replicate only through the lytic cycle of phage reproduction and are incapable of forming a lysogenic association with their host (Chan et al. 2013). In order for phages to encompass all of these desirable properties, they must be extensively genetically characterised and tested for biological efficacy, lysis efficiency and persistence, prior to any inclusion in a phage therapy formulation.

Alternatively, phage-encoded enzymes may be employed as alternatives to conventional antimicrobials (Fenton et al. 2010; O'Mahony et al. 2011; Shen et al. 2012). In nature, phage proteins may have a structural role forming the coat, tail structures or spikes or may also have a functional, enzymatic role (Ackermann and DuBow 1987a). Tailed phages, for example, may have lysozyme located at the tail tip, and the spikes of some capsule-specific phages demonstrate endoglycosidase activity (Murphy et al. 1995) allowing these structures, even when used independantly of complete phage particles, to bind to cell surfaces (Andres et al. 2012). Phages may also encode genes for enzymes with lytic activity, targeting the integrity of the host cell wall, for example, amidase, muramidase and endopeptidase enzymes. Several preparations of phage enzymes have been tested in animal models (O'Flaherty et al. 2009) and have been found to be potent, novel antimicrobials. These enzymes therefore represent a powerful new avenue for the development of phage-based therapies to reduce and control microbial populations of concern.

The future of rumen phage research can be anticipated to rapidly expand, taking advantage of the technological advances in high-throughput sequencing in order to genetically characterise individual rumen phages and benefit from their unique enzymatic and structural properties and biological attributes, such as host specificity and lytic potential. Adoption of new sequence-based technologies will also enable researchers to more fully elucidate the roles phages play in maintaining rumen microbial population balance and genetic transfer and determine their overall contribution to rumen bacterial lysis and ruminant nutrition.

Dedication The authors would like to dedicate this chapter to the memory of Emeritus Professor Nancy Millis (1922–2012) who made considerable contributions to rumen phage research and the development of microbiology in Australia.

References

Abedon ST (2012) Phage therapy best practices. In: Hyman P, Abedon ST (eds) Bacteriophages in health and disease. CABI Press, Wallingford, pp 256–272

Ackermann HW, Dubow MS (1987a) General properties of bacteriophages. CRCPress, Boca Raton

Ackermann HW, DuBow MS (1987b) Natural groups of bacteriophages. CRC Press, Boca Raton

Adams JC, Gazaway JA, Brailsford MD et al (1966) Isolation of bacteriophages from the bovine rumen. Experentia 22:717–718

Ambrozic J, Ferme D, Grabnar M et al (2001) The bacteriophages of ruminal prevotellas. Folia Microbiol 46:37–39

Andres D, Roske Y, Doering C et al (2012) Tail morphology controls DNA release in two Salmonella phages with one lipopolysaccharide receptor recognition system. Mol Microbiol 83:1244–1253

Angly F, Rodriguez-Brito B et al (2005) PHACCS, an online tool for estimating the structure and diversity of uncultured viral communities using metagenomic information. BMC Bioinf 6:41

Angly FE, Felts B et al (2006) The marine viromes of four oceanic regions. PLoS Biol 4:2121–2131

Annison EF, Lewis D (1959) Metabolism in the rumen. Methuen, London

Attwood GT, Kelly WJ, Altermann EH et al (2008) Analysis of the *Methanobrevibacter ruminantium* draft genome: understanding methanogen biology to inhibit their action in the rumen. Aust J Exp Agric 48:83–88

Aziz RK, Bartels D et al (2008) The RAST server: rapid annotations using subsystems technology. BMC Genomics 9:75

Baresi L, Bertani G (1984) Isolation of a bacteriophage for a methanogenic bacterium. Abstr Ann Meet Am Soc Microbiol 28:133

Barnet YM (1972) Bacteriophages of *Rhizobium trifolii* I morphology and host range. J Gen Virol 15:1–15

Barrangou R (2013) CRISPR-Cas systems and RNA-guided interference. Wiley Interdiscip Rev-RNA 4:267–278

Barrangou R, Fremaux C, Deveau H et al (2007) CRISPR provides acquired resistance against viruses in prokaryotes. Science 315:1709–1712

Bentley SD, Parkhill J (2004) Comparative genomic structure of prokaryotes. Annu Rev Genet 38:771–792

Berg Miller ME, Yeoman CJ et al (2012) Phage–bacteria relationships and CRISPR elements revealed by a metagenomic survey of the rumen microbiome. Environ Microbiol 14:207–227

Bird SH, Romulo B, Leng RA (1994) Effects of lucerne supplementation and defaunation on feed intake, digestibility, N-retention and productivity of sheep fed straw based diets. Anim Feed Sci Technol 45:119–129

Bland C, Ramsey TL, Sabree F et al (2007) CRISPR Recognition Tool (CRT): a tool for automatic detection of clustered regularly interspaced palindromic repeats. BMC Bioinf 8:209

Bobay L-M, Rocha EPC, Touchon M (2013) The adaptation of temperate bacteriophages to their host genomes. Mol Biol Evol 30:737–751

Bose M, Barber RD (2006) Prophage Finder: a prophage loci prediction tool for prokaryotic genome sequences in silico. Biol 6:223–227

Brailsford MD, Hartman PA (1968) Characterisation of *Streptococcus durans* bacteriophages. Can J Microbiol 14:397–402

Brenner S, Horne RW (1959) A negative staining method for high resolution electron microscopy of viruses. Biochim Biophys Acta 34:103–110

Brulc JM, Antonopoulos DA et al (2009) Gene-centric metagenomics of the fiber-adherent bovine rumen microbiome reveals forage specific glycoside hydrolases. Proc Natl Acad Sci U S A 106:1948–1953

Callaway TR, Edrington TS et al (2008) Bacteriophage isolated from feedlot cattle can reduce *Escherichia coli* O157: H7 populations in ruminant gastrointestinal tracts. Foodborne Pathog Dis 5:183–191

Callaway TR, Edrington TS et al (2011) Evaluation of phage treatment as a strategy to reduce salmonella populations in growing swine. Foodborne Pathog Dis 8:261–266

Campbell AM (1992) Chromosomal insertion sites for phages and plasmids. J Bacteriol 174:7495–7499

Chan BK, Abedon ST, Loc-Carrillo C (2013) Phage cocktails and the future of phage therapy. Future Microbiol 8:769–783

Cheong JP, Brooker JD (1998) Lysogenic bacteriophage M1 from *Selenomonas ruminantium*: isolation, characterisation and DNA sequence analysis of the integration site. Microbiology 144:2195–2202

Clokie MRJ, Millard AD, Letarov AV et al (2011) Phages in nature. Bacteriophage 1:31–45

Cottle DJ, Nolan JV, Wiedemann SG (2011) Ruminant enteric methane mitigation: a review. Anim Prod Sci 51:491–514

D'Herelle F (1918) Technique de la recherche du microbe filtrant bacteriophage (*Bacteriophagum intestinale*). C R Seances Soc Biol Fil 81:1160–1162

Dixon RM, Nolan JV (1986) Nitrogen and carbon flows between the caecum, blood and rumen in sheep given chopped lucerne (*Medicago sativa*) hay. Br J Nutr 55:313–332

Donlan RM (2009) Preventing biofilms of clinically relevant organisms using bacteriophage. Trends Microbiol 17:66–72

Duckworth DH (1987) History and basic properties of bacterial viruses. In: Goyal SM, Gerba CP, Bitton G (eds) Phage ecology. John Wiley and Sons Inc, New York, pp 1–44

Fenton M, Ross P, McAuliffe O et al (2010) Recombinant bacteriophage lysins as antibacterials. Bioeng Bugs 1:9–16

Firkins JL, Weiss WP, Piwonka EJ (1992) Quantification of intraruminal recycling of microbial nitrogen using nitrogen-15. J Anim Sci 70:3223–3233

Flewett TH, Bryden AS, Davies H (1974) Diagnostic electron microscopy of faeces. J Clin Pathol 27:603–614

Gantner S, Andersson AF, Alonso-Saez L et al (2011) Novel primers for 16S rRNA-based archaeal community analyses in environmental samples. J Microbiol Methods 84:12–18

Gill JJ, Hyman P (2010) Phage choice, isolation, and preparation for phage therapy. Curr Pharm Biotechnol 11:2–14

Gregg K, Kennedy BG, Klieve AV (1994) Cloning and DNA sequence analysis of the region containing AttP of the temperate phage AR29 of *Prevotella ruminicola* AR29. Microbiology 140:2109–2114

Grissa I, Vergnaud G (2007) CRISPRFinder: a web tool to identify clustered regularly interspaced short palindromic repeats. Nucleic Acids Res 35:W52–W57

Grissa I, Vergnaud G, Pourcel C (2007) The CRISPRdb database and tools to display CRISPRs and to generate dictionaries of spacers and repeats. BMC Bioinf 8:172

Hatfull GF (2008) Bacteriophage genomics. Curr Opin Microbiol 11:447–453

Hazlewood GP, Munn EA, Orpin CG (1983) Temperate bacteriophages of *Selenomonas ruminantium* and a Fusobacterium sp isolated from the ovine rumen. In: Abstracts Canadian society for microbiology 33rd annual meeting, Winnipeg, Manitoba, Canada. p 76

Hendrickson H (2012) The lion and the mouse: how bacteriophages create, liberate and decimate bacterial pathogens. In: Hyman P, Abedon S (eds) Bacteriophages in health and disease. CABI, Oxford

Hess M, Sczyrba A et al (2011) Metagenomic discovery of biomass-degrading genes and genomes from cow rumen. Science 331:463–467

Hong SH, Kim JS et al (2004) The genome sequence of the capnophilic rumen bacterium *Mannheimia succiniciproducens*. Nat Biotechnol 22:1275–1281

Hoogenraad NJ, Hird FJR, Holmes I et al (1967) Bacteriophages in rumen contents of sheep. J Gen Virol 1:575–576

Hurwitz BL, Sullivan MB (2013) The Pacific Ocean Virome (POV): a marine viral metagenomic dataset and associated protein clusters for quantitative viral ecol. PLoS One 8:e57355

Iverson WJ, Millis NF (1976a) Characterisation of *Streptococcus bovis* bacteriophages. Can J Microbiol 22:847–852

Iverson WJ, Millis NF (1976b) Lysogeny in *Streptococcus bovis*. Can J Microbiol 22:853–857

Iverson WG, Millis NF (1977) Succession of *Streptococcus bovis* strains with differing bacteriophage sensitivities in the rumens of two fistulated sheep. Appl Environ Microbiol 33:810–813

Jiang WH, Patterson JA, Steenson LR (1995) Isolation and characterisation of a temperate bacteriophage from a ruminal acetogen. Curr Microbiol 31:336–339

Kellenberger E (1995) History of phage research as viewed by a European. FEMS Microbiol Rev 17:7–24

Kim MS, Park EJ, Roh SW et al (2011) Diversity and abundance of single-stranded DNA viruses in human feces. Appl Environ Microbiol 54:1637–1641

Klieve AV (1988) Bacteriophages of rumen bacteria. PhD thesis, University of New England

Klieve AV (2005) Part 2.2. Bacteriophages. In Makkar HPS, McSweeney CS (eds) Methods in gut microbial ecology for ruminants. International Atomic Energy Agency/Springer/Academic, Dordrecht, The Netherlands, pp 39–46

Klieve AV, Bauchop T (1988) Morphological diversity or ruminal bacteriophages from sheep and cattle. Appl Environ Microbiol 54:1637–1641

Klieve AV, Bauchop T (1991) Phage resistance and altered growth habit in a strain of *Streptococcus bovis*. FEMS Microbiol Lett 80:155–160

Klieve AV, Gilbert RA (2005) Part 4.2. Bacteriophage populations. In: Makkar HPS, McSweeney CS (eds) Methods in gut microbial ecology for ruminants. International Atomic Energy Agency/Springer/ Academic, Dordrecht, The Netherlands, pp 129–137

Klieve AV, Hegarty R (1999) Opportunities for biological control of ruminal methanogenesis. Aust J Agr Res 50:1315–1319

Klieve AV, Swain RA (1993) Estimating ruminal bacteriophage numbers using pulsed field gel electrophoresis and laser densitometry. Appl Environ Microbiol 59:2299–2303

Klieve AV, Hudman JF, Bauchop T (1989) Inducible bacteriophages from ruminal bacteria. Appl Environ Microbiol 55:1630–1634

Klieve AV, Gregg K, Bauchop T (1991) Isolation and characteristics of lytic phages from *Bacteroides ruminicola* ss *brevis*. Curr Microbiol 23:183–187

Klieve AV, Turner AF, Heck GL (1998) Dietary influences on bacteriophage numbers in the rumen. Proc Aust Soc Anim Prod 22:341

Klieve AV, Heck GL, Prance MA et al (1999) Genetic homogeneity and phage susceptibility of ruminal strains of *Streptococcus bovis* isolated in Australia. Lett Appl Microbiol 29:108–112

Klieve AV, Bain PA, Yokoyama MT et al (2004) Bacteriophages that infect the cellulolytic ruminal bacterium *Ruminococcus albus* AR67. Lett Appl Microbiol 38:333–338

Krupovic M, Prangishvili D, Hendrix RW et al (2011) Genomics of bacterial and archaeal viruses: dynamics within the prokaryotic virosphere. Microbiol Mol Biol Rev 75:610

Kuhl S, Hyman P, Abedon ST (2012) Diseases caused by phages. In: Hyman P, Abedon ST (eds) Bacteriophages in health and disease. CABI Press, Wallingford, pp 21–32

Leahy SC, Kelly WJ et al (2010) The genome sequence of the rumen methanogen *Methanobrevibacter ruminantium* reveals new possibilities for controlling ruminant methane emissions. Plos One 5:e8926

Leedle JAZ, Bryant MP, Hespell RB (1982) Diurnal variations in bacterial numbers and fluid parameters in ruminal contents of animals fed low- or high-forage diets. Appl Environ Microbiol 44:402–412

Leng RA (1990) Factors affecting the utilisation of 'poor-quality' forages by ruminants particularly under tropical conditions. Nutr Res Rev 3:277–303

Lima-Mendez G, Van Helden J, Toussaint A et al (2008) Prophinder: a computational tool for prophage prediction in prokaryotic genomes. Bioinformatics 24:863–865

Lockington RA, Attwood GT, Brooker JD (1988) Isolation and characterisation of a temperate bacteriophage from the ruminal anaerobe *Selenomonas ruminantium*. Appl Environ Microbiol 54:1575–1580

Lood R, Collin M (2011) Characterization and genome sequencing of two *Propionibacterium acnes* phages displaying pseudolysogeny. BMC Genomics 12:198

Markowitz VM, Chen IMA et al (2012) IMG: the integrated microbial genomes database and comparative analysis system. Nucleic Acids Res 40:D115–D122

Matteuzzi D, Sozzi T (1971) Bifidobacterium bacteriophage from calf rumen Zeitschrift fur Allgemeine. Mikrobiologie 11:57–58

Monk AB, Rees CD, Barrow P (2010) Under the microscope, bacteriophage applications: where are we now? Lett Appl Microbiol 51:363–369

Morrison M, Mackie RI (1996) Nitrogen metabolism by ruminal microorganisms: current understanding and future perspectives. Aust J Agr Res 47:227–246

Mumm JP, Landy A, Gelles J (2006) Viewing single lambda site-specific recombination events from start to finish. EMBO J 25:4586–4595

Murphy FA, Fauquet CM, Bishop DHL et al (1995) Virus taxonomy, 6th report of the international committee on taxonomy of viruses. Springer-Verlag, New York

Nagaraja TG, Titgemeyer EC (2007) Ruminal acidosis in beef cattle: the current microbiological and nutritional outlook. J Dairy Sci 90:E17–E38

Nagaraja TG, Newbold CJ, Van Nevel CJ et al (1997) Manipulation of ruminal fermentation. In: Hobson PN, Stewart CS (eds) The rumen microbial ecosystem. Chapman and Hall, London

Nemcova R, Styriak I, Stachova M et al (1993) Isolation and partial characterisation of three rumen *Lactobacillus plantarum* bacteriophages. Microbiologica 16:177–180

Nolan JV, Leng RA (1972) Dynamic aspects of ammonia and urea metabolism in sheep. Br J Nutr 27:177–194

O'Flaherty S, Ross RP, Coffey A (2009) Bacteriophage and their lysins for elimination of infectious bacteria. FEMS Microbiol Rev 33:801–819

O'Mahony J, Fenton M, Henry M et al (2011) Lysins to kill – a tale of viral weapons of mass destruction. Bioeng Bugs 2:306–308

Orpin CG, Munn EA (1974) The occurrence of bacteriophages in the rumen and their influence on rumen bacterial populations. Experentia 30:1018–1020

Paynter MJB, Ewert DL, Chalupa W (1969) Some morphological types of bacteriophages in bovine rumen contents. Appl Microbiol 18:942–943

Porter K, Russ BE, Dyall-Smith ML (2007) Virus-host interactions in salt lakes. Curr Opin Microbiol 10:418–424

Purushe J, Fouts DE, Morrison M et al (2010) Comparative genome analysis of *Prevotella ruminicola* and *Prevotella bryantii*: insights into their environmental niche. Microb Ecol 60:721–729

Refardt D (2011) Within-host competition determines reproductive success of temperate bacteriophages. ISME J 5:1451–1460

Ritchie AE, Robinson IM, Allison MJ (1970) Rumen bacteriophage: survey of morphological types. In: Societie

Francaise de Microscopie Electronique. Grenoble pp 333–334

Ronning CM, Losada L, Brinkac L et al (2010) Genetic and phenotypic diversity in Burkholderia: contributions by prophage and phage-like elements. BMC Microbiol 10:202

Rosario K, Breitbart M (2011) Exploring the viral world through metagenomics. Curr Opin Virol 1:289–297

Ross EM, Moate PJ, Bath CR et al (2012) High throughput whole rumen metagenome profiling using untargeted massively parallel sequencing. BMC Genet 13:53

Roux S, Faubladier M, Mahul A et al (2011) Metavir: a web server dedicated to virome analysis. Bioinformatics 27:3074–3075

Roux S, Enault F, Robin A et al (2012) Assessing the diversity and specificity of two freshwater viral communities through metagenomics. PLoS One 7:33641

Sampson TR, Saroj SD, Llewellyn AC et al (2013) A CRISPR/Cas system mediates bacterial innate immune evasion and virulence. Nature 497:254

Seet SGM (2005) Genome sequence of bacetriophage Phi-AR29: a basis for integrative plasmid vectors. PhD thesis, Murdoch University

Shen Y, Mitchell M, Donovan DM et al (2012) Phage-based enzybiotics. In: Abedon S, Hyman P (eds) Bacteriophages in health and disease. CABI Press, Wallingford, Oxfordshire, UK pp 217–239

Sorek R, Kunin V, Hugenholtz P (2008) CRISPR – a widespread system that provides acquired resistance against phages in bacteria and archaea. Nat Rev Microbiol 6:181–186

Sorek R, Lawrence CM, Wiedenheft B (2013) CRISPR-mediated adaptive immune systems in bacteria and archaea. Annu Rev Biochem 82:237–266

Stewart CS, Flint HJ, Bryant MP (1997) The rumen bacteria. In: Hobson PN, Stewart CS (eds) The rumen microbial ecosystem. Chapman and Hall, London

Styriak I, Kmet V, Spanova A (1989) Isolation and characterisation of two rumen *Streptococcus bovis* bacteriophages. Microbiologica 12:317–322

Styriak I, Galfi P, Kmet V (1991) Preliminary observations of interaction between bacteriophages and *Streptococcus bovis* bacteria on ruminal epithelium primoculture. Vet Microbiol 29:281–287

Styriak I, Galfi P, Kmet V (1994a) The adherence of 3 *Streptococcus-bovis* strains to cells of rumen-epithelium primoculture under various conditions. Arch Anim Nutr 46:357–365

Styriak I, Spanova A, Montagova H et al (1994b) Isolation and characterisation of a new ruminal bacteriophage lytic to *Streptococcus bovis*. Curr Microbiol 28:355–358

Styriak I, Pristas P, Javorsky P (1998) Lack of surface receptors not restriction-modification system determines F4 phage resistance in *Streptococcus bovis* II/1. Folia Microbiol 43:35–38

Styriak I, Pristas P, Javorsky P (2000) Lack of GATC sites in the genome of *Streptococcus bovis* bacteriophage F4. Res Microbiol 151:285–289

Swain RA, Nolan JV, Klieve AV (1996) Natural variability and diurnal fluctuations within the bacteriophage population of the rumen. Appl Environ Microbiol 62:994–997

Tamada H, Harasawa R, Shinjo T (1985) Isolation of a bacteriophage in *Fusobacterium necrophorum*. Jpn J Vet Sci 47:483–486

Tarakanov BV (1974) Lysogenic cultures of *Streptococcus bovis* isolated from rumen of cattle and sheep. Microbiologica 43:375–377

Tarakanov BV (1976) Biological properties of *Streptococcus bovis* bacteriophages isolated from lysogenic cultures and sheep rumen. Microbiologiia 45:695–700

Tarakanov BV (1994) Regulation of microbial processes in the rumen by bacteriophages of *Streptococcus bovis*. Microbiology 63:373–378 (translated from Mikrobiologiya 363, 657–667)

Tarakanov BV (1996) Biology of lysogenic strains of *Streptococcus bovis* and virulent mutants of their temperate phages. Microbiology 65:575–580

Tarakanov BV (2006) The phenomenon of bacteriophagy in the rumen of ruminants. Nauchny mir, Moscow

Thurber RV, Haynes M, Breitbart M et al (2009) Laboratory procedures to generate viral metagenomes. Nat Protoc 4:470–483

Torrella F, Morita RY (1979) Evidence by electron micrographs for a high incidence of bacteriophage particles in the waters of Yaquina Bay, oregon: ecological and taxonomical implications. Appl Environ Microbiol 37:774–778

Twort FW (1915) An investigation on the nature of ultra-microscopic viruses. Lancet 2:1241–1243

Tyutikov FM, Bespalova IA, Rebentish BA et al (1980) Bacteriophages of methanotrophic bacteria. J Bacteriol 144:375–381

Van Soest PJ (1994) Nutritional ecology of the ruminant. Cornell University Press, Ithaca

Wallace RJ, Cotta MA (1988) Metabolism of nitrogen-containing compounds. In: Hobson PN (ed) The rumen microbial ecosystem. Elsevier, London, pp 217–249

Watanabe K, Kodama Y, Harayama S (2001) Design and evaluation of PCR primers to amplify bacterial 16S ribosomal DNA fragments used for community fingerprinting. J Microbiol Methods 44:253–262

Weinbauer MG (2004) Ecology of prokaryotic viruses. FEMS Microbiol Rev 28:127–181

Wells JE, Russell JB (1996) Why do many ruminal bacteria die and lyse so quickly? J Dairy Sci 79:1487–1495

Williams AG, Coleman GS (1997) The rumen protozoa. In: Hobson PN, Stewart CS (eds) The rumen microbial ecosystem. Chapman and Hall, London, pp 73–139

Zhou Y, Liang Y, Lynch KH et al (2011) PHAST: a fast phage search tool. Nucleic Acids Res 39:W347–W352

Rumen Methanogens 10

Laura M. Cersosimo and André-Denis G. Wright

Abstract

Presently, there are 120 species of methanogens, representing 33 genera. Rumen methanogens produce methane that contributes to both global enteric methane emissions and to a dietary energy loss to the host. Methanogens are strict anaerobes, are difficult to grow *in vitro*, require different substrates (hydrogen, carbon dioxide, acetate, methanol, methylamines, and formate) for methane production, and exist synergistically with bacteria and symbiotically with rumen ciliate protozoa. The differences in morphology and utilization of different substrates by different methanogens enable them to be found in a diverse number of habitats. With the introduction of new technologies, rumen methanogen research has greatly expanded. Future scientific advancements will contribute more insight into decreasing dietary energy losses and enteric methane emissions.

Keywords

Enteric methane • Rumen methanogens • *Methanobrevibacter*

10.1 Introduction

Until the late 1970s, all living things could be classified into two divisions, the *Bacteria* and *Eukarya* (animals, plants, fungi, and protists). This coexisted with another scheme, which had come to be accepted as the five-kingdom system: *Monera* (bacteria), Protista, *Fungi*, Plantae, and Animalia. Then in 1977, Woese and Fox (1977) at the University of Illinois proposed a new group

L.M. Cersosimo (✉) • A.-D.G. Wright
Department of Animal Science,
College of Agriculture and Life Sciences,
University of Vermont, 570 Main Street,
Burlington, VT 05401, USA
e-mail: lcersosi@uvm.edu; agwright@uvm.edu

A.K. Puniya et al. (eds.), *Rumen Microbiology: From Evolution to Revolution*, DOI 10.1007/978-81-322-2401-3_10

of prokaryotes called the Archaebacteria. More than a decade later, molecular phylogenetic analysis of 16S rRNA (prokaryotes) and 18S rRNA (eukaryotes) indicated that extant organisms fall into three major groups or domains. Woese et al. (1990) suggested a change from the classical five-kingdom system to a three-domain system by grouping together all the eukaryotes into the domain Eukarya and by calling the other two major domains *Bacteria* and Archaea.

Members of the domain Archaea have a cellular composition unlike those of other prokaryotes (e.g., *Bacteria*), the biggest difference being the composition of the cell walls. For example, the cell wall of Archaea is composed of pseudomureins or proteins, not peptidoglycans as observed in bacteria. Other differences include the RNA polymerases present and lipid membrane structure. The domain Archaea is comprised of four phyla: Crenarchaeota, Korarchaeota, Nanoarchaeota, and Euryarchaeota. The Euryarchaeota, the largest phylum, includes the methanogenic archaea, which produce methane; the halophilic archaea, which live and grow in hypersaline; and some extremely thermophilic archaea, which thrive at relatively high temperatures between 45 C and 122 °C. The methanogenic archaea or methanogens are the most frequently studied and observed group of Archaea within the rumen (Paul et al. 2012). Methanogens are strict anaerobes, found in hydrothermal vents, lakes, landfills, and the gastrointestinal tract of animals, including insects and humans. The focus of this chapter will be on the methanogens found in the forestomach (i.e., rumen) of ruminants.

Within the rumen, methanogens represent less than 1 % of the total microbial population and maintain a synergistic relationship with bacteria and a symbiotic relationship with protozoa (Wright and Klieve 2011). Additionally, some methanogens are associated with the ciliated protozoa. The ciliated protozoa, or ciliates, support methanogenesis by transferring hydrogen to the methanogens. The protozoa benefit from the removal of hydrogen that would normally cause an inhibitory effect on their metabolism. The methanogens then use the hydrogen to reduce carbon dioxide to methane.

10.2 Significance of Rumen Methanogens

10.2.1 Contribution to Greenhouse Gas Emissions

After methane is produced in the rumen, roughly 89 % of it is expelled from the host to the environment via eructation and exhalation (Murray et al. 1976). Once methane is in the environment, it contributes to greenhouse gas (GHG) emissions. Methane has a global warming potential (i.e., the relative measure of how much heat a GHG can trap in the atmosphere) 25 times that of carbon dioxide (IPCC 2006).

Because excess methane impacts the environment, rumen methanogens contribute greatly to enteric methane emissions. In the USA, enteric methane emissions account for 21 % of total methane emissions and 73 % of the emissions from the agricultural sector (*EPA*2014). In comparison with the USA, Australia and Venezuela's ruminant livestock populations contribute to 71–91 % of all agricultural methane emissions, respectively (PCNCCV 2005; Eckard 2013). These statistics further implicate the call for rumen methanogen mitigation strategies and the overall necessity for more insight into decreasing methane emissions within the agricultural sector, while maintaining the milk, meat, or fiber demands in an expanding world population.

10.2.2 Effect of Methane Loss on Dietary Energy

Although the production of enteric methane contributes to the production of GHG, it also leads to a 2–12 % loss of energy intake in domestic ruminants (Johnson and Johnson 1995). This loss of energy intake implies that domestic ruminants will need to consume more feed, which in turn increases farm expenditures. By examining the methanogen community structures within both domestic and wild ruminants along with new methane measurement technologies, new insights into how to mitigate the loss of gross energy in the form of methane could be gained.

10.3 Common Methanogen Taxa Within the Rumen

10.3.1 The Order Methanobacteriales

The order Methanobacteriales is home to the predominant methanogens found in the rumen (Wright et al. 2004, 2008; Sundset et al. 2009a; Sirohi et al. 2010; King et al. 2011; St-Pierre and Wright 2012a, b). They can have short lancet cocci to long, filamentous rod shapes that are typically Gram-positive, but some can be Gram-negative (Sirohi et al. 2010). The order Methanobacteriales is further divided into two families: Methanothermaceae and Methanobacteriaceae (Boone et al. 1993). The family Methanothermaceae contains the genus *Methanothermus*, while the family Methanobacteriaceae contains the genera *Methanobacterium*, *Methanobrevibacter*, *Methanosphaera*, and *Methanothermobacter*.

Methanobrevibacter is the most abundant methanogen found in the rumen (Wright et al. 2004, 2008; King et al. 2011; St-Pierre and Wright 2012a, b). Species belonging to the genus *Methanobrevibacter* (*Mbr*) demonstrate poor to no mobility, appear as short rods or lancet shaped cocci, and prefer temperatures between 37 C and 39 °C (Sirohi et al. 2010). Species of *Methanobrevibacter* include *Mbr. smithii*, *Mbr. gottschalkii*, *Mbr. thaueri*, *Mbr. ruminantium*, *Mbr. olleyae*, *Mbr. millerae*, *Mbr. wolinii*, *Mbr. woesei*, and *Mbr. arboriphilus*.

Methanobacterium (*Mb*) is not as speciose as *Methanobrevibacter* and is a minor representative of rumen methanogen populations (St-Pierre and Wright 2012a). The genus *Methanobacterium* consists of methanogens with varying shapes (curved, crooked to straight), preferred temperatures (i.e., mesophiles that grow at moderate temperatures and extreme thermophiles that thrive at temperatures greater than 45 °C), and motilities (nonmotile to motile) (Sirohi et al. 2010). *Methanobacterium* species observed include *Mb. aarhusense*, *Mb. alcaliphilum*, *Mb. bryantii*, and *Mb. formicicum* (Janssen and Kirs 2008; St-Pierre et al. 2012a).

The genus *Methanosphaera* has also been observed in the rumen (King et al. 2011; St-Pierre and Wright 2012a; Wright et al. 2004). *Methanosphaera stadtmanae* is a nonmotile methanogen that expresses similar characteristics in cell wall structure to species belonging to the family Methanobacteriaceae.

10.3.2 The Order Methanomicrobiales

The order Methanomicrobiales contains four families, Methanoregulaceae, Methanospirillaceae, Methanocorpusculaceae, and, the largest family, the Methanomicrobiaceae. Only members of the Methanospirillaceae and Methanomicrobiaceae families have been found in the rumen (Schauer et al. 1982; Singh et al. 2010). The cell wall composition of Methanomicrobiaceae differs from that of other methanogens in that it contains proteins (i.e., glycoproteins) instead of pseudomureins (Garcia et al. 2006). Furthermore, the cell morphology varies from cocci to short rods to plate-shaped cells (Boone et al. 1993; Ferry and Kastead 2007).

Although species belonging to the family Methanomicrobiaceae are not typically the most predominant methanogens found within the rumen, *Methanomicrobium mobile* and *Methanoculleus olentangyi* have been observed in the rumen (Wright et al. 2006; Williams et al. 2009; Kumar et al. 2011). In contrast to other methanogens, *Methanomicrobium mobile* requires complex nutrients similar to those found in the rumen ecosystem. Described by Hungate (1966) as a major methanogen within the rumen, in vitro culturing of *Methanomicrobium mobile* was very difficult. To thrive, *Methanomicrobium mobile* requires a heat-stable cofactor derived from a rumen fluid called mobile factor. Without it, *Methanomicrobium mobile* will lyse within 2–4 days (Tanner and Wolfe 1988). However, in Murrah buffalo from India, *Methanomicrobium mobile* was the dominant species identified in 94.4 % of clones, while members of the order Methanomicrobiales were observed in 54.4 % of the operational taxonomic units (OTU) identified from Surti buffalo from India (Chaudhary and Sirohi 2009; Singh et al. 2012). Although species belonging to the order Methanomicrobiales are less common in other ruminants, it is thought that the geographical location, diet, and species of ruminant, such as the buffalo, may affect the outcome of the methanogen community (Singh et al. 2012).

10.3.3 The Order Methanosarcinales

Members of the order Methanosarcinales contain cell walls that are devoid of pseudomurein, but composed of proteins. In some instances, the cells form large aggregates via a heteropolysaccharide sheath (Ferry and Kastead 2007). The cell morphology ranges from large spherical to pleomorphic cells that typically stain Gram-positive (Sirohi et al. 2010). The order Methanosarcinales is further divided into two families, the Methanosaetaceae and the Methanosarcinaceae, the latter of which has been found in the rumen. Within the family Methanosarcinaceae, *Methanosarcina* is the genus that includes the species, *Methanosarcina barkeri* and *Methanosarcina mazei* (Hook et al. 2010; Singh et al. 2010).

10.3.4 The Order Methanoplasmatales

A recently proposed seventh order of methanogens is the Methanoplasmatales. Methanogens belonging to this order have been found in marine environments, soil, and the intestines of termites and mammals (Paul et al. 2012). This new order now contains methanogens that were referred as belonging to the rumen cluster C (RCC) or the Thermoplasmatales-affiliated Lineage C (TALC). This new order consists mainly of uncultured archaea representing novel lineages of methanogens that are most distantly related to uncultured Thermoplasmatales that are part of the order Methanoplasmatales (Wright et al. 2004, 2006; Janssen and Kirs 2008; Paul et al. 2012).

The predominance of these uncultured novel archaea was observed in the rumen of Svalbard reindeer (53 % of clones), sheep in Queensland, Australia (80 %), and cattle consuming a potato by-product diet in Prince Edward Island (50 % of clones) (Wright et al. 2006, 2007; Sundset et al. 2009b).

10.4 SGMT and RO Clades

Species belonging to the genus *Methanobrevibacter* are the predominant methanogens in the rumen; both the phylogenetic distribution and representation of different *Methanobrevibacter* species have been reported (King et al. 2011). It was observed that *Methanobrevibacter*-related sequences were distributed between two major clades, the *smithii-gottschalkii-millerae-thaueri* (*SGMT*) clade and the *ruminantium-olleyae* (RO) clade (King et al. 2011; St-Pierre and Wright 2012b). Interestingly, sequences within a host tend to phylogenetically group within either the *SGMT* or *RO* clade. For example, if a host has a large representation of the *SGMT* group, then the *RO* group is typically lower, or vice versa.

Both the *SGMT* and *RO* clades have been observed in several ruminant hosts (Table 10.1). A greater representation of the *RO* clade was seen in lactating Holstein cows (59.0 %) and in feedlot cattle (i.e., Hereford and Hereford-Angus cross) on both a corn-based (48.0 %) and potato-based (21.1 %) diets in Ontario and Prince Edward Island, respectively (Whitford, et al. 2001; Wright et al. 2007). Moreover, the largest representation of the *RO* clade was observed in the hoatzin, a South American bird with a crop that has a high capacity for fermentation, and its existence dates back to 50 million years before the arrival of ruminants (Wright et al. 2009). Due to the prominence of the *RO* clade in the hoatzin and the ancestry of this unique folivorous bird, it was hypothesized that the *RO* clade is likely to be more ancestral than members in the *SGMT* clade (Wright et al. 2009).

Table 10.1 Distribution of methanogens belonging to the *SGMT* and *RO* clades

Host species	*SGMT* clade (%)	*RO* clade (%)	Total clones	References
Corn-fed Mediterranean water buffalo	55.3	36.1	467	Franzolin et al. (2012)
Grazing Mediterranean water buffalo	81.5	6.7	104	Franzolin et al. (2012)
Sugar cane-fed Mediterranean water buffalo	57.7	33.7	119	Franzolin et al. (2012)
Alpaca	51.6	15.0	947	St-Pierre and Wright (2012b)
Norwegian reindeer	24.1	0	54	Sundset et al. (2009a)
Svalbard reindeer	26.8	0	97	Sundset et al. (2009b)
Corn-fed cattle (Ontario)	4.0	48.0	127	Wright et al. (2007)
Potato-fed cattle (PEI)	9.7	21.1	114	Wright et al. (2007)
Sheep (Venezuela)	62.5	32.7	104	Wright et al. (2007)
Holstein dairy cattle	36.0	59.0	180	King et al. (2011)
Jersey dairy cattle	53.0	44.0	185	King et al. (2011)

Source: St-Pierre et al. (2012)

In contrast, dominance in *SGMT* clade distribution was revealed in lactating Jersey cows (53.0 %), alpacas (51.6 %), sheep from Venezuela (62.5 %), both Svalbard and Norwegian reindeer (26.8, 24.1 %), and water buffalo fed either corn (55.3 %), pasture (81.5 %), or sugar cane (57.7 %) (Wright et al. 2007, 2008; Sundset et al. 2009a, b; King et al. 2011; Franzolin et al. 2012; St-Pierre and Wright 2012b). The disparity among host species between the representation of *SGMT* and *RO* is speculated to be due to the conditions in which each of these two major phylogenetic groupings thrives, but further experimentation is needed.

10.5 Methanogenesis

Methanogenesis is performed by all methanogens and transpires in several anaerobic environments, including the lower intestinal tract, landfills, rice paddies, hydrothermal vents, and the rumen. Since methanogens cannot degrade complex molecules (i.e., glucose), they depend on other microorganisms (i.e., protozoa and bacteria) in the rumen to provide them with substrates for methanogenesis that may include hydrogen, carbon dioxide, methylamines, methanol, formate, or acetate (Ferry and Kastead 2007).

10.5.1 Hydrogen-Utilizing Methanogens

The production of methane is a normal product of rumen fermentation that acts as a pathway for the deposition of metabolic hydrogen produced (Kumar et al. 2011). Methane generated in the rumen is a significant electron sink that generates an electrochemical gradient across the cell membrane to produce ATP (Klieve et al. 2012; Stewart and Bryant 1988). Since hydrogen is utilized by the methanogens to produce methane, the concentration of hydrogen is maintained at low levels in the rumen. Hydrogen is typically derived as a catabolic product from both bacteria and protozoa where it is utilized by methanogens to reduce carbon dioxide to methane. The intricate pathway of methanogenesis via carbon dioxide involves the activation of carbon dioxide by the cofactor, methanofuran, to the formation of formyl-methanofuran. The methyl group from 5-methyl-tetrahydromethanopterin is transferred to coenzyme M, resulting in methyl-CoM producing methane and carbon dioxide (Caspi et al. 2010).

Most methanogens belonging to the following genera are capable of using hydrogen to reduce carbon dioxide to make methane: *Methanobrevibacter*, *Methanobacterium*, *Methanothermobacter*, *Methanothermus*, and some members of the genus *Methanosarcina* (Ferry and Kastead 2007).

10.5.2 Formate-Utilizing Methanogens

In the rumen, cellulolytic bacteria, fungi, and ciliate protozoa release formate (HCOO-) into the rumen fluid where it is utilized by methanogens (Ellis et al. 1990). According to Schauer et al. (1982), 18 % of the methane produced in the rumen is derived from formate. Hungate et al. (1970) further demonstrated this derivation by displaying that the majority of the formate is converted to hydrogen and carbon dioxide via formate hydrogenases and then used in the production of methane by methanogens.

Although *Methanobrevibacter* species tend to utilize hydrogen and carbon dioxide for the production of methane, some species are also capable of using formate plus carbon dioxide. The following species also demonstrate the capacity to utilize formate: *Methanobacterium formicicum*, *Methanomicrobium mobile*, *Methanobrevibacter olleyae*, *Methanobrevibacter millerae*, *Methanobrevibacter smithii*, *Methanobrevibacter ruminantium*, *Methanospirillum hungatei*, and *Methanobrevibacter woesei* (Balch 1979; Hook et al. 2010).

10.5.3 Methanol- and Methylamine-Utilizing Methanogens

Stadtman and Barker (1951) first observed the use of methanol and methylamines for the production of methane in *Methanosarcina barkeri*. In fact, the majority of the members of the order Methanosarcinales (e.g., *Methanosarcina barkeri* and *Methanosarcina mazei*) use methanol and methylamines as substrates (Ferry and Kastead 2007). Methanol and methylamines are C-1 substrates (carbon-containing compounds lacking C-C bonds) that transfer a methyl group to a coenzyme, forming methyl-CoM, which eventually becomes methane and carbon dioxide (Caspi et al. 2010). One in four of the methyl-CoM molecules is oxidized to provide six electrons for the reduction of three methyl-CoM molecules to methane.

Methanosphaera stadtmanae (order Methanobacteriales) cannot reduce carbon dioxide to methane but instead uses hydrogen to reduce methanol to methane (Miller and Wolin 1985). Typically, in ruminants, *Methanosphaera stadtmanae* phylotypes were found in less than 2 % of the total clones in both Jersey and Holstein dairy cattle, Western Australian sheep, Norwegian and Svalbard reindeer, and alpacas (King et al. 2011; St-Pierre and Wright 2012b; Sundset et al. 2009a, b). However, it was identified as the prominent methanogen found in fecal samples from ten Sumatran orangutans in captivity on a diet composed primarily of fruit (Facey et al. 2012) and has also been identified in the human intestine. Facey et al. (2012) suggested that the degradation of pectin found in fruit by anaerobic bacteria to methanol allowed *Methanosphaera stadtmanae* to thrive, whereas a typical ruminant diet lacks pectin, thus significantly lowering methanol levels and the prevalence of this methanogen.

Poulsen et al. (2013) challenged the concept that the Methanosarcinales and some members of the Methanobacteriales were the only methanogens to utilize methylamines as substrates for methanogenesis. Rumen fluid from lactating Holstein dairy cows supplemented with rapeseed oil was collected and used in vitro for the identification of transcripts for the utilization of both methylamines and methanol, demonstrating that members of the TALC/RCC clade can use both substrates (Poulsen et al. 2013).

10.5.4 Acetogen- and Acetate-Utilizing Methanogens

Because enteric methane emissions are affecting both the energy requirements of ruminants and the environment, several mitigation strategies have been explored. One such strategy is to use an alternative hydrogen sink where acetate is produced by acetogenic bacteria instead of methane by methanogens. Acetogenic bacteria, or acetogens, participate in reductive acetogenesis where hydrogen from the fermentation of feed is converted into acetate (Joblin 1999). Acetogens have been found in deer, cattle, and sheep, but methanogens remain the prevalent hydrogen sink.

Although acetate is an end product of reductive acetogenesis, acetate can be utilized by some methanogens to be reduced to methane. Acetate is generally metabolized to methane and carbon dioxide in freshwater, sewage, and marine mud via acetate fermentation. Acetate's involvement in methanogenesis includes the transfer of a methyl group to tetrahydrosarcinapterin and then to coenzyme M to produce methyl-CoM that is demethylated to produce methane (Caspi et al. 2010). Although acetate is involved in methanogenesis in natural environments, it is not metabolized greatly within the rumen because the rate of passage of ruminal contents, such as saliva and feedstuffs, is greater than the growth rate of acetate-utilizing methanogens (Janssen and Kirs 2008). The reduction of carbon dioxide to methane is also more thermodynamically sound and yields more energy than acetate (Joblin 1999). Knowing this, it is less likely that a strict acetate using methanogen will be found within the rumen. Some rumen methanogens such as those belonging to the genus *Methanosarcina* (*Methanosarcina barkeri* and *Methanosarcina mazei*) have the capacity to metabolize acetate, but also use methanol and methylamines to generate methane (Caspi and Pelligrini-Toole 2006; Sirohi et al. 2010). *Methanosphaera stadtmanae* requires acetate for growth, but will use methanol for methanogenesis.

10.6 Concluding Remarks

Methanogens are strict anaerobes, with fastidious growth requirements. They require different substrates (hydrogen, carbon dioxide, acetate, methanol, methylamines, and formate) for methane production and exist synergistically with bacteria and symbiotically with rumen ciliate protozoa. The differences in morphology and utilization of different substrates by different methanogens enable them to be found in a diverse number of habitats. Knowledge of this ecosystem is rapidly accumulating, particularly with the advent of molecular biology and culture-independent technologies.

References

Balch W (1979) Methanogens: reevaluation of a unique biological group. Microbiol Rev 43:260–296

Boone DR, Whitman WB, Rouviére P (1993) Diversity and taxonomy of methanogens. In: Ferry JG (ed) Methanogenesis: ecology physiology biochemistry and genetics. Chapman and Hall, New York, pp 35–80

Caspi R, Pelligrini-Toole A (2006) MetaCyc pathway: methanogenesis from acetate

Caspi R, Altman T, Dale JM et al (2010) The MetaCyc database of metabolic pathways and enzymes and the BioCyc collection of pathway/genome databases. Nucl Acid Res 38:473–479

Chaudhary PP, Sirohi SK (2009) Dominance of *Methanomicrobium* phylotype in methanogen population present in Murrah buffaloes (*Bubalus bubalis*). Lett Appl Microbiol 49(2):274–277

Eckard R (2013) Greenhouse emissions from Australian agriculture. Retrieved 12 July 2013 from http://www.dpivicgovau/agriculture/dairy/emissions-in-dairy/methane-research2

Ellis J, Williams AG, Lloyd D (1990) Formate and glucose stimulation of methane and hydrogen production in rumen liquor. Curr Microbiol 20(4):251–254

EPA (2014) Inventory of US Greenhouse gas emissions and sinks: 1990–2010: Ch 6 Agriculture

Facey HV, Northwood KS, Wright ADG (2012) Molecular diversity of methanogens in fecal samples from captive Sumatran orangutans (*Pongo abelii*). Am J Primat 74(5):408–413

Ferry JG, Kastead KA (2007) Methanogenesis. In: Cavicchioli R (ed) Archaea: molecular and cellular biology. ASM Press, Washington, DC, pp 288–314

Franzolin R, St-Pierre B, Northwood K et al (2012) Analysis of rumen methanogen diversity in water buffaloes (*Bubalus bubalis*) under three different diets. Microb Ecol 64(1):131–139

Garcia J, Ollivier B, Whitman WB (2006) The Prokaryotes (M Dworkin, S Falkow, E Rosenberg, K-H Schleifer, E Stackebrandt eds) Prokaryotes (3):pp 208–230. Springer: New York

Hook SE, Wright ADG, McBride BW (2010) Methanogens: methane producers of the rumen and mitigation strategies. Archaea, Vancouver, pp 1–11

Hungate RE (1966) The rumen and its microbes. Academic, New York/London. ISBN 9780123616500

Hungate RE, Smith W, Bauchop T et al (1970) Formate as an intermediate in the bovine rumen fermentation. J Bacteriol 102:389–397

IPCC (Intergovernmental Panel on Climate Change) (2006) IPCC guidelines for national greenhouse gas inventories prepared by the national greenhouse gas inventories programme. In Eggleston S, Buendia L, Miwa K, Ngara T, Tanabeed K. Institute for global environmental strategies (IGES), Hayama

Janssen PH, Kirs M (2008) Structure of the archaeal community of the rumen. Appl Environ Microbiol 74(12):3619–3625

Joblin KN (1999) Ruminal acetogens and their potential to lower ruminant methane emissions. Aust J Agric Res 50:1307–1313

Johnson KA, Johnson DE (1995) Methane emissions from cattle. J Anim Sci 73(8):2483–2492

King EE, Smith RP, St-Pierre B et al (2011) Differences in the rumen methanogen populations of lactating Jersey and Holstein dairy cows under the same diet regimen. Appl Environ Microbiol 77(16):5682–5687

Klieve AV, Ouwerkerk D, Maguire AJ (2012) Archaea in the foregut of macropod marsupials: PCR and amplicon sequence-based observations. J Appl Microbiol 113(5):1065–1075

Kumar S, Dagar SS, Puniya AK (2011) Isolation and characterization of methanogens from rumen of Murrah buffalo. Ann Microbiol 62(1):345–350

Miller TL, Wolin MJ (1985) *Methanosphaera stadtmaniae* gen nov sp nov: a species that forms methane by reducing methanol with hydrogen. Arch Microbiol 141:116–122

Murray RM, Bryant AM, Leng RA (1976) Rates of production of methane in the rumen and large intestine of sheep. Br J Nutr 36:1–14

Paul K, Nonoh JO, Mikulski L et al (2012) “Methanoplasmatales” thermoplasmatales-related archaea in termite guts and other environments are the seventh order of methanogens. Appl Environ Microbiol 78(23):8245–8253

PCNCCV (Primera Comunicación Nacional en Cambio Climático de Venezuela) (2005) Caracas Republica Bolivariana de Venezuela

Poulsen M, Schwab C, Jensen BB et al (2013) Methylotrophic methanogenic thermoplasmata implicated in reduced methane emissions from bovine rumen. Nat Comm 4:e1428

Schauer NL, Brown DP, Ferry JG (1982) Kinetics of formate metabolism in *Methanobacterium formicicum* and *Methanospirillum hungatei*. Appl Environ Microbiol 44(3):549–554

Singh KM, Pandya PR, Parnerkar S et al (2010) Methanogenic diversity studies within the rumen of Surti buffaloes based on methyl coenzyme M reductase A (mcrA) genes point to Methanobacteriales. Pol J Microbiol 59(3):175–178

Singh KM, Tripathi AK, Pandya PR et al (2012) Methanogen diversity in the rumen of Indian Surti buffalo (*Bubalus bubalis*) assessed by 16S rDNA analysis. Res Vet Sci 92(3):451–455

Sirohi SK, Pandey N, Singh B et al (2010) Rumen methanogens: a review. Indian J Microbiol 50(3):253–262

Stadtman TJ, Barker HA (1951) Studies on the methane fermentation IX the origin of methane in the acetate and methanol fermentations by *Methanosarcina*. Bacteriol 61(1):81–86

Stewart CS, Bryant MP (1988) The rumen bacteria. In: Hobson PN (ed) The rumen microbial ecosystem. Elsevier, London, pp 21–75

St-Pierre B, Wright ADG (2012a) Diversity of gut methanogens in herbivorous animals. Animal1: 49–56

St-Pierre B, Wright ADG (2012b) Molecular analysis of methanogenic archaea in the forestomach of the alpaca (*Vicugna pacos*). BMC Microbiol 12:1–12

Sundset M, Edwards JE, Cheng YF et al (2009a) Molecular diversity of the rumen microbiome of Norwegian reindeer on natural summer pasture. Microb Ecol 57(2):335–348

Sundset MA, Edwards JE, Cheng YF et al (2009b) Rumen microbial diversity in Svalbard reindeer with particular emphasis on methanogenic archaea. FEMS Microb Ecol 70(3):553–562

Tanner RS, Wolfe RS (1988) Nutritional requirements of *Methanomicrobium mobile*. Appl Environ Microbiol 54(3):625–628

Whitford MF, Teather RM, Forster RJ (2001) Phylogenetic analysis of methanogens from the bovine rumen. BMC Microbiol 1:1–5

Williams YJ, Popovski S, Rea SM et al (2009) A vaccine against rumen methanogens can alter the composition of archaeal populations. Appl Environ Microbiol 75(7):1860–1866

Woese CR, Fox GE (1977) Phylogenetic structure of the prokaryotic domain: the primary kingdoms. Proc Natl Acad Sci U S A 74(11):5088–5090

Woese CR, Kandler O, Wheelis ML (1990) Towards a natural system of organisms: proposal for the domains Archaea bacteria and Eucarya. Proc Natl Acad Sci U S A 87:4576–4579

Wright ADG, Klieve AV (2011) Does the complexity of the rumen microbial ecology preclude methane mitigation? Anim Feed Sci Technol 166–167: 248–253

Wright ADG, Williams AJ, Christophersen CT et al (2004) Molecular diversity of rumen methanogens from sheep in Western Australia. Appl Environ Microbiol 70(3):1263–1270

Wright ADG, Toovey AF, Pimm CL (2006) Molecular identification of methanogenic archaea from sheep in Queensland Australia reveal more uncultured novel archaea. Anaerobe 12(3):134–139

Wright ADG, Auckland CH, Lynn DH (2007) Molecular diversity of methanogens in feedlot cattle from Ontario and Prince Edward Island Canada. Appl Environ Microbiol 73(13):4206–4210

Wright ADG, Ma X, Obispo NE (2008) *Methanobrevibacter* phylotypes are the dominant methanogens in sheep from Venezuela. Microb Ecol 56(2):390–394

Wright ADG, Northwood KS, Obispo NE (2009) Rumen-like methanogens identified from the crop of the folivorous South American bird the hoatzin (*Opisthocomus hoazin*). ISME J 3(10):1120–1126

Part III

Rumen Manipulation

11 Plant Secondary Metabolites

Serena Calabrò

Abstract

Secondary metabolites are those compounds which are produced by certain microorganisms in a phase of subsequent to growth, may have some survival function, have unusual chemical structures, and are formed as mixtures of closely related members of a chemical family. This chapter deals with the plant secondary metabolites and their application in rumen biology.

Keywords

Rumen fermentation • Methane production • Microbial activity

11.1 Introduction

The rumen microbiome is an efficient anaerobic fermentation system that confers specific advantages on ruminants, e.g., efficient digestion of large amount of lignocellulosic compounds, capability to use nonprotein nitrogen sources, and detoxification of toxic ingredients. The study of the rumen microbiome has been approached by numerous researchers over the years. The results of their research have helped to improve the feed efficiency guaranteeing better animal performance. Currently, these kinds of studies are primarily aimed at ensuring the health of the gastrointestinal tract and, therefore, the health of entire animal body. Another much discussed topic today is the environmental impact of livestock that may in part be contained also with proper nutrition management, through a manipulation of rumen fermentation. The plant secondary metabolites (PSM) are the focus of interest of many research groups worldwide. Although in the past, their presence in feed was considered only negative, today numerous studies have confirmed that PSM can even improve animal performance and health and limit environmental impact. Certainly, the effects of PSM are very variable in function of their type and concentrations in the diets.

The objective of this chapter is to introduce plant secondary metabolites, their classification, and methods to detect and to describe how to use them to manipulate rumen fermentation.

S. Calabrò (✉)
Department of Veterinary Medicine and Animal Production, University of Napoli Federico II, via F Delpino 1, Naples 80137, Italy
e-mail: serena.calabro@unina.it

A.K. Puniya et al. (eds.), *Rumen Microbiology: From Evolution to Revolution*, DOI 10.1007/978-81-322-2401-3_11

11.1.1 Rumen Manipulation

The rumen microbial ecosystem is highly stable and optimized due to natural selection of microbes. However, it is not completely efficient for the presence of a significant portion of undigested feed in the feces and for the loss of energy through CH_4 emission. Nowadays, the improvement of its efficiency through its manipulation is assumed to be essential for many reasons, for example, for the high genetic manipulation made in plants. Moreover, forages from tropical and subtropical area have some undesirable limitation for animal feeding like low energy value [due to the presence of high amount of lignin, silica, and cutin in the cell walls limiting structural carbohydrate fermentation, volatile fatty acid (VFA) and microbial biomass production, reducing feed intake and prolonged retention time], deficiency of essential nutrients (energy, protein, minerals), imbalanced end products (high acetate, low propionate), and the presence of anti-nutritional compounds (Santra and Karim 2003). Therefore, many reasons exist for manipulating rumen fermentation in order to improve feed utilization, feed conversion efficiency, and animal performance. The manipulating techniques might include dietary intervention and the use of suitable chemicals, probiotics, or feed additives, such as plant secondary metabolites. Bioactive plant metabolites are an important contemporary research area to create substitutes for chemical feed additives due to their potential to modify rumen fermentation, mainly in terms of reducing methane (CH_4) production and selecting rumen microorganisms (bacteria, protozoa, fungi) to increase feed utilization and VFA production.

11.1.2 Plant Secondary Metabolites

The plant secondary metabolites are so defined because they are opposed to the primary plant structural components (i.e., protein, carbohydrates, lipids), which are involved in the primary biochemical processes in the plant body such as cell division, growth and respiration, and storage of nutrients. These molecules are characterized by low concentration (less than 1 % of the total carbon), even if their levels vary temporally and spatially among and within species, and their proportions among other compounds are in constant flux. Moreover, their presence and concentration in a given plant are influenced by genetics, stage of life cycle, and environmental factors (Estell 2010). The improvement of analytical techniques (i.e., chromatography, biochemical techniques, molecular biology) allowed the detection of more than 200,000 defined PSM molecules with important role in the adaptation of plants to the environment (Bourgaud et al. 2001). Most PSM possess biological activity on other living organisms (i.e., they affect some animal metabolic processes and/or the growth rate of some microorganisms), thus protecting the plants from the predation of insects or the grazing by herbivores. While the PSM role in herbivore is quite clear, not so well known is the mechanisms by which ruminants cope with PSM and their effects on livestock production (Bodas et al. 2012). Due to their large biological activities, PSM have been used for centuries in traditional medicine and, nowadays, in pharmaceutics, cosmetics, and nutraceutics. For the same reason, animal nutrition companies are screening bioactive compounds of plants in order to obtain feed additives or rumen manipulators.

11.2 Classification of Plant Secondary Metabolites

The plant secondary compounds are difficult to classify because their metabolic pathways of synthesis and their properties and mechanisms of action often overlap (Bodas et al. 2012). The PSM classification is usually made according to their biosynthetic pathways in three large molecule categories: phenolics, terpenes and steroids, and alkaloids.

11.2.1 Phenolics

A good example of a widespread metabolite family is given by phenolics: because these molecules are involved in lignin synthesis, they are common to all higher plants. Phenols are characterized by having at least one aromatic ring with one or more hydroxyl group and are classified on their biosynthetic origin basis: from acetyl-CoA units, from shikimic acid, and from gallic and ellagic acids. Condensed tannins, or proanthocyanidins, belong to this group and are large polymers of flavonoid molecules joined by covalent bonds. Flavonoids have different functions: act in regulation of primary metabolism, provide attractive colors to flowers, and have antimicrobial properties (Crozier et al. 2006).

11.2.2 Terpenes and Steroids

Terpenes are synthesized from isoprene and classified on the number of its units. *Monoterpenes* are the most representative molecules constituting 90 % of essential oils. *Triterpenes* are one of the most important subclasses due to their wide distribution in the plant kingdom and their applications in pharmacology. Important compounds in this category are steroids that stimulate plant growth and saponins that are used in pharmacological industry for their anti-inflammatory and antimicrobial actions (Grassmann 2005).

11.2.3 Alkaloids

These compounds are sparsely distributed in the plant kingdom and are much more specific to defined plant genus and species. Alkaloids contain nitrogen, and most of them are extremely toxic or teratogenic to other organisms, but some of them have pharmacological effects. They can be classified by the common natural source basis, for the similarity of the carbon skeleton, or on their biogenetic precursor basis (Bruneton 1999).

11.3 Effect of Plant Secondary Metabolites on Rumen Activity

Plant secondary compounds have been shown to express action on rumen activity as antimicrobial agents against bacteria, protozoa, archaea, and fungi. The most active substances are phenolic compounds, but some nonphenolic substances have also demonstrated to be active. The effects on rumen microorganisms' activity depend on plant species and are modulated by rumen pH, diet administered, and methods of compounds' extraction. The microcidal or microstatic action mainly comes from the capacity of such molecules in intruding into the microorganism cell membrane and disintegrate its structures (Bodas et al. 2012). The antimicrobial activity of these compounds is highly specific and may be used to manipulate rumen fermentation by selective inhibition of a microbial group of the ecosystem.

11.3.1 Antimicrobial Activity

Wallace et al. (1994) reported that saponins stimulated growth of *Prevotella ruminicola*, whereas growth of *Butyrivibrio fibrisolvens* and *Streptococcus bovis* was inhibited. Tannins inhibit growth of cellulolytic and proteolytic bacteria, and this effect is correlated with molecular weight (McSweeney et al. 2001). The inhibitory effect of essential oils on Gram-positive bacteria varies as function of their chemical composition: the same extract can have stimulatory or inhibitory effects when obtained from different plant species of the same genus (Ferme et al. 2004).

11.3.2 Antimethanogenic and Antiprotozoal Activity

Some groups of plants, rich in saponins, tannins, and essential oils, are recently largely reevaluated for their antimethanogenic and antiprotozoal activity in the rumen. Saponins are high molecular weight glycosides in which sugars are linked to a triterpene or steroidal aglycone moiety. The effect of saponins varies in function of the chemical composition and level of inclusion in the plant (Kamra et al. 2006). Some of these studies are listed in Table 11.1. However, it has been reported that the rumen microbes are able to adapt to saponins by prolonged feeding of such feeds (Wallace et al. 2002). Tannins are nutritionally important plant secondary metabolites and are complex phenolic organic molecules. The literature indicates that phenolic acids have been found to be toxic for many rumen microbes, especially ciliate protozoa, fiber-degrading microbes, and methanogens. As a result of this property, the methanogenesis in the rumen is also inhibited. The in vivo and in vitro investigation (Table 11.1) showed that the significant effect depends on the extraction type (i.e., water, methanol, ethanol). The presence of a variety of essential oils in some plant (i.e., spices) is responsible for specific antimicrobial activity and may affect rumen fermentation significantly. Many investigations have recently carried out to evaluate the effect of water, methanol, and ethanol extracts of some spices on the in vitro and in vivo rumen fermentation (Table 11.1).

11.3.3 Effects on Rumen Fermentation

The inclusion of saponins in diet of animals changes the site of digestion of organic matter and fiber, from rumen to hindgut (Patra and Saxena 2009). Tannins reduce ruminal protein digestibility and plant cell wall digestion, slowing their digestion rate, due to the binding to dietary protein and structural polysaccharides (i.e., cellulose, hemicelluloses, and pectin) (McSweeney et al. 2001). Essential oils suppress colonization and digestion of readily degradable substrates by amylolytic and proteolytic bacteria without affecting fiber digestion (Wallace et al. 2002).

Table 11.1 The effect of plant secondary compounds on methane production and protozoal counts

Plant	Effect	References
Saponins		
Sapindus mukorossi	Decreases CH_4 production and digestibility	Agarwal et al. (2006)
Soapnut fruit	Increases body weight and efficiency of feed utilization, reduces protozoa count	Kamra et al. (2000)
Sapindus rarak fruit	Reduces protozoa and N-NH_3, increases bacteria, improves daily body weight gain and feed conversion efficiency	Thalib et al. (1996)
Sapindus saponaria	Reduces CH_4 emission by sheep fed on grass-based or grass-legume-based diets	Hess et al. (2004)
Tannins		
Acacia concinna fruit	Inhibits methanogenesis, affects protozoal counts	Patra (2004)
Yucca extract	Reduces protozoa, does not affect bacterial activity	Wang et al. (1998)
Terminalia belerica and *chebula* fruit pulp	Inhibit CH_4 emission, decrease number of protozoa	Patra et al. (2006)
Psidium guajava leaves	Exhibit antimethanogenic activity	Ushida et al. (1989)
Chestnut, mimostatin, and quebracho	Inhibit DM digestibility, CH_4 and total gas production	Roth et al. (2001)
Essential oils		
Foeniculum vulgare, *Syzygium aromaticum*, *Zingiber officinale*	Inhibit in vitro CH_4 emission	Patra (2004)
Allium sativum	Suppresses in vitro CH_4 emission without affecting fiber-degrading enzymes and DM degradability of feeds	Patra (2004)
Blend of essential oils (thymol, guaiacol, and limonene)	Reduces crude protein degradation of soybean meal	Molero et al. (2004)
40 ppm essential oils	Affect growth of *Clostridium sticklandii* and *Prevotella ruminicola*	Silva et al. (1987)
Essential oils	Do not influence ruminal VFA, ammonia-N, and protozoal population	Hobson (1969)

Saponins, tannins, and essential oils usually reduce the amount of N-NH_3 produced in the rumen, which improves assimilation of feed amino acid nitrogen by ruminants (Spanghero et al. 2008). The effects on rumen N-NH_3 concentrations are probably related to a reduction in protozoal numbers (Newbold et al. 1997), which play a key role in ruminal feed protein degradation. The lower microbial activity and substrate degradation due to reduction in total and cellulolytic bacterial numbers in the rumen, in response to the use of tannins in the diet, is usually linked to a reduction in VFA production (Bhatta et al. 2009) and changing in the proportions of acetate (increase) and propionate (decrease). Effects of saponins seem to be pH dependent (i.e., the lower the pH, the more pronounced the effects) and, so, are influenced by the diet to which saponins are added (Bodas et al. 2012). Effects of essential oils on VFA production in the rumen are inconsistent: Spanghero et al. (2008) found decreases in VFA concentrations, Newbold et al. (2004) reported a tendency in stimulating VFA production, and Beauchemin and McGinn (2006) reported no changes in VFA production.

11.4 Rumen Manipulation with Plant Secondary Metabolites

The numerous results reported in literature indicate that plant secondary compounds seem to have a potential to be used as feed additives for rumen manipulation, in particular for defaunation, and to reduce methane emission (Kamra et al. 2006).

11.4.1 Plant Secondary Metabolites and Methanogenesis

The effects of tannin content on in vitro fermentation characteristics and kinetics and methane production have been studied by many authors. For example, sainfoin (*Onobrychis viciifolia*. Scop.) does not induce bloating in ruminants, and in comparison with other forage legumes, it allows a higher absorption of protein in the small intestine. These effects may be explained by the presence of moderate amounts of condensed tannins that are negatively correlated to in vitro CH_4 production (Guglielmelli et al. 2011; Calabrò et al. 2012). The inhibition of methanogenesis is due to reduction in fiber degradation that limits the acetate production through reduction in cellulolytic numbers and/or the poor availability of compounds sequestered by tannins. However, the direct effect of CTs on methanogenic bacteria should thus be considered. Moreover, the amount of CH_4 produced is due to a series of interactions between the nutritive characteristics and the presence of tannins, being both variable with advancing growth of plant.

11.4.2 Defaunation

Rumen protozoa are the largest in size among rumen microbes and contribute 40–50 % of the total microbial biomass and enzyme activities in the rumen. Nevertheless, it is clear that protozoa may be nonessential for the ruminant; recently, it has been hypothesized that they should play a significant role in the rumen metabolism specially to stabilize the rumen pH. The process of making the rumen free of protozoa is called defaunation. Some of the methanogenic archaea have a symbiotic relationship with ciliate protozoa present in the rumen and remain either inside the body of the protozoa or are attached to their surface. Probably this is the reason why defaunation is usually associated with reduced production of methane in the rumen as the methanogens lose their symbiotic partner resulting in their reduced biological activity (Santra and Karim 2003).

Defaunation decreased energy losses through methanogenesis by 5.5–7.9 % of gross energy intake (Mathieu et al. 1996). Stimulatory effects of some plant extracts on some bacterial populations seem to be a consequence of the inhibition of protozoa and fungi. As a consequence of the decrease in the protozoal numbers, an increase in the total bacterial population is expected as there is no predation of bacteria by the protozoa.

11.5 Methods to Detect PMS in Plants

Plant secondary metabolites provide reliable assays to meet the challenge of fulfilling the huge demand for feed; therefore, they are currently the subject of much research interest. Makkar et al. (2007) published a volume which contained methods for analysis of some important PMS, useful for animal feed additives as well as human food. Because their extraction presents specific challenges that must be addressed throughout the solvent extraction process, successful extraction begins with careful selection and preparation of plant samples and thorough review of the appropriate literature for indications of which protocols are suitable for a particular class of compounds or plant species. Jones and Douglas Kinghorn (2005) present an overview of the process of plant extraction, with an emphasis on common problems encountered and methods for reducing or eliminating these problems. During the extraction of plant material, it is important to minimize interference from compounds that may coextract with the target compounds and to avoid contamination of the extract, as well as to prevent decomposition of important metabolites or artifact formation as a result of extraction conditions or solvent impurities.

11.6 Conclusions

The use of plant secondary metabolites as rumen manipulators will be the research issue for many rumen microbiology laboratories. The large amount of literature generated on this topic also indicates the extent of interest of the researchers. Many plants are already screened for their secondary metabolites that have been extracted in different solvents. Most of them have interesting rumen activity in terms of antimethanogenic and antiprotozoal. Further detailed screening and identification of the active principles are essential to know the mechanism of action. These metabolites appear to have a potential for commercial feed additives in livestock production.

References

Agarwal N, Kamra ND, Chaudhary LC et al (2006) Effect of *Sapindus mukorossi* extracts on in vitro methanogenesis and fermentation characteristics in buffalo rumen liquor. J Appl Anim Res 30(1):1–4

Beauchemin KA, McGinn S (2006) Methane emissions from beef cattle: effects of fumaric acid essential oil and canola oil. J Anim Sci 84:1489–1592

Bhatta R, Uyeno Y, Tajima K et al (2009) Difference in the nature of tannins on in vitro ruminal methane and volatile fatty acid production and on methanogenic archaea and protozoal populations. J Dairy Sci 92:5512–5522

Bodas R, Prietoa N, García-González R et al (2012) Manipulation of rumen fermentation and methane production with plant secondary metabolites. Anim Feed Sci Technol 176:78–93

Bourgaud F, Gravot A, Milesi S et al (2001) Production of plant secondary metabolites: a historical perspective. Plant Sci 161(5):839–851

Bruneton J (1999) Pharmacognosy phytochemistry medicinal plants, 2nd edn. Intercept Ltd, London

Calabrò S, Guglielmelli A, Iannaccone F et al (2012) Fermentation kinetics of Sainfoin hay with and without PEG. J Anim Physiol Anim Nutr 96(5):842–849

Crozier A, Jaganath IB, Clifford MN (2006) Phenols polyphenols and tannins: an overview. In: Crozier A, Clifford MN, Ashihara H (eds) Plant secondary metabolites occurrence structure and role in the human diet. Blackwell, Chennai, pp 1–17

Estell RE (2010) Coping with shrub secondary metabolites by ruminants. Small Rumin Res 94(1–3):1–9

Ferme D, Banjac M, Calsamiglia S (2004) The effects of plant extracts on microbial community structure in a rumen-simulating continuous-culture system as revealed by molecular profiling. Folia Microbiol 49:151–155

Grassmann J (2005) Terpenoids as plant antioxidants. Vitam Horm 72:505–535

Guglielmelli A, Calabrò S, Primi R et al (2011) In vitro fermentation patterns and methane production of sainfoin (*Onobrychis viciifolia* Scop) hay with different condensed tannin contents. Grass Forage Sci 66:488–500

Hess HD, Beuret RA, Lötscher M et al (2004) Ruminal fermentation methanogenesis and nitrogen utilization of sheep receiving tropical grass hay-concentrate diets offered with Sapindus saponaria fruits and *Cratylia argentea* foliage. Anim Sci 79:177–189

Hobson PN (1969) Rumen bacteria. In: Norris JR, Ribbons DW (eds) Methods in microbiology, vol 3B. Academic, London, pp 133–149

Jones WP, Douglas K (2005) A extraction of plant secondary metabolites. In: Sarker SD, Zahid L, Gray AI (eds) Natural products isolation, vol 20. Humana Press, Totowa, pp 323–351

Kamra DN, Singh R, Agarwal N et al (2000) Soapnut (Reetha) as natural defaunating agent – its effect on rumen fermentation and *in sacco* degradability of jowar hay in buffaloes. Buffalo J 16:99–104

Kamra ND, Agarwal N, Chaudhary LC (2006) Inhibition of ruminal methanogenesis by tropical plants containing secondary compounds. Int Congr Ser 1293:156–163

Makkar HPS, Becker K, Sidhuraju P (2007) Plant secondary metabolites methods in molecular biology. Humana Press, Totowa, p 129

Mathieu F, Jouany JP, Senaud J (1996) The effect of Saccharomyces cerevisiae and *Aspergillus oryzae* on fermentation in the rumen of faunated and defaunated sheep: protozoal and probiotic interaction. Reprod Nutr Dev 36:271–287

McSweeney C, Palmer B, Bunch R et al (2001) Effect of the tropical forage calliandra on microbial protein synthesis and ecology in the rumen. J Appl Microbiol 90:78–88

Molero R, Ibara M, Calsamiglia S et al (2004) Effects of a specific blend of essential oil compounds on dry matter and crude protein degradability in heifers fed diets with different forage to concentrate ratios. Anim Feed Sci Technol 114:91–104

Newbold CJ, El Hassan SM, Wang J et al (1997) Influence of foliage from African multipurpose trees on activity of rumen protozoa and bacteria. Br J Nutr 78:237–249

Newbold CJ, McIntosh M, Williams P (2004) Effects of a specific blend of essential oil compounds on rumen fermentation. Anim Feed Sci Technol 114:105–112

Patra AK (2004) Studies on inhibition of ciliate protozoa and stimulation of fibre degrading microbes in the rumen of buffalo by plant secondary metabolites. PhD thesis, Indian Veterinary Research Institute, Izatnagar, pp 1–234

Patra AK, Saxena J (2009) A review of the effect and mode of action of saponins on microbial population and fermentation in the rumen and ruminant production. Nutr Res Rev 22:204–219

Patra AK, Kamra DN, Agarwal N (2006) Effect of plant extracts on in vitro methanogenesis enzyme activities and fermentation of feed in rumen liquor of buffalo. Anim Feed Sci Technol 128(3–4):276–291

Roth S, Steingaß H, Drochner H (2001) Reducing methane emission and optimization of N-supply in ruminants by treating feeds with tannins. http://www.uni-hohenheimde/i3ve/00068900/28171041htm

Santra A, Karim SA (2003) Rumen manipulation to improve animal productivity. Asian-Aust J Anim Sci 16(5):748–763

Silvia AT, Wallace RJ, Ørskov ER (1987) Use of particle bound microbial activity to predict the rate and extent of fibre degradation in the rumen. Br J Nutr 57:407–415

Spanghero M, Zanfi C, Fabbro E et al (2008) Effects of a blend of essential oils on some end products of in vitro rumen fermentation. Anim Feed Sci Technol 145:364–374

Thalib A, Widiawati Y, Hamid H et al (1996) The effects of saponins from Sapindus rarak fruit on rumen microbes and performance of sheep. J Ilmu Ternak Vet 2:17–21

Ushida K, Tanaka H, Kjima Y (1989) Effect of phenolic acids on gas and volatile fatty acid production by mixed rumen population with or without protozoa. Jpn J Zootech Sci 60:1135–1142

Wallace RJ, McEwan NR, McIntosh FM (2002) Natural products as manipulators of rumen fermentation. Asian-Aust J Anim Sci 15:1458–1468

Wallace RJ, Arthaud L, Newbold CJ (1994) Influence of *Yucca schidigera* extract on ruminal ammonia concentration and ruminal micro-organisms. Appl Environ Microbiol 60:1762–1767

Wang Y, McAllister TA, Newbold CJ et al (1998) Effect of Yucca schidigera extract on fermentation and degradation of steroidal saponins in the rumen simulation technique (RUSITEC). Anim Feed Sci Technol 74:143–153

Microbial Feed Additives

12

Ravinder Nagpal, Bhuvnesh Shrivastava, Nikhil Kumar, Tejpal Dhewa, and Harmesh Sahay

Abstract

The recent global concerns associated with the use of antibiotics and other growth-promoting stimulants in the animal feeds have created a window of opportunities and responsibilities for livestock nutritionists to search for alternative safer feed additives, such as direct-fed microbials based on viable naturally occurring beneficial microorganisms. In the last decade, nutritionists and microbiologists have elucidated comprehensive details related to the normal intestinal microbiota of mammals and also the benefits it provides to the host. In this milieu, numerous species and strains of probiotic bacteria, yeast, and fungal cultures have been isolated and experimented, and a number of these microbes have been found to confer numerous benefits to the host animal when added to a diet. Some of the benefits associated with these microbial supplements are stimulation of beneficial microbial growth in the rumen, stabilization of the rumen pH, improved ruminal fermentation and end-product production, increased nutrient flow and digestibility, alleviated stress, enhanced immune response, reduced pathogens, reduced acidosis, improved weight gain, increased milk yield, etc. Although the research on the use of microbes as animal feed additives has been intensive as well as extensive in the last couple of decades, more investigations, nevertheless, are still needed to ascertain the species- and strain-specific effects associated with these microorganisms, to explicate the molecular mechanisms involved in the animal responses, and also to search for more efficient and steady formulations for maximum and consistent livestock productivity.

R. Nagpal (✉)
Division of Laboratories for Probiotic Research, Juntendo University Graduate School of Medicine, Hongo, Bunkyo-ku, Tokyo, Japan
e-mail: nagpal511@gmail.com

B. Shrivastava • H. Sahay
Dairy Microbiology Division, ICAR-National Dairy Research Institute, Karnal 132001, India

N. Kumar
Basic and Applied Sciences, National Institute of Food Technology Entrepreneurship and Management, Kundli 131028, India

T. Dhewa
Bhaskaraycharya College of Applied Sciences, University of Delhi, New Delhi, India

A.K. Puniya et al. (eds.), *Rumen Microbiology: From Evolution to Revolution*, DOI 10.1007/978-81-322-2401-3_12

Keywords
Anaerobic fungi • Animal nutrition • Direct-fed microbials • Gut microbes • Probiotics • Rumen • Rumen fermentation • Rumen bacteria • Rumen yeasts

Abbreviations

CFU	Colony-forming units
DFM	Direct-fed microbial
FAO	Food and Agriculture Organization of the United Nations
FDA	US Food and Drug Administration
GI	Gastrointestinal
GIT	Gastrointestinal tract
MFA	Microbial feed additives
WHO	World Health Organization

12.1 Introduction

Since the adoption of the term DFM (direct-fed microbial; a formulation of viable naturally occurring bacteria, fungi, or yeast) by FDA about two decades ago, there have been gigantic advances and progress that have established the beneficial outcomes of exploiting microorganisms and other feed additives in animal diets for improving animal's healthiness and productivity. The use of beneficial microbes in animal diets is appearing to be a promising mean for improved nutrient utilization and increased overall productivity of ruminants. A wide array of bacteria and yeasts have been screened out from the pools of naturally occurring microbes which are capable of positively manipulating and improving the production enhancements in ruminants such as nutrient digestion, nutrition, overall health, and productivity. Numerous DFMs are accessible today that have been rigorously documented for their positive effects on animal's weight gain, rumen development and digestion, restoration of intestinal microflora, producing bacteriocins, competitively reducing opportunistic pathogens, etc. In addition, several DFMs could also produce nutrients and stimulatory growth factors which in turn promote other beneficial microorganisms to flourish within the animal's digestive tract and boost the release of digestive enzymes thereby aiding in the feed digestion by the animal. Moreover, some selected DFM formulations have been proposed to perk up the immune responses as well as metabolism of the animals. In this context, this chapter endeavors to review and summarize some important investigations related to the use of microorganisms as feed additives, the animal responses to these additives, and also the mechanisms by which these formulations improve the production efficiency of the animal.

12.2 Rumen: The Dynamic Microbial Ecosystem

The rumen is a dynamic and continuous-culture-type fermentation container with a highly complex and competitive microbial ecosystem within, yet the rumen microbial ecosystem represents and facilitates a classic symbiotic association between the microbes and the host animal. The feedstuff consumed by the animal is fermented to volatile fatty acids and microbial cell proteins to supply energy and proteins (Weimer 1998). In particular, the plant materials such as celluloses, hemicelluloses, pectins, starches and other polysaccharides are hydrolyzed by the rumen microorganisms to monomers or dimers of sugars and are finally fermented to provide a variety of products such as acetic acid, propionic acid, butyric acid, methane, CO_2, etc. (Hungate 1988). Though the majority of rumen microorganisms are strictly anaerobic, some oxygen, however, is endured by facultative anaerobes which consume the traces of oxygen and aid in retaining the low oxygen potential (Van Soest 1994). Generally, the population densities of rumen bacteria are dominating with the magnitude of 10^{10}–10^{11}, with rumen protozoa prevailing in the range of 10^5–10^7 and anaerobic fungi and some facultatively anaerobic bacteria falling in the range of $<10^4$ (Mcallister and Cheng 1996). However, these proportions of microbial population may vary significantly or slightly in response to animal diet types such as hay to concentrate rations, concentrate diets, complexity of fibers, pH fluctuations, etc. (Mackie et al. 1984). Under normal conditions, the internal temperature of rumen is maintained at 37 °C with the stable pH range of 6.8–7.0.

12.3 The Inception of DFM

Enhancing the ruminants' nutrition, nutrient utilization, productivity, and overall health has always been the foremost objective of majority of rumen ecosystem investigations. In this context, numerous kinds of feed additives and supplements have been exploited in an attempt to improve the ruminant nutrition, health, and productivity. These additives have broadly included the use of a variety of antibiotics, enzymes, probiotics, and prebiotic supplements with an ability to influence the rumen fermentation and microbial ecosystem. However, as a consequence of the increased risk and safety concern with the use of antibiotics and other growth promoters in the animal feed industry, there has been a selective increase in the field of alternative potential feed additives, particularly the microbial feed additives (MFAs) or direct-fed microbials (DFMs) based on naturally occurring microbes (Krehbiel et al. 2003; Hong et al. 2005). Consequently, such MFAs and DFMs have been increasingly researched and explored by livestock producers in order to exploit novel avenues to enhance animal performance and health. A wide array of terminologies has frequently been in use to categorize the microbial-based additives or supplements for ruminants. In fact, the terms probiotics and DFMs have also been sporadically used synonymously which appears to be rather conflicting as per the guidelines. Probiotics are generally defined as "live microbial feed supplements which beneficially affects the host animal through improvement of intestinal microbial balance" (Parker 1974). Conversely, DFMs or MFAs are merely defined as "a source of viable naturally-occurring microbes" (FDA 1995, 2003). A DFM is also defined as "a feed additive or supplement that includes one or more viable naturally-occurring microorganisms" (Krehbiel et al. 2003). Kmet et al. (1993) proposed a definition of ruminal probiotics as "live cultures of microorganisms that are intentionally added into the rumen in order to improve the animal health or nutrition." On the other hand, the term "probiotic" is generic and wide-ranging and could comprise microbial cultures, extracts, as well as enzyme preparations (Elam et al. 2003). Therefore, FDA has itself recommended the use of the term "direct-fed microbials" to depict microbial-based feed supplements that contain live, naturally occurring microorganisms including bacteria, fungi, and yeast (FDA 2003). To date, numerous microorganisms have been exploited as DFM for ruminants. When such live microorganisms are incorporated as a supplement or additive, their details and certification are usually mentioned on the label of the commercial products, as per the guidelines of the US Food and Drug Administration (FDA) to use the term "direct-fed microbial." Majority of such DFM mainly include lactic acid bacteria, probiotics, and yeasts, since these microbes confer several benefits when added to the animal's diet. These benefits may include stimulation of beneficial microbial growth in the rumen, decrease in ruminal acidosis, stabilization of the rumen pH, positive alteration of ruminal fermentation and end-product production, increase in ruminal propionate concentrations, increased plasma hormones, increase in nutrient flow, efficient nutrient digestibility, enhanced immune responses, and increased milk production (Yoon and Stern 1995; Francisco et al. 2002; Krehbiel et al. 2003; Nocek and Kautz 2006; Stein et al. 2006; Aleman et al. 2007). Some of the most common microorganisms frequently exploited as DFMs or MFAs are cited in Fig. 12.1, and general mechanisms of action responsible for the benefits conferred by these DFMs to the ruminants' health are presented in Fig. 12.2.

Microbial Feed Additives/ Direct-fed Microbials

Lactobacilli
L. acidophilus, L. plantarum, L. brevis, L. casei, L. reuteri, L. salivarius, L. bulgaricus, L. cellobiosus, L. carvatus, L. delbrueckii, L. fermentum, L. helveticus, L. lactis

Bifidobacteria
B. longum, B. pseudolongum, B. lactis, B. adolescentis, B. animalis, B. bifidum, B. thermophilium, B. infantis

Propionibacteria
P. shermanii, P. acidipropionici, P. freudenreichii, P. jensenii

Pediococci
P. acidilactici, P. cerevisiae (damnosus), P. pentosaceus

Anaerobic fungi
Neocallimastix sp., *Piromyces* sp., *Sphaeromonas* sp., *Orpinomyces* sp.

Saccharomyces
S. cerevisiae, S. boulardii

Aspergillus
A. niger, A. oryzae

Streptococci
S. bovis, S. faecium, S. thermophilus, S. cremoris, S. diacetylactis, S. intermedius, S. lactis

Bacillus
B. subtilis, B. coagulans, B. licheniformis, B. lentus, B. pumilus

Leuconostoc mesenteroides

Megasphaera elsdenii

Enterococci
E. faecalis, E. faecium

Fig. 12.1 Microorganisms frequently used as microbial feed additives/direct-fed microbials (Yoon and Stern 1995; Walker 2007; Seo et al. 2010)

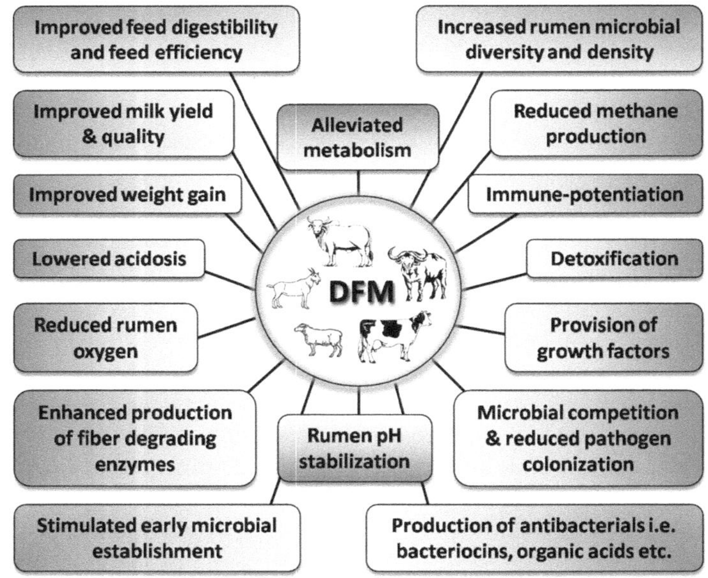

Fig. 12.2 Some benefits associated with different microbial feed additives/direct-fed microbials

12.4 Yeast-Based DFMs

Although the limelight of DFMs has mostly been focused on probiotic bacteria, a limited research has also been conducted on exploring the effects of yeast cultures to the animal diets. Since the majority of these yeast-based products contain either the yeast (*Saccharomyces cerevisiae*) or the aerobic fungus (*Aspergillus oryzae*) or occasionally both in combination, these are usually tagged as fungal DFM, fungal feed additives, or fungal probiotics (Agarwal et al. 2000). Given that the usefulness of these fungi is basically derived from their effects on rumen fermentation, these additives are generally categorized as rumen modifiers or growth promoters (Wallace and Newbold 1992). Though several yeasts and yeast-containing products have long been used in ruminant diets as a source of protein and energy (Eckles and Williams 1925; Carter and Philips 1944), it was since the late 1980s that an increased interest was observed in the yeast-based additives that could enhance the gut functions as those generally conferred by probiotics (Wallace and Newbold 1992). For instance, Linn and Raeth-Knight (2006) recently observed an increase in overall body weight gain and feed efficiency in calves after adding yeast cultures to the animal diet. Martin and Nisbet (1990) observed an about 20 % increase in ammonia production by *A. oryzae* in an in vitro rumen fermentation experiment using a concentrated diet. Newbold et al. (1996) also reported that ammonia concentrations were increased when *S. cerevisiae* was fed to sheep. It has also been proposed that these yeast cultures could stimulate ammonia production by the mixed rumen population, thereby indicating that these may enhance proteolysis. Maybe this increase in ammonia production is possibly caused by the yeasts providing supplementary nutrients to the ruminal microorganisms or maybe due to the endogenous proteolytic activity of yeasts (Martin and Nisbet 1990). Yeasts also provide growth factors, such as malate and vitamins, which encourage the growth of lactate-utilizing bacteria, help in stabilizing the rumen pH, and prevent the risk of acidosis (Rossi et al. 2006). Cellulolytic bacteria in the rumen are particularly stimulated by yeast cultures and improve fiber digestion. Quigley et al. (1992) observed increased ruminal acetate and butyrate and decreased propionate concentrations when calves were given the yeast *Saccharomyces cerevisiae*. Lesmeister et al. (2004) also found an increased daily gain and metabolic index after feeding *S. cerevisiae* to Holstein calves. Recently, Magalhaes et al. (2008) also fed Holstein calves a diet containing *S. cerevisiae* but observed no differences in intake or growth except for a significant decrease in fecal scores. Ruppert et al. (1998) found an increase in average daily gain and feed intake in Holstein calves after feeding a combination of *S. cerevisiae*, *S. faecium*, and *L. acidophilus*. Sretenovic et al. (2008) explored the effects of a commercial preparation containing live yeast cultures of *S. cerevisiae* in combination with probiotic bacteria on Holstein-Friesian cows and reported that the preparation influenced the quantity and composition of the milk and the supplemented group also had a lower somatic cell count which could indicate a better health of the udder. Viable yeast cultures have also been reported to improve the performance with an increase in dry matter intake and milk production in dairy ruminants (Sniffen et al. 2004; Jouany 2006; Stella et al. 2007) and enhance the growth parameters, viz., average daily gain, final weight, feed intake, feed efficiency, etc., in beef cattle (Lesmeister et al. 2004). These improvements have also been correlated to higher total culturable ruminal bacteria (Newbold et al. 1996) and enhanced growth and fiber-degrading activities of cellulolytic microbes (Mosoni et al. 2007), thereby leading to improved fiber digestibility (Guedes et al. 2008; Marden et al. 2008).

Viable yeasts have also been found to stabilize the ruminal pH and reduce the risk of acidosis (Chaucheyras-Durand et al. 2008; Marden et al. 2008). Several studies have observed that viable yeasts influence the balance of lactate-metabolizing bacteria in vitro, possibly by limiting the production of lactic acid by *Streptococci* and by encouraging the uptake of lactic acid by *Megasphaera elsdenii* or *Selenomonas ruminantium* (Chaucheyras et al. 1996; Rossi et al. 2004). Brossard et al. (2006) also found that *S. cerevisiae* could stimulate the populations of ciliated protozoa and prevent pH decrease.

In view of the fact that ruminant methane emission is projected to characterize nearly 5 % of the global warming scenario, mitigating the methane excretion by ruminant livestock is also an escalating concern. In this milieu, probiotics could also be a helpful environmental tool to tackle this global concern (Martin et al. 2006; Newbold and Rode 2006). For instance, addition of specific yeast strains has been reported to improve hydrogen utilization and acetate production by ruminal acetogenic bacteria in vitro (Chaucheyras et al. 1995). Furthermore, since acetate is a source of energy for ruminants and the eructated methane corresponds to a loss of about 5–10 % of gross energy intake, an increase in the proportion of hydrogen utilized for acetate production rather than methane production could also be a fascinating way from an energetic point of view for the animal (Martin et al. 2006). In this perspective, high hydrogen-utilizing bacteria from diverse rumen environments could represent the opportunity to increase the ruminal contribution of alternative reductive acetogenesis (Klieve and Joblin 2007).

However, the overall conclusive results supporting the addition of DFMs to calf diets have been somewhat incoherent and arguable, since some studies reported increased growth performance and intake (Abe et al. 1995; Cruywagen et al. 1995; Ruppert et al. 1998; Lesmeister et al. 2004; Raeth-Knight et al. 2007), while others reported no significant changes (Higginbotham and Bath 1993; Magalhaes et al. 2008). Although increase in the rumen microbial number appears to be the main mechanism of action of the yeasts, however, such effects of yeasts in the rumen may vary based on a number of factors such as animal's age, host species, environmental factors, prevalent rumen microbial ecosystem, composition of diet, etc. Therefore, further studies are awaited to substantiate these effects and aid in envisaging nutritional strategies to extract maximum benefits.

12.5 Anaerobic Ruminal Fungi as Feed Additives

The role of rumen anaerobic fungi in fiber digestion has been studied widely by several researchers (Theodorou et al. 1989; Samanta et al. 2001; Dey et al. 2004; Lee et al. 2004; Paul et al. 2004; Thareja et al. 2006; Dayanand et al. 2007; Tripathi et al. 2007; Saxena et al. 2010; Nagpal et al. 2011). It has been suggested that the rhizoids of their thalli are efficient in penetrating the complex plant tissues than most of the rumen bacteria and protozoa, which could lead to a more rapid degradation of forage entering the rumen (Bauchop and Mountfort 1981; Orpin and Joblin 1988; Nagpal et al. 2008). These fungi could easily and advantageously enter the leaf interiors through stomata in the epidermal layer, thereby degrading the plant fiber and enhancing its utilization by the animal. Species of *Piromyces*, *Neocallimastix*, *Caecomyces*, *Orpinomyces*, and *Anaeromyces* have been reported to degrade the plant fiber to a significant degree and increase the plant-fiber digestibility (Akin et al. 1990; Borneman and Akin 1990; Gordon 1990; Gordon and Phillips 1998; Manikumar et al. 2002, 2003; Dey et al. 2004; Tripathi et al. 2007; Saxena et al. 2010). Various reports have also suggested a positive correlation between anaerobic fungi and voluntary intake of low-digestible herbage diets (Akin and Borneman 1990; Gordon and Phillips 1993). These fungi could also supply proteins to the host animal through the action of proteolytic enzymes and also as a proportion of the microbial proteins synthesized in the rumen that pass to abomasum and intestines for absorption and digestion (Gordon and Phillips 1998). Therefore, increasing the biomass of anaerobic fungi in the rumen is suspected to enhance the supply of high-quality microbial proteins to host ruminant (Gordon and Phillips 1998). Nevertheless, it is now a well-established fact that anaerobic rumen fungi effectively contribute in the hydrolysis of plant biomass in ruminants, owing to their penetration into the plant tissues leading to an enhanced degradation of forage in the rumen. These fungi are well equipped with a wide array of fibrolytic enzymes important for rumen fermentation and hence possess the potential to improve feed intake, body weight, milk output, and overall animal productivity.

12.6 Probiotics as Feed Supplements

The term probiotic is derived by the combination of a Latin preposition "pro" ("for" or "in support") with a Greek noun *bios* ("biotic" or "life") denoting "for life," "in favor of life," or "in support of life." However, ever since the first proposed definition of probiotics as "substances secreted by one microorganism that stimulate another microorganism" by Lilly and Stillwell in 1965, the term has gone through an extensive transformation and variation over the succeeding years. The subsequent chronicles of proposed definitions of the term "probiotics" post-1965 are summarized in Table 12.1. However, this transition was finally concluded till the final and definitive definition laid out by FAO/WHO in 2009 as "live microorganisms which when administered in adequate amounts confer a health benefit on the host." This final definition, though, was recently modified slightly with only a minor grammatical correction as "live microorganisms that, when administered in adequate amounts, confer a health benefit on the host" by Hill et al. (2014) on behalf of the International Scientific Association for Probiotics and Prebiotics consensus statement.

Table 12.1 The chronicles of definition of the term "probiotics"

Proposed definition	Reference
Substances secreted by one microorganism that stimulate another microorganism	Lilly and Stillwell (1965)
Organisms and substances that have a beneficial effect on the host animal by contributing to its intestinal microbial balance	Parker (1974)
A live microbial feed supplement that beneficially affects the host animal by improving its intestinal microbial balance	Fuller (1989)
Live cultures of microorganisms that are intentionally added into the rumen in order to improve the animal health or nutrition (rumen probiotics)	Kmet et al. (1993)
A live microbial feed supplement that improves the intestinal microbial balance of the host animal	Cruywagen et al. (1995)
A live microbial culture of cultured dairy product that beneficially influences the health and nutrition of the host	Salminen (1996)
Living microorganisms that on ingestion in certain numbers exert health benefits beyond inherent basic nutrition	Guarner and Schaafsma (1998)
A preparation of or a product containing viable, defined microorganisms in sufficient numbers that alter the microflora (by implantation or colonization) in a compartment of the host and that exert beneficial health effects in this host	Schrezenmeir and De Vrese (2001)
Natural live organisms either of bacteria or fungal cultures used as feed additives in livestock feeding and in human diets	Todd (2001)
Specific live or inactivated microbial cultures that have documented targets in reducing the risk of human disease or in their nutritional management	Isolauri et al. (2002)
Preparation of viable microorganisms that are consumed by humans or other animals with the aim of inducing beneficial effects by qualitatively or quantitatively influencing their gut microflora and/or modifying their immune status	Fuller (2004)
A preparation or a product containing viable, defined microorganisms in sufficient number that alter the microflora (by implantation or colonization) in a compartment of the host and that exert beneficial health effects on the host	Roselli et al. (2005)
Live microorganisms which when administered in adequate amounts confer a health benefit on the host	FAO/WHO (2009)
Live microorganisms that, when administered in adequate amounts, confer a health benefit on the host	Hill et al. (2014)

Elle Metchnikoff (1908) was the pioneer to propose that consuming specific *Lactobacillus* species can be desirable and may help in preventing intestinal diseases caused by gut pathogens. Notably, the concept of probiotics and their use as feed supplements ignited mainly post World War II as a result of volcanic increase in the use of antibiotics was found to destroy commensal and beneficial gut bacteria and result in increasing cases of acute and chronic diarrhea in humans. However, the extensive use of probiotics for improving the health and performance of ruminants was recognized much lately, mainly in the past couple of decades. So far, probiotics have been exploited to impart several beneficial functions such as preventing ruminants against intestinal disorders, increasing nutrient utilization and feed conversion efficiency, preventing rumen acidosis, enhancing milk production, and increasing weight gain (Fig. 12.2) (Windschitl et al. 1991; Ghorbani et al. 2002; Nocek et al. 2002; Krehbiel et al. 2003). However, these functions may vary from strain to strain as well as from host to host, since for a probiotic to confer a specific effect, there has to be a symbiotic relationship between the probiotic and the host animal in terms of gastrointestinal tract environment of the animal, prevalent conditions for bacterial adherence, growth and reproduction, inhibitory compounds, major target of action, and the efficacy of the probiotic strain. In addition, the microbial symbiosis inside the GI tract is also of particularly importance, since the end effects of probiotics are considered to be mediated not only by their own but also by their influences on other rumen microbes (Van Eys and Den Hartog 2003). In several instances, probiotic bacteria have been observed to cause the rumen microbiota to adapt to the presence of lactic acid within the rumen (Nocek et al. 2002). Some of the most commonly used probiotics comprise, but are not restricted to, *Lactobacilli*, *Bifidobacteria*, *Streptococci*, *Enterococci*, and *Propionibacteria* (Walker 2007) (Fig. 12.1), while the majority of these species has been found to be the most active and effectual in the lower gut of ruminants.

Just as in the case of antimicrobials and growth promoters, the safety of probiotics has been a foremost concern for the consumers and the end users. However, after analysis of the extensive literature, one can instantly and undoubtedly conclude that safety is not a concern with probiotics since these beneficial strains, especially lactic acid bacteria, have been routinely used for many years in the making of numerous food products. Moreover, these probiotic bacteria are a regular part of the normal flora of plants and animals including humans, predominantly in the oral cavity, GI tract, and urinary tract of majority of warm-blooded animals (Fox 1988). In view of the fact that the pathogenicity associated with these probiotics has been very rare, the probiotics are considered as safe to the host animals, to the humans, as well as to the environment. However, it should be noted that not all the strains or species of lactic acid bacteria or other probiotics are equivalent; therefore, while defining probiotics, Fuller (1989) also prudently proposed the following characteristics as a criterion for a good probiotic candidate to be used for any health use, particularly in humans:

- A strain, which is able to exert a beneficial outcome on the host animal, for instance, increased growth or disease resistance
- Able to survive and metabolize in the gut environment, for example, it should be resistant to low pH, bile salts, organic acids, etc.
- Nonpathogenic, nontoxic, nonallergic, and noncarcinogenic
- Present in viable form and, if possible, in large numbers
- Stable and able to remain viable for longer periods under storage and field conditions

In using DFMs, particularly as animal feed additive, specific guidelines have been summarized by the FAO (2002) as follows:

- A DFM has to be viable when administered.
- Documented health benefits of a DFM in the target host must have undergone controlled evaluation.
- Microbe(s) to be used as DFM must be defined taxonomically at genus, species, and strain level.
- DFM must be safe for any proposed use.

The supplementation of probiotic bacterial species in animal feeds has been shown to improve the average daily gain, daily feed intake, rumination efficiency, and feed conversion. For example, Abe et al. (1995) reported a significantly improved body weight gain, feed intake, and feed conversion following the administration of probiotic *Bifidobacterium pseudolongum* or *Lactobacillus acidophilus* to neonatal calves of 7–35 days of age. In addition, they also observed a marked decrease in the occurrence of diarrhea in Holstein calves. Timmerman et al. (2005) also observed that calf-species probiotics could reduce the incidences of diarrhea in veal calves used in their study. Taras et al. (2006) also observed a decrease in the percentage of piglets with post-weaning diarrhea after incorporating *Enterococcus faecium* into the animal feeds. These reports collectively signify that probiotics may reduce the growth or dominance of pathogens that causes diarrhea. Cruywagen et al. (1995) used *L. acidophilus* in milk replacer and found an increase in average daily gain in calves receiving *L. acidophilus*, thereby advocating that adding probiotics to milk replacer may be beneficial in the first few weeks of life.

In ruminants' GI tract, probiotics are mostly selected to target the rumen, the major site of feed digestion. The rumen microbial ecosystem is primarily comprised of an extensive diversity of strictly anaerobic bacteria, anaerobic ruminal fungi, ciliated protozoa, and archaea which accounts for fermentation of more than 70 % of the dietary constituents. However, in young pre-ruminants, probiotics such as lactic acid bacteria (*Lactobacillus* spp., *Bifidobacterium* spp., *Enterococcus* spp., *Propionibacterium* spp.) usually target the small intestine, since the rumen is not yet fully developed, and they could be an interesting approach to stabilize the gut microflora and reduce the risk of pathogen colonization. Viable yeasts distributed from the first days after birth have been found to support microbial colonization and the setup of fermentative capacities in the rumen (Chaucheyras-Durand and Fonty 2002). Also, an improved weight gain and rumen development in young calves has been observed with such microbial additives (Galvao et al. 2005; Adams et al. 2008). Lactate-producing bacteria (*Lactobacillus*, *Enterococcus*) may also characterize a feasible way to limit rumen acidosis in high-concentrate-fed animals (Nocek et al. 2002; Nocek and Kautz 2006). For example, *M. elsdenii* and *Propionibacterium* spp. have been administered as direct-fed microbials to avoid ruminal lactic acidosis (Klieve et al. 2003; Stein et al. 2006). Probiotic strains of *L. acidophilus* have been reported to decrease the population of *E. coli* O157 and *Salmonella enterica* in sheep and cattle feces (Chaucheyras-Durand et al. 2006; Stephens et al. 2007; Tabe et al. 2008). In this context, exploitation of such probiotic strains could be a practical approach to minimize the release of pathogen in the environment, thus limiting the risk of associated foodborne zoonotic infections in humans.

In lactating cattle, Jaquette et al. (1988) reported a significant increase in the milk yield for cows fed a diet containing *L. acidophilus*. In a similar experiment by Gomez-Basauri et al. (2001), cows fed with lactic acid bacteria and oligosaccharide were found to consume less dry matter and more milk yield. Block et al. (2000) observed an increased milk yield when lactating cows were given a combination of *L. acidophilus*, *L. plantarum*, and *S. cerevisiae*. In contrast, Francisco et al. (2002) observed no effect when a strain of *Propionibacterium jensenii* P169 was used for dairy cows on milk production or milk fat and lactose percentage. However, cows fed with P169 exhibited a higher percentage of milk proteins and solids-not-fat and plasma nonesterified fatty acid concentrations. In addition, the body weight and plasma leptin concentrations were found to be higher, indicating that the use of *Propionibacteria* in dairy cattle could modify some characteristics of metabolism during lactation. In another study with same strain P169 in combination with yeast culture, cows had higher milk lactose, milk fat, and solids-not-fat (Stein et al. 2006). Aleman et al. (2007) also reported lower plasma glucose and higher plasma insulin in cows fed with strain P169. Collectively, these reports advocate that DFMs based on bacteria or fungi or their combinations could effectively increase milk production by lactating cows.

Nevertheless, these results may vary depending on the type of microbe, daily concentration, duration of intake period, mode of administration, human handler, physiological and metabolic health of the animal, etc.

12.7 Prebiotics: Favoring Probiotics' Growth

The promising outcomes of prebiotics on gut microflora and metabolic health of humans have encouraged livestock nutritionists to investigate their influence on ruminant as well as non-ruminant herbivores. Recently, there has been a mounting interest in the significance of intestinal microflora due to their considerable involvement in numerous physiological functions such as fiber degradation, nutrient utilization, pathogen prohibition, gastrointestinal development, immunomodulation, etc. Also, the ill concerns of using antibiotics and/or hormones in animal feed have pushed up the need and search for safe and effective alternatives. In this context, the efficacy of prebiotics for positive modulation of the gastrointestinal microbial ecosystem of ruminants as well as non-ruminants has recently been highlighted. It has been observed that several selected prebiotics could confer beneficial alterations in the gut microbiota, decrease rumen ammonia nitrogen and methane emission, and enhance microbial protein synthesis and weight gains in ruminants. It can easily be speculated that prebiotics would soon become a regular part of animal diets as an ecological additive for constructively modulating the GI microflora and animal productivity. Some of the speculated mechanisms of action for benefits associated with probiotics are positive manipulation of gut microflora, immunopotentiation, perked up lipid and mineral metabolism, enhanced nitrogen retention, reduced rumen ammonia, improved fecal consistency, hydrolysis of phytic acid, and some other trophic effects. Prebiotic oligosaccharides are fermented and utilized as source of energy by several rumen bacteria (Samanta et al. 2012), and hence, these could be used as effective alternatives to ameliorate several gut-related maladies of livestock. Several rumen bacteria such as *Butyrivibrio fibrisolvens*, *Eubacterium ruminantium*, and *Ruminococcus albus* have been found to effectively utilize prebiotic oligosaccharides as growth substrate, while the rumen pH remains unaffected (Cota and Whitefield 1998; Santoso et al. 2003; Mwenya et al. 2004). A lower rumen ammonia nitrogen concentration has been observed in animals consuming prebiotics, maybe because of the inhibition of ammonia-producing bacteria or due to the utilization of ammonia for rumen microbial protein synthesis (Mwenya et al. 2004, 2005). Administration of prebiotics to calves has been found to lead to higher nitrogen retention, increased microbial protein synthesis, increased weight gains, and improved feces consistency (Kaufhold et al. 2000; Santoso et al. 2003; Verdonk and Van Leeuwen 2004). The inclusion of prebiotic oligosaccharides into the diet of calves has been shown to reduce the populations of *E. coli* and total anaerobes and increase the numbers of Bifidobacteria (Bunce et al. 1995). Prebiotic oligofructose has also been observed to improve the body weight gain, feed conversion efficiency, and reduce incidence of diarrhea in calves (Mul 1997).

12.8 Future Prospects and Conclusion

Recent researches have confirmed that supplementing ruminants' diets with efficient fiber-degrading microbes has considerable potential to enhance the nutrient utilization and animal nutrition. Such improvements have been attributed to increase in density and diversity of rumen microflora, increase in beneficial microbes and their beneficial by-products, reduction in pathogens, maintenance of favorable rumen pH, increase in ruminal fibrolytic enzymes, and improvement in ruminal fiber digestion and nutrient utilization, thereby resulting in increased feed digestibility and animal productivity. Several microbial formulations are now commercially accessible, and this catalog of microbial feed additives is projected to grow further in the coming years as a consequence of rising alarms about the use of antibiotics and growth promoters in livestock production (Table 12.2). With the advent of new

Table 12.2 Various commercial feed additives available worldwide

Product name	Product details	Manufacturing co./details
Enviva® PRO	A *Bacillus*-based solution comprises of a multi-strain, highly stable solution for poultry, offering superior nutrient absorption and healthy benefits. Based on unique customer-specific inhibitor (CSI) technology, it can be used in diets with or without antibiotics	Danisco Animal Nutrition, PO Box 777 Marlborough, Wiltshire, SN8 1XN, United Kingdom http://animalnutrition.dupont.com/productsservices/feed-probiotic-solutions/
Enviva® MPI	A lactobacillus-based fermented milk product based on a combination of lactobacillus probiotics with a proven mode of action to support piglet gut health and protect the growing pig against common production diseases. This product is available only in certain countries in the EU	
Agmaster®	A silage inoculant-based solution offering probiotic silage inoculants that improve the ensiling of grass, corn, alfalfa haylage, small grains, and high-moisture grains, helping to resolve waste management and pollution issues	
Syncra® AVI	Support broiler nutrition and unlock the full potential of feed with our unique probiotic and enzyme combination	
Jacoby's Synergy	Synergy is the next generation in feed additives for ruminants. It contains a combination of natural plant extracts; live, host-specific strains of beneficial bacteria; active live yeast; *Aspergillus oryzae*; and bypass protein. This product addresses the animal's entire digestive and circulatory systems through an all-natural means that results in increased growth, milk yield, and more consistent feed intake. Synergy improves metabolic efficiency to create a healthier animal	Jacoby Feed and Seeds Ltd. 103 E Mesquite Melvin, Texas 76858 http://jacobyfeed.com/wp-content/uploads/2013/09/SYNERGY.pdf
Levucell® SC	It is rumen-specific live yeast and active dry yeast for use as a direct-fed microbial in ruminant feeds	Lallemand Inc. 1620 rue Prefontaine, Montreal Quebec H1W 2N8, Canada http://www.lallemand.com/our-business/animal-nutrition/products
Levucell® SB	This is active dry yeast for monogastrics and young ruminants	
Bactocell®	This is a lactic acid-producing bacteria *Pediococcus acidilactici* MA18/5M selected specifically for monogastric feeding programs	
MicroCell®	These microbial feed additives are freeze-dried bacteria for use in ruminant feeds	
BioStart LactoPlus	It is a prebiotic digestive enhancer for dairy cows and is powered by Signal Molecule Technology. It contains a concentrated blend of enzymes and secondary metabolites from the fermentation of *Lactobacillus acidophilus*, *Bifidobacterium*, *Streptococcus thermophilus*, and *Propionibacteria*. It aids in maintaining efficient rumen function through periods of stress such as feed changes, mating, and environmental stress	216 Lake Road, Takapuna, Auckland 0622, New Zealand http://www.bio-start.co.nz/products/animal-probiotics/ruminant/
ProP169®	This is a new technology containing 60 billion colony-forming units of *Propionibacterium freudenreichii* strain P169 per 2 g feeding	Bio-Vet Home Office 300 Ernie Drive, Barneveld, WI 53507 http://www.bio-vet.com/AdultCattledfm.htm
Generator™ ELITE Direct Fed Microbial	A microbial supplement for ruminants. Generator™ ELITE provides high levels of microorganisms, including rumen/intestinal origin bacteria, live yeast, digestive enzyme units, and yeast cell walls	
Generator™ Pro-Sacc™ Direct Fed Microbial	Generator™ Pro-Sacc™ now contains Bio-Vet's proprietary Tri-Pro™ (P5-P63-P169) blend of rumen *Propionibacteria*	
Generator™ B Direct Fed Microbial	Generator™ B provides high levels of microorganisms, including rumen/intestinal origin bacteria. Designed for finishing beef rations, especially grain-fed beef. Generator™ B contains Bio-Vet's proprietary Tri-Pro™ (P5-P63-P169) blend of rumen *Propionibacteria*	

tools and technologies, studies on genetic modification of rumen microorganisms and use of genetically engineered microbes for better establishment and fibrolytic enzyme production in the rumen are also appearing. With the importance and potential of improved animal performance by using microbial feed additives/ DFMs, these products are undoubtedly going to play an important role in the future of livestock nutrition and production. However, there have been extensive discrepancies at the level of animal performance in response to various DFMs, and the mechanisms of action underlying these responses remain to be explicated comprehensively. Thus, additional insights into these mechanisms would diminish the inconsistencies in the animal performance responses and would certainly prompt further improvements in the development of new avenues of more efficient and reproducible microbial feed formulations for maximum health effects and production performance of the ruminants.

Conflict of Interest's Statement The first author is postdoc fellow at Juntendo Laboratory. This manuscript, however, does not represent any scientific or monetary viewpoint of Juntendo. The author declares no competing financial interests.

References

Abe F, Ishibashi N, Shimamura S (1995) Effect of administration of Bifidobacteria and lactic acid bacteria to newborn calves and piglets. J Diary Sci 78(912):2838–2846

Adams MC, Luo J, Rayward D et al (2008) Selection of a novel direct-fed microbial to enhance weight gain in intensively reared calves. Anim Feed Sci Technol 145:41–52

Agarwal N, Kamra DN, Chaudhary LC et al (2000) Selection of *Saccharomyces cerevisiae* strains for use as a microbial feed additive. Lett Appl Microbiol 31:270–273

Akin DE, Borneman WS (1990) Role of rumen fungi in fiber degradation. J Dairy Sci 73:3023–3032

Akin DE, Rigsby LL, Lyon CE et al (1990) Relationship of tissue digestion to textural strength in Bermuda grass and alfalfa stems. Crop Sci 30:990–993

Aleman MM, Stein DR, Allen DT et al (2007) Effect of feeding two levels of Propionibacteria to dairy cows on plasma hormones and metabolites. J Dairy Res 74:146–153

Bauchop T, Mountfort DO (1981) Cellulose fermentation by a rumen anaerobic fungus in both the absence and presence of rumen methanogens. Appl Environ Microbiol 42:1103–1110

Block E, Nocek JE, Kautz WP et al (2000) Direct fed microbial and anionic salt supplementation to dairy cows fed 21 days pre- to 70 days post-partum. J Anim Sci 78(1):304

Borneman WS, Akin DE (1990) Lignocellulose degradation by rumen fungi and bacteria: ultrastructure and cell wall degrading enzymes. In: Akin DE, Ljungdahl LG, Wilson JR, Harris PJ (eds) Microbial and plant opportunities to improve lignocellulose utilization by ruminants. Elsevier, New York, pp 325–339

Brossard L, Chaucheyras-Durand F, Michalet-Doreau B et al (2006) Dose effect of live yeasts on rumen microbial communities and fermentations during butyric latent acidosis in sheep: new type of interaction. Anim Sci 82:829–836

Bunce TJ, Howard MD, Kerley MS et al (1995) Protective effect of fructooligosaccharides in prevention of mortality and morbidity from infectious *E. coli* K88. J Anim Sci 73(1):69

Carter HE, Phillips GE (1944) The nutritive value of yeast proteins. Fed Proc 3:123–128

Chaucheyras F, Fonty G, Bertin G et al (1995) *In-vitro* H_2 utilization by a ruminal acetogenic bacterium cultivated alone or in association with an archaea methanogen is stimulated by a probiotic strain of *Saccharomyces cerevisiae*. Appl Environ Microbiol 61:3466–3467

Chaucheyras F, Fonty G, Bertin G et al (1996) Effects of a strain of *Saccharomyces cerevisiae* (Levucell SC), a microbial additive for ruminants, on lactate metabolism *in vitro*. Can J Microbiol 42:927–933

Chaucheyras-Durand F, Fonty G (2002) Influence of a probiotic yeast (*Saccharomyces cerevisiae* CNCM I-1077) on microbial colonization and fermentation in the rumen of newborn lambs. Microb Ecol Health Dis 14:30–36

Chaucheyras-Durand F, Madic J, Doudin F et al (2006) Biotic and abiotic factors influencing *in vitro* growth of *E. coli* O157:H7 in ruminant digestive contents. Appl Environ Microbiol 72:4136–4142

Chaucheyras-Durand F, Walker ND, Bach A (2008) Effects of active dry yeasts on the rumen microbial ecosystem: past, present and future. Anim Feed Sci Technol 145:5–26

Cota MA, Whitefield TR (1998) Xylooligosaccharides utilization by ruminal anaerobic bacterium *Selenomonas ruminantium*. Curr Microbiol 36:183–189

Cruywagen CW, Jordan I, Venter L (1995) Effect of *Lactobacillus acidophilus* supplementation of milk replacer on preweaning performance of calves. J Dairy Sci 79:483–486

Dayanand TL, Nagpal R, Puniya AK et al (2007) *In-vitro* degradation of urea-NH_3 treated wheat straw using anaerobic ruminal fungi. J Anim Feed Sci 16:484–489

Dey A, Sehgal JP, Puniya AK et al (2004) Influence of an anaerobic fungal culture (*Orpinomyces* sp.) administration on growth rate, ruminal fermentation and nutrient digestion in calves. Asian-Aust J Anim Sci 17:733–884

Eckles CH, Williams VM (1925) Yeast as a supplementary feed for lactating cows. J Dairy Sci 8:89–93

Elam NA, Gleghorn JF, Rivera JD et al (2003) Effects of live culture of *Lactobacillus acidophilus* (strains NP45 and NP51) and *Propionibacterium freudenreichii* on performance, carcass and intestinal characteristics, and *Escherichia coli* strain O157 shedding of finishing beef steers. J Anim Sci 81:2686–2698

FAO (Food and Agriculture Organization of the United Nations) (2002) Guidelines for the evaluation of probiotics in food. ftp://ftp.fao.org/es/esn/food/wgreport2.pdf

FAO/WHO (2009) Guidelines for the evaluation of probiotics in foods. Available at: ftp://ftp.fao.org/es/esn/food/wgreport2.pdf

FDA (1995) Direct-fed microbial products. Compliancy Policy Guide, Silver Spring

FDA (2003) Office of Regulatory Affairs: compliance policy guides. Sec. 689.100. Direct-Fed Microbial Products (CPG 7126:41). Available at: http://www.fda.gov/ora/compliance_ref/cpg/cpgvet/cpg689-100.html

Fox SM (1988) Probiotics intestinal inoculants for production animals. Vet Med 83:806–830

Francisco CC, Chamberlain CS, Waldner DN et al (2002) Propionibacteria fed to dairy cows: effects on energy balance, plasma metabolites and hormones, and reproduction. J Dairy Sci 85:1738–1751

Fuller R (1989) Probiotics in man and animals. J Appl Bacteriol 66:365–378

Fuller R (2004) What is a probiotic? Biologist 51:232

Galvao KN, Santos JE, Coscioni A et al (2005) Effect of feeding live yeast products to calves with failure of passive transfer on performance and patterns of antibiotic resistance in fecal *Escherichia coli*. Reprod Nutr Develop 45:427–440

Ghorbani GR, Morgavi DP, Beauchemin KA et al (2002) Effects of bacterial direct-fed microbials on ruminal fermentation, blood variables, and the microbial populations of feedlot cattle. J Anim Sci 80:1977–1985

Gomez-Basauri J, de Ondarza MB, Siciliano-Jones J (2001) Intake and milk production of dairy cows fed lactic acid bacteria and mannanoligosaccharide. J Dairy Sci 84(1):283

Gordon GLR (1990) Selection of anaerobic fungi for better fibre degradation in the rumen. In: Akin DE, Ljungdahl LG, Wilson JR, Harris PJ (eds) Microbial and plant opportunities to improve lignocellulose utilization by ruminants. Elsevier, New York, pp 301–309

Gordon GLR, Phillips MW (1993) Removal of anaerobic fungi from the rumen of sheep by chemical treatment and the effect on feed consumption and *in vivo* fibre digestion. Lett Appl Microbiol 17:220–223

Gordon GLR, Phillips MW (1998) The role of anaerobic gut fungi in ruminants. Nutr Res Rev 11:1–36

Guarner F, Schaafsma GJ (1998) Probiotics. Int J Food Microbiol 39:237–238

Guedes CM, Gonçalves D, Rodrigues MAM et al (2008) Effects of a *Saccharomyces cerevisiae* yeast on ruminal fermentation and fibre degradation of maize silages in cows. Anim Feed Sci Technol 145:27–40

Higginbotham GE, Bath DL (1993) Evaluation of *Lactobacillus* fermentation cultures in calf feeding systems. J Dairy Sci 76:615–620

Hill C, Guarner F, Reid G et al (2014) Expert consensus document: the International Scientific Association for Probiotics and Prebiotics consensus statement on the scope and appropriate use of the term probiotic. Nat Rev Gastroenterol Hepatol 11(8):506–514. doi:10.1038/nrgastro.2014.66

Hong HA, Hong Duc L, Cutting S (2005) The use of bacterial spore formers as probiotics. FEMS Microbiol Rev 29:813–835

Hungate ER (1988) The ruminant and the rumen. In: Hobson PN (ed) The rumen microbial ecosystem. Elsevier Applied Sciences, London, pp 1–19

Isolauri E, Rautava S, Kalliomaki M (2002) Role of probiotics in food hypersensitivity. Curr Opin Allergy Clin Immunol 2:263–271

Jaquette RD, Dennis RJ, Coalson JA et al (1988) Effect of feeding viable *Lactobacillus acidophilus* (BT1386) on performance of lactating dairy cows. J Dairy Sci 71(1):219

Jouany JP (2006) Optimizing rumen functions in the close-up transition period and early lactation to dry matter intake and energy balance in cows. Anim Reprod Sci 96:250–264

Kaufhold JN, Hammon HM, Blum JW (2000) Fructooligosaccharides supplementation effects on metabolic, endocrine and hematological traits in veal calves. J Vet Med A 47:17–29

Klieve A, Joblin KJ (2007) Comparison in hydrogen utilization of ruminal and marsupial reductive acetogens. In: 5-year science progress report 2002–2007. New Zealand Pastoral Greenhouse Gas Research Consortium, Wellington, pp 34–35

Klieve AV, Hennessy D, Ouwerkerk D et al (2003) Establishing populations of *Megasphaera elsdenii* YE 34 and *Butyrivibrio fibrisolvens* YE 44 in the rumen of cattle fed high grain diets. J Appl Microbiol 95:621–630

Kmet V, Flint HJ, Wallace RJ (1993) Probiotics and manipulation of rumen development and function. Arch Anim Nutr 44:1–10

Krehbiel CR, Rust SR, Zhang G et al (2003) Bacterial direct-fed microbials in ruminant diets: performance response and mode of action. J Anim Sci 81:E120–E132

Lee SS, Choi CK, Ahn BH et al (2004) *In vitro* stimulation of rumen microbial fermentation by a rumen anaerobic fungal culture. Anim Feed Sci Technol 115:215–226

Lesmeister KE, Heinrichs AJ, Gabler MT (2004) Effects of supplemental yeast (*Saccharomyces cerevisiae*) culture on rumen development, growth characteristics, and blood parameters in neonatal dairy calves. J Dairy Sci 87:1832–1839

Lilly DM, Stillwell RH (1965) Probiotics: growth promoting factors produced by microorganisms. Science 147:747–748

Linn J, Raeth-Knight M (2006) Yeast in dairy cattle diets. 2006 4-state dairy nutrition and management conference, pp 85–90. Ames, IA, Iowa State University, MWPS

Mackie RI, Gilchrist FMC, Heath S (1984) *In vivo* study of ruminal microorganisms influencing lactate turnover and its contribution to volatile fatty acid production. J Agric Sci 103:37

Magalhaes VJA, Susca F, Lima FS et al (2008) Effect of feeding yeast culture on performance, health, and immunocompetence of dairy calves. J Dairy Sci 91:1497–1509

Manikumar B, Puniya AK, Singh K (2002) Influence of ruminal fungi on degradation of wheat straw *in vitro*. Indian J Microbiol 42:133–136

Manikumar B, Puniya AK, Singh K (2003) Effect of ruminal fungi on *in vitro* degradation of rice straw. Indian J Anim Sci 73:312–314

Marden JP, Julien C, Monteils V et al (2008) How does live yeast differ from sodium bicarbonate to stabilize ruminal pH in high-yielding dairy cows? J Dairy Sci 91:3528–3535

Martin SA, Nisbet DJ (1990) Effects of *Aspergillus oryzae* extract on fermentation of amino acids, bermuda grass and starch by mixed ruminal microorganisms *in vitro*. J Anim Sci 68:2142–2149

Martin C, Morgavi D, Doreau M et al (2006) Comment réduire la production de méthane chez les ruminants? Forages 187:283–300

Mcallister TA, Cheng KJ (1996) Microbial strategies in the ruminal digestion of cereal grain. Anim Feed Sci Technol 62:29–36

Metchnikoff E (1908) The prolongation of life, 1st edn. G.P. Putnam's Sons, New York

Mosoni P, Chaucheyras-Durand F, Bera-Maillet C et al (2007) Quantification by real-time PCR of cellulolytic bacteria in the rumen of sheep after supplementation of a forage diet with readily fermentable carbohydrates: effect of a yeast additive. J Appl Microbiol 103:2676–2685

Mul AJ (1997) Application of oligofructose in animal feeds. In: Proceedings of international symposium non-digestible oligosaccharides: healthy food for colon? Wageningen, p 106

Mwenya B, Santoso B, Sar C et al (2004) Effects of including beta 1,4- galactooligosaccharides, lactic acid bacteria or yeasts culture on methanogenesis as well as energy and nitrogen metabolism in sheep. Anim Feed Sci Technol 115:313–326

Mwenya B, Sar C, Santoso B et al (2005) Comparing the effects of beta 1,4-galactooligosaccharides, and L-cysteine to monensin on energy and nitrogen utilization in steers fed a very high concentrate diets. Anim Feed Sci Technol 118:19–30

Nagpal R, Puniya AK, Griffith G et al (2008) Anaerobic rumen fungi: potential and applications. In: Khachatourians G, Arora DK, Rajendran TP, Srivastava AK (eds) Agriculturally important microorganisms: an international multi-volume annual review series, vol II. Academic World International, Bhopal, pp 375–393

Nagpal R, Puniya AK, Sehgal JP et al (2011) *In-vitro* fibrolytic potential of anaerobic rumen fungi from ruminants and non-ruminant herbivores. Mycoscience 52:31–38

Newbold CJ, Rode LM (2006) Dietary additives to control methanogenesis in the rumen. Int Congr Ser 1293:138–147

Newbold CJ, Wallace RJ, McIntosh FM (1996) Mode of action of the yeast *Saccharomyces cerevisiae* as a feed additive for ruminants. Brit J Nutr 76:249–261

Nocek JE, Kautz WP (2006) Direct-fed microbial supplementation on ruminal digestion, health, and performance of pre- and postpartum dairy cattle. J Dairy Sci 89:260–266

Nocek JE, Kautz WP, Leedle JA et al (2002) Ruminal supplementation of direct-fed microbials on diurnal pH variation and *in situ* digestion in dairy cattle. J Dairy Sci 85:429–433

Orpin CG, Joblin KN (1988) The rumen anaerobic fungi. In: Hobson PN (ed) The rumen microbial ecosystem. Elsevier Applied Science, London, pp 129–150

Parker RB (1974) Probiotics: the other half of the antibiotic story. Anim Nutr Health 29:4–8

Paul SS, Kamra DN, Sastry VRB et al (2004) Effect of administration of an anaerobic gut fungus isolated from wild blue bull to buffaloes on *in vivo* ruminal fermentation and digestion of nutrients. Anim Feed Sci Technol 115:143–157

Quigley JD, Wallis LB, Dowlen HH et al (1992) Sodium bicarbonate and yeast culture effects on ruminal fermentation, growth, and intake in dairy calves. J Dairy Sci 75:3531–3538

Raeth-Knight ML, Linn JG, Jung HG (2007) Effect of direct-fed microbials on performance, diet digestibility, and rumen characteristics of Holstein dairy cows. J Dairy Sci 90:1802–1809

Roselli M, Finamore A, Britti M et al (2005) Alternatives to in-feed antibiotics in pig production: evaluation of probiotics, zinc oxide or organic acids as protective agents for the intestinal mucosa. A comparison of *in vitro* and *in vivo* results. Anim Res 54:203–218

Rossi F, Luccia AD, Vincenti D et al (2004) Effects of peptidic fractions from *Saccharomyces cerevisiae* culture on growth and metabolism of the ruminal bacteria *Megasphaera elsdenii*. Anim Res 53:177–186

Rossi C, Sgoifo A, Dell-Orto V et al (2006) Effects of live yeast in beef cattle studied. Feedstuffs 16:11

Ruppert LD, Gene CM, Hutjens MF (1998) Feeding of probiotics to calves. Illini Dairy Net 06

Salminen S (1996) Uniqueness of probiotic strains. Int Dairy Feed Sand Nutr Newsl 5:16–18

Samanta AK, Walli TK, Singh KK (2001) Role of different groups of microbes on fibre utilization. Ind J Anim Sci 71:497–498

Samanta AK, Senani S, Kolte AP et al (2012) Production and *in vitro* evaluation of xylooligosaccharides generated from corn cobs. Food Bioprod Process 90:466–474

Santoso B, Kume S, Nonaka K et al (2003) Influence of beta galactooligosaccharide supplementation on nitrogen utilization, rumen fermentation, and microbial nitrogen supply in dairy cows fed silage. Asian-Aust J Anim Sci 26:1137–1142

Saxena S, Sehgal JP, Puniya AK et al (2010) Effect of administration of rumen fungi on production performance of lactating buffaloes. Benef Microbes 1(2):183–188

Schrezenmeir J, de Vrese M (2001) Probiotics, prebiotics, and synbiotics: approaching a definition. Am J Clin Nutr 73(2):361S–364S

Seo JK, Seon-Woo K, Kim MH et al (2010) Direct-fed microbials for ruminant animals. Asian-Aust J Anim Sci 23:1657–1667

Sniffen CJ, Chaucheyras-Durand F, De Ondarza MB et al (2004) Predicting the impact of a live yeast strain on rumen kinetics and ration formulation. In: Proceedings of the 19th annual southwest nutrition and management conference, Tempe, pp 53–59, 24–25 Feb 2004

Sretenovic L, Petrovic MP, Aleksic S et al (2008) Influence of yeast, probiotic and enzymes rations on dairy cows. Biotechnol Anim Husb 24:33–43

Stein DR, Allen DT, Perry EB et al (2006) Effects of feeding Propionibacteria to dairy cows on milk yield, milk components and reproduction. J Dairy Sci 89:111–125

Stella AV, Paratte R, Valnegri L et al (2007) Effect of administration of live *Saccharomyces cerevisiae* on milk production, milk composition, blood metabolites, and fecal flora in early lactating dairy goats. Small Rum Res 67:7–13

Stephens TP, Loneragan GH, Karunasena E et al (2007) Reduction of *Escherichia coli* O157 and *Salmonella* in feces and on hides of feedlot cattle using various doses of a direct-fed microbial. J Food Prot 70:1346–1349

Tabe ES, Oloya J, Doetkott DK et al (2008) Comparative effect of direct-fed microbials on fecal shedding of *Escherichia coli* O157:H7 and *Salmonella* in naturally infected feedlot cattle. J Food Prot 71:539–544

Taras D, Vahjen W, Macha M et al (2006) Performance, diarrhea incidence, and occurrence of *Escherichia coli* virulence genes during long-term administration of a probiotic *Enterococcus faecium* strain to sows and piglets. J Anim Sci 84:608–617

Thareja A, Puniya AK, Goel G et al (2006) *In vitro* degradation of wheat straw by anaerobic fungi from small ruminants. Arch Anim Nutr 60:412–417

Theodorou MK, Longland AC, Dhanoa MS et al (1989) Growth of *Neocallimastix* sp strain R1 on Italian ryegrass hay – removal of neutral sugars from plant-cell walls. Appl Environ Microbiol 55:1363–1367

Timmerman HM, Mulder L, Everts H et al (2005) Health and growth of veal calves fed milk replacers with or without probiotics. J Dairy Sci 88:2154–2165

Todd RK (2001) The probiotic concept. In: Michael PD, Larry RB, Thomas JM (eds) Food microbiology, fundamental and frontiers, vol 2. American Society of Microbiology (ASM) Press, Washington, DC, pp 128–139

Tripathi VK, Sehgal JP, Puniya AK et al (2007) Effect of administration of anaerobic fungi isolated from cattle and wild blue bull (*Boselaphus tragocamelus*) on growth rate and fibre utilization in buffalo calves. Arch Anim Nutr 61:416–423

Van Eys J, den Hartog L (2003) Separation of health performance roles of probiotics may lead to understanding of mode of action. Feedstuffs 78:24–29

Van Soest PJ (1994) Nutritional ecology of the ruminant, 2nd edn. Comstock, Ithaca

Verdonk JMAJ, Van Leeuwen P (2004) The application of inulin type fructans in diets for veal calves and broilers. In: Inulin and oligofructose feed good factors for health and wellbeing. 4th ORAFTI research conference, Paris

Walker ND (2007) DFMs: fact, fiction, and future (part 1). Feedstuffs 11 Jun: 12

Wallace RJ, Newbold CJ (1992) Probiotics for ruminants. In: Fuller R (ed) Probiotics: the scientific basis. Chapman and Hall, London, pp 317–353

Weimer PJ (1998) Manipulating ruminal fermentation: a microbial ecological perspective. J Anim Sci 76:3114–3122

Windschitl PM, Randall KM, Brainard DJ (1991) Growth performance of Holstein dairy calves supplemented with a probiotics. Agricultural and Forestry Experiment Station School of Agriculture and Land Resources Management, Apr: 22

Yoon IK, Stern MD (1995) Influence of direct-fed microbials on ruminal microbial fermentation and performance of ruminants: a review. Asian-Aust J Anim Sci 8:533–555

13 Utilization of Organic Acids to Manipulate Ruminal Fermentation and Improve Ruminant Productivity

M.D. Carro and E.M. Ungerfeld

Abstract

Much research has been conducted to develop feed additives that can improve the efficiency of ruminal fermentation and increase the productivity of ruminants. Ionophore antibiotics have been banned in some countries, and organic acids (OA) have been proposed as an alternative. Most research on the use of OA in ruminant feeding has focused on malate and fumarate, which are intermediates in the tricarboxylic acid cycle and in the randomizing pathway of propionate formation in the rumen. The effects of OA addition on ruminal fermentation and animal performance can be based on more than one mechanism of action. The simplest effect of OA can be thought as the supplementation with a rapidly fermentable substrate, but calculations from in vitro rate of disappearance of fumarate and malate suggest that a considerable proportion of these OA may outflow the rumen unfermented. Stabilization of ruminal pH through a stimulation of lactate utilization by *Selenomonas ruminantium* by malate has been proposed as a beneficial effect of this OA. Ruminal pH stabilization as a result of malate addition has been confirmed in most batch and continuous mixed-culture experiments and in vivo. It has also been proposed that OA can decrease methane production in the rumen by competing for metabolic hydrogen. In most batch culture experiments, the decrease in methane production caused by OA has been small, which was explained by simultaneous release of metabolic hydrogen in the conversion of part of

M.D. Carro (✉)
Departamento de Producción Agraria,
E.T.S.I. Agrónomos, Universidad Politécnica de
Madrid, Ciudad Universitaria, 28040 Madrid, Spain
e-mail: mariadolores.carro@upm.es

E.M. Ungerfeld
Instituto de Investigaciones Agropecuarias (INIA)
Carillanca, Camino Cajón km 10, Temuco, Chile

A.K. Puniya et al. (eds.), *Rumen Microbiology: From Evolution to Revolution*, DOI 10.1007/978-81-322-2401-3_13

added OA to acetate, although substantial decreases were achieved using a slow-release form of fumarate. In vivo effects of OA on methane production have been variable and difficult to explain. With regard to animal performance, OA have sometimes been shown to increase daily gain and feed efficiency in beef cattle and lambs and to enhance milk production in dairy cows, but no effects have been observed in other studies. Inconsistency in productive responses has been attributed to the dose of OA, characteristics of the diet (i.e., forage-to-concentrate ratio, forage type, cereal grains, etc.), chemical form in which OA were fed (i.e., free acid, salts, or encapsulated acids), and characteristics of the experimental animals (physiological state, level of production, ruminal populations, etc.). The use of malate and fumarate in animal feeding is considered safe for the animal, the consumer, and the environment, but the main limitation to their use is currently economical. Future research should aim to identify the conditions in which OA have optimal efficacy.

Keywords
Ruminal fermentation • Ruminant • Organic acids • Carboxylic acids • Malate • Fumarate

13.1 Introduction

The major objective in manipulating ruminal fermentation is to increase the efficiency of fermentation, by maximizing processes that produce metabolites that are nutritionally useful to the host animal as energy or precursors for the synthesis of other compounds and by minimizing the processes that result in energy or nitrogen losses or are environmentally damaging (e.g., methane emissions and nitrogen voided to the environment in urine). Much research effort has been dedicated to the development of feed additives that could improve ruminal fermentation and consequently increase productivity and health of domesticated ruminants and ameliorate their environmental impact. One of the commonly used feed additives are ionophore antibiotics, whose efficiency in improving animal health and performance and reducing environmental pollution is well documented (Nagaraja et al. 1997; Callaway et al. 2003; Guan et al. 2006). However, the concern of consumers about antibiotic use in animal feeding has markedly increased, as there is apprehension about the possibility of emergence and spread of microorganisms resistant to antibiotics as a consequence of their widespread use in production (EFSA 2014a). As a consequence, many countries (including those belonging to the EU) have banned the use of antibiotics as feed additives in animal feeding. In this context, organic acids have been regarded as an alternative to ionophore antibiotics in ruminant feeding (Castillo et al. 2004).

Most research on the use of organic acids in ruminant feeding has focused on malate and fumarate, both of which are naturally occurring in biological tissues. Malate and fumarate are intermediates in the tricarboxylic acid cycle, thus being natural components of plants and animals that are consumed daily by humans. In ruminants, carbohydrates are degraded in the rumen resulting in the production of pyruvate, which is further metabolized by rumen microorganisms to produce volatile fatty acids (VFA). Malate and fumarate are intermediates of the randomizing pathway of propionate production (Fig. 13.1). Propionate is absorbed from the rumen and largely transported to the liver, where it becomes the main source of glucose for the ruminant.

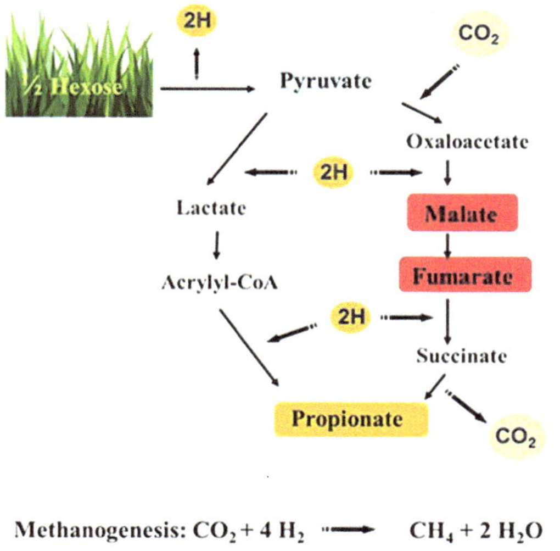

Methanogenesis: $CO_2 + 4\ H_2 \longrightarrow CH_4 + 2\ H_2O$

Fig. 13.1 Pathways of propionate and methane formation in the rumen

When used as feed additives, organic acids may be administered to ruminants as such acids, but its management is problematic as they are corrosive substances and can lower ruminal pH (Asanuma et al. 1999; Liu et al. 2009). Therefore, an alternative is the use of salts (especially sodium salts), which are usually more expensive than free acids, but have the advantages of being nonabrasive and safer to manipulate.

Although several excellent reviews have been previously published on the manipulation of ruminal fermentation by organic acids (Martin 1998; Castillo et al. 2004; Khampa and Wanapat 2007), the purpose of this chapter is to further examine their possible modes of action and to provide an updated overview of published studies on the potential of organic acids as feed additives in practical ruminant feeding.

13.2 Mechanisms of Action of Organic Acids on Ruminal Fermentation

In principle, the effects of organic acid addition on ruminal fermentation and animal performance can be based on more than one mechanism of action.

13.2.1 Energy Density

The simplest effect of organic acids might be thought as the supplementation with a rapidly fermentable substrate that is, in theory, 100 % utilizable, as it does not need to be digested The heats of combustion of malic and fumaric acids are 9.89 and 11.5 kJ/g, respectively. As a comparison, glucose's heat of combustion is 15.6 kJ/g (Domalski 1972), but most glucose-forming part of cellulose in animal feeds is typically not 100 % digested. Therefore, in theory malate and fumarate would yield more digestible energy than glucose polymerized in cellulose that is less than 63 % and 74 % digested, respectively (calculations not shown). Moreover, because a greater proportion of both malate and fumarate are metabolized to propionate and less to acetate compared to glucose (see below), more of their digestible energy is conserved as metabolizable energy, as energy losses as methane (CH_4) are smaller. Also, an increase in propionate supply could be hypothesized to benefit performance of ruminants with high requirements for glucose, like high-producing dairy cows.

However, even though most added malate (Russell and Van Soest 1984; Callaway and Martin 1997; Asanuma and Hino 2000) and fumarate (Callaway and Martin 1997; Asanuma et al. 1999) are metabolized by ruminal mixed cultures within the first 12 h of incubation, if an in vivo ruminal liquid passage rate of 0.129 is assumed (Seo et al. 2007), it can be estimated from the in vitro results reported by Callaway and Martin (1997) and Asanuma et al. (1999) that only between 60 % and 75 % of added fumarate and 35–56 % of added malate might be metabolized before they pass out of the rumen (calculations not shown). This casts doubt on whether organic acids could be considered as energy supplements solely on the basis of their fermentability, except with diets with very low digestibility. The rate of utilization of organic acids has been shown to be stimulated by cellobiose and further stimulated by monensin (Callaway and Martin 1997).

Addition of organic acids to ruminal mixed cultures increased VFA production in many in vitro studies (Carro and Ranilla 2003a, b; Giraldo et al. 2007; Yu et al. 2010). Both malate and fumarate are intermediates in the randomizing pathway of propionate production (Fig. 13.1), which is used by the ruminal bacteria *Selenomonas ruminantium* to produce succinate and propionate (Martin 1998). Therefore, fumarate and malate can decrease the acetate-to-propionate ratio and increase the supply of propionate to the host animal (Mao et al. 2007; Li et al. 2009b; Yu et al. 2010). Much of the added malate and fumarate is indeed metabolized to propionate in vitro (Martin and Streeter 1995; Asanuma et al. 1999), although succinate accumulated at fumarate initial concentration of 30 mM, confirming that fumarate was metabolized mostly through propionate's randomizing pathway (Asanuma et al. 1999).

13.2.2 Control of Lactic Acidosis

Extensive research has been conducted on the effect of organic acids, and in particular malate, on stabilizing ruminal pH through a stimulation of lactate utilization (Martin and Streeter 1995; Callaway and Martin 1996; Martin 2004). *S. ruminantium* is an important lactate utilizer in the rumen. *S. ruminantium* utilization of lactate was strongly stimulated by aspartate and fumarate and especially by malate (Nisbet and Martin 1990; Martin et al. 2000). Addition of malate allowed *S. ruminantium* to grow on lactate at low pH (Evans and Martin 1997), and for some strains of *S. ruminantium*, organic acids are a requirement to grow on lactate (Evans and Martin 1997). It has thus been hypothesized that the administration of fumarate and malate to ruminants could reduce the accumulation of lactic acid in the rumen and thus prevent large drops in pH that cause acidosis.

Aspartate, fumarate, and malate stimulated lactate uptake by *S. ruminantium*, but these organic acids did not promote growth of *S. ruminantium* in the absence of lactate (Nisbet and Martin 1990). Extracts of *Aspergillus oryzae* (Nisbet and Martin 1990) and *Saccharomyces cerevisiae* (Nisbet and Martin 1991) that contained malate also stimulated lactate uptake by *S. ruminantium*. However, stimulation of lactate uptake by malate was considerably smaller in *S. ruminantium* that had been growing on carbohydrates compared to *S. ruminantium* growing on lactate, which suggests that the mechanism of stimulation of lactate uptake is lactate-inducible (Nisbet and Martin 1994). There was evidence of the involvement of membrane proton gradients in lactate uptake by *S. ruminantium*; however, it is unknown if the role of malate on lactate uptake is specifically related to its transport into the cell or to the enhancement of lactate metabolism. It has been suggested that malate and fumarate could help overcome an oxaloacetate deficiency induced by gluconeogenesis (Linehan et al. 1978). Sodium has also been shown to enhance lactate uptake by *S. ruminantium* in the presence of malate, although stimulation by Na was not instantaneous but involved a longer-term cell mechanism (Nisbet and Martin 1994).

S. ruminantium co-metabolizes malate along with lactate. Lactate is metabolized to acetate and propionate, and malate acts as an electron acceptor and is converted to succinate (Martin 1998). When succinate accumulates, malate utilization stops (Evans and Martin 1997). Succinate disappeared after the addition of mixed ruminal bacteria to a *S. ruminantium* culture growing on lactate and malate that had accumulated ~10 mM succinate (Evans and Martin 1997). In contrast, Asanuma et al. (1999) reported accumulation of 17.8 mM succinate in ruminal mixed cultures supplemented with fumarate. However, as stated by Asanuma et al. (1999), 6-h incubations might have not been long enough for the microbiota to adapt to increase the rate of succinate uptake and metabolism to propionate.

Several in vitro mixed-culture experiments showed a decrease in lactate concentrations and increased pH when fumarate and malate were supplemented to batch cultures of ruminal microorganisms (Carro and Ranilla 2003a, b; García-Martínez et al. 2005; Tejido et al. 2005) or Rusitec fermenters (Carro et al. 1999; Gómez et al. 2005). In vitro effects on pH stabilization were confirmed in vivo by Martin et al. (1999), who showed that subclinical acidosis was prevented by malate dosing into the rumen. However, lactate concentration was unaffected and VFA concentration decreased. Lack of change in lactate concentrations along with an increase in pH was also observed in batch cultures by Mao et al. (2007).

13.2.3 Amelioration of Methane Production

Methane (CH_4) production by ruminants is a source of climate change and a loss of energy to the host animal. An approach to decrease CH_4 production in the rumen has been the use of organic acid intermediates of ruminal fermentation that are either electron acceptors (i.e., fumarate) or are metabolized to such compounds (i.e., malate). The rationale to this strategy is to decrease CH_4 production by competing with methanogenesis for metabolic hydrogen, which is instead incorporated into VFA formation. Volatile fatty acids are the main source of energy for ruminants and of glucose in the case of propionate.

The reduction of fumarate to succinate has been estimated to be thermodynamically competitive with methanogenesis (Ungerfeld and Kohn 2006). Because the reduction of 1 mol of fumarate to 1 mol of succinate and subsequent conversion to 1 mol of propionate incorporate 1 mol of reducing equivalents ([2H]), each mole of fumarate metabolized to propionate would be expected to decrease CH_4 production by 0.25 mol (based on $CO_2+4H_2 \rightarrow CH_4+2H_2O$). Based on this stoichiometry, a 10 % decrease in CH_4 production by a dairy cow producing 500 L/d CH_4 would require the animal to ingest 1.4 kg/d of sodium fumarate, which would make sodium fumarate a considerable part of the diet (Newbold et al. 2005).

A meta-analysis of fumarate addition to mixed ruminal batch cultures found a linear decrease of 0.037 mol CH_4/mol of added fumarate, which is considerably lower than the expected 0.25 mol/mol stoichiometry (Ungerfeld et al. 2007). This was explained on the basis of the following: (i) on average, only 0.48 mol of propionate was recovered per each mol of added fumarate, and (ii) for each mol of added fumarate, acetate increased 0.20 mol. Metabolism of 1 mol of fumarate to 1 mol of acetate would result in the release of 2 mol [2H] (in the oxidation of malate to oxaloacetate and in the oxidative decarboxylation of pyruvate to acetyl-CoA). Thus, the release of reducing equivalents in fumarate conversion to acetate almost completely compensated for its incorporation into propionate: 0.48 [2H] − 2×0.20 [2H] mol=0.08 [2H] incorporated per mol of added fumarate. The net result of 0.08 [2H] mol incorporation per mol of added fumarate implies a decrease of 0.08/4=0.02 mol CH_4/mol of added fumarate, which is very close to the regression coefficient of 0.037 mol CH_4/mol of added fumarate (Ungerfeld et al. 2007). Therefore, fumarate's lower than expected efficiency at decreasing CH_4 production was explained by the fact that a substantial proportion of it was metabolized to acetate rather than to propionate. Given that fumarate is a fermentation intermediate of propionate's randomizing pathway, propionate metabolism to acetate might have been at first unforeseen; however, it was estimated that fumarate conversion to acetate would be thermodynamically feasible even at low fumarate concentration and, therefore, expected to occur (Ungerfeld et al. 2007).

Later experiments with slow-release forms of fumaric acid obtained decreases in CH_4 production beyond the decrease achievable through 100 % [2H] incorporation into propionate formation (Wood et al. 2009). In these experiments, the recovery of added fumaric acid as propionate was ~50 %, which is numerically similar to what had been summarized by Ungerfeld et al. (2007). However, in the experiment by Wood et al. (2009), there was no increase in acetate production even with the regular form of fumaric acid, which would have been a source of [2H] available for CH_4 production. Even so, the sole incorporation of [2H] into fumarate conversion to propionate with no fumarate conversion to acetate would have been insufficient to explain the decrease in CH_4 production observed. Perhaps fumarate stimulated other electron sinks, such as microbial biomass. It is of much interest to understand (i) why fumaric acid did not increase acetate production in these experiments and (ii) what other electron sinks might have fumarate stimulated.

The recovery of fumarate as propionate in continuous cultures (López et al. 1999; Kolver et al. 2004; Newbold et al. 2005; Giraldo et al. 2007) has been in general numerically greater than in batch cultures (Carro and Ranilla 2003a; García-Martínez et al. 2005). Perhaps, continuous fumarate supplementation could stimulate amplification over time of microbial populations that convert fumarate to propionate (Ungerfeld et al. 2007). Hattori and Matsui (2008) reported the isolation of 39 different fumarate reducers in the rumen, all of which had highly homologous 16S rRNA gene sequences to known species. Two clone libraries with a total of 74 clones were constructed for fumarate reductase gene. These clones had only between 58 % and 85 % sequence similarity to fumarate reductases from known species (Hattori and Matsui 2008). Mao et al. (2008) reported an increase in the succinate producer *Succinivibrio dextrinisolvens* in the rumen of goats fed fumarate. Yang et al. (2012) observed that feeding fumaric acid to goats increased the ruminal abundance of *S. ruminantium* and reduced the number of methanogens, although *Fibrobacter succinogenes* numbers were not affected. Zhou et al. (2012) also determined changes in the microbial community as a consequence of feeding fumarate to goats. Numbers of succinate producer *Fibrobacter succinogenes* increased, but of *Ruminococcus flavefaciens* decreased, which would be unexpected, as *R. flavefaciens* also produces succinate as a main fermentation product in pure culture (Stewart et al. 1997). Changes in the bacterial community induced by fumarate supplementation can depend on the diet. Mao et al. (2007) found little change in the bacterial community when supplementing fumarate to in vitro cultures when using forage as substrate, but the bacterial community grew increasingly different as the percentage of concentrate in the substrate increased. Also, bacterial species with a high capacity to reduce fumarate to succinate responded to added fumarate by increasing their synthesis of fumarate reductase between 4- and 15-fold (Asanuma and Hino 2000). This shorter-term adaptation mechanism would be unlikely responsible for the greater recovery of fumarate as propionate in continuous cultures compared to batch cultures.

It is estimated that protozoa-associated methanogens are responsible for a substantial proportion of CH_4 produced in the rumen (Finlay et al. 1994; Newbold et al. 1995). Associating with protozoa may then give methanogens an advantage when competing with fumarate reducers for H_2. Protozoa tend to be lost in most continuous culture fermenters (Sharp et al. 1998; Hristov et al. 2012), which may be another reason for the numerically greater effectiveness of fumarate at decreasing CH_4 production in continuous fermenters compared to batch cultures. In agreement with this hypothesis, fumarate was more effective in protozoa-removed than in protozoa-enriched ruminal fluid (Asanuma et al. 1999).

In vivo results of fumarate and malate supplementation on methane mitigation have been variable, and variation could not be consistently related to the diet or the animal species (Ungerfeld et al. 2007). Some studies have found decreases in CH_4 production as a consequence of feeding fumarate (Bayaru et al. 2001; Newbold et al. 2001; Wallace et al. 2006; Wood et al. 2009; Yang et al. 2012), whereas others (McGinn et al. 2004; Beauchemin and McGinn 2006; Kolver and Aspin 2006; McCourt et al. 2008) observed no effect. An increase in CH_4 production when feeding fumarate was also reported (Moss and Newbold 2006). Molano et al. (2008) reported a linear decrease in CH_4 production when supplementing fumarate to wether lambs, but this was explained by a concomitant decrease in dry matter (DM) intake. The greatest in vivo success was obtained by Wood et al. (2009), who fed a regular and a slow-release form of fumarate to lambs and observed supra-stoichiometric responses of 60 % and a 76 % decrease in CH_4 emission, respectively.

An explanation to variation in responses in vivo of CH_4 decrease to fumarate addition was proposed by Wood et al. (2009) based on reported differences among fumarate reducers in their capacity to compete with methanogens for H_2 or formate (Asanuma et al. 1999). These authors suggested that the response to fumarate may depend on which methanogens and/or fumarate fermenters predominate under a given set of conditions (Wood et al. 2009). In agreement with this hypothesis, Mamuad et al. (2014) recently showed that adding the fumarate reducting bacteria

Mitsuokella jalaludinni to in vitro cultures of ruminal microorganisms reduced methane production and increased succinate concentration.

The same as fumarate, malate has been also studied in vitro and in vivo as an alternative electron acceptor to decrease CH_4 production while increasing propionate. Because 1 mol of malate is dehydrated to 1 mol of fumarate (Russell and Wallace 1997), the same expected stoichiometric relationship between fumarate addition and CH_4 decrease holds for malate, i.e., the addition of 1 mol of malate is expected to result in a decrease of 0.25 mol of CH_4. A similar meta-analysis to the one conducted for fumarate was conducted on the effects of malate on ruminal fermentation in batch cultures (Ungerfeld and Forster 2011). Although the decrease in CH_4 was numerically greater with malate than with fumarate, it was more variable and nonsignificant. Malate recovery as propionate and acetate was numerically similar as for fumarate, and therefore the same response in CH_4 decrease would have been expected. It was speculated that malate might have stimulated microbial growth as an alternative electron sink, resulting in a numerically greater decrease in CH_4 production compared to fumarate (Ungerfeld and Forster 2011).

In Rusitec fermenters, the addition of malate slightly increased CH_4 production, although, because it promoted organic matter digestion, malate supplementation decreased CH_4 production expressed on a fermented organic matter basis (Carro et al. 1999), in agreement with later findings (Gómez et al. 2005). The same as with fumarate, in vivo results have been variable, although there have been fewer animal trials compared to fumarate. Decreases in CH_4 production of 6 % and 16 % were reported when supplementing beef heifers with malic acid at 3.75 % and 7.5 %, respectively (Foley et al. 2009a), whereas lack of effect was reported for dairy cows (Cobb et al. 2009; Foley et al. 2009b).

Other propionate precursors have been evaluated as alternative electron acceptors to decrease CH_4 formation. Oxaloacetate is reduced to malate in propionate's randomizing pathway (Russell and Wallace 1997). Addition of oxaloacetate to in vitro cultures of ruminal microorganisms resulted in a small increase in CH_4 production, which can be explained by the fact that most oxaloacetate was recovered as acetate rather than propionate. Oxaloacetate conversion to acetate would involve release of reducing equivalents that would be available for CH_4 production (Ungerfeld et al. 2003). Acrylyl-CoA is a metabolic intermediate in propionate's nonrandomizing pathway (Russell and Wallace 1997). Newbold et al. (2005) obtained moderate decreases in CH_4 production by adding acrylate to batch cultures of ruminal microorganisms, although there was no effect in continuous culture fermenters. Except for fumarate and oxoglutarate, other compounds evaluated in batch cultures were less effective, and conversion of acrylate to propionate was generally similar to fumarate and malate.

Butyrate precursors have also been studied as a means of decreasing CH_4 production by acting as alternative electron acceptors (Ungerfeld et al. 2003). Most acetoacetate and β-hydroxybutyrate, however, were recovered as acetate. More crotonate was metabolized to butyrate, but still much of it was recovered as acetate. Metabolism of much of these carboxylic acids to acetate explains why they had small or no effect on CH_4 production (Ungerfeld et al. 2003).

Other organic acids that are not intermediates of ruminal fermentation, but isomers or double- or triple-bond analogs of them have been evaluated in batch cultures or ruminal microorganisms. Crotonate's position isomer 3-butenoic acid had small effect on CH_4 production. It was hypothesized that propionate and butyrate's triple-bond analogs, propynoic acid and 2-butynoic acid, would decrease CH_4 production by competing for reducing equivalents to reduce their triple bond (Ungerfeld et al. 2003). Effects of 2-butynoic acid on CH_4 production were small, whereas propynoic acid was a potent methanogenesis inhibitor. The triple bond ester ethyl-2-butynoate also strongly inhibited CH_4 production. Because both propynoic acid and ethyl-2-butynoate caused the accumulation of H_2 and other atypical reduced products of ruminal fermentation like formate and ethanol, it was thought that they inhibited methanogens directly rather than compete with methanogenesis for reducing equivalents

(Ungerfeld et al. 2003). Direct toxicity of these compounds to methanogens was subsequently confirmed (Ungerfeld et al. 2004).

Direct inhibition of methanogenesis using chemical compounds results in accumulation of H_2 (Janssen 2010). There have been attempts to use carboxylic acids as electron acceptors to incorporate accumulated H_2 into VFA. Crotonate and 3-butenoic acid were ineffective at decreasing H_2 accumulation caused by propynoic acid and ethyl-2-butynoate (Ungerfeld et al. 2006). On the other hand, Mohammed et al. (2004), Tatsuoka et al. (2008), and Ebrahimi et al. (2011) had some success at using fumarate and malate to decrease H_2 accumulation caused by inhibition of methanogenesis. Thus, even though decreases in CH_4 production by organic acids are generally small, they could perhaps be used to help incorporate electrons into VFA when methanogenesis is inhibited through other means.

13.2.4 Other Effects

Because low pH inhibits fibrolytic organisms, it was hypothesized that malate could improve the digestibility of forages incubated with added lactate. This was confirmed for bermudagrass but not for alfalfa hay (Martin 2004). In batch cultures of mixed ruminal microorganisms, Newbold et al. (2005) found an increase in DM digestibility with disodium fumarate, but not with fumaric acid, and García-Martínez et al. (2005) reported a tendency to increase organic matter digestibility by supplementing disodium fumarate to three substrates differing in their forage-to-concentrate ratio. Supplementing mixed batch cultures with sodium fumarate, Mao et al. (2007) found an increase in DM digestibility of roughage substrates, and Yu et al. (2010) observed an increase in neutral detergent fiber digestibility, with concomitant increases in endoglucanase and xylanase activities. In contrast, Li et al. (2009b) observed a decrease in DM digestibility without changes in fiber digestibility. In Rusitec fermenters, López et al. (1999) found an increase in DM digestibility of a mixed substrate by supplementing fumarate, which was accompanied with an increase in the number of cellulolytic bacteria; however, these results on DM digestibility and cellulolytic bacteria were not confirmed in other studies with Rusitec fermenters also using mixed substrates (Giraldo et al. 2007).

Fumarate supplementation has sometimes decreased NH_4^+ concentration (Bayaru et al. 2001; Li et al. 2009b; Yu et al. 2010; Zhou et al. 2012), which might be related to an increased incorporation of NH_4^+ into microbial amino acids but also to less deamination, which agrees with observation of diminished branched-chain VFA concentration with fumarate supplementation (Yu et al. 2010). Lower NH_4^+ and branched-chain VFA concentrations in the rumen of cattle supplemented with fumaric acid were also observed by Bayaru et al. (2001). Because protozoal numbers are not affected by fumarate supplementation (López et al. 1999; Bayaru et al. 2001; Newbold et al. 2005), decreased NH_4^+ concentration does not seem to relate to less intraruminal N recycling. Effects of fumarate on NH_4^+ are not consistent though, and in other in vitro (López et al. 1999; Carro and Ranilla 2003a; Newbold et al. 2005; Giraldo et al. 2007; Mao et al. 2008) and in vivo (Beauchemin and McGinn 2006) studies, no effects of fumarate on NH_4^+ have been found. Contradictory results were interpreted by Yu et al. (2010) as related to the proportion of concentrates in the diet. High-concentrate diets are associated with greater NH_4^+ incorporation into carbon chains and less deamination, which would make the effect of fumarate on NH_4^+ less apparent (Yu et al. 2010). However, Callaway and Martin (1996) found a decrease in NH_4^+ with fumarate supplementation in the presence of monensin but not in its absence; monensin is thought to inhibit deamination (Nagaraja et al. 1997; Wallace et al. 1997), so a greater effect of fumarate on NH_4^+ might actually be expected in the absence of monensin, if fumarate effect on NH_4^+ depends on background deamination as proposed by Yu et al. (2010). Furthermore, no effects of fumarate on NH_4^+ were found by García-Martínez et al. (2005) in batch cultures with three substrates differing in their forage-to-concentrate ratio, but increases in NH_4^+ were reported for a roughage and a mixed substrate by Mao et al. (2007).

Supplementation with fumarate or malate decreased the in vitro biohydrogenation of linoleic (Li et al. 2010) and linolenic (Li et al. 2009a) acids, increasing the production of *c*9, *t*11-conjugated linoleic acid, a nutraceutical fatty acid beneficial for human health. These authors concluded that malate and fumarate may compete with biohydrogenation of C18:2 in the utilization of metabolic H_2. This mechanism of action of these acids deserves further investigation given the increasing concern of consumers about the intake of saturated fat.

13.3 Effects of Organic Acids on Feed Intake, Diet Digestibility, in vivo Ruminal Fermentation, and Plasma Metabolites

13.3.1 Malate

There are inconsistencies in the literature regarding the effects of malate supplementation on feed intake in ruminants (Tables 13.1 and 13.2), but some of these inconsistent effects seem to be due to differences in rate of supplementation. Malic acid or malate salts at inclusion rates up to 25 g per kg of DM diet had no negative effects on feed intake in beef cattle (Castillo et al. 2007; Liu et al. 2009; Carrasco et al. 2012), dairy cows (Kung et al. 1982; Sniffen et al. 2006), dairy goats (Salama et al. 2002), and lambs (Carro et al. 2006). Devant et al. (2007) observed that inclusion of disodium/calcium malate in the concentrate offered to dairy cows at 20 g/kg DM increased concentrate intake, indicating that there was no negative effect of malate on concentrate palatability. In contrast, Flores et al. (2003) observed a reduction in DM intake of lambs when disodium/calcium malate was included in the diet at rates as low as 2 and 4 g/kg DM. High doses of malic acid have been reported to depress feed intake; thus, Foley et al. (2009a) found that malic acid supplemented at 50 and 75 g per kg of DM diet reduced DMI in steers fed a 40:60 grass silage: concentrate diet by 3.1–4.3 %, respectively.

Most studies have shown a lack of effect of malate supplementation on diet digestibility determined either using digesta markers (Montaño et al. 1999; Foley et al. 2009a) or total feces collection (Carro et al. 2006). Liu et al. (2009) used chromic oxide as digesta marker in steers observed that DM, organic matter, neutral detergent fiber, and acid detergent fiber digestibilities were linearly increased with increasing doses of malic acid supplementation (7.8, 15.6, and 23.4 g/kg DM), but crude protein and ether extract digestibilities were not affected. Similarly, Flores et al. (2003) observed an increase in diet digestibility by supplementing Manchega and Lacaune fattening lambs with disodium–calcium malate at 2 and 4 kg/kg DM. In contrast, Carro et al. (2006) administered equal doses of the same product to Merino fattening lambs and observed no differences in diet digestibility, and Mungói et al. (2012), using the same product at 1 g/kg DM in the diet of Manchega fattening lambs, observed a reduction of diet digestibility. These discrepancies in the results may be related to differences in the composition of the concentrate used in the different trials, as in vitro studies have shown that effects of malic acid or malate salts on ruminal fermentation are influenced by the type of cereal grain in the incubation substrate (Carro and Ranilla 2003b; Sirohi et al. 2012). In milk-producing animals, malate supplementation either had no effects (Khampa et al. 2006b; Sniffen et al. 2006) or increased diet digestibility (Khampa et al. 2006a).

Compared to in vitro studies, less research has been conducted to evaluate malate and fumarate effects on ruminal fermentation in vivo, and the results have been variable. Montaño et al. (1999) in steers, Carrasco et al. (2012) in heifers, Carro et al. (2006) in lambs, and Khampa et al. (2006b) in dairy cows reported no effects of malic acid or malate salt supplementation (up to 26.4 g/kg DM) on ruminal VFA concentration. In contrast, increases in ruminal VFA concentration were observed by others (Kung et al. 1982; Martin et al. 1999; Khampa et al. 2006a; Liu et al. 2009) when supplementing steers and dairy cows with malic acid or its sodium salt at doses ranging from 2.8 to 26.0 g/kg DM; however, VFA con-

Table 13.1 Effects of malate supplementation on feed intake, digestion, and performance of meat-producing ruminants

Reference	Additive and dose	Animal and duration	Diet	Comments
Kung et al. (1982)	Malic acid, 13.2 or 26.0 g/kg DM	Steers (Holstein), 47 days	49:49:3 whole-shelled corn: corn silage: supplements	No effects on DMI, diet AD, plasma concentrations of urea–N, and NH_3–N. Malic acid increased ruminal concentrations of total VFA and propionate
Martin et al. (1999)	Malic acid, 5.2 or 10.4 g/kg DM	Beef steers (angus and crossbred), 10-day step-up period, 52-day finishing study, 113-day study	Rolled corn-based diet	Malic acid linearly increased ADG and F:G ratio in a 10-day period, but had no effects in the 52-day finishing study and the 113-day study, no effects on carcass characteristics and plasma metabolites (glucose, urea–N, cholesterol, triglyceride, and lactate)
Montaño et al. (1999)	Malic acid, 26.4 g/kg DM	Holstein steers, 15 days	Steam-flaked barley-based diet	No effects on DMI, diet AD, rumen concentrations of total and individual VFA, plasma lactate concentrations, and microbial N flow to the duodenum malic acid increased rumen pH at 2 h post-feeding
Flores et al. (2003)	Disodium/calcium malate, 2 or 4 g/kg DM	Fattening lambs (Manchega and Lacaune), 31 days	Barley-based concentrate ad libitum and straw corn-based concentrate ad libitum and straw	Malate increased ADG, diet AD and ruminal pH at slaughter Malate decreased total DMI and F:G ratio There were malate by cereal interactions for most variables. Effects of malate were more marked in the barley-based concentrates than in the corn-based ones
Carro et al. (2006)	Disodium/calcium malate; 2 or 4 g/kg DM	Fattening lambs (Merino), 35 days	Barley-based concentrate ad libitum and straw	No effects on DMI, ADG, F:G ratio, rumen parameters (pH, VFA, lactate, and NH_3–N concentrations), blood metabolites (glucose, urea, and lactate), diet AD, and carcass characteristics
Castillo et al. (2007)	Malic acid, 2 g/kg DM Disodium/calcium malate, 2 g/kg DM	Beef calves (Belgian Blue), 148 days	Barley-based concentrate ad libitum and straw	No effects on ADG and F:G ratio Both malate chemical forms decreased plasma lactate concentrations in the first week of the study

Khampa et al. (2006a)	Sodium malate, 1.7, 3.4, or 5.1 g/kg DM	Dairy steers (Holstein–Friesian crossbred), 21 days	Urea-treated rice straw ad libitum and concentrate (1.5 % of BW)	No effects on DMI and diet AD Malate linearly increased N retention and rumen pH, rumen concentrations of ammonia–N and total VFA, and propionate proportion Malate linearly decreased proportion of acetate and acetate-to-propionate ratio
Foley et al. (2009a)	Malic acid, 25, 37.5, 50, or 75 g/kg DM	Beef heifers (Charolais cross), 28 days Beef steers (Friesian), 28 days	Grass silage and concentrate (40:60)	Malic acid linearly decreased DMI and methane production in beef heifers No effects on diet AD in any experiment
Liu et al. (2009)	Malic acid, 7.8, 15.6, or 23.4 g/kg DM	Beef steers (Chinese Simmental), 21 days	Corn stover and concentrate (60:40)	No effects on DMI Malic acid linearly increased diet AD, urinary excretion of purine derivatives, and total VFA concentrations in the rumen Malic acid linearly decreased rumen pH and NH_3-N and lactate concentrations
Carrasco et al. (2012)	Malic acid, 1 g/kg DM. Disodium–calcium malate; 1 g/kg DM	Beef heifers (Charolais and Limousin crossbreed, Charolais), 141 days	Barley-based concentrate ad libitum and straw	No effects of malate chemical form on DMI, ADG, F:G ratio, rumen fermentation variables (pH, VFA, lactate, and NH_3–N concentrations), blood metabolites (glucose, urea, and lactate), and carcass characteristics
Mungói et al. (2012)	Disodium–calcium malate, 1 g/kg DM	Fattening lambs (Manchega), 21 days	Wheat-based concentrate *ad libitum* and straw Barley–manioc-based concentrate *ad libitum* and straw	No effects on DMI Malate decreased AD of dry matter, organic matter, and acid detergent fiber No effects on diet ADG or F:G ratio There were malate by source of starch interactions for DMI and N balance

Only studies describing diet composition and malic acid or malate dose are included
AD apparent digestibility, *ADG* average daily gain, *BW* body weight, *DMI* dry matter intake, *F:G ratio* feed-to-gain ratio, *VFA* volatile fatty acids

Table 13.2 Effects of malate supplementation on feed intake, digestion, and performance of milk-producing ruminants

Reference	Additive and dose	Animal and duration	Diet	Comments
Kung et al. (1982)	Malic acid, 3.8, 5.5, or 7.7 g/kg DM	Dairy cows (Holstein), 100 days	Corn silage *ad libitum*, 2.2 kg hay, and concentrate at 1 kg/2.5 kg of milk	Malic acid increased persistency of milk production, conversion of feed, and ruminal concentrations of total VFA, acetate, and butyrate No effect on milk yield, but fat and total solids percentage tended to increase No effects on plasma urea and glucose concentrations
Salama et al. (2002)	Disodium/calcium malate, 1 g/kg DM (mixed with a *Saccharomyces cerevisiae* culture)	Dairy goats, 8 weeks	Forage (alfalfa hay and maize whole plant) *ad libitum* and concentrate (0.6 kg/day)	No effects on DMI and milk yield and composition Malate increased BW
Sniffen et al. (2006)	Malic acid, 2 g/kg DM	Dairy cows, 28 days	Silage (corn, alfalfa, and grass) and concentrate diet (49:51)	No effects on DMI, diet AD, and BW change Increased production of milk, milk true protein, and milk lactose
Khampa et al. (2006b)	Sodium malate, 3.4 or 6.9 g/kg DM	Dairy cows (Holstein–Friesian crossbred), 21 days	Urea-treated rice straw ad libitum and concentrate (1.3 kg/day)	No effects on DMI, rumen fermentation variables (pH, VFA, lactate, and NH_3–N concentrations), milk production, and milk composition Malate decreased rumen lactate concentration and increased AD of crude protein, organic matter, and fiber
Devant et al. (2007)	Disodium/calcium malate, 84 g per cow (no data on total DMI intake given)	Dairy cows (Holstein–Friesian), 14 weeks	Total mixed diet *ad libitum* and concentrate (3 kg/day)	Malate increased concentrate DMI (no data available on total DMI) No effects on milk production and composition, rumen pH, and VFA concentrations
Foley et al. (2009b)	Malic acid, 26 g/kg DM	Dairy cows (Holstein–Friesian), 6 weeks	Grazing (mixed-species grass sward) and concentrate (6 kg/day)	No effects on DMI (herbage and total), BW change, milk yield and composition, and methane production

Only studies describing diet composition and malic acid or malate dose are included
AD apparent digestibility, *ADG* average daily gain, *BW* body weight, *DMI* dry matter intake, *F:G ratio* feed-to-gain ratio, *VFA* volatile fatty acids

centration need not reflect changes in VFA production if there are concomitant changes in VFA absorption and/or liquid outflow rates. Malate supplementation in vivo has generally resulted in increased propionate molar proportion (Kung et al. 1982; Khampa et al. 2006a; Liu et al. 2009) and sometimes butyrate (Liu et al. 2009). It has been suggested that malic acid and malate salts could selectively stimulate or inhibit the activity of specific ruminal microbes at some concentrations (Liu et al. 2009).

Only few studies have been conducted on the influence of malate supplementation on plasma metabolites. Martin et al. (1999) and Montaño et al. (1999) in steers, Carrasco et al. (2012) in heifers, Kung et al. (1982) in cows, and Carro et al. (2006) in lambs found no effects of concentrations of malic acid or malate salts included between 3.7 and 26 g/kg DM on plasma concentrations of glucose, L-lactate, and urea–N. In contrast, Castillo et al. (2007) found that supplementing malic acid or a mixture of disodium and calcium malate at 4 g/kg concentrate DM fed to feedlot bulls in the finishing period lowered plasma L-lactate concentration compared to unsupplemented controls. Similarly, Hernández et al. (2011) observed that supplementing finishing bull calves with free malic acid or disodium–calcium malate salts at 4 g/kg DM diet lowered plasma concentrations of L-lactate, urea–N, and creatinine. Reasons for the discrepancies among studies are unknown, but dose of malate may not be the only factor involved.

13.3.2 Fumarate

The effects of fumarate supplementation on feed intake reported in the literature are inconsistent (Table 13.3). Some authors have reported reductions in feed intake (McGinn et al. 2004; Beauchemin and McGinn 2006; Molano et al. 2008), but others have observed no effect (McCourt et al. 2008; Yu et al. 2010; Wood et al. 2009). McCourt et al. (2008) and Wood et al. (2009) administered to growing/fattening lambs up to 113 and 100 g of fumaric/kg DM, respectively, without observing decreased feed intake. In those studies, fumaric acid was encapsulated in oil, under the hypothesis that this would result in slow release and prevent negative effects on feed intake and ruminal pH. Molano et al. (2008) observed a decrease in DM intake when supplementing wether lambs with free fumaric acid supplemented at 80 or 100 g/kg DM. These results suggest that encapsulation may be a useful procedure for administering high doses of organic acids to ruminants avoiding negative effects on feed intake.

Several studies reported no effect of fumarate supplementation at doses ranging from 12 to 50 g/kg DM on DM, energy, and fiber digestibility in steers (Bayaru et al. 2001; McGinn et al. 2004; Beauchemin and McGinn 2006) and dairy cows (Kolver and Aspin 2006). In contrast, Yu et al. (2010) observed greater energy, cellulose, and N digestibility in goats receiving sodium fumarate at 10 g/kg DM compared with unsupplemented animals, and Isobe and Shibata (1993) found that adding fumaric acid (5 g/kg DM) to the diet of goats increased the in situ disappearance of the diet at short incubation times.

Kolver and Aspin (2006) reported no effects of fumaric acid supplementation on plasma metabolites. Yu et al. (2010) on the other hand observed increased plasma glucose concentrations in dairy goats fed sodium fumarate at 10 g/kg DM.

13.4 Effects of Organic Acids on Animal Production

13.4.1 Beef Production

Malate supplementation to meat ruminants has been shown to have a favorable impact on average daily gain in some, but not all, studies. Martin et al. (1999) reported that supplementing malic acid at rates of 3.7 or 7.1 g/kg DM to feedlot steers fed a corn-based diet significantly improved average daily gain and feed-to-gain ratio, and Flores et al. (2003) found similar results in lambs supplemented with malate salts at 2 or 4 g/kg DM. On the contrary, no effects of malic acid or malate salts on daily gain or feed-to-gain ratio were observed by Castillo et al. (2007) and

Table 13.3 Effects of fumarate supplementation on feed intake, digestion, and ruminant productivity

Reference	Additive and dose	Animal and duration	Diet	Comments
Isobe and Shibata (1993)	Fumaric acid. 2.5 or 5 g/kg DM	Goats	80:20 hay: crashed corn 50:50 hay: crashed corn	Fumaric acid increased ruminal pH, acetate and propionate concentrations, and in situ dietary substrate disappearance
Bayaru et al. (2001)	Fumaric acid. 20 g/kg DM	Holstein steers	Sorghum silage	Fumaric acid decreased methane and CO_2 production by 23.0 % and 20.5 %, respectively No effects on AD of dry matter and neutral detergent fiber and ruminal protozoal population Fumaric acid decreased ruminal NH_3–N concentrations and tended to increase total VFA concentration
McGinn et al. (2004)	Fumaric acid, 12 g/kg DM	Steers, 21 days	Barley silage-based diet	No effects on rumen variables (pH, VFA, lactate, and NH_3–N concentrations), methane production and DM, energy, and fiber AD Fumaric acid decreased NDF and ADF intake
Beauchemin and McGinn (2006)	Fumaric acid, 24.4 g/kg DM (plus 75 g sodium bicarbonate/day)	Beef cattle (Angus heifers), 21 days	Whole-crop barley silage: steam rolled barley, concentrate–mineral–vitamin supplement (75:19:6)	No effects on DM, energy and fiber AD, and methane production Fumaric acid decreased DMI and neutral detergent fiber intake, but increased total VFA concentrations and molar proportions of acetate and propionate
Kolver and Aspin (2006)	Sodium fumarate, 50 g/kg DM	Dairy cows	Fresh pasture (low quality), 15 days	No effects on DMI, diet AD, methane production, milk yield and composition, ruminal fermentation, and plasma metabolites

McCourt et al. (2008)	Encapsulated fumaric acid, 59 or 113 g/kg DM	Holstein dairy cows, 21 days	Grazing and concentrate (5 kg/day)	No effects on DMI, methane production, BW change, and milk yield and composition No effects on numbers of ciliates, anaerobic fungi, archaea, *Butyrivibrio*, and *Fibrobacter succinogenes*, but numbers of *Ruminococcus flavefaciens* were reduced
Molano et al. (2008)	Fumaric acid, 40, 60, 80, or 100 g/kg DM	Wether lambs	Dried ground lucerne	Fumaric acid decreased DMI and increased ruminal pH No effects on methane production per kg DMI
Wood et al. (2009)	Fumaric acid, 100 g/kg DM, either encapsulated or free from	Fattening lambs, 56 days	Wheat- and barley-based concentrate and straw	Both forms of fumaric acid resulted in improved ADG, decreased F:G ratio, and decreased methane production
Yu et al. (2010)	Sodium fumarate, 10 g/kg DM	Dairy goats (non-lactating), 21 days	Bermuda grass hay ad libitum and concentrate (0.24 kg/day)	No effects on DMI and BW change Fumarate increased plasma glucose concentrations and AD of crude protein and cellulose
Yang et al. (2012)	Fumaric acid, 15.4 g/kg DM	Goats, 21 days	41:59 Chinese wild rye hay: concentrate 58:42 Chinese wild rye hay: concentrate	No effects on DMI (herbage and total), BW change, milk yield and composition, and methane production

Only studies describing diet composition and fumaric acid or fumarate dose are considered
ADG average daily gain, *DMI* dry matter intake, *F:G ratio* feed-to-gain ratio, *VFA* volatile fatty acids, *AD* apparent digestibility

Carrasco et al. (2012) in beef steers or heifers and by Carro et al. (2006) and Mungói et al. (2012) in fattening lambs using lower doses of malic acid or malate salts (1–4 g/kg DM). No effects of malic acid or malate salts on hot carcass weight and dressing proportion have been observed in steers (Martin et al. 1999; Carrasco et al. 2012) and lambs (Carro et al. 2006).

13.4.2 Milk Production

Responses in milk production to malate supplementation have not been consistent. Kung et al. (1982) observed that feeding 140 g of malic acid per day to dairy cows in early lactation did not improve milk yield, but increased persistency of milk production, and Stallcup (1979) reported that a daily dose of 70 g of malic acid increased both milk yield and total solids and fat milk content. Sniffen et al. (2006) also observed an increase in milk production and milk true protein and lactose in dairy cows consuming 50 g of malic acid per day. Some of the observed increases in milk yield have been explained by concomitant increases in DM intake or diet digestibility. However, in the trial of Sniffen et al. (2006), neither DM intake nor diet digestibility was affected by malic acid supplementation, and these authors attributed the increase in milk yield to an increase in microbial efficiency, as observed in several in vitro experiments (Tejido et al. 2005; Gómez et al. 2005; Sniffen et al. 2006). In contrast, in other studies no effects of malic acid or malate salts on milk yield have been noticed (Salama et al. 2002; Khampa et al. 2006b; Foley et al. 2009b). Only few studies have investigated the effects of fumarate on milk production, but no effects on milk yield or composition have been detected (Kolver and Aspin 2006; McCourt et al. 2008) when supplementing fumaric acid up to 113 g/kg DM.

In general, inconsistent results found in the literature for the effects of malate and fumarate supplementation on feed intake, digestion, and ruminant performance have been attributed to differences in the diet (i.e., forage-to-concentrate ratio, forage type, cereal grains, etc.), dose of organic acids, chemical form in which organic acids were fed (i.e., free acid, salts, or encapsulated acids), and characteristics of the experimental animals (physiological state, level of production, ruminal populations, etc.). Few studies have compared the effects of malic acid with those of malate salts under the same feeding conditions, but no differences between both forms have been found in animal performance (Carrasco et al. 2012), fermentation variables (Carrasco et al. 2012), or plasma metabolites (Castillo et al. 2007; Hernández et al. 2011; Carrasco et al. 2012). Another source of variability in the results could be the basal content of malate or fumarate in the diets fed in the different experiments. For example, Salama et al. (2002) attributed the lack of a positive effect of malate supplementation in their study to the amount of alfalfa (rich in malate) included in the diet. According to their calculations, the control group of goats consumed an average of 12.3 g of malate daily, which was considered adequate to stimulate lactate utilization by *S. ruminantium* (Nisbet and Martin 1991) and therefore could have prevented positive effects of 2 g of added malate/d in the supplemented group.

13.5 Legal Aspects and Future Prospects

As stated by the EU already in 1980, the use of fumaric and malic acids or their salts as additives in ruminant nutrition does not constitute any risk for animal or human health and does not have any unfavorable effects on the quality of animal products. Currently, both acids and their salts are authorized by EU legislation as animal feed additives in the category of "technological additives," and their use is allowed in all livestock species. Setting a maximum content of malate and fumarate in the diet was not considered as necessary (EFSA 2013, 2014b). The use of malic acid and its sodium and calcium salts in animal nutrition is considered safe for the consumer and the environment, but both substances are strongly irritant to skin, eyes, and mucosa, and exposure via inhalation for those handling these additives is considered to present a risk (EFSA 2014b).

In the USA, both malic acid and fumaric acid are on the GRAS (Generally Recognized as Safe) list maintained by the USA Food and Drug Administration, and both acids and their salts are food additives permitted for direct addition to food for human consumption.

The main limitation presented by the use of malic and fumaric acids as feed additives for ruminants is economical, since their high cost makes unprofitable their use in ruminant feeding at effective doses. In the EU, USA, and other countries, the current alternative might be combining these acids or their salts (at low doses, to lower the price) with other additives that have similar actions in the digestive tract of ruminants, such as probiotics or plant extracts. Currently, there are commercial products made by combinations of these products that are marketed for their use in the feeding of both ruminants and non-ruminants. An alternative option for ruminant feeding is to select forages with high content of organic acids. Callaway et al. (1997) showed that malate content of forage varies with forage type (greater in legumes than in grasses), forage variety, maturity (greater in immature than in mature forage), and processing (decreases with haymaking or pelleting), although it is difficult to conclude whether differences in organic acid levels among forages and cultivars are sufficient enough to affect ruminal fermentation and animal performance.

13.6 Conclusions

The main rationales to organic acid supplementation to ruminant diets are (i) to increase digestible energy content and propionate supply of low digestible diets, (ii) to decrease CH_4 production, and (iii) to enhance lactate utilization and diminish the risk of lactic acidosis. Inclusion of organic acids solely to increase dietary energy may not be profitable in most cases, in comparison to supplementing with concentrates, which are highly digestible and considerably cheaper. Furthermore, even though organic acids are theoretically 100 % utilizable, actual utilization is likely to be lower due to a non negligible proportion of the compounds outflowing the rumen before they can be metabolized.

Expected decrease in CH_4 production by supplementing with organic acids is in principle limited by the stoichiometry of electron incorporation into propionate production and by the fact that, at least in vitro, a considerable amount of these propionate precursors seems to be metabolized to acetate, resulting in the release of reducing equivalents available to CH_4 production. However, even though most in vitro and in vivo responses in CH_4 production to organic acid supplementation have been small, larger decreases in CH_4 were reported in some cases with slow-release fumarate formulations both in vitro and in vivo. To an extent, in vitro responses in CH_4 decrease in these experiments could be explained by no acetate increase associated with the addition of fumarate. However, both in vitro and in vivo responses in CH_4 decrease to slow-release fumarate were supra-stoichiometric in relation to the maximum [2H] that could be incorporated into fumarate conversion to propionate, and it would be of much interest to mechanistically understand how those occurred.

Pure culture studies have shown that organic acids and malate in particular stimulate lactate utilization by *S. ruminantium*. In vitro batch and continuous mixed-culture studies have confirmed a decrease in lactate concentration and pH stabilization with malate or fumarate supplementation. In vivo results have confirmed pH stabilization, although without a decrease in lactate concentration and with lower VFA concentration. Contradictory results are reported for plasma L-lactate concentration. The doses of acids and their form (free acids, salts, encapsulated, etc.), the diet fed to the animals (forage-to-concentrate ratio, forage type, cereal grains, etc.), the animal physiological state, and even the natural content of organic acids in the diet are factors that may influence the efficacy of these additives in improving animal performance. The use of organic acids as feed additives is considered safe for the animals, consumers, and the environment, but future research should identify the conditions in which these additives have optimal efficacy.

References

Asanuma N, Hino T (2000) Activity and properties of fumarate reductase in ruminal bacteria. J Gen Appl Microbiol 46:119–125

Asanuma N, Iwamoto M, Hino T (1999) Effect of the addition of fumarate on methane production by ruminal microorganisms in vitro. J Dairy Sci 82:780–787

Bayaru E, Kanda S, Kamada T et al (2001) Effect of fumaric acid on methane production, rumen fermentation and digestibility of cattle fed roughage alone. Anim Sci J 72:139–146

Beauchemin KA, McGinn SM (2006) Methane emissions from beef cattle: effects of fumaric acid, essential oil, and canola oil. J Anim Sci 84:1489–1496

Callaway TR, Martin SA (1996) Effects of organic acid and monensin treatment on in vitro mixed ruminal microorganism fermentation of cracked corn. J Anim Sci 74:1982–1989

Callaway TR, Martin SA (1997) Effects of cellobiose and monensin on in vitro fermentation of organic acids by mixed ruminal bacteria. J Dairy Sci 80:1126–1135

Callaway TR, Martin SA, Wampler JL et al (1997) Malate content of forage varieties commonly fed to cattle. J Dairy Sci 80:1651–1655

Callaway TR, Edrington TS, Rychlik JL et al (2003) Ionophores: their use as ruminant growth promotants and impact on food safety. Curr Issues Intest Microbiol 4:43–51

Carrasco C, Medel P, Fuentetaja A et al (2012) Effect of malate form (acid or disodium/calcium salt) supplementation on performance, ruminal parameters and blood metabolites of feedlot cattle. Anim Feed Sci Technol 176:140–149

Carro MD, Ranilla MJ (2003a) Influence of different concentrations of disodium fumarate on methane production and fermentation of concentrate feeds by rumen micro-organisms *in vitro*. Br J Nutr 90:617–623

Carro MD, Ranilla MJ (2003b) Effect of the addition of malate on *in vitro* rumen fermentation of cereal grains. Br J Nutr 89:279–288

Carro MD, López S, Valdés C et al (1999) Effect of DL-malate on mixed ruminal microorganism fermentation using the rumen simulation technique (RUSITEC). Anim Feed Sci Technol 79:279–288

Carro MD, Ranilla MJ, Giráldez FJ et al (2006) Effects of malate on diet digestibility, microbial protein synthesis, plasma metabolites, and performance of growing lambs fed a high-concentrate diet. J Anim Sci 84:405–410

Castillo C, Benedito JL, Méndez J et al (2004) Organic acids as a substitute for monensin in diets for beef cattle. Anim Feed Sci Technol 115:101–116

Castillo C, Benedito JL, Pereira V et al (2007) Malic acid supplementation in growing/finishing feedlot bull calves: influence of chemicals form on blood acid–base balance and productive performance. Anim Feed Sci Technol 135:222–235

Cobb SM, Michal JJ, Johnson KA (2009) The effect of malic acid supplementation on diet digestibility and methane production by beef cattle fed a forage diet. J Dairy Sci 92E-Suppl. 1:292

Devant M, Bach A, García JA (2007) Effect of malate supplementation to dairy cows on rumen fermentation and milk production in early lactation. J Appl Anim Res 31:169–172

Domalski ES (1972) Selected values of heats of combustion and heats of formation of organic compounds containing the elements C, H, N, O, P, and S. J Phy Chem Ref Data 1:221–277

Ebrahimi SH, Mohini M, Singhal KM et al (2011) Evaluation of complementary effects of 9,10-anthraquinone and fumaric acid on methanogenesis and ruminal fermentation *in vitro*. Archiv Anim Nutr 65:267–277

EFSA (European Food Safety Authority) (2013) Scientific opinion on the safety and efficacy of fumaric acid as a feed additive for all animal species. EFSA J 11(2):3102 (14 pp)

EFSA (European Food Safety Authority) (2014a) Antimicrobial resistance. http://www.efsa.europa.eu/en/topics/topic/amr.htm. Accessed 20 May 2014

EFSA (European Food Safety Authority) (2014b) Scientific opinion on the safety and efficacy of malic acid and a mixture of sodium and calcium malate when used as a technological additives for all animal species. EFSA J 12(2):3563 (18 pp)

Evans JD, Martin SA (1997) Factors affecting lactate and malate utilization by *Selenomonas ruminantium*. Appl Environ Microbiol 63:4853–4858

Finlay BJ, Esteban G, Clarke KJ et al (1994) Some rumen ciliates have endosymbiotic methanogens. FEMS Microbiol Lett 117:157–162

Flores C, Caja G, Romero R et al (2003) Malate in concentrate improves growth performance and digestibility of intensively fattened lambs. J Dairy Sci 86(Suppl 1):149

Foley PA, Kenny DA, Callan JJ et al (2009a) Effect of dl-malic acid supplementation on feed intake, methane emission, and rumen fermentation in beef cattle. J Anim Sci 87:1048–1057

Foley PA, Kenny DA, Lovett DK et al (2009b) Effect of dl-malic acid supplementation on feed intake, methane emissions, and performance of lactating dairy cows at pasture. J Dairy Sci 92:3258–3264

García-Martínez R, Ranilla MJ, Tejido ML et al (2005) Effects of disodium fumarate on *in vitro* rumen microbial growth, methane production and fermentation of diets differing in their forage:concentrate ratio. Br J Nutr 94:71–77

Giraldo LA, Ranilla MJ, Tejido ML et al (2007) Influence of exogenous fibrolytic enzymes and fumarate on methane production, microbial growth and fermentation in Rusitec fermenters. Br J Nutr 98:753–761

Gómez JA, Tejido ML, Carro MD (2005) Influence of disodium malate on microbial growth and fermentation in rumen-simulation technique fermenters receiving medium- and high-concentrate diets. Br J Nutr 93:479–484

Guan H, Wittenberg KM, Ominski KH et al (2006) Efficacy of ionophores in cattle diets for mitigation of enteric methane. J Anim Sci 84:1896–1906

Hattori K, Matsui H (2008) Diversity of fumarate reducing bacteria in the bovine rumen revealed by culture dependent and independent approaches. Anaerobe 14:87–93

Hernández J, Castillo C, Méndez J et al (2011) The influence of chemical form on the effects of supplementary malate on serum metabolites and enzymes in finishing bull calves. Livest Sci 137:260–263

Hristov AN, Lee C, Hristova R et al (2012) A meta-analysis of variability in continuous-culture ruminal fermentation and digestibility data. J Dairy Sci 95:5299–5307

Isobe Y, Shibata F (1993) Rumen fermentation in goats administered fumaric acid. Anim Sci Technol 64:1024–1030

Janssen PH (2010) Influence of hydrogen on rumen methane formation and fermentation balances through microbial growth kinetics and fermentation thermodynamics. Anim Feed Sci Technol 160:1–22

Khampa S, Wanapat M (2007) Manipulation of rumen fermentation with organic acids supplementation in ruminants raised in the tropics. Pak J Nutr 6:20–27

Khampa S, Wanapat M, Wachirapakorn C et al (2006a) Effect of levels of sodium DL-malate supplementation on ruminal fermentation efficiency of concentrates containing high levels of cassava chip in dairy steers. Asian-Aust J Anim Sci 19:837–844

Khampa S, Wanapat M, Wachirapakorn C et al (2006b) Effects of urea level and sodium dl-malate in concentrate containing high cassava chip on ruminal fermentation efficiency, microbial protein synthesis in lactating dairy cows raised under tropical condition. Asian-Aust J Anim Sci 19:368–375

Kolver ES, Aspin PW (2006) Supplemental fumarate did not influence milk solids or methane production from dairy cows fed high quality pasture. Proc N Z Soc Anim Prod 66:409–415

Kolver ES, Aspin PW, Jarvis GN (2004) Fumarate reduces methane production from pasture fermented in continuous culture. Proc N Z Soc Anim Prod 64:155–159

Kung L Jr, Huber JT, Krummrey JD (1982) Influence of adding malic acid to dairy cattle rations on milk production, rumen volatile acids, digestibility, and nitrogen utilization. J Dairy Sci 65:1170–1174

Li XZ, Choi SH, Jin GL et al (2009a) Linolenic acid in association with malate or fumarate increased CLA production and reduced methane generation by rumen microbes. Asian-Austr J Anim Sci 22:819–826

Li XZ, Yan CG, Choi SH et al (2009b) Effects of addition level and chemical type of propionate precursors in dicarboxylic acid pathway on fermentation characteristics and methane production by rumen microbes *in vitro*. Asian-Austr J Anim Sci 22:82–89

Li XZ, Long RJ, Yan CG et al (2010) Rumen microbial responses in fermentation characteristics and production of CLA and methane to linoleic acid in associated with malate or fumarate. Anim Feed Sci Technol 155:132–139

Linehan B, Scheifinger CC, Wolin MJ (1978) Nutritional requirements of *Selenomonas ruminantium* for growth on lactate, glycerol, or glucose. Appl Environ Microbiol 35:317–322

Liu Q, Wang C, Yang WZ et al (2009) Effects of malic acid on rumen fermentation, urinary excretion of purine derivatives and feed digestibility in steers. Animal 2:32–39

López S, Valdés C, Newbold CJ et al (1999) Influence of sodium fumarate on rumen fermentation in vitro. Br J Nutr 81:59–64

Mamuad L, Kim SH, Jeong CD et al (2014) Effect of fumarate reducing bacteria on in vitro rumen fermentation, methane mitigation and microbial diversity. J Microbiol 52:120–128

Mao SY, Zhang G, Zhu YW (2007) Effect of disodium fumarate on *in vitro* rumen fermentation of different substrates and rumen bacterial communities as revealed by denaturing gradient gel electrophoresis analysis of 16S ribosomal DNA. Asian-Austral J Anim Sci 20:543–549

Mao SY, Zhang G, Zhu YW (2008) Effect of disodium fumarate on ruminal metabolism and rumen bacterial communities as revealed by denaturing gradient gel electrophoresis analysis of 16S ribosomal DNA. Anim Feed Sci Technol 140:293–306

Martin SA (1998) Manipulation of ruminal fermentation with organic acids: a review. J Anim Sci 76:3123–3132

Martin SA (2004) Effects of DL-malate on in vitro forage fiber digestion by mixed ruminal microorganisms. Curr Microbiol 48:27–31

Martin SA, Streeter MN (1995) Effect of malate on in vitro mixed ruminal microorganism fermentation. J Anim Sci 73:2141–2145

Martin SA, Streeter MN, Nisbet DJ et al (1999) Effects of DL-malate on ruminal metabolism and performance of cattle fed a high-concentrate diet. J Anim Sci 77:1008–1115

Martin SA, Bertrand JA, Sauls B (2000) In situ disappearance of malate from alfalfa and bermudagrass hay. J Dairy Sci 83:308–312

McCourt AR, Yan T, Mayne S et al (2008) Effect of dietary inclusion of encapsulated fumaric acid on methane production from grazing dairy cows. Proc Br Soc Anim Sci Annu Conf: 64. Br Soc Anim Sci, Scarborough

McGinn SM, Beauchemin KA, Coates T et al (2004) Methane emissions from beef cattle: effects of monensin, sunflower oil, enzymes, yeast, and fumaric acid. J Anim Sci 82:3346–3356

Mohammed N, Lila ZA, Ajisaka N et al (2004) Inhibition of ruminal microbial methane production by b-cyclodextrin iodopropane, malate and their combination *in vitro*. J Anim Physiol Anim Prod 88:188–195

Molano G, Knight TW, Clark H (2008) Fumaric acid supplements have no effect on methane emissions per unit of feed intake in wether lambs. Aust J Exp Agric 48:165–168

Montaño MF, Chai W, Zinn-Ware TE et al (1999) Influence of malic acid supplementation on ruminal pH, lactic acid utilization, and digestive function in steers fed high concentrate finishing diets. J Anim Sci 77:780–784

Moss AR, Newbold CJ (2006) Novel feed additives for decreasing methane emissions from ruminants. Objective 2 (Study 1) propionate precursors and other metabolic intermediates as possible alternative electron acceptors to methanogenesis in ruminal fermentation in vitro. Final Project Report CSA 4320, Ministry of Agriculture, Fisheries and Food, London

Mungói M, Flores C, Casals R et al (2012) Effect of malate and starch source on digestibility and nutrient balance of growing-fattening lambs. Anim Feed Sci Technol 174:154–162

Nagaraja TG, Newbold CJ, Van Nevel CJ et al (1997) Manipulation of ruminal fermentation. In: Hobson PN, Stewart CS (eds) The rumen microbial ecosystem, 2nd edn. Blackie Academic, New York

Newbold CJ, Lassalas B, Jouany JP (1995) The importance of methanogens associated with ciliate protozoa in ruminal methane production *in vitro*. Lett Appl Microbiol 21:230–234

Newbold CJ, Ouda JO, López S et al (2001) Propionate precursors as possible electron acceptors to methane in ruminal fermentation. In: First international conference on greenhouse gases and animal agriculture, Obihiro

Newbold CJ, López S, Nelson N et al (2005) Propionate precursors and other metabolic intermediates as possible alternative electron acceptors to methanogenesis in ruminal fermentation *in vitro*. Br J Nutr 94:27–35

Nisbet DJ, Martin SA (1990) Effect of dicarboxylic acids and *Aspergillus oryzae* fermentation extract on lactate uptake by the ruminal bacterium *Selenomonas ruminantium*. Appl Environ Microbiol 56:3515–3518

Nisbet DJ, Martin SA (1991) Effect of a Saccharomyces cerevisiae culture on lactate utilization by the ruminal bacterium *Selenomonas ruminantium*. J Anim Sci 69:4628–4633

Nisbet DJ, Martin SA (1994) Factors affecting L-lactate utilization by *Selenomonas ruminantium*. J Anim Sci 72:1355–1361

Russell JB, Van Soest PJ (1984) In vitro ruminal fermentation of organic acids common in forage. Appl Environ Microbiol 47:155–159

Russell JB, Wallace RJ (1997) Energy-yielding and energy-consuming reactions. In: Hobson PN, Stewart CS (eds) The rumen microbial ecosystem. Blackie Academic & Professional, London

Salama AAK, Caja G, Garín D et al (2002) Effects of adding a mixture of malate and yeast culture (*Saccharomyces cerevisiae*) on milk production of Murciano-Granadina dairy goats. Anim Res 51:295–303

Seo S, Lanzas C, Tedeschi LO, Fox DG (2007) Development of a mechanistic model to represent the dynamics of liquid flow out of the rumen and to predict the rate of passage of liquid in dairy cattle. J Dairy Sci 90:840–855

Sharp R, Ziemer CJ, Stern MD et al (1998) Taxon-specific associations between protozoal and methanogen populations in the rumen and a model rumen system. FEMS Microbiol Ecol 26:71–78

Sirohi SK, Pandey P, Singh B et al (2012) Effect of malic acid supplementation on rumen fermentation, digestibility and methanogenesis in wheat straw sorghum based total mixed diets in vitro. Indian J Anim Sci 82:1038–1045

Sniffen CJ, Ballard CS, Carter MP et al (2006) Effects of malic acid on microbial efficiency and metabolism in continuous culture of rumen contents and on performance of mid-lactation dairy cows. Anim Feed Sci Technol 127:13–31

Stallcup OT (1979) Influence of addition of dl-malic acid to diets of lactating dairy cows. J Dairy Sci 62(Suppl 1):225–226

Stewart CS, Flint HJ, Bryant MP (1997) The rumen bacteria. In: Hobson PN, Stewart CS (eds) The rumen microbial ecosystem. Blackie Academic & Professional, London

Tatsuoka N, Hara K, Mikuni K et al (2008) Effects of the essential oil cyclodextrin complexes on ruminal methane production *in vitro*. Anim Sci J 79:68–75

Tejido ML, Ranilla MJ, García-Martínez R et al (2005) *In vitro* microbial growth and rumen fermentation of different diets as affected by the addition of disodium malate. Anim Sci 81:31–38

Ungerfeld EM, Forster RJ (2011) A meta-analysis of malate effects on methanogenesis in ruminal batch cultures. Anim Feed Sci Technol 166–167:282–290

Ungerfeld EM, Kohn RA (2006) The role of thermodynamics in the control of ruminal fermentation. In: Sejrsen K, Hvelplund T, Nielsen MO (eds) Ruminant physiology. Wageningen Academic Publishers, Wageningen

Ungerfeld EM, Rust SR, Burnett R (2003) Use of some novel alternative electron sinks to inhibit ruminal methanogenesis. Reprod Nutr Develop 43:189–202

Ungerfeld EM, Rust SR, Boone DR et al (2004) Effects of several inhibitors on pure cultures of ruminal methanogens. J Appl Microbiol 97:520–526

Ungerfeld EM, Rust SR, Burnett R (2006) Effects of butyrate precursors on electron relocation when methanogenesis is inhibited in ruminal mixed cultures. Lett Appl Microbiol 42:567–572

Ungerfeld EM, Kohn RA, Wallace RJ et al (2007) A meta-analysis of fumarate effects on methane production in ruminal batch cultures. J Anim Sci 85:2556–2563

Wallace RJ, Onodera R, Cotta MA (1997) Metabolism of nitrogen-containing compounds. In: Hobson PN, Stewart CS (eds) The rumen microbial ecosystem. Blackie Academic & Professional, London

Wallace RJ, Wood TA, Rowe A, Price J, Yañez DR, Williams SP, Newbold CJ (2006) Encapsulated fumaric acid as a means of decreasing ruminal methane emissions. Int Congr Ser 1293:148–151

Wood TA, Wallace RJ, Rowe A et al (2009) Encapsulated fumaric acid as a feed ingredient to decrease ruminal methane emissions. Anim Sci Technol 152:62–71

Yang CJ, Mao SY, Long LM et al (2012) Effect of disodium fumarate on microbial abundance, ruminal fermentation and methane emission in goats under different forage:concentrate ratios. Animal 6:1788–1794

Yu CW, Chen YS, Cheng YH et al (2010) Effects of fumarate on ruminal ammonia accumulation and fiber digestion in vitro and nutrient utilization in dairy does. J Dairy Sci 93:701–710

Zhou YW, McSweeney CS, Wang JK et al (2012) Effects of disodium fumarate on ruminal fermentation and microbial communities in sheep fed on high-forage diets. Animal 6:815–823

14 Selective Inhibition of Harmful Rumen Microbes

Yutaka Uyeno

Abstract

The microbial ecology of the rumen is extremely complex, allowing the efficient conversion of carbohydrates of plant origin to organic acids. Disturbance of this ecosystem might lead to impairment of host productivity or sometimes to disease in the host. For many years, researchers have attempted to manipulate the numbers and/or activities of rumen microbes, including methanogens and protozoa, to improve the efficiency of ruminant production. However, such ruminal interventions sometimes have unintended consequences.

This chapter presents the background on the diversity and functionality of the rumen microbial communities and then describes practical means to amend ruminant health and productivity by limiting the activity of inherent rumen microbes under specific conditions, as well as the benefits and current applications of probiotics/prebiotics in terms of nutrition and health for the ruminants.

Keywords

Rumen acidosis • Methanogen • Probiotics and prebiotics • Plant secondary metabolites

Y. Uyeno (✉)
Frontier Agriscience and Technology Center, Shinshu University, 8304 Minami-minowa, Nagano 399-4598, Japan
e-mail: ytkuyeno@shinshu-u.ac.jp

14.1 Rumen Microbial Community Overview

Ruminants harbor a complex microbial community comprising a diverse array of bacteria, archaea, and eukaryotes (protozoa and fungi) in the rumen, by which metabolic and physiological functions in the organ are secured (Dehority 2003; Stewart et al. 1988). The extreme bacterial molecular

A.K. Puniya et al. (eds.), *Rumen Microbiology: From Evolution to Revolution*, DOI 10.1007/978-81-322-2401-3_14

diversity reflects the complex metabolic network among microbes. These microorganisms interact with one another and take part in systematic digestion of large amounts of fibrous plant material that cannot be digested by the host. They anaerobically ferment these materials into end products that are in turn used as energy sources by the host. Among ruminal microbes, bacteria take part in decomposing the feed into short-chain fatty acids (SCFA), amino acids, hydrogen, carbon dioxide, etc. Environmental and stochastic factors, such as diet composition, feeding practices, and farm management, can strongly affect the composition and functions of the microbiota and sometimes lead to impaired health and performance in the host animal. The effects of diet on changes in the diversity and numbers of a wide range of bacterial species in the rumen are known (Kocherginskaya et al. 2001; Russell and Rychlik 2001; Tajima et al. 2001a).

Molecular-based monitoring of the intestinal bacterial communities of cattle has revealed that the community dynamics are much greater than previously thought based on cultivation methods (Whitford et al. 1998; Tajima et al. 1999; Uyeno et al. 2010). The majority of the rumen bacterial community is affiliated with two phyla, *Bacteroidetes* and *Firmicutes* (Whitford et al. 1998; Uyeno et al. 2007). The other groups, such as *Ruminococcus flavefaciens* and *Ruminococcus albus* (Krause et al. 1999, 2001), and the genus *Fibrobacter* (Stahl et al. 1988; Ziemer et al. 2000), all of which are fibrolytic bacteria and thus contribute to the organ's characteristics, also have niches in the community. The detection coverage, i.e., the sum of the populations that can be identified, is at most around 80 % of the total bacteria. These numbers include uncultured rumen bacterial groups (i.e., functionally unknown organisms), whose relative proportions are found to be 1–3 % each in the rumen community of around 1-year-old heifers (Uyeno et al. 2010). Accordingly, a certain proportion of the bacterial community has yet to be identified, which implies the difficulty of producing a complete description of the whole community, mainly due to the genetically diverse structure of the bacteria in rumen ecosystems.

The archaeal component of the ecosystem, which is thought to be represented exclusively by methanogens, is implicated in the removal of hydrogen. This makes the rumen archaea an indispensable part of this organ (St-Pierre and Wright 2012). Cultivation-based analyses have identified *Methanobrevibacter*, *Methanomicrobium*, and *Methanobacterium* as predominant methanogenic archaeal genera in the rumen (Stewart et al. 1988, 1997; Kumar et al. 2009). A previous study of the molecular diversity of rumen methanogens in sheep in Australia revealed those belonging to the order *Methanobacteriales*, especially the genus *Methanobrevibacter* (Wright et al. 2004a), and putative novel species in the rumen (Tajima et al. 2001b; Yanagita et al. 2000). The population of these methanogens has been estimated by various molecular studies and was within the range of 0.3–3.3 % (Lin et al. 1997; Sharp et al. 1998; Uyeno et al. 2007; Kong et al. 2010).

As the rumen community structure is robust, colonization by microbes that are inherently harmful for the community or for the host is usually unsuccessful. However, specific microbial groups that normally colonize the rumen are regarded as harmful in some situations, for example:

- Accumulation of organic acids which are little absorbed via the rumen epithelium and subsequent decrease in rumen pH. This is often observed in cattle under conditions of heat stress or in feedlot cattle that are provided excess concentrates.
- Methane generation from the rumen. Of course, this is inevitable due to the normal function of the rumen, but efforts to mitigate the generation of methane have been made to increase energy deposition in the hosts' body and to lower greenhouse gas emission from ruminants.

Approaches that work on specific microbes may reduce their influence for maintaining the function of the community in the rumen.

14.2 Problems in Ruminant Health and Productivity Due to Overactivity of a Specific Microbe Inherent in the Rumen

14.2.1 Heat Stress and Rumen Acidosis

Heat stress occurs when animals are exposed to environmental temperatures in excess of the upper critical temperature, particularly in combination with high relative humidity or sunshine (West 2003). A large quantity of metabolic heat mainly generated from rigorous fermentation in the rumen causes heat load in cattle and thus increases body temperature, which is then maintained by adjusting heat loss to the environment. Cattle respond physiologically by reducing exercise and feed intake during periods of heat stress. Greater maintenance costs associated with heat stress reduce the efficiency of energy utilization by cattle.

Based on current knowledge, a combination of various practices may be necessary to optimize production of dairy cattle in hot, humid climates. Nutritional management is the simplest approach. To prevent elevation of body temperature, reduction of forage in the diet in hot climates is primarily implemented to increase dietary nutrient density and to alleviate the negative effects on an animal's productivity. Consequently, feed efficiency often increases, and even an increase in milk yield of heat-stressed cattle may be observed, which can be explained by a reduction of metabolic heat production compared to feeding on forage resulting from increased dietary energy concentration (Ominski et al. 2002; Mader and Davis 2004). However, this dietary change sometimes leads to a drop in ruminal pH caused by faster growth of lactate-producing bacteria, such as *Streptococcus bovis*, as lactate is much less absorbed in the rumen epithelium than volatile fatty acids, such as acetate, propionate, and butyrate. The dietary change is also associated with risks of developing subacute rumen acidosis (SARA) and physical disorders such as acute and chronic acidosis, which often occur with accompanying laminitis and liver abscesses (Stone 2004).

Maintenance of normal rumen function is necessary regardless of changes in the diet. To accomplish this, it is necessary to characterize bacterial populations in the rumen of Holstein heifers maintained under different environmental conditions. Our feeding trials under different environmental temperatures using four rumen-fistulated healthy Holstein heifers (9 and 15 months old) resulted in a decrease of dry matter intake (DMI) especially in forage consumption and body weight gain with the rise in environmental temperature from 20 °C to 33 °C (Tajima et al. 2007; Uyeno et al. 2010). These changes were accompanied by a substantial decrease in SCFA concentration, which can be a contributing factor to impaired animal growth performance under more severe conditions. We profiled the changes in the ruminal bacterial community structure of Holstein heifers during heat stress. In response to the rise in environmental temperature, major fibrolytic bacteria (i.e., *Fibrobacter*) decreased by 50–80 %, while major saccharolytic bacteria (i.e., *Streptococcus*) increased by up to 2.5–10 times. Consistent population-level changes of certain bacterial groups with heat stress were observed. It is reasonable that fibrolytic bacterial activity decreases and saccharolytic bacterial activity increases as a consequence of preferential consumption of concentrates. The effects of heat stress on the rumen fermentation characteristics and on the impaired growth in dairy heifers can be explained indirectly by changes in the bacterial population, as a decline in ruminal pH was observed with the rapid growth of starch-degrading bacteria in the rumen. Temperature shifts seemed to have a profound effect on the ruminal bacterial population. Although the relative populations of these genera are relatively small in the rumen bacterial community, the changes in the relative population (relative activity) of these bacteria may be important in the ruminal adaptive response. It is noteworthy that no defined group diminished, suggesting that the overall bacterial diversity (richness) at the genus and species levels was maintained even under conditions of heat stress.

Thus, the presence of ruminal bacterial community members can be maintained, and a group that is decreased in the population under one set of conditions may bloom again when the conditions change.

Ruminal acidosis is one of the most common nutritional disorders in feedlot cattle, which usually consume large amounts of fermentable carbohydrates and limited amounts of forage. Certain feeding regimens in ruminants can lead to overgrowth of *S. bovis* in the rumen. Along with dietary management, e.g., gradual introduction of grain-based diet and feeding of coarsely chopped roughage, ionophores are extensively used to increase animal performance by modifying ruminal bacterial populations to obtain more efficient fermentation (Herrera et al. 2009). On the other hand, SARA is a metabolic disease of dairy cattle that occurs during early and mid-lactation and has traditionally been characterized by low pH that is not related to accumulation of lactic acid but to other factors, such as microbial population shifts and immune responses. It has been hypothesized that factors such as moderately low pH increase production of bacterial lipopolysaccharides (LPS) and gut permeability of LPS, which could trigger systemic inflammation, acting as endotoxins (Gozho et al. 2005; Plaizier et al. 2012). Khafipour et al. (2009) analyzed rumen samples from eight lactating Holstein dairy cattle to examine the microbiome structure using two molecular techniques. The most prominent shift occurred during SARA was a decline in Gram-negative *Bacteroidetes* organisms, but severe grain-induced SARA was dominated by *S. bovis* and *Escherichia coli*. These results suggested that the severity of SARA and degree of inflammation were highly correlated with the abundance of *E. coli* and not with lipopolysaccharide in the rumen. Intrinsically, *E. coli* is regarded to be absent or a minor constituent of the rumen community, but this group may be a contributing factor to the onset of disease.

Feed management to minimize undesirable changes in rumen fermentation characteristics can be implemented for maintenance of host health and productivity under severe conditions (Lettat et al. 2012). This includes use of feed supplements and live microbes to manipulate ruminal microbial functions, e.g., sodium bicarbonate as dietary buffer and antibiotics (monensin and lasalocid) (Shwartz et al. 2009; Baumgard et al. 2011), which inhibit microbial strains that produce lactate, and stimulation of lactate activity using bacteria or starch-engulfing ruminal protozoa (Al-Saiady et al. 2004; Stone 2004). Other methods of controlling lactic acid bacteria in the ruminal environment are also available. Inoculation with microorganisms capable of hydrolyzing starch or of metabolizing lactate at low pH may also help alleviate the occurrence of rumen acidosis.

14.2.2 Rumen Methane Emission and Its Mitigation

The complexity of the rumen microbial ecosystem supports the ability to efficiently convert various carbohydrates to volatile fatty acids for host energy via stepwise disposal of hydrogen through reduction of carbon dioxide to methane. Cattle with higher feed efficiencies are reported to produce 20–30 % less methane (Buddle et al. 2011; Eckard et al. 2010), clearly indicating that large amounts of ingested energy are usually discarded as methane in ruminant production. In addition, methane production in ruminants has attracted a great deal of attention in relation to its contribution to the greenhouse gas effect and global warming. Given their critical role in interspecies hydrogen transfer, a more intricate issue is how to counteract hydrogen in the rumen when methanogens are inhibited. A principal concept of the manipulation of methane production is a means of diverting hydrogen away from methane formation (Wright and Klieve 2011). For example, propionate production from succinate is an alternative metabolic pathway to dispose of reducing power. Reducing available metabolic hydrogen for methanogenesis (e.g., with alternative electron sinks other than disposal of hydrogen) is another useful method.

To determine effective means without impairing ruminal fermentation, various approaches have been evaluated as shown in Table 14.1. Various strategies aim at inhibition of enteric methane generation either directly or indirectly.

Table 14.1 Approaches aiming at inhibition of enteric methane generation

Mode of inhibition	Compounds	References
To limit methanogens (direct approach)	Chemical inhibitors (ionophores [monensin, lasalocid])	Russell and Strobel (1989), Callaway and Martin (1997), Ipharraguerre and Clark (2003), Odongo et al. (2007)
	Biological approaches (vaccine, bacteriophage, bacteriocins)	Dumitru et al. (2003), Sar et al. (2004), Williams et al. (2008), Buddle et al. (2011)
	Hydrolyzed tannins	McSweeney et al. (2001), Hook et al. (2010), Krueger et al. (2010), Patra and Saxena (2011), Bhatta et al. (2012)
To reduce hydrogen supply from rumen metabolic pathways (indirect approach)	Organic acids (fumarate)	Mohammed et al. (2004), van Zijderveld et al. (2011)
	Unsaturated lipids (linoleic acid)	Yang et al. (2009), Li et al. (2010), Shinkai et al. (2012)
	Condensed tannins	Puchala et al. (2005), Animut et al. (2008), Guglielmelli et al. (2011), Liu et al. (2011), Mohammed et al. (2011)

Compounds that inhibit the activity of methanogens are likely to reduce or eliminate methane production. Researchers have mainly focused on manipulating the numbers and/or activities of rumen methanogens. Approaches are promising, but the diversity and plasticity of functions of the rumen bacterial and methanogenic communities may be limiting factors for their successful application. Substantial decreases in methane reduction by these approaches would be difficult to achieve without compromising production. However, accompanying benefits (e.g., decrease in rumen protein degradability and increase in post-rumen protein availability) are expected. A number of abatement options can fit animal production systems in the immediate or near future, many of which are likely to be cost effective.

14.3 Current Measures for Selective Inhibition of Rumen Microbes Under Specific Conditions

14.3.1 Ionophores

Ionophores, which have been used to enhance feed conversion efficiency and growth rate in cattle, have been shown to inhibit the growth of lactic acid bacteria in the rumen. Monensin is a class of ionophores, which have been proposed to act primarily by selective inhibition of Gram-positive bacteria, to which *S. bovis* belongs, because Gram-negative bacteria have an outer lipopolysaccharide layer that limits penetration of monensin across the bacterial cell membrane (Russell and Strobel 1989). In addition, reduction of the Gram-positive population is thought to reduce available H_2 and thus suppress methanogenesis and increase propionate supply, improving the energy balance for ruminant animals. It is typically added to ruminant diets to increase the efficiency of feed utilization.

It should be noted that under certain production conditions, monensin feeding can occasionally lead to undesirable effects on milk composition (McIntosh et al. 2003). In particular, feeding monensin to lactating cows sometimes causes a reduction of milk fat content or milk fat yield (Sauer et al. 1997; Benchaar et al. 2006), possibly by inhibition of acetate-producing bacteria, which also belong to Gram-positive and biohydrogenating species (*Megasphaera elsdenii* and *Butyrivibrio fibrisolvens*) which displayed strong suppression upon monensin feeding (Weimer et al. 2008).

14.3.2 Plant Secondary Metabolites

There is an increasing interest in use of plant secondary metabolites (PSMs) in many parts of the tropics to mitigate enteric ruminal methane emissions and thereby improve animal performance. The PSMs are well recognized as antimicrobial agents, which act against bacteria, protozoa, and fungi (i.e., they do not necessarily target archaea) as a substitute for chemical feed additives. Among these, tannins and saponins, which are widely distributed in the plant kingdom, constitute the major classes of PSMs that are currently under research in a number of laboratories (Martin et al. 2010; Bodas et al. 2012). Tannins are polyphenol substances of diverse molecular weights and variable complexity, which can form complexes with feed and microbial proteins. The tannins are classified into two groups based on their structure and chemical properties: hydrolysable tannins (HT) and condensed tannins (CT). Tannins have been shown to favorably modulate rumen fermentation, such as reducing protein degradation in the rumen, prevention of bloating, and inhibition of methanogenesis. Saponins are chemically defined as high molecular weight glycosides in which sugars are linked to a triterpene or steroidal aglycone moiety. In addition to having the potential to depress protozoa by forming complexes with sterols in protozoan cell membranes (Guo et al. 2008), saponins show greater antibacterial activity against Gram-positive bacteria than Gram-negative bacteria. SCFA production often results in accumulation of propionate in the rumen by limited hydrogen generation. It may therefore be useful as a supplement to indirectly inhibit methane production by limiting the substrate for methanogens in ruminants without a deleterious effect on rumen function (Goel and Makkar 2012).

The mode of action of tannins has been discussed with various schemes; for example, it can form complexes with proteins, enzymes, or other macromolecules thereby inducing membrane disruption. Tannin complexes with the cell wall membrane of bacteria cause morphological changes and secretion of extracellular enzymes. Either interaction is likely to restrict the transport of nutrients into the cell and retard or prevent the growth of the organism. Tannins are therefore regarded as natural antimicrobial compounds that exert inhibitory effects on either methanogenic archaea or fiber-degrading bacteria, although this has yet to be described completely.

The antimethanogenic activities of HT and CT have been extensively demonstrated in several in vitro and in vivo studies (Hess et al. 2004; Tan et al. 2011). Methane production from ruminal fermentation decreases markedly when ruminants are fed tannin-containing forage and tannin extracts (Patra and Saxena 2011; Ramirez-Restrepo and Barry 2005; Waghorn 2008). HT appears to act more through inhibition of the growth and/or activity of methanogens and/or hydrogen-producing microbes (direct effect), whereas CT can decrease methane more through reduction of fiber digestion (indirect effect) that limits H_2 derived from synthesis of acetate (McSweeney et al. 2001; Carulla et al. 2005). In relation to this, feeding tannins from quebracho, mimosa, sumac, and chestnut normally result in a decrease in ruminal acetate concentration (Beauchemin et al. 2007; Deaville et al. 2010; Castro-Montoya et al. 2011). Instead, the molar proportion of propionate increases, which contributes to a decrease in hydrogen supply for methane production. However, it should be noted that not all types of tannins produce beneficial nutritional and environmental responses (Goel and Makkar 2012). The effects of tannins on ruminal bacteria were reported to be dependent on their nature, activity, and concentration in a plant or plant product that have differences in affinity for bacterial and plant proteins (Min et al. 2003; Patra et al. 2012). Generally, tannins with low molecular weight have greater inhibitory effects on rumen microbes because of their higher protein-precipitating capacities than high molecular weight polymeric tannins. On the other hand, there have been some reports that CT fractions with higher molecular weight have greater inhibitory effects on methane production (Tavendale et al. 2005; Huang et al. 2011).

There have been very few studies regarding the effects of tannins on rumen archaea populations. Bhatta et al. (2009) evaluated the effects of six commercially available natural sources of tannins (three sources of HT and three sources of CT in different combinations) on total archaea expressed as % of total archaeal 16S rRNA over total 16S rRNA in the sample, using mixed continuous cultures (RUSITEC system). The total archaeal population was 12 % lower with the combination of HT and CT than with HT alone, with reduced methane production by 5.5 %. In addition, SCFA composition varied with CT in the materials tested, each of which may have a stimulatory effect on specific microbes, an inhibitory effect on other microbes, or both. The different modes of action of two types of tannins may explain why the effects of HT+CT on total gas and methane production were greater compared with HT only. On the other hand, the deliberate ejection of some methanogens would only provide an opportunity for others to become members of the consortium, so the population changed less.

Rumen protozoa are also involved in methanogenesis because of their ecto- and endosymbiotic relationships with methanogenic archaea, which utilize hydrogen produced by the protozoa (McAllister and Newbold 2008; Kumar et al. 2009). Defaunation (removing protozoa) is a method to lower the supply of metabolic hydrogen to methanogens. As reported by Tan et al. (2011), CT in feed components is suggested to have an adverse effect on the numbers of ruminal protozoa, resulting in a decrease in archaeal count. Selective suppression of protozoa has been suggested to be a promising approach to reduce methane production (Cieslak et al. 2012), but it has not yet been put into practice. Even animals can be successfully defaunated; protozoa coming from other faunated subjects are able to reestablish themselves in the rumen.

The effects of tannins with rumen nutritional conversion are not limited to carbohydrate digestion systems. The effects on ruminal nitrogen metabolism are well documented. Multiple phenolic hydroxyl groups of tannins can react with proteins predominantly via hydrogen bonds, forming tannin–protein complexes to prevent breakdown by proteases and bind proteins at ruminal pH, which affects the metabolism of rumen microbes. Some tannins are known to produce excessive amounts of rumen escape CP (Rubanza et al. 2005; Ben Salem et al. 2005; Kariuki and Norton 2008). The reversible nature of the protein–HT complex, which can be released at the abomasum level, is also considered to contribute to improvement of protein availability for the host.

14.3.3 Probiotics/Prebiotics

Probiotics and prebiotics can modulate the balance and activities of the gastrointestinal microbiota and are therefore considered beneficial to the host animal and have been used as functional foods.

Dietary supplementation, for example, with probiotics, prebiotics, and organic acids, has attracted a great deal of attention with continuing demand to optimize animal health and growth performance while reducing feed costs (Allen et al. 2013). Probiotics can enhance intestinal health by stimulating the development of a healthy microbiota (dominated by beneficial bacteria) and improving mucosal immunity. The mechanism includes a selective decrease of harmful bacteria.

Prebiotics are nondigestible food ingredients, such as dietary fibers and oligosaccharides, that when consumed in sufficient amounts are selectively fermentable by a limited number of microbes in the gut to stimulate growth and/or activity, resulting in health benefits. Prebiotics, especially in combination with probiotics (symbiotic), participate to create an environment where useful bacteria are predominant, eventually benefiting host health, such as by competitive exclusion of pathogens or the stimulation of health-promoting metabolites. Application of probiotics and prebiotics is a means of redirecting the autonomic change toward a desirable community.

Both concepts of probiotics and prebiotics are applicable to rumen microbiota as well as to human intestinal microbiota, but they may require different types of application from those to the community of monogastric animals. Probiotics for adult ruminants have mostly been selected to target the rumen compartment. Such probiotics have a positive effect on various digestive processes, especially cellulolysis and synthesis of microbial protein. The main form of probiotic commonly used in dairy cattle consists of various strains of yeast (*Saccharomyces cerevisiae*), as has been tested as an additive to ruminant diets (Lascano and Heinrichs 2009; Moya et al. 2009). The most consistent effects following addition of yeast culture to the diet include improved productivity in both lactating and growing animals. The mode of action of yeast products has not been completely determined, but changes in rumen fermentation rate and patterns are generally involved. The metabolic potential and effect of yeast are largely dependent on the strain, and kinetics differ according to whether the yeast is active (live) or inactive (dead). Certain strains of active dry yeast are particularly effective at raising and stabilizing ruminal pH by stimulating certain populations of ciliate protozoa, which rapidly engulf starch and thereby effectively compete with amylolytic lactate-producing bacteria (Stone 2004; Brossard et al. 2006; Nocek and Kautz 2006). The beneficial effects of yeast are greater in forage-based diets than in high-concentrate diets. A less acidic ruminal environment benefits growth and fiber-degrading activities of cellulolytic microorganisms (Callaway and Martin 1997; Beauchemin et al. 2003; Chung et al. 2011). In beef cattle, growth parameters (average daily gain, final weight, intake, feed to gain ratio) have been reported to be improved by continuous live yeast supplementation (Chaucheyras-Durand and Durand 2010), possibly due to stabilizing ruminal pH as the result of revitalizing fibrolytic bacteria may also be effective when beef cattle are fed a high readily fermentable diet. Yeast also has the potential to affect the fermentation process in a manner that reduces methane gas formation (McGinn et al. 2004; Chung et al. 2011). Scavenging oxygen within the rumen is also a function of yeast, creating a more anaerobic environment, which is required by ruminal microorganisms (Dehority 2003), and may function as a barrier for pathogenic anaerobic bacteria. In this context, yeast itself acts as not only a probiotic, but also helps other rumen community members grow and thus acts as a type of prebiotic.

With regard to bacterial probiotics for adult ruminants, lactate-producing bacteria (*Enterococcus*, *Lactobacillus*), which would sustain lactic acid at a constant level compared to *S. bovis*, may represent a possible means of limiting acidosis in high-concentrate-fed animals (Nocek et al. 2002; Nocek and Kautz 2006;). *Megasphaera elsdenii* and *Propionibacterium* spp., which utilize lactate, have also been administered as direct-fed microbials to avoid ruminal lactate accumulation (Stein et al. 2006; Vasconcelos et al. 2008). As feedlot cattle are fed more high-grain diets compared to lactating cows, they are at increased risk of subclinical ruminal acidosis. Direct-fed bacteria may decrease the risk of acidosis in feedlot cattle fed highly fermentable diets, to counter the class of saccharolytic bacteria that contribute to a decrease in rumen pH.

Mitigating methane excretion by ruminants as an increasing target of probiotics may represent an interesting ecological tool. For example, hydrogen utilization and acetate production by a ruminal acetogenic bacterial isolate have been shown to be improved in vitro by the addition of a yeast strain, even in the presence of methanogens (Chaucheyras-Durand et al. 2008). High hydrogen-utilizing bacterial species isolated from diverse gut environments could also be used to increase reductive acetogenesis (McAllister and Newbold 2008). Redirection of hydrogen for acetate production instead of methane production would also be favorable for energy generation for the host, as acetate is an important energy source.

14.3.4 Immunization

Immunological approaches to manipulating rumen microbial populations have been investigated. Some vaccination approaches target methanogens, other bacteria, and protozoa. Providing oral doses of an avian-derived polyclonal antibody against lactic acid-producing bacteria was effective in decreasing ruminal lactate concentration and target bacteria and therefore in preventing the onset of acidosis in cattle and sheep fed high-grain diets (Shu et al. 1999; Gill et al. 2000). Di Lorenzo and coworkers (2006) determined that feeding a polyclonal antibody preparation (PAP) made from hen eggs immunized against *S. bovis* was successful in improving the rumen environment (decreasing ruminal counts of target bacteria and increasing pH). In their subsequent in vivo trial, the PAP was still effective in enhancing the gain: feed ratio of feedlot cattle possibly related to changes in ruminal *S. bovis* counts and in the fermentation products (DiLorenzo et al. 2008). These PAPs could therefore be effective in preventing the deleterious effects associated with these bacteria and possibly in enhancing animal performance, improving feed efficiency. This technology could be utilized to target other microorganisms of interest, such as methanogenic bacteria.

Vaccination against rumen methanogens can reduce methane emissions (Wright and Klieve 2011). However, trials using this strategy have provided inconsistent results, in part because of the need for more consideration of the composition, function, and microbial interactions within the ecosystem. A preliminary study for vaccination targeting approximately 20 % of the methanogen population reduced methane production (per kg/DMI) by 7.7 %, although the results were not repeatable with subsequent vaccine preparations (Wright et al. 2004b). The same research group also developed a vaccine based on five methanogen strains that was administered in three steps to sheep, but the vaccination failed to demonstrate any methane abatement with a mixture of five methanogens (Williams et al. 2009). Currently, production of effective vaccines to reduce methane emissions in ruminants based on crude whole-cell preparations is not easy to achieve. Intrinsically, a highly specific vaccine can be made to target specific strains of methanogens, but removal of some methanogens may also allow for other methane-producing microbes that are unaffected by the vaccine to take their place. Wedlock et al. (2010) also performed an in vivo experiment to introduce their immunization concept. Whole cells and four subcellular fractions (cytoplasmic, two cell wall preparations, and cell wall-derived proteins) prepared from a *Methanobrevibacter ruminantium* strain were used for repeated vaccination of 10-month-old sheep. While rumen ecology of the sheep did not change, vaccination with antigenic fractions induced strong antibody responses in serum. Antisera from sheep vaccinated with fractions of methanogens induced cell agglutination and decreasing growth of methanogens and production of methane in an in vitro assay, demonstrating the feasibility of a semi-active immunization to mitigate methane emission.

Limited genera and/or species comprising the methanogenic ecology may contribute significantly to the difference in methane gas production between cattle with different feed efficiencies. Immunization approaches have been shown to favorably modulate rumen fermentation, but consistent beneficial effects of rumen modulation and animal performance have not been observed. Alternatively, for establishing more efficient ways to mitigate methane emission, systematic intervention in the rumen microbial populations by combination with vaccination and other chemical means may also be feasible.

14.4 Concluding Remarks

It has been demonstrated that rumen microbial composition can be altered by various factors, including diet, age, and stress. As the rumen is an open system, these interventions can be achieved but sometimes can result in unintended consequences. Any changes in the microbiota composition have the potential to influence energy expenditure, satiety, and food intake of the host. Host health and productivity may therefore be determined as the ability to maintain a balance within the rumen ecosystem. Functional interactions between microbes and relations between microbes and host cells are warranted as a fundamental aspect of future research. For example, beneficial effects of nutritional supplementation have been observed by in vitro studies, and in vivo studies with animal hosts are required to determine the feasibility of these materials as feed additives or ingredients in ruminant diets to mitigate methane production or lactic acid accumulation without detrimental effects on the animal.

The same is true for probiotics, where these beneficial effects are still observed mainly in vitro. Comprehensive in vivo studies with animal hosts are required to confirm the potential and feasibility as feed additives or ingredients in ruminant diets. Moreover, further efforts and combined physiological and ecological approaches in addition to nutritional approaches are required for setting up suitable conditions for domestic ruminants for their best performance.

References

Allen HK, Levine UY, Looft T et al (2013) Treatment, promotion, commotion: antibiotic alternatives in food-producing animals. Trend Microbiol 21:114–119

Al-Saiady MY, Al-Shaikh MA, Al-Mufarrej SI et al (2004) Effect of chelated chromium supplementation on lactation performance and blood parameters of Holstein cows under heat stress. Anim Feed Sci Technol 117:223–233

Animut G, Puchala R, Goetsch AL et al (2008) Methane emission by goats consuming different sources of condensed tannins. Anim Feed Sci Technol 144:228–241

Baumgard LH, Wheelock JB, Sanders SR et al (2011) Postabsorptive carbohydrate adaptations to heat stress and monensin supplementation in lactating Holstein cows. J Dairy Sci 94:5620–5633

Beauchemin KA, Yang WZ, Morgavi DP et al (2003) Effects of bacterial direct-fed microbials and yeast on site and extent of digestion, blood chemistry, and subclinical ruminal acidosis in feedlot cattle. J Anim Sci 81:1628–1640

Beauchemin KA, McGinn S, Martinez T et al (2007) Use of condensed tannin extract from quebracho trees to reduce methane emissions from cattle. J Anim Sci 85:1990–1996

Ben Salem H, Makkar H, Nefzaoui A et al (2005) Benefit from the association of small amounts of tannin-rich shrub foliage (*Acacia cyanophylla* Lindl) with soya bean meal given as supplements to Barbarine sheep fed on oaten hay. Anim Feed Sci Technol 122:173–186

Benchaar C, Petit H, Berthiaume R et al (2006) Effects of addition of essential oils and monensin premix on digestion, ruminal fermentation, milk production, and milk composition in dairy cows. J Dairy Sci 89:4352–4364

Bhatta R, Uyeno Y, Tajima K et al (2009) Difference in the nature of tannins on in vitro ruminal methane and volatile fatty acid production and on methanogenic archaea and protozoal populations. J Dairy Sci 92:5512–5522

Bhatta R, Saravanan M, Baruah L et al (2012) Nutrient content, in vitro ruminal fermentation characteristics and methane reduction potential of tropical tannin-containing leaves. J Sci Food Agric 92:2929–2935

Bodas R, Prieto N, García-González R et al (2012) Manipulation of rumen fermentation and methane production with plant secondary metabolites Anim. Feed Sci Technol 176:78–93

Brossard L, Chaucheyras-Durand F, Michalet-Doreau B et al (2006) Dose effect of live yeasts on rumen microbial communities and fermentations during butyric latent acidosis in sheep: new type of interaction. J Anim Sci 82:829–836

Buddle BM, Denis M, Attwood GT et al (2011) Strategies to reduce methane emissions from farmed ruminants grazing on pasture. Vet J 188:11–17

Callaway TR, Martin SA (1997) Effects of cellobiose and monensin on in vitro fermentation of organic acids by mixed ruminal bacteria. J Dairy Sci 80:1126–1135

Carulla J, Kreuzer M, Machmüller A et al (2005) Supplementation of *Acacia mearnsii* tannins decreases methanogenesis and urinary nitrogen in forage-fed sheep. Aust J Agric Res 56:961–970

Castro-Montoya J, Makkar H, Becker K (2011) Chemical composition of rumen microbial fraction and fermentation parameters as affected by tannins and saponins using an in vitro rumen fermentation system. Can J Anim Sci 91:433–448

Chaucheyras-Durand F, Durand H (2010) Probiotics in animal nutrition and health. Benef Microbes 1:3–9

Chaucheyras-Durand F, Walker N, Bach A (2008) Effects of active dry yeasts on the rumen microbial ecosystem: past, present and future. Anim Feed Sci Technol 145:5–26

Chung YH, Walker ND, McGinn SM et al (2011) Differing effects of 2 active dried yeast (*Saccharomyces cerevisiae*) strains on ruminal acidosis and methane production in nonlactating dairy cows. J Dairy Sci 94:2431–2439

Cieslak A, Zmora P, Pers-Kamczyc E et al (2012) Effects of tannins source (*Vaccinium vitis idaea* L) on rumen microbial fermentation in vivo. Anim Feed Sci Technol 176:102–106

Deaville ER, Givens DI, Mueller-Harvey I (2010) Chestnut and mimosa tannin silages: effects in sheep differ for apparent digestibility, nitrogen utilisation and losses. Anim Feed Sci Technol 157:129–138

Dehority BA (2003) Numbers, factors affecting the population and distribution of rumen bacteria. In: Dehority BA (ed) Rumen microbiology. Nottingham University Press, Nottingham, pp 265–294

DiLorenzo N, Diez-Gonzalez F, DiCostanzo A (2006) Effects of feeding polyclonal antibody preparations on ruminal bacterial populations and ruminal pH of steers fed high-grain diets. J Anim Sci 84:2178–2185

DiLorenzo N, Dahlen CR, Diez-Gonzalez F et al (2008) Effects of feeding polyclonal antibody preparations on

rumen fermentation patterns, performance, and carcass characteristics of feedlot steers. J Anim Sci 86:3023–3032
Dumitru R, Palencia H, Schroeder SD et al (2003) Targeting methanopterin biosynthesis to inhibit methanogenesis. Appl Environ Microbiol 69:7236–7241
Eckard RJ, Grainger C, de Klein CAM (2010) Options for the abatement of methane and nitrous oxide from ruminant production: a review. Livest Sci 130:47–56
Gill HS, Shu Q, Leng RA (2000) Immunization with *Streptococcus bovis* protects against lactic acidosis in sheep. Vaccine 18:2541–2548
Goel G, Makkar HS (2012) Methane mitigation from ruminants using tannins and saponins. Trop Anim Health Prod 44:729–739
Gozho G, Plaizier J, Krause D et al (2005) Subacute ruminal acidosis induces ruminal lipopolysaccharide endotoxin release and triggers an inflammatory response. J Dairy Sci 88:1399–1403
Guglielmelli A, Calabro S, Primi R et al (2011) In vitro fermentation patterns and methane production of sainfoin (*Onobrychis viciifolia* Scop) hay with different condensed tannin contents. Grass Forage Sci 66:488–500
Guo YQ, Liu JX, Lu Y et al (2008) Effect of tea saponin on methanogenesis, microbial community structure and expression of mcrA gene, in cultures of rumen micro-organisms. Lett Appl Microbiol 47:421–426
Herrera P, Kwon YM, Ricke SC (2009) Ecology and pathogenicity of gastrointestinal *Streptococcus bovis*. Anaerobe 15:44–54
Hess HD, Beure RA, Lötscher M et al (2004) Ruminal fermentation, methanogenesis and nitrogen utilization of sheep receiving tropical grass hay-concentrate diets offered with *Sapindus saponaria* fruits and *Cratylia argentea* foliage. Anim Sci 79:177–189
Hook SE, Wright A-DG, McBride BW (2010) Methanogens: methane producers of the rumen and mitigation strategies. Archaea 2010:Article ID 945785
Huang XD, Liang JB, Tan HY et al (2011) Effects of Leucaena condensed tannins of differing molecular weights on in vitro CH_4 production. Anim Feed Sci Technol 166–167:373–376
Ipharraguerre IR, Clark JH (2003) Usefulness of ionophores for lactating dairy cows: a review. Anim Feed Sci Technol 106:39–57
Kariuki I, Norton B (2008) The digestion of dietary protein bound by condensed tannins in the gastro-intestinal tract of sheep. Anim Feed Sci Technol 142:197–209
Khafipour E, Li S, Plaizier JC et al (2009) Rumen microbiome composition determined using two nutritional models of subacute ruminal acidosis. Appl Environ Microbiol 75:7115–7124
Kocherginskaya SA, Aminov RI, White BA (2001) Analysis of the rumen bacterial diversity under two different diet conditions using denaturing gradient gel electrophoresis, random sequencing, and statistical ecology approaches. Anaerobe 7:119–134
Kong Y, He M, McAlister T et al (2010) Quantitative fluorescence in situ hybridization of microbial communities in the rumens of cattle fed different diets. Appl Environ Microbiol 76:6933–6938
Krause DO, Dalrymple BP, Smith WJ et al (1999) 16S rDNA sequencing of *Ruminococcus albus* and *Ruminococcus flavefaciens*: design of a signature probe and its application in adult sheep. Microbiology 145:1797–1807
Krause DO, Bunch RJ, Conlan LL et al (2001) Repeated ruminal dosing of *Ruminococcus* spp does not result in persistence, but changes in other microbial populations occur that can be measured with quantitative 16S-rRNA-based probes. Microbiology 147:1719–1729
Krueger WK, Gutierrez-Bañuelos H, Carstens GE et al (2010) Effects of dietary tannin source on performance, feed efficiency, ruminal fermentation, and carcass and non-carcass traits in steers fed a high-grain diet. Anim Feed Sci Technol 159:1–9
Kumar S, Puniya A, Puniya M et al (2009) Factors affecting rumen methanogens and methane mitigation strategies. World J Microbiol Biotechnol 25:1557–1566
Lascano GJ, Heinrichs AJ (2009) Rumen fermentation pattern of dairy heifers fed restricted amounts of low, medium, and high concentrate diets without and with yeast culture. Livest Sci 124:48–57
Lettat A, Noziere P, Silberberg M et al (2012) Rumen microbial and fermentation characteristics are affected differently by bacterial probiotic supplementation during induced lactic and subacute acidosis in sheep. BMC Microbiol 12:142
Li XZ, Long RJ, Yan CG et al (2010) Rumen microbial responses in fermentation characteristics and production of CLA and methane to linoleic acid in associated with malate or fumarate. Anim Feed Sci Technol 155:132–139
Lin C, Raskin L, Stahl DA (1997) Microbial community structure in gastrointestinal tracts of domestic animals: comparative analyses using rRNA-targeted oligonucleotide probes. FEMS Microbiol Ecol 22:281–294
Liu H, Vaddella V, Zhou D (2011) Effects of chestnut tannins and coconut oil on growth performance, methane emission, ruminal fermentation, and microbial populations in sheep. J Dairy Sci 94:6069–6077
Mader TL, Davis MS (2004) Effect of management strategies on reducing heat stress of feedlot cattle: feed and water intake. J Anim Sci 82:3077–3087
Martin LM, Wood KM, McEwen PL et al (2010) Effects of feeding corn naturally contaminated with Fusarium mycotoxins and/or a modified yeast cell wall extract on the performance, immunity and carcass characteristics of grain-fed veal calves. Anim Feed Sci Technol 159:27–34
McAllister TA, Newbold CJ (2008) Redirecting rumen fermentation to reduce methanogenesis. Aust J Exp Agric 48:7–13
McGinn SM, Beauchemin KA, Coates T et al (2004) Methane emissions from beef cattle: effects of monensin,

sunflower oil, enzymes, yeast, and fumaric acid. J Anim Sci 82:3346–3356

McIntosh F, Williams P, Losa R et al (2003) Effects of essential oils on ruminal microorganisms and their protein metabolism. Appl Environ Microbiol 69:5011–5014

McSweeney CS, Palmer B, McNeill DM et al (2001) Microbial interactions with tannins: nutritional consequences for ruminants. Anim Feed Sci Technol 91:83–93

Min B, Barry T, Attwood G et al (2003) The effect of condensed tannins on the nutrition and health of ruminants fed fresh temperate forages: a review. Anim Feed Sci Technol 106:3–19

Mohammed N, Ajisaka N, Lila ZA et al (2004) Effect of Japanese horseradish oil on methane production and ruminal fermentation in vitro and in steers. J Anim Sci 82:1839–1846

Mohammed R, Zhou M, Koenig KM et al (2011) Evaluation of rumen methanogen diversity in cattle fed diets containing dry corn distillers grains and condensed tannins using PCR-DGGE and qRT-PCR analyses. Anim Feed Sci Technol 166–167:122–131

Moya D, Calsamiglia S, Ferret A et al (2009) Effects of dietary changes and yeast culture (*Saccharomyces cerevisiae*) on rumen microbial fermentation of Holstein heifers. J Anim Sci 87:2874–2881

Nocek JE, Kautz WP (2006) Direct-fed microbial supplementation on ruminal digestion, health, and performance of pre- and postpartum dairy cattle. J Dairy Sci 89:260–266

Nocek JE, Kautz WP, Leedle JAZ et al (2002) Ruminal supplementation of direct-fed microbials on diurnal pH variation and in situ digestion in dairy cattle. J Dairy Sci 85:429–433

Odongo NE, Bagg R, Vessie G et al (2007) Long-term effects of feeding monensin on methane production in lactating dairy cows. J Dairy Sci 90:1781–1788

Ominski KH, Kennedy AD, Wittenberg KM et al (2002) Physiological and production responses to feeding schedule in lactating dairy cows exposed to short-term, moderate heat stress. J Dairy Sci 85:730–737

Patra AK, Saxena J (2011) Exploitation of dietary tannins to improve rumen metabolism and ruminant nutrition. J Sci Food Agric 91:24–37

Patra AK, Min B-R, Saxena J (2012) Dietary tannins on microbial ecology of the gastrointestinal tract in ruminants. In: Patra AK (ed) Dietary phytochemicals and microbes. Springer, Dordrecht, pp 237–262

Plaizier J, Khafipour E, Li S et al (2012) Subacute ruminal acidosis (SARA), endotoxins and health consequences. Anim Feed Sci Technol 172:9–21

Puchala R, Min B, Goetsch A et al (2005) The effect of a condensed tannin-containing forage on methane emission by goats. J Anim Sci 83:182–186

Ramirez-Restrepo C, Barry T (2005) Alternative temperate forages containing secondary compounds for improving sustainable productivity in grazing ruminants. Anim Feed Sci Technol 120:179–201

Rubanza C, Shem M, Otsyina R et al (2005) Polyphenolics and tannins effect on in vitro digestibility of selected *Acacia* species leaves. Anim Feed Sci Technol 119:129–142

Russell JB, Rychlik JL (2001) Factors that alter rumen microbial ecology. Science 292:1119–1122

Russell JB, Strobel H (1989) Effect of ionophores on ruminal fermentation. Appl Environ Microbiol 55:1–6

Sar C, Santoso B, Mwenya B et al (2004) Manipulation of rumen methanogenesis by the combination of nitrate with b-1-4 galacto-oligosaccharides or nisin in sheep. Anim Feed Sci Technol 115:129–142

Sauer FD, Fellner V, Kinsman R et al (1997) Methane output and lactation response in Holstein cattle with monensin or unsaturated fat added to the diet. J Anim Sci 76:906–914

Sharp R, Ziemer CJ, Stern MD et al (1998) Taxon-specific associations between protozoal and methanogen populations in the rumen and a model rumen system. FEMS Microbiol Ecol 26:71–78

Shinkai T, Enishi O, Mitsumori M et al (2012) Mitigation of methane production from cattle by feeding cashew nut shell liquid. J Dairy Sci 95:5308–5316

Shu Q, Gill H, Hennessy D et al (1999) Immunisation against lactic acidosis in cattle. Res Vet Sci 67:65–71

Shwartz G, Rhoads ML, VanBaale MJ et al (2009) Effects of a supplemental yeast culture on heat-stressed lactating Holstein cows. J Dairy Sci 92:935–942

Stahl DA, Flesher B, Mansfield HR et al (1988) Use of phylogenetically based hybridization probes for studies of ruminal microbial ecology. Appl Environ Microbiol 54:1079–1084

Stein DR, Allen DT, Perry EB et al (2006) Effects of feeding Propionibacteria to dairy cows on milk yield, milk components, and reproduction. J Dairy Sci 89:111–125

Stewart CS, Fonty G, Gouet P (1988) The establishment of rumen microbial communities. Anim Feed Sci Technol 21:69–97

Stewart CS, Flint HJ, Bryant MP (1997) The rumen bacteria. In: Hobson PN, Stewart CS (eds) The rumen microbial ecosystem, 2nd edn. Chapman and Hall, London, pp 10–72

Stone WC (2004) Nutritional approaches to minimize subacute ruminal acidosis and laminitis in dairy cattle. J Dairy Sci 87:E13–E26

St-Pierre B, Wright ADG (2012) Diversity of gut methanogens in herbivorous animals. Animal 7(supplement s1):49–56

Tajima K, Aminov RI, Nagamine T et al (1999) Rumen bacterial diversity as determined by sequence analysis of 16S rDNA libraries. FEMS Microbiol Ecol 29:159–169

Tajima K, Aminov RI, Nagamine T et al (2001a) Diet-dependent shifts in the bacterial population of the rumen revealed with real-time PCR. Appl Environ Microbiol 67:2766–2774

Tajima K, Nagamine T, Matsui H et al (2001b) Phylogenetic analysis of archaeal 16S rRNA libraries from the rumen suggests the existence of a novel group of archaea not associated with known methanogens. FEMS Microbiol Lett 200:67–72

Tajima K, Nonaka I, Higuchi K et al (2007) Influence of high temperature and humidity on rumen bacterial diversity in Holstein heifers. Anaerobe 13:57–64

Tan HY, Sieo CC, Abdullah N et al (2011) Effects of condensed tannins from Leucaena on methane production, rumen fermentation and populations of methanogens and protozoa in vitro. Anim Feed Sci Technol 169:185–193

Tavendale MH, Meagher LP, Pacheco D et al (2005) Methane production from in vitro rumen incubations with *Lotus pedunculatus* and *Medicago sativa*, and effects of extractable condensed tannin fractions on methanogenesis. Anim Feed Sci Technol 123–124(Part 1):403–419

Uyeno Y, Sekiguchi Y, Tajima K et al (2007) Evaluation of group-specific, 16S rRNA-targeted scissor probes for quantitative detection of predominant bacterial populations in dairy cattle rumen. J Appl Microbiol 103:1995–2005

Uyeno Y, Sekiguchi Y, Tajima K et al (2010) An rRNA-based analysis for evaluating the effect of heat stress on the rumen microbial composition of Holstein heifers. Anaerobe 16:27–33

Van Zijderveld SM, Fonken B, Dijkstra J et al (2011) Effects of a combination of feed additives on methane production, diet digestibility, and animal performance in lactating dairy cows. J Dairy Sci 94:1445–1454

Vasconcelos JT, Elam NA, Brashears MM et al (2008) Effects of increasing dose of live cultures of *Lactobacillus acidophilus* (Strain NP 51) combined with a single dose of *Propionibacterium freudenreichii* (Strain NP 24) on performance and carcass characteristics of finishing beef steers. J Anim Sci 86:756–762

Waghorn G (2008) Beneficial and detrimental effects of dietary condensed tannins for sustainable sheep and goat production – progress and challenges. Anim Feed Sci Technol 147:116–139

Wedlock D, Pedersen G, Denis M et al (2010) Development of a vaccine to mitigate greenhouse gas emissions in agriculture: vaccination of sheep with methanogen fractions induces antibodies that block methane production in vitro. N Z Vet J 58:29–36

Weimer PJ, Stevenson DM, Mertens DR et al (2008) Effect of monensin feeding and withdrawal on populations of individual bacterial species in the rumen of lactating dairy cows fed high-starch rations. Appl Microbiol Biotechnol 80:135–145

West JW (2003) Effects of heat-stress on production in dairy cattle. J Dairy Sci 86:2131–2144

Whitford MF, Forster RJ, Beard CE et al (1998) Phylogenetic analysis of rumen bacteria by comparative sequence analysis of cloned 16S rRNA genes. Anaerobe 4:153–163

Williams YJ, Rea SM, Popovski S et al (2008) Reponses of sheep to a vaccination of entodinial or mixed rumen protozoal antigens to reduce rumen protozoal numbers. Brit J Nutr 99:100–109

Williams YJ, Popovski S, Rea SM et al (2009) A vaccine against rumen methanogens can alter the composition of archaeal populations. Appl Environ Microbiol 75:1860–1866

Wright ADG, Klieve AV (2011) Does the complexity of the rumen microbial ecology preclude methane mitigation? Anim Feed Sci Technol 166–167:248–253

Wright ADG, Williams AJ, Winder B et al (2004a) Molecular diversity of rumen methanogens from sheep in western Australia. Appl Environ Microbiol 70:1263–1270

Wright ADG, Kennedy P, O'Neill CJ et al (2004b) Reducing methane emissions in sheep by immunization against rumen methanogens. Vaccine 22:3976–3985

Yanagita K, Kamagata Y, Kawaharasaki M et al (2000) Phylogenetic analysis of methanogens in sheep rumen ecosystem and detection of Methanobacterium mobile by fluorescence in situ hybridization. Biosci Biotechnol Biochem 64:1737–1742

Yang SL, Bu DP, Wang JQ et al (2009) Soybean oil and linseed oil supplementation affect profiles of ruminal microorganisms in dairy cows. Animal 3:1562–1569

Ziemer CJ, Sharp R, Stern MD et al (2000) Comparison of microbial populations in model and natural rumens using 16S ribosomal RNA-targeted probes. Environ Microbiol 2:632–643

15 'Omics' Approaches to Understand and Manipulate Rumen Microbial Function

Bhuvnesh Shrivastava, Kavish Kumar Jain, Ravinder Kumar, Sonali Prusty, Sandeep Kumar, Subhojit Chakraborty, Hotam Singh Chaudhary, Monica Puniya, and R.C. Kuhad

Abstract

Diverse populations of rumen microorganisms in gut contribute to develop ability of breaking down fibrous foods, which are mostly unusable by humans (Owens FN, Goetsch AL (1988) Ruminal fermentation. In: Church DC (ed) The ruminant animal, digestive physiology and nutrition. Prentice-Hall, Englewood Cliffs, p 160). Rumen is having a larger population of microorganisms, more than a trillion organisms and wide diversity (hundreds of species and thousands of subspecies), per ounce of rumen contents (Xu et al., J Anim Sci 85:1024–1029, 2007). There are various traditional approaches through which overall performance of the rumen has been attempted to improve, e.g. plant secondary metabolites, microbial feed additives, chemical feed additives, selective stimulation of beneficial rumen microbes and selective inhibition of harmful rumen microbes. In spite of these, nowadays various new approaches are being used to improve our understanding of the relationships among the various rumen microorganisms and towards how they interact with their hosts (Chaucheyras-Durand and Ossa, Prof Anim Sci 30:1–12, 2014). To better characterize species in the rumen, new advanced technological aids such as gene sequencing and study of gene (genomics), protein (proteomics) and metabolite (metabolomics) expression are being frequently used. This chapter will majorly emphasize on recent tools used in exploring the diversity of rumen.

B. Shrivastava (✉) • R. Kumar • M. Puniya
Dairy Microbiology Division, ICAR-National Dairy Research Institute, Karnal 132001, India
e-mail: bhuvneshbiotech@gmail.com

K.K. Jain • S. Kumar • S. Chakraborty • R.C. Kuhad
Lignocellulose Biotechnology Lab, University of Delhi South Campus, New Delhi 110021, India

S. Prusty
Dairy Cattle Nutrition Division, ICAR-National Dairy Research Institute, Karnal, Haryana 132001, India

H.S. Chaudhary
Department of Biotechnology, Madhav Institute of Technology and Science, Gwalior 474005, India

A.K. Puniya et al. (eds.), *Rumen Microbiology: From Evolution to Revolution*, DOI 10.1007/978-81-322-2401-3_15

Keywords
Reumen • Metabolomics • Genomics • Proteomics

15.1 Introduction

The tremendous importance of domestic ruminants is majorly due to their unique ability of converting the low-quality forages into high-quality, energy-rich, high-protein products (especially milk and meat), utmost needed for human consumption and survival. There are various different known and unknown metabolic pathways or microorganisms, and the specific factors exist, which control the growth and metabolic activity of the microorganisms in the rumen. The latest assessment of microbial diversity in the rumen ecosystem using DNA- and RNA-based sequence analysis has further paved the way for 'omics' technologies, and thereby new insights into the structure and functions of these complex microbial communities are being sought constantly (Fig. 15.1). This chapter will give major emphasis on reviewing recent advances in exploring the diversity of rumen through various advanced molecular techniques. Majority of these works involve exploring genetic manipulation of rumen microflora which facilitates control in specific metabolic pathways by induction or removal of certain intermediate molecules or expression of function-specific genes which provide certain functions. Certain extended approaches include the creation of synthetic genes for expression in rumen bacteria, unravelling the rumen diversity through proteomic approaches and studying and manipulating metabolic pathways for rumen improvement (Teather 1985; Chaucheyras-Durand and Ossa 2014).

15.2 Rumen Manipulation Using Genomic Approaches

Exploring the diversity of rumen through various fast and high-throughput molecular techniques can be used in maintaining the diversity in rumen subsequently to improve the performance of the cattle rumen. These techniques may employ DNA profiling techniques (PCR based), quantitative PCR, sequencing, fluorescent in situ hybridization (FISH), DNA microarrays and flow cytometry (Gong and Yang 2012). PCR-based DNA profiling techniques using PCR primers which target the specific genes to amplify the 16S/ITS rDNA sequence template targeted DNA sample (Snel et al. 2002; Li et al. 2003), while, on the other hand, bacterial population size of gut microbiota could be measured using qPCR method (Sekhavati et al. 2009). The quantification is achieved by measuring fluorescence emitted by the fluorophore. The DNA-binding dye (SYBR® green) binds non-specifically with the double-stranded DNA (dsDNA) during the PCR amplification process of the target gene, and as the

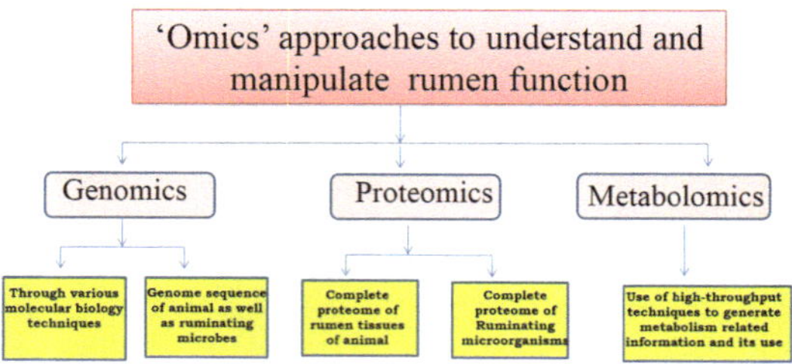

Fig. 15.1 Various 'omics' technologies to understand and manipulate rumen function

formation of dsDNA increases, the fluorescence intensity increases giving us the quantification of DNA. Similarly, DNA sequencing has been proved to be a fast and powerful tool. Next-generation have increased the pace of genomic and metagenomic studies, but sometimes it generates fragments of the short reads. To address this issue, nowadays the third-generation sequencing platform (PacBio RS of Pacific Biosciences) also appeared in the sequencing market (Gong and Yang 2012). Eckburg et al. (2005) have used the sequencing and fluorescent in situ hybridization (FISH) together to study diversity of gut microbiota. Among these, FISH has been found advantageous as of being quantitative or automated and can be combined with flow cytometer or an image analyzer providing information about the bacterial distribution in gut (Palma et al. 2009). FISH, however, has low sensitivity but has many additional advantages of detecting uncultured bacteria without enrichment, whereas flow cytometry is a powerful technique for studying gut microbiota. Both of these techniques have been used together successfully ranging from analysing mucosa-associated microbiota in the gut of pigs and chickens (Collado and Sanz 2007a), healthy lambs and calves (Collado and Sanz 2007b) to gut microbiota during pregnancy in women (Collado et al. 2008).

Palmer et al. 2007 extensively used DNA microarrays to detect various members of the gut microbiota to explore gut diversity. In their study, chips which contained probes targeting up to 359 species simultaneously were developed on an Agilent platform. Recently, phylogenetic microarrays which employ high density of 16S rRNA genes (Paliy et al. 2009) and small-subunit (SSU) rRNA genes (Rajilic-Stojanovic et al. 2010) have also been found to be useful in evaluating the microbial diversity of the human gut. In this kind of phylogenetic microarrays, many microbial phylotypes (more than 1,000) were accessed to study the human gut microbiota (Rajilic-Stojanovic et al. 2010). These results can be used in various clinical and medical applications to support the various findings. Jami and Mizrahi (2012) used molecular techniques to characterize the rumen bacterial population of 16 individual lactating cows using tag amplicon pyrosequencing. Results obtained were analysed using Bray–Curtis metric in terms of abundance and occurrence and showed 51 % similarity in bacterial taxa across samples. By adding taxon phylogeny to the analysis using a weighted UniFrac metric, the similarity increased to 82 %. They found 32 genera that were common in all samples but exhibit high variability in abundance across samples. In another approach, Li et al. (2013) studied the expression profile of the major xylanase of the glycoside hydrolase family 10 (GH 10) from the rumen samples at the time of feeding cycles of a small-tail Han sheep. Evidence of the presence of 44 distinct GH 10 xylanase gene fragments at both the genomic and transcriptional levels was found. The data of their relative abundance on comparison revealed that results from the evaluation of functional genes at the transcriptional level are more reliable indicators for understanding fluctuations in xylanase levels. Dynamic expression of xylanase gene was observed when investigated and observed up to 24 h, a complex trend of six xylanase genes, detected at all-time points of the feeding cycle. Correlation analysis indicated that the rumen is a dynamic ecosystem where the transcript profiles of xylanase genes are closely related to ruminal conditions, especially rumen pH and bacterial population. Recently, Wang et al. (2014) studied the effect of Zn source on expression of ZIPII transporter genes in Guanzhong dairy goats. They found that it changes the levels of expression of the genes involved in Zn metabolism in the tissues relevant to digestion or metabolism which were tissue specific and varied with the zinc source, indicating that the source of zinc affects the SLC39A1-3 mRNA expression. Mohammadzadeh et al. (2014) also compared the biomass and community structure of bacterial, protozoan and archaeal communities in samples of rumen and faeces of goats and to what extent the diet (alfalfa hay with or without supplemented oats) offered to them exerts an influence. Diversity, biomass and community structure of bacterial, protozoan and archaeal communities was assessed by real-time PCR (qPCR) and denaturing gradient gel electrophoresis (DGGE), respectively. The number of archaea and bacteria in both rumen and faeces was higher and lower, respectively, in animals fed AH diet ($P<0.005$). Contrary to that, protozoan numbers were not affected by the diet but were lower ($P<0.001$) in faeces than in rumen. The analysis of the community structure revealed that consistently different populations in structure in rumen and faeces for the three studied microbial groups and that supplementing alfalfa hay with oats led to a decrease in the similarity between sites in the rumen and faeces.

In addition to these, the genome sequence approach also generates a lot of information that can be used to improve the rumen performance. Jiang et al. (2014) have developed and analysed a high-quality reference sheep genome and transcriptomes from 40 different tissues. Various genes encoding keratin cross-linking proteins associated with rumen evolution have been mapped and found that these are overexpressed. Further, genes involved in lipid metabolism have been selected and were amplified which showed differential or altered pattern of tissue expression. This may be because of changes in the barrier lipids of the skin, an interaction between lipid metabolism and wool synthesis, and an increased role of volatile fatty acids in ruminants compared with nonruminant animals. Similarly genome of various gut microfloras has been sequenced. The draft genome of six isolates *Coxiella burnetii* European isolates from ruminants has been recently sequenced (Sidi-Boumedine et al. 2014). The availability of these genomes will help in understanding the potential host specificity and performance. Thus, by having this accurate and fast information, manipulation in rumen can be designed which can further improve the performance of the rumen ecosystem in digesting plant fibre.

15.3 Rumen Manipulation Using Proteomic Approaches

For rumen microbial diversity, proteomics is another useful approach through which the real expression pattern can be obtained. It is a powerful tool for the identification of proteins and to study their localization, functions, modifications and possible interactions or complexes they can form. This has been nicely reviewed recently giving an overview of proteome studies of various tissues and biological fluids in different farm animals (Jayasri et al. 2014). Applying these techniques, the exact enzyme machinery or the proteins involved in fibre digestion as well as other physiological activities in the rumen can be analysed. Wang et al. (2012) investigated the changes of rumen papillae proteome affected by dietary change. In their study twenty-four primiparous Holstein cows (*Bos taurus*) with similar day in milk and body weight were randomly assigned two treatments (12 cows fitted with ruminal cannulas). (1) Cows were fed high-concentrate diet consisted of corn (*Zea mays*) stalk and concentrate, while (2) cows were fed low-concentrate diet consisted of Chinese wild rye (*Elymus chinensis*), corn silage, alfalfa (*Medicago sativa*) hay and concentrate in a 4-week trial. Rumen papillae were collected from ventral rumen wall of cows after a 4-week trial. Alteration of protein were detected and identified using two-dimensional gel electrophoresis in combination with matrix-assisted laser desorption/ionization time-of-flight mass spectrometry (MALDI-TOF). The results showed that acyl-CoA synthetase family member 2, hydroxymethylglutaryl-CoA synthase, peroxiredoxin-2 and voltage-dependent anion-selective channel protein 1 were upregulated in response to high-concentrate diet, while keratin 6A and larva-specific keratin (RLK) were upregulated in response to low-concentrate diet. The identified proteins were mainly associated with functions related to stress, metabolism and signal transduction. Based on these findings, it was concluded that the changes of rumen papillae proteins were affected by dietary composition that mediate rumen epithelial adaptation to dietary change. Similarly, Holligan et al. (2013) investigated proteins which play regulatory roles in the pancreas of domestic cattle (*Bos taurus*) and that were associated with differences in pancreatic α-amylase activity. To study the proteomic profile, two groups of crossbred steers were kept on a high-moisture corn-based diet, and their α-amylase activity was noted and ranked accordingly.

Similarly, proteome of rumen epithelial tissue was analysed by SDS-PAGE coupled with LC–MS/MS. Eight hundred and thirteen nonredundant proteins were identified of which 7.4 % featured membrane-spanning domains and 15.4 % harboured a signal peptide (Yang et al. 2013). According to the gene ontology annotation, the most abundant proteins which exhibited binding activities related to their molecular functions were proteins of cellular components or belonged to various metabolic processes. A predominant group of canonical pathways in the rumen epithelial tissue was identified using the IPA software. These findings have paved the way for future research and provide a useful resource for further experiments aiming at investigation of alterations of proteins in response to diet changes.

In the similar context, further analysis of secretome of some rumen-associated microflora also may advance the existing knowledge and can be managed to use efficiently. Wang et al. (2011) created an effective platform which uses a combination of transcriptomic and proteomic approaches to *Neocallimastix patriciarum* to accelerate gene identification, enzyme classification and application in rice straw degradation. By conducting complementary studies of transcriptome (Roche 454 GS and Illumina GA IIx) and secretome (ESI–trap LC–MS/MS), they identified 219 putative GH contigs and classified them into 25 GH families. Through this approach they recognized four major enzymes involved in rice straw degradation including β-glucosidase, endo-1,4-b xylanase, xylanase B and Cel48A exoglucanase. From the sequences of assembled contigs, 19 putative cellulase genes, including the GH1, GH3, GH5, GH6, GH9, GH18, GH43 and GH48 gene families, have been cloned, which were highly expressed in *N. patriciarum* cultures grown on different feedstocks. These GH genes were expressed in *Pichia pastoris* and/or *Saccharomyces cerevisiae* for functional characterization, which can be further administered in the rumen as per requirement. In a similar approach, extracellular polysaccharide-degrading proteome of *Butyrivibrio proteoclasticus* was analysed for the growth phase-dependent abundance patterns of secreted proteins that recovered from cells grown in vitro with xylan or xylose as sole carbon source (Dunne et al. 2012). A total number of five polysaccharidases and two carbohydrate-binding proteins (CBPs) were found among 30 identified secreted proteins. The overall finding of this experiment established (1) the 17.5-fold abundance of endo-1,4-β-xylanase Xyn10B suggesting its major role in hemicellulose degradation and (2) the modulation of the secretion of hemicellulose-degrading enzymes and ATP-dependent sugar uptake systems in *B. proteoclasticus* advocating the notion about this organism as a major contributor to polysaccharide degradation in the rumen. In another study, Bond et al. (2012) studied membrane proteins of the rumen bacterium *Butyrivibrio proteoclasticus* and identified 13 membrane proteins predicted to function as carbohydrate transporters when cells were grown with fructose or xylan as the sole substrate. In another study to get insights into *Escherichia coli* O157:H7 (O157), survival mechanisms in the bovine rumen, growth characteristics and proteome of O157 cultured in rumen fluid (RF; pH 6.0–7.2 and low volatile fatty acid content) obtained from rumen-fistulated cattle fed low-protein-content 'maintenance diet' under diverse in vitro conditions were characterized (Kudva et al. 2014). Bottom-up proteomics (LC–MS/MS) of whole cell lysates of O157 cultured under anaerobic conditions in filter-sterilized RF (fRF; devoid of normal ruminal microbiota) and nutrient-depleted and filtered RF (dRF) resulted in an anaerobic O157 fRF and dRF proteome comprising 35 proteins functionally associated with cell structure, motility, transport, metabolism and regulation, but interestingly not with O157 virulence.

In silico, secretome proteins can be predicted from completely sequenced genomes using various available algorithms that identify membrane-targeting sequences. However, this approach is impractical in case of metasecretome (collection of surface, secreted and transmembrane proteins from environmental microbial communities). The major limitation is due to poor representation of metasecretome open reading frames (ORFs) in the dataset which comprise only 10–30 % of total metagenome causing an overall low coverage even in large-scale projects (Ciric et al. 2014). To address the issue Ciric et al. (2014) combined secretome-selective phage display and next-generation sequencing and carried out the sequence analysis of complex rumen microbial community on the metasecretome component of the metagenome. Using this strategy they not only achieved 29-fold higher enrichment of secreted fibrolytic enzymes from the plant-adherent microbial community of the bovine rumen but also identified hundreds of heretofore rare modules belonging to cellulosomes, cell-surface complexes specialized for recognition and degradation of the plant fibre. It infers that metasecretome phage display combined with next-generation sequencing can successfully be employed in the study of diversity of low-abundance surface and secreted proteins.

15.4 Rumen Manipulation Using Metabolomic Approaches

Metabolomics is a recent approach and emerging field of 'omics' science that uses various high-throughput approaches, such as ^{1}H-NMR spectroscopy, DNA sequencing, RNA sequencing and protein sequencing coupled with multivariate analysis, which extract a lot of amount of data related to metabolic phenotypes in mammals, plants and microbes (Vinayavekhin et al. 2010). Metabolomics has opened new avenues in the field of nutrition research, allowing scientists to explore the complex metabolic pathways in response to diet (Wishart et al. 2008). Rumen microfloras are easy to manipulate, so it has also been a viable approach to the pathways in these organisms and synthesize the desired product. For example, lactate dehydrogenase (LdhA), pyruvate lyase (PflB), phosphotransacetylase (Pta) and acetate kinase (AckA) were deleted from *Mannheimia succiniciproducens* MBEL55E to engineer it which resulted in a succinic acid-overproducing variant, *M. succiniciproducens* LPK7 (Lee et al. 2006). However, the slower cell growth resulting from the gene deletions led to decrease in the overall volumetric succinic acid productivity (Lee et al. 2006). In order to better understand the physiological changes in the LPK7 strain compared with the MBEL55E strain, the proteomes of *M. succiniciproducens* LPK7 at the exponential and stationary phases were analysed and were compared with those obtained with MBEL55E. This makes it understandable why pyruvic acids were excreted in the LPK7 strain differently from the MBEL55E strain and were able to solve the problem by the supplementation of pantothenate and L-cysteine as a proof of the hypothesis (Lee and Lee 2009). Metabolomics was successfully employed by Ametaj et al. (2010) to evaluate changes in rumen metabolites of dairy cows fed increasing proportions of barley grain. ^{1}H-NMR spectroscopy was used to analyse rumen fluid samples representing four different diets, and an increase was observed in the concentrations of rumen methylamine as well as glucose, alanine, maltose, propionate, uracil, valerate, xanthine, ethanol and phenylacetate for cows fed 30 % and 45 % grain. In another attempt made by the similar group, Saleem et al. (2012) have tried to understand why high-grain diets during early lactation have a high incidence of metabolic disorders. Rumen fluid samples from dairy cows fed four different diets were taken, and metabolomic analysis was carried out by an integrated approach using proton nuclear magnetic resonance spectroscopy, gas chromatography–mass spectrometry and direct flow injection tandem mass spectroscopy. A total number of 93 metabolites were identified and quantified in rumen samples from dairy cows fed graded amounts of barley grain (i.e. 0 %, 15 %, 30 % and 45 % of diet dry matter). Rumen metabolites arising from the diet containing 45 % barley grain were found clearly different from those containing 0 %, 15 % and 30 % barley grain. This metabolic profile revelled the effect of high-grain diets (>30 %) which resulted in increased concentrations of several toxic, inflammatory and unnatural compounds in rumen fluid and perturbations in several amino acids. This established the utility of multiple metabolomic platforms for understanding the metabolic causes and effects. Recently, the microbial metabolite profile in dairy cows by the inclusion of roughage of different types in diet was investigated (Zhao et al. (2014). Diet containing four types of roughage (corn stover (CS group) or a mixture of alfalfa hay, *Leymus chinensis* hay and corn silage (MF group)) was fed to the dairy cows, and rumen fluid was analysed using nuclear magnetic resonance (NMR) spectroscopy. A significant difference between the ruminal metabolome of the CS and MF groups at both time points was noticed, while the MF group has higher levels of acetate, valerate, hydrocinnamate and methylamine and lower levels of glucose, glycine, propionate and isovalerate compared to the CS group. This study further established that different types of roughage can influence the metabolite profile in the rumen specially the content and pattern of amino acid, organic acid, etc. Further, Liang et al. (2014) studied the role of fatty acid-binding protein 3 (FABP3) in milk fat synthesis using quantitative reverse transcriptase (qRT)-PCR, Western blotting and fluorescent immune-staining techniques. It was found that FABP3 upregulated the expression of sterol regulatory element-binding protein and proliferator-activated receptor gamma to increase lipid droplet accumulation. These results helped us in understanding the mechanism of FABP3 in regulating milk fat synthesis by throwing new insights.

15.5 Conclusion

With the help of these recent 'omics' techniques, fast and accurate information can be generated which may give several immediate solutions to improve the rumen performance. It can also be used to find out or map various genomic, proteomic and metabolic pathways that makes easier to manipulate the rumen performance by just diet or other means.

References

Ametaj BN, Zebeli Q, Saleem F et al (2010) Metabolomics reveals unhealthy alterations in rumen metabolism with increased proportion of cereal grain in the diet of dairy cows. Metabolomics 6:583–594

Bond JJ, Dunne JC, Kwan FYS et al (2012) Carbohydrate transporting membrane proteins of the rumen bacterium, *Butyrivibrio proteoclasticus*. J Proteomics 75:3138–3144

Chaucheyras-Durand F, Ossa F (2014) The rumen microbiome: composition, abundance, diversity, and new investigative tools. Prof Anim Sci 30:1–12

Ciric M, Moon DC, Leahy LC et al (2014) Metasecretome-selective phage display approach for mining the functional potential of a rumen microbial community. BMC Genomics 15:356

Collado MC, Sanz Y (2007a) Quantification of mucosa-adhered microbiota of lambs and calves by the use of culture methods and fluorescent in situ hybridization coupled with flow cytometry techniques. Vet Microbiol 121:299–306

Collado MC, Sanz Y (2007b) Characterization of the gastrointestinal mucosa associated microbiota of pigs and chickens using culture-based and molecular methodologies. J Food Prot 70:2799–2804

Collado MC, Isolauri E, Laitinen K et al (2008) Distinct composition of gut microbiota during pregnancy in overweight and normal-weight women. Am J Clin Nutr 88:894–899

Dunne JC, Li D, Kelly WJ et al (2012) Extracellular polysaccharide-degrading proteome of *Butyrivibrio proteoclasticus*. J Proteome Res 11(1):131–142

Eckburg PB, Bik EM, Bernstein CN et al (2005) Diversity of the human intestinal microbial flora. Science 308:1635–1638

Gong J, Yang C (2012) Advances in the methods for studying gut microbiota and their relevance to the research of dietary fiber functions. Food Res Int 48:916–929

Habib G, Raza M, Saleem M (2008) Effect of tree leaves with or without urea as a feed supplement on nutrient digestion and nitrogen balance in sheep. Anim Feed Sci Technol 144:335–343

Holligan S, Wang J, Cant JP et al (2013) A proteomics approach to detect tissue-wide adaptive changes in the pancreas associated with increased pancreatic α-amylase activity in domestic cattle (*Bos taurus*). Comp Biochem Physiol Part D Genomics Proteomics 8:65–71

Jami E, Mizrahi I (2012) Composition and similarity of bovine rumen microbiota across individual animals. PLoS One 7(3):e33306

Jayasri K, Padmaja K, Prasad PE (2014) Proteomics in animal health and production. IOSR J Agric Vet Sci 7:50–56

Jiang Y, Xie M, Chen W et al (2014) The sheep genome illuminates biology of the rumen and lipid metabolism. Science (New York, NY) 344:1168–1173

Kudva IT, Stanton TB, Lippolis JD (2014) The *Escherichia coli* O157:H7 bovine rumen fluid proteome reflects adaptive bacterial responses. BMC Microbiol 14:48

Lee JW, Lee SY (2009) Proteome-based physiological analysis of the metabolically engineered succinic acid producer *Mannheimia succiniciproducens* LPK7. Bioprocess Biosyst Eng. doi:10.1007/s00449-009-0339-4

Lee SJ, Song H, Lee SY (2006) Genome-based metabolic engineering of *Mannheimia succiniciproducens* for succinic acid production. Appl Environ Microbiol 72:1939–1948

Li M, Gong J, Cottrill M et al (2003) Evaluation of QIAamp® DNA Stool Mini Kit for ecological studies of gut microbiota. J Microbiol Methods 54:13–20

Li Z, Zhao H, Yang P et al (2013) Comparative quantitative analysis of gene expression profiles of glycoside hydrolase family 10 xylanases in the sheep rumen during a feeding cycle. Appl Environ Microbiol 79:1212–1220

Liang M-Y, Hou X-M, Qu B et al (2014) Functional analysis of FABP3 in the milk fat synthesis signaling pathway of dairy cow mammary epithelial cells. In Vitro Cell Dev Biol Anim. doi:10.1007/s11626-014-9780-z

Mohammadzadeh H, Yáñez-Ruiz DR, Martínez-Fernandez G et al (2014) Molecular comparative assessment of the microbial ecosystem in rumen and faeces of goats fed alfalfa hay alone or combined with oats. Anaerobe 29:52–58

Paliy O, Kenche H, Abernathy F et al (2009) High-throughput quantitative analysis of the human intestinal microbiota with a phylogenetic microarray. Appl Environ Microbiol 75:3572–3579

Palma DG, Nadal I, Collado MC et al (2009) Effects of a gluten-free diet on gut microbiota and immune function in healthy adult human subjects. Br J Nutr 102:1154–1160

Palmer C, Bik EM, DiGiulio DB et al (2007) Development of the human infant intestinal microbiota. PLoS Biol 5:e177

Rajilic-Stojanovic M, Maathuis A, Heilig HGHJ et al (2010) Evaluating the microbial diversity of an in vitro model of the human large intestine by phylogenetic microarray analysis. Microbiology 156:3270–3281

Saleem F, Ametaj BN, Bouatra S et al (2012) A metabolomics approach to uncover the effects of grain diets on rumen health in dairy cows. J Dairy Sci 95:6606–6623

Sekhavati MH, Mesgaran MD, Nassiri MR et al (2009) Development and use of quantitative competitive PCR assays for relative quantifying rumen anaerobic fungal population s in both in vitro and in vivo systems. Mycol Res 113:1146–1153

Sidi-Boumedine K, Ellis RJ, Adam G et al (2014) Draft genome sequences of six ruminant *Coxiella burnetii* isolates of European origin. Genome Announc 2:e00285-14. doi:10.1128/genomeA. 00285-14

Snel J, Harmssen HJM, van der Wielen PWJJ et al (2002) Dietary strategies to influence the gastrointestinal microflora of young animals, and its potential to improve intestinal health. In: Blok MC et al (eds) Nutrition and health of the gastrointestinal tract. Wageningen Academic Publishers, Wageningen, pp 37–69

Teather RM (1985) Application of gene manipulation to rumen microflora. Can J Anim Sci 65:563–574

Vinayavekhin N, Homan EA, Saghatelian A (2010) Exploring disease through metabolomics. J Agric Food Chem 5:91–103

Wang T-Y, Chen H-L, Lu M-Y J et al (2011) Functional characterization of cellulases identified from the cow rumen fungus *Neocallimastix patriciarum* W5 by transcriptomic and secretomic analyses. Biotechnol Biofuel 4:24

Wang et al (2012) Effects of dietary composition on the rumen papillae tissue proteome of cows by the combination of two-dimensional gel electrophoresis and mass spectrometry. J Agri Biotechnol 20:1042–1047. ISSN 1674–7968 CN: 11-3342/S

Wang LF, Yang GY, Yang GQ et al (2014) Effect of zinc source on the expression of ZIPII transporter genes in Guanzhong dairy goats. Anim Feed Sci Technol. http://dx.doi.org/10.1016/j.anifeedsci.2014.06.006

Wishart DS, Lewis MJ, Morrissey JA et al (2008) The human cerebrospinal fluid metabolome. J Chromatogr B Anal Technol Biomed Life Sci 871:164–173

Yang Y, Wang J, Yuan T et al (2013) Proteome profile of bovine ruminal epithelial tissue based on GeLC–MS/MS. Biotechnol Lett 35:1831–1838

Zhao J, Bu D, Sun P et al (2014) Metabolomics analysis reveals large effect of roughage types on rumen microbial metabolic profile in dairy cows. Lett Appl Microbiol 59:79–85

Zoetendal EG, Collier CT, Koike S et al (2004) Molecular ecological analysis of the gastrointestinal microbiota: a review. J Nutr 134:65–472

Part IV

Exploration and Exploitation of Rumen Microbes

Rumen Metagenomics

16

Robert W. Li

Abstract

The rumen microbiome plays a critical role in normal physiology and nutrition of ruminants. Alterations in the rumen microbiome have important physiological and pathological implications. The advent of next-generation sequencing technologies and rapid development of computational tools and reference databases provide powerful tools in rumen microbiome studies. Rumen metagenomics enables studies on the collective genetic structure and functional composition of the rumen microbial community in a culture-independent manner and can be simply divided into functional metagenomics and sequencing-based computational metagenomics. Recent progresses in mining the rumen microbial community for novel enzymes, such as fibrolytic enzymes, or other biomolecules for industry and biotechnology applications using functional screening are discussed. Rapid advances in computational metagenomic tools and methods are summarized. Metagenomics has provided novel insights into the structure and function of the rumen microbiome. Recent efforts suggest that the core rumen microbiome consists of 8 phyla and 15 families, which likely contribute to the basic function of the rumen. Systematic investigations of the rumen microbiome, including its viral (virome) and plasmid (plasmidome) fractions, have revealed previously unrecognized biodiversity in the rumen. Resistance and resilience of the rumen microbial community in response to perturbation is also discussed. Moreover, the need for mechanistic models and applications of general ecological theories and principles in rumen metagenomic studies is emphasized.

R.W. Li (✉)
United States Department of Agriculture, Agriculture Research Service, Animal Genomics and Improvement Laboratory, 10300 Baltimore Avenue, Beltsville, MD 20705, USA
e-mail: Robert.Li@ars.usda.gov

A.K. Puniya et al. (eds.), *Rumen Microbiology: From Evolution to Revolution*, DOI 10.1007/978-81-322-2401-3_16

Keywords
Rumen • Microbiome • Microbiota, metagenomics • 16S rRNA gene • Functional screening • Ruminant • Microbial • Ecosystem • Resilience • Next-generation sequencing • Assembly

16.1 Introduction

The rumen microbiome represents the totality of rumen microorganisms, their genetic elements, and environmental interactions. The rumen microbiome plays an essential role in ruminant physiology and nutrition and pathology as well as host immunity. Rumen microorganisms convert plant fiber to short-chain fatty acids (SCFA), which contribute up to 75 % of the total metabolizable energy in ruminants. In addition to the fibrolytic capacity, rumen microorganisms also participate in ruminal nitrogen metabolism, including dietary protein degradation. However, nitrogen losses during protein degradation and methane produced during rumen fermentation are substantial contributors to water and air pollution as well as global warming. Rumen microorganisms produce a large amount of vitamins. As a result, ruminants generally do not need dietary supplement of water-soluble vitamins and vitamin K. Rumen microorganisms are able to modulate nutrient absorption and may be among the major determinants of nutrient utilization efficiency (Li et al. 2012a; Jami et al. 2014). Moreover, ruminal biohydrogenation, the saturation process of dietary unsaturated fatty acids controlled by rumen microorganisms, can be manipulated for healthier meat products (Jenkins et al. 2008). It is well known that rumen microbes play a key role in detoxifying plant secondary compounds (Wallace 2008). The involvement of the rumen microbiome in xenobiotic metabolism has been well documented (Li et al., 2014). Previous studies have identified rumen microbes responsible for the degradation of nitroaromatic explosive compounds, such as 2,4,6-trinitrotoluene (De Lorme and Craig 2009) and hexahydro-1,3,5-trinitro-1,3,5-triazine (Eaton et al. 2011).

The complexity of the rumen microbiome has long been appreciated, as evidenced by the presence of myriad microbial interactions (Li et al. 2012a). One of the major obstacles hindering our understanding of the structure and function of the rumen microbiome is that only approximately 11 % of rumen bacteria appear to be culturable (Edwards et al. 2004). DNA fingerprinting techniques widely used in earlier studies, such as terminal-restriction fragment length polymorphism (t-RFLP) and single-strand conformation polymorphism (SSCP), have limited throughput and low resolution and are therefore unable to provide a holistic view of the structure and function of the rumen microbiome. Furthermore, the rumen microbiome functions as a tightly integrated system in which all resident species interacts closely to contribute to its emergent properties. Predominant species perform all major microbial conversions in this ecosystem. Nevertheless, numerically minor species also play an important role in maximizing rumen ecosystem outputs. Disruption of one species could cause a chain reaction and result in undesired or unpredicted consequences. These properties call for a move from studies of individual rumen microorganisms in isolation or in pure culture to community-level studies, especially in their natural habitats.

Metagenomics has emerged in the past few years as a powerful tool for studying the rumen microbiome, thanks to the advent of next-generation sequencing (NGS) technologies and rapid progress in reference databases and bioinformatic tools. Metagenomics addresses the collective genetic structure and functional composition of a microbial community without the bias or necessity for culturing its individual inhabitants (Galbraith et al. 2004). Rumen metagenomics enables comprehensive studies of the structure and function of the rumen microbiome using culture-independent approaches. Rumen metagenomics generally includes two major arenas: high-throughput screening of cloned expression libraries made from rumen metagenome DNA for gene products of interest (functional metagenomics) and sequencing-based characterization of the aggregate collection of genomes and genes present in rumen microbial communities, at both DNA (metagenomics) and RNA levels (metatranscriptomics). Functional screening technology was first applied to rumen materials to mine novel enzymes in 2005 (Ferrer et al. 2005), whereas the first publication using next-generation sequencing-based rumen computational metagenomics can be traced back to 2009 (Brulc et al. 2009). Since then, metagenomic technologies have been extensively utilized to investigate rumen microbial communities. The rumen of an individual animal is believed to harbor hundreds and up to 1,000 microbial species. Therefore, microarrays (such as PhyloChips), DNA fingerprinting techniques, or traditional Sanger sequencing-based methods that are unable to provide a sequencing depth of greater than 1,000 sequence reads will be excluded for discussion in this chapter. We will summarize recent advances in metagenomic technologies and novel metagenomic insights into the structure and function of the rumen microbiome.

16.2 Functional Metagenomics

Functional metagenomics is the study of the collective genome of a microbial community by expressing it in a foreign host (Ekkers et al. 2012). The vast majority of enzymes that catalyze biochemical reactions are encoded by genes present in microbial communities under various environmental conditions. For example, a recently developed database lists 510 commercially useful enzymes used in various sectors, including agriculture, energy, and biomedicine (Sharma et al. 2010). Therefore, functional screening has become an increasingly important field for discovering novel biomolecules for applications in biotechnology and medicine. This approach relies on cloning of vast genetic diversity from a target habitat in various vectors (e.g., plasmids, cosmids, fosmids, or bacterial artificial chromosomes) and then expressing cloned metagenome libraries in foreign host systems (e.g., *E. coli*) followed by detection and characterization of desired functional activities in the expression libraries using various strategies (Simon and Daniel 2009). Functional screening provides direct access to largely unexploited microbial genetic diversity in the environment. Lignocellulose biomass, including cellulose, hemicellulose, pectin, and lignin, is the most abundant source of organic carbon on the planet. Efficient enzymatic conversion of biomass into biofuel has been of great interest recently. The complete degradation of lignocelluloses requires a concerted action of dozens of enzymes from various families, such as endo-β-1,4-glucanases, cellobiohydrolases, β-glucosidases, endoxylanases, β-xylosidases, α-l-arabinofuranosidases, acetyl xylan esterases, feruloyl esterases, and α-glucuronidases. Previous studies suggest that the rumen microbial ecosystem harbors a dazzling array of microbial diversity (Li et al. 2012a, b, c; Sparks et al. 2012) and is a rich source of efficient fibrolytic enzymes. A relatively small fraction of rumen microorganisms have been successfully cultured to date. The largely unexplored ruminal microbial diversity represents an untapped source of unique lignocellulose-digesting enzymes, especially those with multiple functions. Numerous efforts have been made to isolate fiber-digesting enzymes from the rumen, including various hydrolases from at least 8 glycosyl hydrolase families, such as GH3, GH5, GH8, GH9, GH10, GH13, GH26, GH43, GH48, and GH57. Morgavi et al. (2013) summarized the screening results prior to 2012. Results from functional screening of the rumen microbiome since 2012 are listed in Table 16.1.

Despite the fact that the huge potential of functional screening in mining genetic diversity for biotechnology applications has been demonstrated by the abovementioned case studies,

Table 16.1 Lignocellulose-digesting enzymes mined from the rumen using functional metagenomic approaches since 2012

Enzyme (family)	Rumen	Reference
Cellulases (GH5)	Buffalo	Nguyen et al. (2012)
Cellulases (GH5)	Buffalo	Cheema et al. (2012)
Cellulases (GH5, GH9, GH45, GH48)	Yak	Dai et al. (2012)
Endocellulase/xylanase (GH5)	Bovine	Rashamuse et al. (2013)
Endoglucanase cellulases	Cow	Gong et al. (2012)
Endohemicellulases (GH8, GH10, GH11, GH26, GH28, GH53)	Yak	Dai et al. (2012)
Esterase	Cow	Kim et al. (2012a)
Exocellulase (GH48)	Cow	Ko et al. (2013)
Feruloyl esterase	Calf	Ferrer et al. (2012)
Feruloyl esterase	Cow	Cheng et al. (2012a)
Glycosyl hydrolases (GH43)	Calf	Ferrer et al. (2012)
Glycosyl hydrolases (GH5)	Reindeer	Pope et al. (2012)
Xylanase (GH10)	Bovine	Cheng et al. (2012b)
Xylanase (GH10)	Bovine	Gong et al. (2012)
α-Glucuronidase (GH67)	Cow	Lee et al. (2012a)
β-Glucosidase (GH3)	Bovine	Gruninger et al. (2014)

numerous challenges remain. First, only a small fraction of functional diversity is captured in expression libraries, partially due to the difficulty in expressing target genes in a foreign host. Moreover, current methods to detect desired function or enzyme activities are less sensitive; and the throughput of screening methods is relatively low. Novel strategies, such as fractionation of the microbial community using habitat biasing methods to reduce the complexity of the microbiome or to enrich desired activities, have been developed to overcome these limitations (Ekkers et al. 2012). Furthermore, the potential power of novel technology using in vitro compartmentalization (IVC) in combination with fluorescent-activated cell sorting (FACS) in aiding functional screening of complex microbial ecosystems has been recognized (Ferrer et al. 2009). It is foreseeable that in combination with rapid advances in directed evolution techniques and methods (Dalby 2011), more enzymes and biomolecules with improved activities will be isolated using functional screening from the rumen microbiome for a wide range of applications.

16.3 Computational Metagenomics: Methods and Approaches

The advent of ultrahigh-throughput next-generation sequencing technologies and rapid development of computational tools and resources have stimulated computational metagenomic studies. As a result, computational metagenomics provides novel insights into the structure and function of microbial communities of host-associated habitats or from environmental samples at unprecedented resolution. The approach targeting small subunits (SSU) of rRNA genes (16S or 18S) allows us to interrogate the microbial composition and structure of the rumen microbiome. The whole-genome shotgun (WGS) approach provides unique opportunities to gain novel insights into the protein repertoire and metabolic potential of the microbiome, which lead to biological pathway reconstruction. Moreover, WGS approach enables taxonomical assignment to understand the microbial composition and structure of the rumen microbiota.

16.3.1 Ribosomal RNA Gene-Based Analysis

SSU ribosomal RNA genes, such as 16S rRNA for prokaryotes and 18S rRNA genes for eukaryotes, can be amplified from metagenomic DNA of various fractions of rumen materials. These two genes are most frequently used for phylogenetic analysis and microbial diversity studies in the rumen. Taxonomic informativeness of 9 well-defined hypervariable regions (V1 to V9) of the 16S rRNA gene varies tremendously (Chakravorty et al. 2007). As a result, effects of various primer combinations on classification accuracies have been designed and compared (Nossa et al. 2010; Soergel et al. 2012). The position of primers and amplicon length are major determinants of taxonomic precision. Most importantly, taxonomic informativeness of primers is habitat dependent. No primers are truly universal and work best in all environments (Soergel et al. 2012). For example, primer pairs 343 F and 798R, targeting on hypervariable regions V3 to V4, produce maximal classification accuracy under the current limitation of NGS platforms and may be the most suitable for human foregut microbiome studies (Nossa et al. 2010). Primer pairs targeting on V1 to V3, V3 to V5, and V6 to V9 generally result in overall similar and yet accurate classification with minor bias (Vilo and Dong 2012). Indeed, primers targeting V1 to V3 and V3 to V5 regions are commonly used in rumen microbiome studies (Table 16.2).

The amplicons from target regions of 16S (18S) rRNA genes are sequenced using next-generation sequencers, such as 454 FLX or Illumina sequencers. While barcoded pyrosequencing has been the mainstay in sequencing the 16S amplicons of the rumen samples (Jami et al. 2013; Li et al. 2012c; Wu et al. 2012b), Illumina-based sequencing technology is increasingly gaining attention. The newly launched Illumina MiSeq sequencer with version 3 reagent kits enables generation of up to 25 million sequences with a length up to 2×300 bp (pair end). The reagent cost for such a run is approximately $1400.

Raw sequence quality needs to be checked and then filtered and trimmed. Sequencing error and PCR single-base substitutions from the 454 platform can be removed using AmpliconNoise, a development of the PyroNoise algorithm that is capable of separately removing 454 sequencing errors and PCR single-base errors (Quince et al. 2011). The Perseus program can be used to remove chimeras. Processed 16S sequence reads are then analyzed using taxonomy-dependent and taxonomy-independent approaches. The taxonomy-dependent approach generally assigns 16S sequences to various levels of taxa based on sequence similarities to annotated sequences deposited in public databases. The commonly used SSU databases include EzTaxon-e (Kim et al. 2012b), Greengenes (DeSantis et al. 2006), RDP (Cole et al. 2009), and SILVA (Quast et al. 2013). The algorithm RDP Classifier (Wang et al. 2007) is among the most frequently used programs for taxonomic classification and has resulted in the publication of more than 400 articles since its launch. However, inherent limitations of this approach are obvious: (1) inability to assign novel sequences from previously undescribed species that have no matches in existing reference databases, (2) accuracy and robustness of taxonomic classification that is dependent on the coverage and quality of the database used, and (3) low resolution. This approach is often unable to assign input query sequences to species or strain levels. These limitations become more serious for rumen microbiome studies because the SSU sequences of rumen origin are particularly underrepresented in public databases. To overcome these limitations, the taxonomy-independent clustering approach has been developed. This approach uses various clustering algorithms to assign query sequences into operational taxonomic units (OTUs) based on a distance matrix at a specified threshold (Chen et al. 2013). Its independence from existing databases allows the analysis of novel sequences. More than 15 taxonomy-independent algorithms, such as CD-HIT-OTU (Fu et al. 2012; Li and Godzik 2006), CROP (Hao et al. 2011), ESPRIT (Sun et al. 2009) and ESPRIT-tree (Cai and Sun 2011),

Table 16.2 A summary of metagenomic studies in ruminants using next-generation sequencing technologies

Sequencing	Species	Reference
16S hypervariable regions targeted		
V1–V3 (and V4–V5; V6–V8)	Buffalo	Pitta et al. (2014)
V1–V3	Cattle (Angus heifers)	Petri et al. (2013a, b)
V1–V3	Cattle (cows)	Mao et al. (2013)
V1–V2	Cattle (cows)	de Menezes et al. (2011)
V6–V8	Cattle (cows)	Hess et al. (2011)
V1–V3	Cattle (Friesian cows)	Sandri et al. (2014)
V3–V5	Cattle (Holstein calves)	Li et al. (2012a)
V1–V3	Cattle (Holstein calves)	Malmuthuge et al. (2014)
V3–V5	Cattle (Holstein cows and Angus beef steers)	Wu et al. (2012a)
V2–V3	Cattle (Holstein cows)	Jami et al. (2013)
V2–V3	Cattle (Holstein cows)	Jami and Mizrahi (2012)
V2–V3	Cattle (Holstein cows)	Jami et al. (2014)
V3–V5	Cattle (Holstein cows)	Li et al. (2012b)
V1–V2	Cattle (Holstein cows)	Pinloche et al. (2013)
V3–V4	Cattle (Holstein cows)	Thoetkiattikul et al. (2013)
V6–V8	Cattle (Nellore steer)	de Oliveira et al. (2013)
V1–V3, V4 (archaeal V3, V6–V8), 18S, ITS1	*Cervus*, red deer, sheep, cattle	Kittelmann et al. (2013)
V1–V3	Goats	Lee et al. (2012b)
V1–V3	Goats	Huo et al. (2014)
V1–V3	Reindeer	Pope et al. (2012)
V1–V2	Sheep	Castro-Carrera et al. (2014)
V6–V8	Sheep (West African Dwarf)	Omoniyi et al. (2014)
Whole-genome shotgun (WGS)		
WGS	Buffalo	Singh et al. (2012a, b)
WGS	Camels	Bhatt et al. (2013)
WGS	Cattle (Angus steers)	Brulc et al. (2009)
WGS	Cattle (cows)	Ferrer et al. (2012)
WGS	Cattle (cows)	Hess et al. (2011)
WGS	Cattle (Holstein calves)	Li et al. (2012b)
WGS	Cattle (Jersey cows)	Wang et al. (2013)
WGS	Goats	Lim et al. (2013)
WGS	Reindeer	Pope et al. (2012)
WGS	Sheep	Ellison et al. (2014); Li et al. (2014)

MOTHUR (Schloss et al. 2009), UClust (Edgar 2010), and UPARSE (Edgar 2013), have been published to date. A novel algorithm, TBC, incorporating the basic concept of taxonomy into clustering has been published (Lee et al. 2012c). Recently, the relative performance and parameters of some of these algorithms have been compared (Chen et al. 2013; Wu et al. 2012a). For the rumen data, CD-HIT-OTU performs well in our hands (Wu et al. 2012a; Li et al. 2012d). This algorithm, which uses a greedy incremental clustering process to identify OTUs from 16S rRNA gene sequences, is able to assign millions of reads in a relatively short time. Most importantly, the program avoids overestimation of OTUs, a common problem associated with many existing programs, and results in accurate estimation of microbial diversity. In addition to these algorithms, publicly accessible pipelines, such as MOTHUR and QIIME (www.qiime.org), are very popular in analyzing SSU sequences and have been widely used to analyze the rumen datasets (Castro-Carrera et al. 2014; Lee et al. 2012b; Omoniyi et al. 2014; Pitta et al. 2014; Pope et al. 2012). QIIME also wraps other applications, such as FastTree, PyNAST, RDP Classifier, and UClust. The microbial community structure between different samples can then be compared and visualized using UniFrac (Lozupone and Knight 2005) and Fast UniFrac (Hamady et al. 2010).

16.3.2 Whole-Genome Shotgun Approach

WGS sequencing provides an opportunity to analyze both microbial diversity and functionality encoded in the genomes of rumen microbial communities. Driven by its application potential in metagenomics, numerous tools have been developed to analyze WGS data (Fig. 16.1). NGS technologies, such as those from Roche 454 pyrosequencing, Illumina, Ion Torrent, and PacBio platforms, have significantly reduced the time and cost of metagenome sequencing, which are revolutionizing metagenomic studies. Unique features and advantages of various NGS platforms, including future DNA sequencing technologies, have been extensively reviewed (Zhang

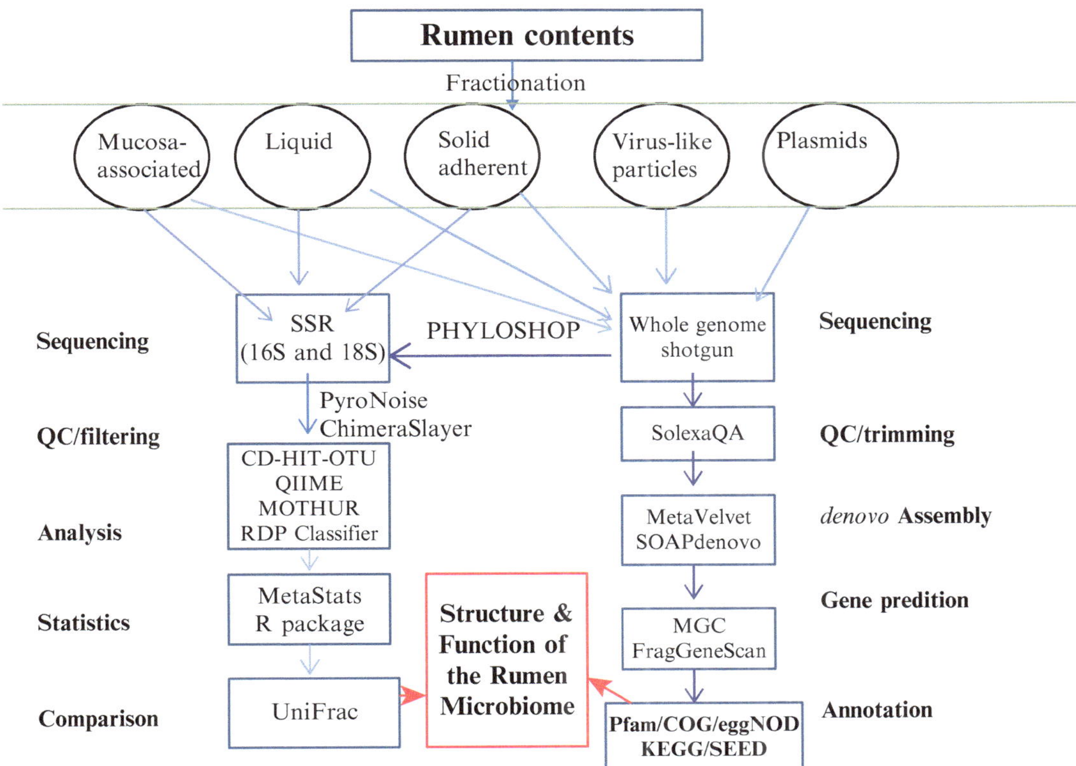

Fig. 16.1 A typical workflow and computational tools for rumen metagenomics

et al. 2012). NGS generally relies on either synthesis or hybridization at a massively parallel scale. For example, Illumina HiSeq 2500, in conjunction with improved version 4 chemistry, generates up to 1 terabase (1 trillion base) of sequence data in a single run (~167 Gb per day). The ultralow cost, enormous throughput, and extreme convenience of these NGS technologies have directly contributed to their instant acceptance and utility in the metagenomic community. Sequences generated by NGS technologies have some unique characteristics, such as short read lengths, platform-specific biases, and relatively high error rates, which could have a significant impact on downstream analyses. The computational challenges in handling short sequence reads have been extensively discussed (Pop 2009; Pop and Salzberg 2008). The challenges generally include difficulties in dealing with repetitive sequences as well as the need to modify existing algorithms to solve platform-specific errors and high error rates (Pop 2009). Furthermore, the production of billions of reads in a single sequencing run by such as Illumina HiSeq 2500 sequencers poses a tremendous challenge on computational resources.

Bioinformatic pipelines for NGS shotgun sequences generally include six steps: raw read quality control (QC) and trimming, assembly, functional annotation and metagenomic pathway reconstruction, taxonomy assignment, statistical analysis, and global network inference (Fig. 16.1). These processes have been extensively reviewed (Kim et al. 2013; Luo et al. 2013). The first step in dealing with WGS sequences involves QC, filtering, and trimming processes. Host sequence contamination can be removed using DeconSeq (Schmieder and Edwards 2011) or BLAST/Blat. Sequencing errors and PCR single-base substitutions as well as chimeras from the 454 platform can be removed using AmpliconNoise. Raw sequences generated by the Illumina platform can be trimmed using SolexaQA (Cox et al. 2010).

WGS sequences after these QC steps will generally need to be assembled into longer contigs for downstream applications. Assembly improves functional annotation. The basic framework of NGS assembly includes 4 stages: a preprocessing filtering, a graph construction process, a graph simplification process, and a post-processing filtering (El-Metwally et al. 2013). More than a dozen short-read assemblers have been developed to facilitate the analysis of short WGS sequences (Huang et al. 2012; Miller et al. 2010). The de Bruijn graph-based approach is among the most commonly used in short-read de novo assemblers, such as ABySS (Simpson et al. 2009), EULER-USR (Chaisson et al. 2009), SOAP*denovo* and its memory-efficient version, SOAP*denovo*2 (http://soap.genomics.org.cn/soapdenovo.html), and Velvet (Zerbino and Birney 2008). Recently, efforts have been made to understand the unique features and limitations of these short-read assemblers (Zhang et al. 2011; Huang et al. 2012; Mende et al. 2012; Vazquez-Castellanos et al. 2014). For 454 pyrosequencing data, Newbler is among the widely used assemblers and has been used in the analysis of rumen WGS sequences (Li et al. 2012b). Genovo, a de novo assembler specifically designed for 454-based metagenomic sequences using a generative probabilistic model (Laserson et al. 2011), and its extended version, Xgenovo (Afiahayati et al. 2013), perform well and are able to generate long contigs (Vazquez-Castellanos et al. 2014). Our results using simulated metagenomic datasets show that ABySS and SOAP*denovo* produce higher N50 and require relatively low memory usage, while Velvet and SOAP*denovo* produce higher genome coverage (Huang et al. 2012). Both Velvet and SOAP*denovo* result in a lower percentage of contig chimerism; while not specifically designed for metagenomic datasets, de Bruijn graph-based assemblers have been proven appropriate for large datasets with hundreds of millions of short reads (Zhang et al. 2011) and have been nevertheless extensively used in metagenomic studies. Recently, by making use of abundance differences and graph connectivity for the decomposition of the de Bruijn graph, an extended version of Velvet, MetaVelvet, has been developed to handle metagenomic data (Namiki et al. 2012). MetaVelvet is able to generate significantly higher N50 scores than other short-read assemblers, leading to an increased number of predicted genes in our hands. Similarly, a de novo metagenomic assembler, Meta-IDBA (Peng et al. 2011), and its revised version, IDBA-UD (Peng et al. 2012), have been shown to generate longer contigs with high accuracy.

Genes or open reading frames (ORF) are then predicted from assembled contigs using a variety of gene prediction (gene-calling) algorithms. A dozen gene-calling programs have been developed for the metagenomic datasets, such as FragGeneScan (Rho et al. 2010), Glimmer-MG (Kelley et al. 2012), MetaGene (Noguchi et al. 2006) and MetaGeneAnnotator (Noguchi et al. 2008), MetaGeneTack (Tang et al. 2013), and Orphelia (Hoff et al. 2009). Gene prediction is an essential step for WGS metagenome data analysis for two reasons: (1) it is necessary for functional annotation and pathway reconstruction, and (2) gene prediction reduces the computational burden of protein similarity searches by nearly a factor of 6, compared to BLASTX (Trimble et al. 2012). In a direct comparison of 5 commonly used gene-calling algorithms, it is found that FragGeneScan performs better than MetaGeneAnnotator, MetaGeneMark, or Orphelia, especially for short reads (<1,000 bp) with sequence errors (Trimble et al. 2012). FragGeneScan combines sequencing error models and codon usages in a hidden Markov model to predict ORF in short reads (Rho et al. 2010) and has been used in the publically available MG-RAST pipeline (Wilke et al. 2013). Recently, a newly improved algorithm, MGC, has been published (El Allali and Rose 2013). This program relies on different models for different regions with various GC-contents and includes amino acid usage features to improve overall accuracy. It performs better in terms of sensitivity and specificity than both FragGeneScan and Orphelia in simulated metagenomic data (El Allali and Rose 2013).

To gain insights into functional potentials, predicted genes are further annotated against various public databases using a homology-based approach. For example, COG (Tatusov et al. 2000) and eggNOG (Powell et al. 2014) databases can be used to classify functional categories of predicted genes. Pfam (Punta et al. 2012), TIGRfam (Haft et al. 2003), and FIGfam (Meyer et al. 2009) databases can be used for protein family analysis. For rumen metagenomic data, the Carbohydrate-Active enZYmes database (CAZy), a database collecting and annotating the families of catalytic and carbohydrate-binding modules of enzymes that degrade, modify, or create glycosidic bonds, has been frequently used to mine fibrolytic enzymes in the rumen (Brulc et al. 2009; Hess et al. 2011). Pfam is a widely used database for protein family analysis and includes more than 14,800 annotated protein families in its latest release (v27.0). These families are also organized into groupings of related families (clans) based on similarity of sequences and structures. Furthermore, Gene Ontology (GO) can be extracted from these protein families using the Pfam2GO program (Hayete and Bienkowska 2005). The metagenomic features, such as COG functional profiles and metabolic subsystem data, between samples from two treatment groups can be analyzed using statistical packages, such as MetaStats (White et al. 2009). Metabolic pathways can then be reconstructed using databases such as the Kyoto Encyclopedia of Genes and Genomes (KEGG) (Kanehisa 2002) and BRENDA, the enzyme database (Schomburg et al. 2004). Metabolic pathways that differ between samples from two treatment groups can then be analyzed for statistical significance using MetaPath (Liu and Pop 2011) and Metagenomic Annotation Networks (Vey and Moreno-Hagelsieb 2012). Furthermore, network analysis tools and algorithms can be used to infer global co-occurrence patterns for the microbiome-wide microbial interactions (Faust and Raes 2012; Faust et al. 2012). The tools available for co-occurrence and association network analysis in metagenomic databases include CoNet (http://psbweb05.psb.ugent.be/conet/download.php), extended Local Similarity Analysis (eLSA) (Xia et al. 2011), QIIME, and a metagenome-wide association study (Qin et al. 2012).

Recently, using statistical distribution methods that ignore known biological processes to handle metagenomic data has been questioned (Liberles et al. 2013). These authors suggest that mechanism models based on ecological relationships, such as predator–prey dynamics and competitive relationships that widely exist in the rumen microbial community, should be incorporated in metagenomic data analysis for ecological inference.

In addition to the individual bioinformatic tools and resources described above, several publicly available Web-based pipelines have been developed for metagenomic data analysis. These online platforms, such as CAMERA (http://camera.calit2.net/), IMG/M (http://img.jgi.doe.gov/), METAREP (http://jcvi.org/metarep/), and MG-RAST (http://metagenomics.anl.gov/), have integrated various tools for gene prediction, functional or protein family assignment, and protein interaction and metabolic pathway inference in a user-friendly format. For example, the latest version of the MG-RAST server integrates data uploading, QC, and annotation and analysis of various datasets, such as 16S or amplicon sequences and WGS metagenome and metatranscriptome sequences (Wilke et al. 2013). MG-RAST relies on FragGeneScan as a gene-calling program and BLAT for homology-based similarity search. Since its launch in 2007, more than 108,500 datasets have been analyzed.

WGS metagenomic sequence data not only enable analysis of functionality and metabolic potential of the microbial community but also provide a means for taxonomical assignment (binning). Numerous tools, such as metaBEETL (Ander et al. 2013), have been developed for taxonomic classification of WGS data from microbial communities. These tools are generally divided into 3 major categories: homology or similarity-based, composition-based, and hybrid approach. The lowest common ancestor (LCA) algorithm has formed a base for many of the similarity-based classification methods, such as in WebCARMA (Gerlach et al. 2009), CloudLCA (Zhao et al. 2012), MEGAN (Huson et al. 2007), and DiScRIBinATE (Ghosh et al. 2010). It is shown that the latter significantly reduces binning time with superior assignment accuracy. Composition-based methods, which exploit the uniqueness of DNA base composition in genomes of different taxonomic entities, have been implemented in programs such as PhyloPythia (McHardy et al. 2007), Phymm (Brady and Salzberg 2009), TACOA (Diaz et al. 2009), and TaxSOM (Weber et al. 2011). The hybrid approach for binning generally combines both similarity- and composition-based methods, such as PhymmBL (Brady and Salzberg 2009) and RITA (MacDonald et al. 2012), for better accuracy. However, these two methods tend to be computationally time-consuming. To overcome this problem, a new algorithm, MetaPhlAn, has been developed (Segata et al. 2012). MetaPhlAn estimates the relative abundance of microbial cells by mapping short reads against a set of 400,141 clade-specific marker genes and allows for more accurate taxonomical assignment down to a species level in minutes of computational times for millions of WGS reads (Segata et al. 2012).

16.3.3 Stable Isotope Labeling

Stable or "heavy" isotopes, such as ^{13}C and ^{15}N, can be used to label various substrates, either small molecules (e.g., glucose) and polysaccharides (such as inulin) or even whole plants. Microorganisms that utilize these labeled substrates will most likely incorporate the heavy or stable isotope more efficiently into their biomolecules, including DNA or RNA. The labeled DNA or RNA can be readily separated from unlabeled, normal "light" DNA or RNA by isopycnic density-gradient ultracentrifugation, in combination with magnetic-bead capture techniques and isotope ratio mass spectrometry. The enriched SIP-RNA/DNA can be then studied using standard molecular technologies. For example, SIP has been widely used to investigate community function in microbial ecosystems or genes responsible for bioremediation (Uhlik et al. 2013). The potential of SIP technology in metagenomic studies is immediately recognized. The combined approaches not only permit the detection of low-abundance species in a complex microbial community but also facilitate the discovery of novel enzymes and bioactive compounds (Chen and Murrell 2010). Moreover, SIP-metagenomic technologies enable the enrichment of metabolically active fractions of microorganisms from environmental samples or the gut microbiome, which can provide a powerful link between microbial phylogeny and metabolic functionality and activity. Such a link is important in understanding the role of rumen microbiota in normal physiology and pathogenesis. SIP-RNA technology has demonstrated that changes in the functional activity of the human gut microbiota are associated with nutrient sources and medium types (Reichardt et al. 2011). In ruminants, compounds labeled by SIP have been widely used in nutrition studies. For example, ^{13}C-labeled n-alkanes of plant origin are used as an internal tracer to assess digesta passage kinetics through the gastrointestinal tract (Warner et al. 2013). In addition, a stable isotope tracer, ^{13}C-linolenic acid, has been used to investigate the biohydrogenation process of linolenic acid in a bovine rumen microbial community (Lee and Jenkins 2011). Using a small-scale repeated batch culture model of cattle fecal microbial communities, ^{13}C-labeled fructose in combination with a modified t-RFLP molecular fingerprinting identifies *Streptococcus bovis* as the most dominant and *Lactobacillus vitulinus* and *Megasphaera elsdenii* as minor fructose fermenter, while several species of *Clostridium* cluster IV are non-fermenters of fructose (Michinaka and Fujii 2012). It is conceivable that the importance of SIP-RNA/DNA technology in rumen metagenomic studies will soon be recognized.

16.3.4 Gnotobiotic Rumen Models

Gnotobiotic, including germ-free, animals, which have well-defined microbial composition, provide an elegant model system to study myriad interactions between individual microbes and between microbes and the host. The microbial communities of varying complexity and origin can be then sequentially introduced to gnotobiotic animals to examine the effects of genetic background, dietary conditions, and physiological stages on the microbial community structure and dynamics. Synthetic gut microbiota with known microbial composition and abundance can be created in germ-free animals. When the complete genome and transcriptome of these introduced microbial species become known, these systems can then be used to measure perturbation dynamics of the entire microbial community and to refine tools and algorithms using computational metagenomics.

Gnotobiotic lambs have been used to study rumen microbial establishment sequences and interactions of microbes of different functional groups for more than a decade (Fonty et al. 1989). The rumen of these animals harbors a defined microbial community. Rumen microbial species with known function can be sequentially added to the defined community. Therefore, gnotobiotic lambs are an ideal model to study the role of specific microorganisms and their interactions with other species in the rumen, such as the relationship between H_2-producing and H_2-consuming communities. Early results show that lambs lacking methanogens can be raised to adulthood, although their feed intake is lower compared to conventional lambs with functional methanogens (Fonty et al. 2007). A concomitant reduction in SCFA production as well as overall microbial complexity in these lambs is also observed. Moreover, acetogens can colonize and become rapidly established in the rumen of methanogen-free lambs, suggesting their establishment is independent of other microbes, unlike cellulolytic bacteria that generally require the presence of a diverse microbiota for establishment. On the other hand, methanogen colonization in the rumen does not substantially affect acetogen diversity (Gagen et al. 2012). Recently, interactions between fibrolytic species and methanogens have been examined using the gnotobiotic model (Chaucheyras-Durand et al. 2010). Methane emission is reduced when the dominant fibrolytic species in the rumen is a non-H_2 producer, such as *Fibrobacter succinogenes*, compared to the rumen with H_2-producing fibrolytic species, such as ruminococci and anaerobic fungi. These results suggest that dietary intervention strategies to promote non-H_2-producing fibrolytic species may represent a novel approach to mitigate methane production in farm animals. Utilization of metagenomic tools in the gnotobiotic rumen model will facilitate our understanding of microbial establishment sequence and succession of the rumen microbiome.

16.3.5 Metatranscriptomics, Metaproteomics, and Metabolomics

Sequencing-based metagenomics addresses the collective genetic structure and functional composition at the DNA level of a microbial community in a culture-independent manner. While vitally important, metagenome characterization using the DNA-based approach does not itself reveal how the genetic information of a given microbiota is actually expressed. To characterize how genes in the metagenome are expressed and regulated, metatranscriptomics, a comprehensive measure of mRNA transcript abundance, dynamics, and regulation under a variety of environmental conditions or developmental and physiological/pathological stages, is developed (Lim et al. 2013; Qi et al. 2011). While metatranscriptomic analysis provides insights into how the metagenome is expressed and regulated, metaproteomics allows comprehensive characterization of the gene products (proteins) encoded in the metagenome and their posttranslational modifications and turnover. Metaproteomics has recently been applied to analyze the human salivary supernatant (Jagtap et al. 2012). However, due to limitations in accurate detection and mass measurement of peptides and their annotation, metaproteomics currently allow characterization of only a relatively small fraction of the gene products in a complex gut microbiota (Wilmes and Bond 2006). Similarly, a comprehensive survey of metabolites in the host, diet (forage), and its rumen microbiome, metabolomics, has been conducted to provide information on key players responsible for the microbiome function (Lee et al. 2012b; Kingston-Smith et al. 2013). Metabolomic profiling using nuclear magnetic resonance spectroscopy, in combination with 454 pyrosequencing, demonstrates uniqueness of the microbial composition and metabolites in the rumen of Korean native goats (Lee et al. 2012b). Metabolic data facilitate the study of interactions between bacteria-specific metabolites and host proteins (Jacobsen et al. 2013). Together, rapid integrations of these OMIC technologies, including metagenomics for metagenomic DNA, metatranscriptomics for RNA, metaproteomics for proteins and peptides, and meta-metabolomics for metabolites, will provide a holistic insight into the structure and function of the rumen microbiota.

16.4 Metagenomic Insights into the Structure and Function of the Rumen Microbiome

16.4.1 Microbial Establishment and Succession During Rumen Development

Microbial products, such as SCFAs, play a critical role in rumen development. As a result, numerous attempts have been made to understand microbial establishment sequences and ecological succession of the developing rumen microbiome. It is generally accepted that the rumen is sterile at birth. Earlier studies demonstrate that bacteria start colonization in the rumen within the first 24 h of life and strictly anaerobic bacteria become predominant by the second day after birth (Fonty et al. 1989). Major functional groups of microorganisms, including fibrolytic bacteria and methanogens, become established in the rumen within the first week of life, followed by protozoa (Morvan et al. 1994; Quigley et al. 1985). Our knowledge of the microbial establishment and succession in the developing rumen and during the transition to mature rumen has been significantly expanded (Li et al. 2012b; Jami et al. 2013; Malmuthuge et al. 2014), largely due to the advent of metagenomic technologies. For example, it is generally accepted that methanogens start colonizing the rumen 3–4 days after birth (Fonty et al. 1989). A recent study shows that methanogens can be detected in the ovine rumen 17 h after birth (Gagen et al. 2012). A systematic cataloging of microbial diversity and functionality in the developing rumen using both

16S rRNA gene-based and WGS approaches has been attempted (Li et al. 2012a, b, c). Sequences from more than 24 prokaryotic phyla and 22 eukaryotic phyla are identified in rumen microbial communities of preruminant (14-day-old and 42-day-old) calves fed the same milk diet. Furthermore, the rumen microbiome of preruminant calves harbors considerable functional diversity, as evidenced by the existence of 8,298 Pfam protein families. A total of 156 and 120 genera are identified in the rumen microbiota of 14-day-old and 42-day-old calves, respectively. Fibrolytic bacteria and glycoside hydrolase protein families are abundant in the developing rumen, before the calves are fed a solid diet. Moreover, the fibrolytic capability of the developing rumen increases as calves' age. Interestingly, rumen development has a significant impact on microbial diversity. Genus-level richness indices ACE and Chao1 are significantly higher in the rumen of 14-day-old calves than that of 42-day-old calves. The rumen microbiome of younger calves displays a more heterogeneous microbial composition and harbors a greater number of bacterial genera (many of them may be transient) than the older calves fed the same milk replacer diet (Li et al. 2012b). This observation is in agreement with a general ecological theory that biodiversity tends to increase during early succession as new species arrive but may decline in later succession as competition eliminates opportunistic species. Rumen microbial composition changes from birth to adulthood have been monitored using pyrosequencing (Jami et al. 2013). This study shows that the fibrolytic species, such as *Ruminococcus albus*, is detectable in the developing rumen as early as one day of age while another major fibrolytic species, *Fibrobacter succinogenes*, appears much later. The observation that the presence of fibrolytic capacity in the developing rumen prior to exposure of solid diet is in agreement with previous studies (Li et al. 2012b) and in human infants (Koenig et al. 2011). Developmental stages appear to be one of the major determinants of the rumen microbial establishment as evidenced by the significant differences observed in rumen microbial composition between 14-day-old and 42-day-old calves (Li et al. 2012b) and between 6-month-old and 2-year-old cattle (Jami et al. 2013) that are fed the same diet. The direct comparison of microbial establishment sequences in the bovine rumen (Jami et al. 2013) and the hindgut of the human infants (Koenig et al. 2011) identifies similar wavelike patterns, coincident with critical life events such as diet, development, and health status, suggesting that similar forces may drive the establishment of microbial communities in two different habitats (Jami et al. 2013).

16.4.2 Rumen Microbial Diversity and the Core Rumen Microbiome

The collective microbial diversity in the rumen has been illustrated by a meta-analysis of all curated 16S rRNA gene sequences of rumen origin (13,478 bacterial and 3,516 archaeal sequences) deposited into the RDP database (Kim et al. 2011). This analysis has identified a total of 19 phyla and 5,271 and 942 OTUs for rumen bacteria and archaea, respectively. Approximately 1,000 OTUs are likely present in the rumen of fistulated cows and 587 OTUs are detected in all 4 samples from these two cows (Hess et al. 2011). In our studies, a total of 21 phyla are collectively identified from the rumen microbiome of dairy cows (Li et al. 2012c), with the mean of 16 phyla for the mature rumen of dairy cattle (Wu et al. 2012a). The mean number of genera in the rumen of individual cattle is 79.9 ± 14.0 ($\pm$ sd). In the mature rumen of dairy and beef cattle, the mean numbers of OTUs identified are 512 and 343, respectively. Together, these results suggest that the number of microbial species in a typical rumen of individual animals will likely be in the range of several hundreds.

The rumen microbiome is highly responsive to diet (Ellison et al. 2014), developmental stage (Li et al. 2012b; Jami et al. 2013), genetics, and numerous environmental factors. Substantial variations exist in microbial compositions and functionality among individual rumen samples within a species and between ruminant species (Jami and Mizrahi 2012). It is probable that a set of core taxa or OTUs (species) are shared by the rumen microbiome of individual animals in all ruminant species within the large context of variation. The core rumen microbiome, consisting of a common set of microbial taxa that are shared by all individual samples, may contribute to basic rumen function. Defining the core rumen microbiome is of significance in understanding basic structure and function of the rumen microbiome and has been recently attempted (Li et al. 2012b, c; Wu et al. 2012a; Petri et al. 2013b). The core rumen microbiome of the bovine rumen microbiome, both the developing and mature rumen of dairy and beef cattle, consists of 8 phyla, 11 classes, and 15 families (Wu et al. 2012a). These 8 phyla, accounting for 99.5 % of input 16S sequences, are *Bacteroidetes*, *Firmicutes*, *Proteobacteria*, *Spirochaetes*, *Fibrobacteres*, *Verrucomicrobia*, *Synergistetes*, and *Actinobacteria*, in descending order of relative abundance. The core bovine rumen microbiome likely represents minimal components of the rumen microbial community. As Table 16.2 shows, only a small number of approximately 150 ruminant species have been systematically studied for rumen microbial diversity using metagenomic tools. The rumen microbiome composition between ruminant species has been investigated (Kittelmann et al. 2013). Recently, we have compared microbial community compositions of the bovine ($N=8$), caprine ($N=10$), and ovine ($N=10$) rumen in order to define the core rumen microbiome using deep 16S sequencing. The mean number of 16S rRNA gene sequences per sample is 79,213.0 ± 11,682.2 (mean ± SD; $N=28$; the mean read length = 300 bp), a sequencing depth estimated to reach 99.9 % coverage (Kim et al. 2011). Our preliminary results show that collectively, 22 phyla and 94 families are detected in the rumen microbiome of cattle (cows), goats, and sheep. The mean number of the phyla per animal in the rumen microbiome of cattle, goats, and sheep is 16.8, 11.8, and 16.9, respectively. The caprine rumen has a significantly smaller number of phyla than the bovine and ovine rumen ($P<10^{-5}$) in this study. The family-level composition follows the similar trend. The core rumen microbiome consists of 8 phyla, *Actinobacteria*, *Bacteroidetes*, Euryarchaeota, *Firmicutes*, *Proteobacteria*, *Spirochaetes*, *Synergistetes*, and *Verrucomicrobia*, which is consistent with our previous study of the core bovine microbiome (Wu et al. 2012a).

The 15 families consisting of the core rumen microbiome (Table 16.3), representing >95 % of assigned 16S sequences in each sample, will

Table 16.3 The relative abundance of the 15 families consisting of the core rumen microbiome

Family	Cattle ($N=8$)	Goats ($N=10$)	Sheep ($N=10$)
Prevotellaceae	51.12 (±5.99)	24.36 (±16.68)	35.16 (±11.56)
Lachnospiraceae	22.01(±3.56)	24.35 (±13.25)	23.97 (±11.87)
Ruminococcaceae	8.18 (±3.18)	4.43 (±4.83)	19.74 (±9.60)
Veillonellaceae	4.33 (±3.90)	13.51(±14.96)	10.09 (±6.78)
Acidaminococcaceae	4.21 (±1.26)	0.73 (±0.64)	2.39 (±1.37)
Porphyromonadaceae	1.76 (±1.28)	0.90 (±1.33)	1.21 (±0.99)
Spirochaetaceae	1.40 (±0.44)	0.45 (±0.65)	0.73 (±0.65)
Succinivibrionaceae	1.26 (±0.68)	29.84 (±20.46)	0.36 (±0.30)
Erysipelotrichaceae	0.88 (±0.32)	0.24 (±0.32)	1.73 (±1.05)
Synergistaceae	0.07 (±0.04)	0.03 (±0.02)	0.09 (±0.06)
Coriobacteriaceae	0.06 (±0.02)	0.05 (±0.06)	0.24 (±0.15)
Desulfobulbaceae	0.06 (±0.04)	0.00 (±0.00)	0.21 (±0.12)
Moraxellaceae	0.03 (±0.03)	0.02 (±0.01)	0.02 (±0.06)
Clostridiales incertae sedis XIII	0.03 (±0.02)	0.05 (±0.04)	0.16 (±0.11)
Campylobacteraceae	0.01 (±0.01)	0.05 (±0.04)	0.12 (±0.08)

The number represents the percentage of the abundance (± SD). The abundance = the number of 16S sequences assigned to this particular family/the total number of 16S sequences assigned to all families at 80 % confidence cutoff using RDP Classifier likely contribute to the basic function of the rumen microbial ecosystem. However, the relative abundance of 15 families consisting of the core rumen microbiome varies significantly among the 3 host species, despite their high prevalence. For example, the abundance of the family *Prevotellaceae*, the most abundant family in the rumen of all 3 species, in the caprine rumen (24.26 %) is significantly lower than in the bovine (51.12 %) and ovine (35.16 %) rumen, while the abundance of 4 families, *Acidaminococcaceae*, *Desulfobulbaceae*, *Campylobacteraceae*, and *Succinivibrionaceae*, is significantly different between species. The abundance of *Ruminococcaceae* is significantly higher in the ovine rumen (19.74 %) than in the bovine (8.18 %) and the caprine (4.43 %) rumen. On the other hand, the abundance of *Lachnospiraceae* is relatively stable in the rumen of the 3 species (between 22.00 and 24.35 %).

16.4.3 Rumen Virome and Plasmidome

Previous studies have focused on bacterial and archaeal diversity in the rumen. Lack of conserved proteins and genes, such as 16S rRNA genes, among viruses or phages has hindered their discovery. The advent of computational metagenomics has made possible systematically cataloging viruses in the rumen and surveying the phage diversity. A recent study shows that more than 20,000 viral genotypes exist in the rumen of 2 of the 3 cattle analyzed (Berg Miller et al. 2012). While over 70 % of viral sequences have no significant matches to sequences in public databases, sequences associated with prophages outnumber those lytic phages approximately 2:1. Moreover, rumen viral sequences carry functional genes; and the majority of these genes belong to phages, prophages, transportable elements, and plasmid subsystem according to the SEED database, as expected. In another study, 14 putative viral sequences over 30 kb are identified (Ross et al. 2013). Cows housed together and fed the same diet display similar taxonomical virome profiles than those housed separately. Intriguingly, these two cohorts have similar functional characterizations, suggesting the rumen virome appears to be functionally conserved (Ross et al., 2013). Together, these results provide further evidence that viruses play an important role in horizontal gene transfer between various microorganisms, spreading antibiotic resistance genes, controlling bacterial population dynamics, and affecting animal nutrition and protein metabolism in the rumen.

Due to a relatively low copy number per bacterial cell and difficulty in distinguishing them from chromosomal DNA, rumen plasmids have not been systematically studied until recently (Brown Kav et al. 2012; Mizrahi 2012). Brown Kav et al. (2013) recently reported a method to enrich plasmid DNA from the rumen. The method takes advantage of an exonuclease that is able to digest chromosomal DNA that is sheared and linearized during extraction procedures. The resultant circular plasmid DNA is amplified using a phi29 DNA polymerase and further sequenced using an Illumina sequencer. This study provides a first glimpse of the diversity and function of the rumen plasmidome (i.e., the collective plasmid population of rumen origin). Notably, while the rumen microbial hosts can be assigned to the three major phyla, *Firmicutes*, *Bacteroidetes*, and *Proteobacteria*, using rumen plasmid sequences, a significant difference in the relative abundance is evident compared to the phylogenetic assignment based on the 16S sequences from the very same rumen source. For example, *Proteobacteria* account for approximately 20 % of rumen plasmid sequences, whereas its abundance appears to be significantly lower (~5 %) according to the 16S approach (Brown Kav et al. 2012). Functional analysis using the SEED database suggests that, in addition to the intrinsic plasmid-coding functions, the rumen plasmidome shares similarities with the rumen metagenomes and displays a significantly higher representation of the functional categories, such as "amino acids," "cell wall and capsule," "cofactors, vitamins, etc.," and "protein metabolism." These results demonstrate that rumen plasmids may play an important role in lateral gene transfer between rumen microorganisms.

16.4.4 Resistance and Resilience of the Rumen Microbiome

One of the primary functions of the rumen is to degrade lignocellulosic fiber to produce short-chain fatty acids for energy. The relative stability of structure and composition of the rumen microbiome becomes a prerequisite for such functions. The stability can be defined as (1) the ability to return to an equilibrium state following perturbation and (2) the ability to resist changes (resistance) or the rate of return to an equilibrium following perturbation or overall system variability (Robinson et al. 2010). Therefore, the stability of the rumen microbiome imparts resilience to perturbation, ensuring continued rumen function.

The rumen microbiome is susceptible to both natural and anthropogenic stresses and is highly responsive to changes in environmental conditions and host factors, such as critical physiological or pathological events, resulting in the creation of novel niche for other microbial species. The microbial diversity of the rumen is a reflection of the coevolution between microbial communities and their host and represents equilibrium between functional redundancy of a stable community and niche specialization. Although a few dominant microorganisms are likely to be responsible for the majority of the metabolic activity and energy influx in the rumen, it is well known that uncommon species often serve as a reservoir of genetic and functional diversity, playing key roles in microbial ecosystems. These species can become numerically important if environmental conditions change. Recently, scientific communities have begun to study the extent of temporal and spatial shifts in functionality and phylogenic composition of the rumen microbiome in response to various stresses, such as diet, critical life events (such as weaning and acidosis), and antibiotic usage, as well as their ecological and physiological implications (Li et al. 2012b, c; Jami et al. 2013; Petri et al. 2013a, b).

We have characterized temporal changes of the rumen microbiome of dairy cows in response to an exogenous butyrate disturbance (Li et al. 2012c). We reanalyzed the raw data using improved bioinformatic tools for this chapter. Our results indicate that in the rumen microbiome of dairy cows in their mid-lactation, the five most abundant phyla, *Bacteroidetes*, *Firmicutes*, *Proteobacteria*, *Fibrobacteres*, and *Spirochaetes* in this order, account for >99 % of observed 16S sequences. A 168-h exogenous butyrate perturbation results in significant changes in abundance of 4 of the 5 most abundant phyla. The relative abundance of *Bacteroidetes* and *Fibrobacteres* is significantly decreased, from 68.20 % to 56.74 % (at the basal level to after perturbation) and 1.43 % to 0.82 %, respectively. On the other hand, the abundance of *Firmicutes* and *Spirochaetes* is significantly increased, from 23.34 % to 30.32 % and 0.98 % to 2.17 %, respectively. The phylum *Firmicutes* includes a majority of butyrate-producing bacteria. The observation of exogenous butyrate increasing the relative abundance of *Firmicutes* suggests that butyrate itself may be butyrogenic. Indeed, a readily available energy source (exogenous butyrate) reduces the need of fibrolytic capacity of the rumen, resulting in a decreased abundance of *Fibrobacteres*. The abundance of these 5 phyla returns to a pre-disturbance level 168 h after perturbation, suggesting the resilient nature of the rumen microbiome. The rumen microbiome is also recovered 1 week after diet-induced acidosis challenge (Petri et al. 2013a). The analysis at family level demonstrates the same trend. The perturbation with exogenous butyrate significantly impacts 7 of the 20 most abundant families (accounting for 99.8 % of the sequences), including the 3 most abundant families, *Prevotellaceae*, *Lachnospiraceae*, and *Ruminococcaceae*. After the perturbation withdrawal, the abundance of these families returns to their pre-disturbed levels. Among the 20 top families, the long-lasting impact of the disturbance is observed for only 2 minor families: *Anaeroplasmataceae* remains elevated, while *Acetobacteraceae* is further repressed, 168 h after perturbation withdrawal. Our data demonstrate that the rumen microbial ecosystem displays substantial resilience to short-term disturbances. Furthermore, considerable hysteresis of the rumen microbiome is observed. The ecological role and consequences of the two families, *Anaeroplasmataceae* and *Acetobacteraceae*, in the new established rumen microbial community are worthy of further scientific attention.

16.5 Conclusions

Metagenomics has significantly expanded our knowledge of the rumen microbial diversity and the structure and function of the rumen microbiome, thanks to rapid advances in next-generation sequencing technologies and computational tools and resources. The rumen microbiome, consisting of hundreds of microbial species and myriad microbial interactions, plays a critical role in nutrition and normal physiology of ruminants. It is highly responsive to changes in diet, development, environmental factors, and host genetics. Alterations in the rumen microbiome have important implications for animal well-being and production efficiency. While metagenomics has proven to be a powerful tool in rumen microbiome studies, numerous challenges remain. Temporal and spatial fluctuations as well as intra- and interindividual variations of ruminal microbial composition have yet to be assessed. Current development of bioinformatic tools and resources that cope with fragmental and voluminous next-generation sequencing data is still in its infancy. Notably, the lack of fully assembled and annotated reference genomes of rumen origin in public databases has hindered functional annotation of metagenomic data. The lack of commonly accepted data analysis pipelines and standardized report formats makes direct comparisons of various metagenomic studies difficult, if not impossible. Most importantly, general theories and principles of microbial ecology have yet to be fully applied to rumen metagenomic studies. Mechanistic models are still needed to aid the interpretation of biological relevance of metagenomic data. The recent launch of the Hungate1000 community sequencing project, which aims to sequence up to 1,000 microbial genomes of rumen origin (http://www.hungate1000.org.nz/), representing a broad spectrum of rumen microbial taxa, will undoubtedly facilitate the assembly and annotation of WGS metagenomic data. It is conceivable that comprehensive studies of the rumen microbiome using metagenomic tools will broaden our understanding of the structure and function of the rumen microbiome and its role in normal physiology and pathology of ruminants, which in turn should guide our efforts to develop optimal rumen manipulation strategies for more efficient fiber digestion as well as mitigation of environmental footprints of animal farming.

Acknowledgments Mention of trade names or commercial products in this publication is solely for the purpose of providing specific information and does not imply recommendation or endorsement by the US Department of Agriculture. The USDA is an equal opportunity provider and employer.

References

Afiahayati, Sato K, Sakakibara Y (2013) An extended genovo metagenomic assembler by incorporating paired-end information. Peer J 1:e196

Ander C, Schulz-Trieglaff OB, Stoye J et al (2013) MetaBEETL: high-throughput analysis of heterogeneous microbial populations from shotgun DNA sequences. BMC Bioinfo 14(Suppl 5) :S2

Berg Miller ME, Yeoman CJ, Chia N et al (2012) Phage-bacteria relationships and CRISPR elements revealed by a metagenomic survey of the rumen microbiome. Environ Microbiol 14:207–227

Bhatt VD, Dande SS, Patil NV et al (2013) Molecular analysis of the bacterial microbiome in the forestomach fluid from the dromedary camel (*Camelus dromedarius*). Mol Biol Rep 40:3363–3371

Brady A, Salzberg SL (2009) Phymm and PhymmBL: metagenomic phylogenetic classification with interpolated Markov models. Nat Methods 6:673–676

Brown Kav A, Sasson G, Jami E et al (2012) Insights into the bovine rumen plasmidome. Proc Natl Acad Sci U S A 109:5452–5457

Brown Kav A, Benhar I, Mizrahi I (2013) A method for purifying high quality and high yield plasmid DNA for metagenomic and deep sequencing approaches. J Microbiol Methods 95:272–279

Brulc JM, Antonopoulos DA, Miller ME, Wilson MK, Yannarell AC et al (2009) Gene-centric metagenomics of the fiber-adherent bovine rumen microbiome reveals forage specific glycoside hydrolases. Proc Natl Acad Sci U S A 106:1948–1953

Cai Y, Sun Y (2011) ESPRIT-tree: hierarchical clustering analysis of millions of 16S rRNA pyrosequences in quasilinear computational time. Nucleic Acids Res 39:e95

Castro-Carrera T, Toral PG, Frutos P et al (2014) Rumen bacterial community evaluated by 454 pyrosequencing and terminal restriction fragment length polymorphism analyses in dairy sheep fed marine algae. J Dairy Sci 97:1661–1669
Chaisson MJ, Brinza D, Pevzner PA (2009) De novo fragment assembly with short mate-paired reads: does the read length matter? Genet Res 19:336–346
Chakravorty S, Helb D, Burday M et al (2007) A detailed analysis of 16S ribosomal RNA gene segments for the diagnosis of pathogenic bacteria. J Microbiol Methods 69:330–339
Chaucheyras-Durand F, Masseglia S, Fonty G et al (2010) Influence of the composition of the cellulolytic flora on the development of hydrogenotrophic microorganisms hydrogen utilization and methane production in the rumens of gnotobiotically reared lambs. Appl Environ Microbiol 76:7931–7937
Cheema TA, Jirajaroenrat K, Sirinarumitr T (2012) Isolation of a gene encoding a cellulolytic enzyme from swamp buffalo rumen metagenomes and its cloning and expression in Escherichia coli. Anim Biotechnol 23:261–277
Chen Y, Murrell JC (2010) When metagenomics meets stable-isotope probing: progress and perspectives. Trends Microbiol 18:157–163
Chen W, Zhang CK, Cheng Y et al (2013) A comparison of methods for clustering 16S rRNA sequences into OTUs. PLoS One 8:e70837
Cheng F, Sheng J, Cai T et al (2012a) A protease-insensitive feruloyl esterase from China Holstein cow rumen metagenomic library: expression characterization and utilization in ferulic acid release from wheat straw. J Agric Food Chem 60:2546–2553
Cheng F, Sheng J, Dong R, Men Y, Gan L, Shen L (2012b) Novel xylanase from a holstein cattle rumen metagenomic library and its application in xylooligosaccharide and ferulic Acid production from wheat straw. J Agric Food Chem 60:12516–12524
Cole JR, Wang Q, Cardenas E, Fish J, Chai B, Farris RJ, Kulam-Syed-Mohideen AS, McGarrell DM, Marsh T, Garrity GM, Tiedje JM (2009) The ribosomal database project: improved alignments and new tools for rRNA analysis. Nucleic Acids Res 37:D141–D145
Cox MP, Peterson DA, Biggs PJ (2010) SolexaQA: at-a-glance quality assessment of Illumina second-generation sequencing data. BMC Bioinf 11:485
Dai X, Zhu Y, Luo Y, Song L, Liu D, Liu L, Chen F, Wang M, Li J, Zeng X, Dong Z, Hu S, Li L, Xu J, Huang L, Dong X (2012) Metagenomic insights into the fibrolytic microbiome in yak rumen. PLoS One 7:e40430
Dalby PA (2011) Strategy and success for the directed evolution of enzymes. Curr Opin Struct Biol 21:473–480
De Eaton HL, Lorme M, Chaney RL, Craig AM (2011) Ovine ruminal microbes are capable of biotransforming hexahydro-135-trinitro-135-triazine (RDX). Microb Ecol 62:274–286
De Lorme M, Craig M (2009) Biotransformation of 246-trinitrotoluene by pure culture ruminal bacteria. Curr Microbiol 58:81–86
de Menezes AB, Lewis E, O'Donovan M, O'Neill BF, Clipson N, Doyle EM (2011) Microbiome analysis of dairy cows fed pasture or total mixed ration diets. FEMS Microbiol Ecol 78:256–265
de Oliveira MN, Jewell KA, Freitas FS, Benjamin LA, Totola MR, Borges AC, Moraes CA, Suen G (2013) Characterizing the microbiota across the gastrointestinal tract of a Brazilian Nelore steer. Vet Microbiol 164:307–314
DeSantis TZ, Hugenholtz P, Larsen N, Rojas M, Brodie EL, Keller K, Huber T, Dalevi D, Hu P, Andersen GL (2006) Greengenes a chimera-checked 16S rRNA gene database and workbench compatible with ARB. Appl Environ Microbiol 72:5069–5072
Diaz NN, Krause L, Goesmann A, Niehaus K, Nattkemper TW (2009) TACOA: taxonomic classification of environmental genomic fragments using a kernelized nearest neighbor approach. BMC Bioinf 10:56
Edgar RC (2010) Search and clustering orders of magnitude faster than BLAST. Bioinformatics 26:2460–2461
Edgar RC (2013) UPARSE: highly accurate OTU sequences from microbial amplicon reads. Nat Methods 10:996–998
Edwards JE, McEwan NR, Travis AJ, Wallace RJ (2004) 16S rDNA library-based analysis of ruminal bacterial diversity. Antonie Van Leeuwenhoek 86:263–281
Ekkers DM, Cretoiu MS, Kielak AM, Elsas JD (2012) The great screen anomaly–a new frontier in product discovery through functional metagenomics. Appl Microbiol Biotechnol 93:1005–1020
El Allali A, Rose JR (2013) MGC: a metagenomic gene caller. BMC bioinformatics 14(Suppl 9):S6
Ellison MJ, Conant GC, Cockrum RR, Austin KJ, Truong H, Becchi M, Lamberson WR, Cammack KM (2014) Diet alters both the structure and taxonomy of the ovine gut microbial ecosystem. DNA Res 21(5):115–125
El-Metwally S, Hamza T, Zakaria M, Helmy M (2013) Next-generation sequence assembly: four stages of data processing and computational challenges. PLoS Comput Biol 9:e1003345
Faust K, Raes J (2012) Microbial interactions: from networks to models. Nat Rev Microbiol 10:538–550
Faust K, Sathirapongsasuti JF, Izard J, Segata N, Gevers D, Raes J, Huttenhower C (2012) Microbial co-occurrence relationships in the human microbiome. PLoS Comput Biol 8:e1002606
Ferrer M, Golyshina OV, Chernikova TN, Khachane AN, Reyes-Duarte D, Santos VA, Strompl C, Elborough K, Jarvis G, Neef A, Yakimov MM, Timmis KN, Golyshin PN (2005) Novel hydrolase diversity retrieved from a metagenome library of bovine rumen microflora. Environ Microbiol 7:1996–2010
Ferrer M, Beloqui A, Vieites JM, Guzzaroni ME, Berger I, Aharoni A (2009) Interplay of metagenomics and

in vitro compartmentalization. J Microbial Biotechnol 2:31–39

Ferrer M, Ghazi A, Beloqui A, Vieites JM, Lopez-Cortes N, Marin-Navarro J, Nechitaylo TY, Guazzaroni ME, Polaina J, Waliczek A, Chernikova TN, Reva ON, Golyshina OV, Golyshin PN (2012) Functional metagenomics unveils a multifunctional glycosyl hydrolase from the family 43 catalysing the breakdown of plant polymers in the calf rumen. PLoS One 7:e38134

Fonty G, Gouet P, Nebout JM (1989) Development of the cellulolytic microflora in the rumen of lambs transferred into sterile isolators a few days after birth. Can J Microbiol 35:416–422

Fonty G, Joblin K, Chavarot M, Roux R, Naylor G, Michallon F (2007) Establishment and development of ruminal hydrogenotrophs in methanogen-free lambs. Appl Environ Microbiol 73:6391–6403

Fu L, Niu B, Zhu Z, Wu S, Li W (2012) CD-HIT: accelerated for clustering the next-generation sequencing data. Bioinformatics 28:3150–3152

Gagen EJ, Mosoni P, Denman SE, Al Jassim R, McSweeney CS, Forano E (2012) Methanogen colonisation does not significantly alter acetogen diversity in lambs isolated 17 h after birth and raised aseptically. Microb Ecol 64:628–640

Galbraith EA, Antonopoulos DA, White BA (2004) Suppressive subtractive hybridization as a tool for identifying genetic diversity in an environmental metagenome: the rumen as a model. Environ Microbiol 6:928–937

Gerlach W, Junemann S, Tille F, Goesmann A, Stoye J (2009) WebCARMA: a web application for the functional and taxonomic classification of unassembled metagenomic reads. BMC Bioinf 10:430

Ghosh TS, Monzoorul Haque M, Mande SS (2010) DiScRIBinATE: a rapid method for accurate taxonomic classification of metagenomic sequences. BMC Bioinf 11(Suppl 7):S14

Gong X, Gruninger RJ, Qi M, Paterson L, Forster RJ, Teather RM, McAllister TA (2012) Cloning and identification of novel hydrolase genes from a dairy cow rumen metagenomic library and characterization of a cellulase gene. BMC Res Notes 5:566

Gruninger RJ, Gong X, Forster RJ, McAllister TA (2014) Biochemical and kinetic characterization of the multifunctional beta-glucosidase/beta-xylosidase/alpha-arabinosidase Bgxa1. Appl Microbiol Biotechnol 98(7):3003–3012

Haft DH, Selengut JD, White O (2003) The TIGRFAMs database of protein families. Nucleic Acids Res 31:371–373

Hamady M, Lozupone C, Knight R (2010) Fast UniFrac: facilitating high-throughput phylogenetic analyses of microbial communities including analysis of pyrosequencing and PhyloChip data. ISME J 4:17–27

Hao X, Jiang R, Chen T (2011) Clustering 16S rRNA for OTU prediction: a method of unsupervised Bayesian clustering. Bioinformatics 27:611–618

Hayete B, Bienkowska JR (2005) Gotrees: predicting go associations from protein domain composition using decision trees. Pac Symp Biocomput 10:127–138

Hess M, Sczyrba A, Egan R, Kim TW, Chokhawala H, Schroth G, Luo S, Clark DS, Chen F, Zhang T, Mackie RI, Pennacchio LA, Tringe SG, Visel A, Woyke T, Wang Z, Rubin EM (2011) Metagenomic discovery of biomass-degrading genes and genomes from cow rumen. Science 331:463–467

Hoff KJ, Lingner T, Meinicke P, Tech M (2009) Orphelia: predicting genes in metagenomic sequencing reads. Nucleic Acids Res 37:W101–W105

Huang Y, Li W, Perkins D, Li RW (2012) Comparison of de novo short read assemblers on simulated metagenomic data (chapter 5). In: Li RW (ed) Metagenomics and its applications in agriculture biomedicine and environmental studies. Nova Science, New York

Huo W, Zhu W, Mao S (2014) Impact of subacute ruminal acidosis on the diversity of liquid and solid-associated bacteria in the rumen of goats. World J Microbiol Biotechnol 30:669–680

Huson DH, Auch AF, Qi J, Schuster SC (2007) MEGAN analysis of metagenomic data. Genome Res 17:377–386

Jacobsen UP, Nielsen HB, Hildebrand F, Raes J, Sicheritz-Ponten T, Kouskoumvekaki I, Panagiotou G (2013) The chemical interactome space between the human host and the genetically defined gut metabotypes. ISME J 7:730–742

Jagtap P, McGowan T, Bandhakavi S, Tu ZJ, Seymour S, Griffin TJ, Rudney JD (2012) Deep metaproteomic analysis of human salivary supernatant. Proteomics 12:992–1001

Jami E, Mizrahi I (2012) Composition and similarity of bovine rumen microbiota across individual animals. PLoS One 7:e33306

Jami E, Israel A, Kotser A, Mizrahi I (2013) Exploring the bovine rumen bacterial community from birth to adulthood. ISME J 7:1069–1079

Jami E, White BA, Mizrahi I (2014) Potential role of the bovine rumen microbiome in modulating milk composition and feed efficiency. PLoS One 9:e85423

Jenkins TC, Wallace RJ, Moate PJ, Mosley EE (2008) Board-invited review: recent advances in biohydrogenation of unsaturated fatty acids within the rumen microbial ecosystem. J Anim Sci 86:397–412

Kanehisa M (2002) The KEGG database. Novartis Found Symp 247:91–101; discussion 101–103 119–128 244–152

Kelley DR, Liu B, Delcher AL, Pop M, Salzberg SL (2012) Gene prediction with Glimmer for metagenomic sequences augmented by classification and clustering. Nucleic Acids Res 40:e9

Kim M, Morrison M, Yu Z (2011) Status of the phylogenetic diversity census of ruminal microbiomes. FEMS Microbiol Ecol 76:49–63

Kim MK, Kang TH, Kim J, Kim H, Yun HD (2012a) Cloning and identification of a new group esterase

(Est5S) from noncultured rumen bacterium. J Microbiol Biotechnol 22:1044–1053

Kim OS, Cho YJ, Lee K, Yoon SH, Kim M, Na H, Park SC, Jeon YS, Lee JH, Yi H, Won S, Chun J (2012b) Introducing EzTaxon-e: a prokaryotic 16S rRNA gene sequence database with phylotypes that represent uncultured species. Int J Syst Evol Microbiol 62:716–721

Kim M, Lee KH, Yoon SW, Kim BS, Chun J, Yi H (2013) Analytical tools and databases for metagenomics in the next-generation sequencing era. Genomics Inform 11:102–113

Kingston-Smith AH, Davies TE, Rees Stevens P, Mur LA (2013) Comparative metabolite fingerprinting of the rumen system during colonisation of three forage grass (*Lolium perenne* L) varieties. PLoS One 8:e82801

Kittelmann S, Seedorf H, Walters WA, Clemente JC, Knight R, Gordon JI, Janssen PH (2013) Simultaneous amplicon sequencing to explore co-occurrence patterns of bacterial archaeal and eukaryotic microorganisms in rumen microbial communities. PLoS One 8:e47879

Ko KC, Han Y, Cheong DE, Choi JH, Song JJ (2013) Strategy for screening metagenomic resources for exocellulase activity using a robotic high-throughput screening system. J Microbiol Methods 94:311–316

Koenig JE, Spor A, Scalfone N, Fricker AD, Stombaugh J, Knight R, Angenent LT, Ley RE (2011) Succession of microbial consortia in the developing infant gut microbiome. Proc Natl Acad Sci U S A 108(Suppl 1):4578–4585

Laserson J, Jojic V, Koller D (2011) Genovo: de novo assembly for metagenomes. J Comput Biol 18:429–443

Lee YJ, Jenkins TC (2011) Biohydrogenation of linolenic acid to stearic acid by the rumen microbial population yields multiple intermediate conjugated diene isomers. J Nutr 141:1445–1450

Lee CC, Kibblewhite RE, Wagschal K, Li R, Orts WJ (2012a) Isolation of alpha-glucuronidase enzyme from a rumen metagenomic library. Protein J 31:206–211

Lee HJ, Jung JY, Oh YK, Lee SS, Madsen EL, Jeon CO (2012b) Comparative survey of rumen microbial communities and metabolites across one caprine and three bovine groups using bar-coded pyrosequencing and (1)H nuclear magnetic resonance spectroscopy. Appl Environ Microbiol 78:5983–5993

Lee JH, Yi H, Jeon YS, Won S, Chun J (2012c) TBC: a clustering algorithm based on prokaryotic taxonomy. J Microbiol 50:181–185

Li W, Godzik A (2006) Cd-hit: a fast program for clustering and comparing large sets of protein or nucleotide sequences. Bioinformatics 22:1658–1659

Li RW, Sparks ME, Connor EE (2012a) Dynamics of the rumen microbiota. In: Li RW (ed) Metagenomics and its applications in agriculture biomedicine and environmental studies. Nova Science, New York

Li RW, Connor EE, Li C, Baldwin Vi RL, Sparks ME (2012b) Characterization of the rumen microbiota of pre-ruminant calves using metagenomic tools. Environ Microbiol 14:129–139

Li RW, Wu S, Baldwin RLVI, Li W, Li C (2012c) Perturbation dynamics of the rumen microbiota in response to exogenous butyrate. PLoS One 7:e29392

Li RW, Wu S, Li W, Navarro K, Couch RD, Hill D, Urban JF Jr (2012d) Alterations in the porcine colon microbiota induced by the gastrointestinal nematode Trichuris suis. Infect Immun 80:2150–2157

Li RW, Giarrizzo JG, WU S, Li W, Duringer JM, Craig AM (2014) Metagenomic insights into the RDX-degrading potential of the ovine rumen microbiome. PLoS One 9(11), e110505

Liberles DA, Teufel AI, Liu L, Stadler T (2013) On the need for mechanistic models in computational genomics and metagenomics. Genome Biol Evol 5:2008–2018

Lim S, Seo J, Choi H, Yoon D, Nam J, Kim H, Cho S, Chang J (2013) Metagenome analysis of protein domain colection within cellulase genes of goat rumen microbes. Asian-Aust J Anim Sci 26(8):1144–1151

Lim YW, Schmieder R, Haynes M, Willner D, Furlan M, Youle M, Abbott K, Edwards R, Evangelista J, Conrad D, Rohwer F (2013) Metagenomics and metatranscriptomics: windows on CF-associated viral and microbial communities. J Cyst Fibros 12(2):154–164

Liu B, Pop M (2011) MetaPath: identifying differentially abundant metabolic pathways in metagenomic datasets. BMC Proc 5(Suppl 2):S9

Lozupone C, Knight R (2005) UniFrac: a new phylogenetic method for comparing microbial communities. Appl Environ Microbiol 71:8228–8235

Luo C, Rodriguez RL, Konstantinidis KT (2013) A user's guide to quantitative and comparative analysis of metagenomic datasets. Methods Enzymol 531:525–547

MacDonald NJ, Parks DH, Beiko RG (2012) Rapid identification of high-confidence taxonomic assignments for metagenomic data. Nucleic Acids Res 40:e111

Malmuthuge N, Griebel PJ, Guan le L (2014) Taxonomic identification of commensal bacteria associated with the mucosa and digesta throughout the gastrointestinal tracts of preweaned calves. Appl Environ Microbiol 80:2021–2028

Mao S, Huo W, Zhu W (2013) Use of pyrosequencing to characterize the microbiota in the ileum of goats fed with increasing proportion of dietary grain. Curr Microbiol 67:341–350

McHardy AC, Martín HG, Tsirigos A, Hugenholtz P, Rigoutsos I (2007) Accurate phylogenetic classification of variable-length DNA fragments. Nat Methods 4:63–72

Mende DR, Waller AS, Sunagawa S, Jarvelin AI, Chan MM, Arumugam M, Raes J, Bork P (2012) Assessment of metagenomic assembly using simulated next generation sequencing data. PLoS One 7:e31386

Meyer F, Overbeek R, Rodriguez A (2009) FIGfams: yet another set of protein families. Nucleic Acids Res 37:6643–6654

Michinaka A, Fujii T (2012) Efficient and direct identification of fructose fermenting and non-fermenting bacteria from calf gut microbiota using stable isotope probing and modified t-RFLP. J Gen Appl Microbiol 58:297–307

Miller JR, Koren S, Sutton G (2010) Assembly algorithms for next-generation sequencing data. Genomics 95:315–327

Mizrahi I (2012) The rumen plasmidome: a genetic communication hub for the rumen microbiome. Mob Genet Elements 2:152–153

Morgavi DP, Kelly WJ, Janssen PH, Attwood GT (2013) Rumen microbial (meta)genomics and its application to ruminant production. Animal 7(Suppl 1):184–201

Morvan B, Dore J, Rieu-Lesme F, Foucat L, Fonty G, Gouet P (1994) Establishment of hydrogen-utilizing bacteria in the rumen of the newborn lamb. FEMS Microbiol Lett 117:249–256

Namiki T, Hachiya T, Tanaka H, Sakakibara Y (2012) MetaVelvet: an extension of Velvet assembler to de novo metagenome assembly from short sequence reads. Nucleic Acids Res 40:e155

Nguyen NH, Maruset L, Uengwetwanit T, Mhuantong W, Harnpicharnchai P, Champreda V, Tanapongpipat S, Jirajaroenrat K, Rakshit SK, Eurwilaichitr L, Pongpattanakitshote S (2012) Identification and characterization of a cellulase-encoding gene from the buffalo rumen metagenomic library. Biosci Biotechnol Biochem 76:1075–1084

Noguchi H, Park J, Takagi T (2006) MetaGene: prokaryotic gene finding from environmental genome shotgun sequences. Nucleic Acids Res 34:5623–5630

Noguchi H, Taniguchi T, Itoh T (2008) MetaGeneAnnotator: detecting species-specific patterns of ribosomal binding site for precise gene prediction in anonymous prokaryotic and phage genomes. DNA Res 15:387–396

Nossa CW, Oberdorf WE, Yang L, Aas JA, Paster BJ, Desantis TZ, Brodie EL, Malamud D, Poles MA, Pei Z (2010) Design of 16S rRNA gene primers for 454 pyrosequencing of the human foregut microbiome. World J Gastroenterol 16:4135–4144

Omoniyi LA, Jewell KA, Isah OA, Neumann AP, Onwuka CF, Onagbesan OM, Suen G (2014) An analysis of the ruminal bacterial microbiota in West African Dwarf sheep fed grass- and tree-based diets. J Appl Microbiol 116(5):1094–1105

Peng Y, Leung HC, Yiu SM, Chin FY (2011) Meta-IDBA: a de Novo assembler for metagenomic data. Bioinformatics 27:i94–i101

Peng Y, Leung HC, Yiu SM, Chin FY (2012) IDBA-UD: a de novo assembler for single-cell and metagenomic sequencing data with highly uneven depth. Bioinformatics 28:1420–1428

Petri RM, Schwaiger T, Penner GB, Beauchemin KA, Forster RJ, McKinnon JJ, McAllister TA (2013a) Changes in the rumen epimural bacterial diversity of beef cattle as affected by diet and induced ruminal acidosis. Appl Environ Microbiol 79:3744–3755

Petri RM, Schwaiger T, Penner GB, Beauchemin KA, Forster RJ, McKinnon JJ, McAllister TA (2013b) Characterization of the core rumen microbiome in cattle during transition from forage to concentrate as well as during and after an acidotic challenge. PLoS One 8:e83424

Pinloche E, McEwan N, Marden JP, Bayourthe C, Auclair E, Newbold CJ (2013) The effects of a probiotic yeast on the bacterial diversity and population structure in the rumen of cattle. PLoS One 8:e67824

Pitta DW, Kumar S, Veiccharelli B, Parmar N, Reddy B, Joshi CG (2014) Bacterial diversity associated with feeding dry forage at different dietary concentrations in the rumen contents of Mehshana buffalo (*Bubalus bubalis*) using 16S pyrotags. Anaerobe 25:31–41

Pop M (2009) Genome assembly reborn: recent computational challenges. Brief Bioinform 10:354–366

Pop M, Salzberg SL (2008) Bioinformatics challenges of new sequencing technology. Trends Genet 24:142–149

Pope PB, Mackenzie AK, Gregor I, Smith W, Sundset MA, McHardy AC, Morrison M, Eijsink VG (2012) Metagenomics of the Svalbard reindeer rumen microbiome reveals abundance of polysaccharide utilization loci. PLoS One 7:e38571

Powell S, Forslund K, Szklarczyk D, Trachana K, Roth A, Huerta-Cepas J, Gabaldon T, Rattei T, Creevey C, Kuhn M, Jensen LJ, von Mering C, Bork P (2014) eggNOG v40: nested orthology inference across 3686 organisms. Nucleic Acids Res 42:D231–D239

Punta M, Coggill PC, Eberhardt RY, Mistry J, Tate J, Boursnell C, Pang N, Forslund K, Ceric G, Clements J, Heger A, Holm L, Sonnhammer EL, Eddy SR, Bateman A, Finn RD (2012) The Pfam protein families database. Nucleic Acids Res 40:D290–D301

Qi M, Wang P, O'Toole N, Barboza PS, Ungerfeld E, Leigh MB, Selinger LB, Butler G, Tsang A, McAllister TA, Forster RJ (2011) Snapshot of the eukaryotic gene expression in muskoxen rumen – a metatranscriptomic approach. PLoS One 6(5), e20521

Qin J, Li Y, Cai Z, Li S, Zhu J, Zhang F, Liang S, Zhang W, Guan Y, Shen D, Peng Y, Zhang D, Jie Z, Wu W, Qin Y, Xue W, Li J, Han L, Lu D, Wu P, Dai Y, Sun X, Li Z, Tang A, Zhong S, Li X, Chen W, Xu R, Wang M, Feng Q, Gong M, Yu J, Zhang Y, Zhang M, Hansen T, Sanchez G, Raes J, Falony G, Okuda S, Almeida M, LeChatelier E, Renault P, Pons N, Batto JM, Zhang Z, Chen H, Yang R, Zheng W, Li S, Yang H, Wang J, Ehrlich SD, Nielsen R, Pedersen O, Kristiansen K, Wang J (2012) A metagenome-wide association study of gut microbiota in type 2 diabetes. Nature 490:55–60

Quast C, Pruesse E, Yilmaz P, Gerken J, Schweer T, Yarza P, Peplies J, Glockner FO (2013) The SILVA ribosomal RNA gene database project: improved data processing and web-based tools. Nucleic Acids Res 41:D590–D596

Quigley JD 3rd, Schwab CG, Hylton WE (1985) Development of rumen function in calves: nature of protein reaching the abomasum. J Dairy Sci 68:694–702

Quince C, Lanzen A, Davenport RJ, Turnbaugh PJ (2011) Removing noise from pyrosequenced amplicons. BMC Bioinf 12:38

Rashamuse KJ, Visser DF, Hennessy F, Kemp J, der Merwe MP R-v, Badenhorst J, Ronneburg T, Francis-Pope R, Brady D (2013) Characterisation of two bifunctional cellulase-xylanase enzymes isolated from a bovine rumen metagenome library. Curr Microbiol 66:145–151

Reichardt N, Barclay AR, Weaver LT, Morrison DJ (2011) Use of stable isotopes to measure the metabolic activity of the human intestinal microbiota. Appl Environ Microbiol 77:8009–8014

Rho M, Tang H, Ye Y (2010) FragGeneScan: predicting genes in short and error-prone reads. Nucleic Acids Res 38:e191

Robinson CJ, Bohannan BJ, Young VB (2010) From structure to function: the ecology of host-associated microbial communities. Microbiol Mol Biol Rev 74:453–476

Ross EM, Petrovski S, Moate PJ, Hayes BJ (2013) Metagenomics of rumen bacteriophage from thirteen lactating dairy cattle. BMC Microbiol 13:242

Sandri M, Manfrin C, Pallavicini A, Stefanon B (2014) Microbial biodiversity of the liquid fraction of rumen content from lactating cows. Animal 8(4):572–579

Schloss PD, Westcott SL, Ryabin T, Hall JR, Hartmann M, Hollister EB, Lesniewski RA, Oakley BB, Parks DH, Robinson CJ, Sahl JW, Stres B, Thallinger GG, Van Horn DJ, Weber CF (2009) Introducing mothur: open-source platform-independent community-supported software for describing and comparing microbial communities. Appl Environ Microbiol 75:7537–7541

Schmieder R, Edwards R (2011) Fast identification and removal of sequence contamination from genomic and metagenomic datasets. PLoS One 6:e17288

Schomburg I, Chang A, Ebeling C, Gremse M, Heldt C, Huhn G, Schomburg D (2004) BRENDA the enzyme database: updates and major new developments. Nucleic Acids Res 32:D431–D433

Segata N, Waldron L, Ballarini A, Narasimhan V, Jousson O, Huttenhower C (2012) Metagenomic microbial community profiling using unique clade-specific marker genes. Nat Methods 9:811–814

Sharma VK, Kumar N, Prakash T, Taylor TD (2010) MetaBioME: a database to explore commercially useful enzymes in metagenomic datasets. Nucleic Acids Res 38:D468–D472

Simon C, Daniel R (2009) Achievements and new knowledge unraveled by metagenomic approaches. Appl Microbiol Biotechnol 85:265–276

Simpson JT, Wong K, Jackman SD, Schein JE, Jones SJ, Birol I (2009) ABySS: a parallel assembler for short read sequence data. Genome Res 19:1117–1123

Singh KM, Ahir VB, Tripathi AK, Ramani UV, Sajnani M, Koringa PG, Jakhesara S, Pandya PR, Rank DN, Murty DS, Kothari RK, Joshi CG (2012a) Metagenomic analysis of Surti buffalo (*Bubalus bubalis*) rumen: a preliminary study. Mol Biol Rep 39:4841–4848

Singh KM, Jakhesara SJ, Koringa PG, Rank DN, Joshi CG (2012b) Metagenomic analysis of virulence-associated and antibiotic resistance genes of microbes in rumen of Indian buffalo (Bubalus bubalis). Gene 507:146–151

Soergel DA, Dey N, Knight R, Brenner SE (2012) Selection of primers for optimal taxonomic classification of environmental 16S rRNA gene sequences. ISME J 6:1440–1444

Sparks ME, Huang Y, Baldwin RL VI, Li W, Connor EE, Li C, Sonstegard TS, Schroeder SG, Bequette BJ, Li RW (2012) Detection of functional shifts in the rumen microbiota in response to propionate intake in cattle. In: Li RW (ed) Metagenomics and its applications in agriculture biomedicine and environmental studies. Nova Science, New York

Sun Y, Cai Y, Liu L, Yu F, Farrell ML, McKendree W, Farmerie W (2009) ESPRIT: estimating species richness using large collections of 16S rRNA pyrosequences. Nucleic Acids Res 37:e76

Tang S, Antonov I, Borodovsky M (2013) MetaGeneTack: ab initio detection of frameshifts in metagenomic sequences. Bioinformatics 29:114–116

Tatusov RL, Galperin MY, Natale DA, Koonin EV (2000) The COG database: a tool for genome-scale analysis of protein functions and evolution. Nucleic Acids Res 28:33–36

Thoetkiattikul H, Mhuantong W, Laothanachareon T, Tangphatsornruang S, Pattarajinda V, Eurwilaichitr L, Champreda V (2013) Comparative analysis of microbial profiles in cow rumen fed with different dietary fiber by tagged 16S rRNA gene pyrosequencing. Curr Microbiol 67:130–137

Trimble WL, Keegan KP, D'Souza M, Wilke A, Wilkening J, Gilbert J, Meyer F (2012) Short-read reading-frame predictors are not created equal: sequence error causes loss of signal. BMC Bioinf 13:183

Uhlik O, Leewis MC, Strejcek M, Musilova L, Mackova M, Leigh MB, Macek T (2013) Stable isotope probing in the metagenomics era: a bridge towards improved bioremediation. Biotechnol Adv 31:154–165

Vazquez-Castellanos JF, Garcia-Lopez R, Perez-Brocal V, Pignatelli M, Moya A (2014) Comparison of different assembly and annotation tools on analysis of simulated viral metagenomic communities in the gut. BMC Genomics 15:37

Vey G, Moreno-Hagelsieb G (2012) Metagenomic annotation networks: construction and applications. PLoS One 7:e41283

Vilo C, Dong Q (2012) Evaluation of the RDP classifier accuracy using 16S rRNA gene variable regions. Metagenomics 1:e235551 doi: 10.4303/mg/235551

Wallace RJ (2008) Gut microbiology – broad genetic diversity yet specific metabolic niches. Animal 2:661–668

Wang Q, Garrity GM, Tiedje JM, Cole JR (2007) Naive Bayesian classifier for rapid assignment of rRNA sequences into the new bacterial taxonomy. Appl Environ Microbiol 73:5261–5267

Wang L, Hatem A, Catalyurek UV, Morrison M, Yu Z (2013) Metagenomic insights into the carbohydrate-active enzymes carried by the microorganisms adhering to solid digesta in the rumen of cows. PLoS One 8:e78507

Warner D, Dijkstra J, Hendriks WH, Pellikaan WF (2013) Passage kinetics of 13C-labeled corn silage components through the gastrointestinal tract of dairy cows. J Dairy Sci 96:5844–5858

Weber M, Teeling H, Huang S, Waldmann J, Kassabgy M, Fuchs BM, Klindworth A, Klockow C, Wichels A, Gerdts G, Amann R, Glockner FO (2011) Practical application of self-organizing maps to interrelate biodiversity and functional data in NGS-based metagenomics. ISME J 5:918–928

White JR, Nagarajan N, Pop M (2009) Statistical methods for detecting differentially abundant features in clinical metagenomic samples. PLoS Comput Biol 5:e1000352

Wilke A, Glass EM, Bartels D, Bischof J, Braithwaite D, D'Souza M, Gerlach W, Harrison T, Keegan K, Matthews H, Kottmann R, Paczian T, Tang W, Trimble WL, Yilmaz P, Wilkening J, Desai N, Meyer F (2013) A metagenomics portal for a democratized sequencing world. Methods Enzymol 531:487–523

Wilmes P, Bond PL (2006) Metaproteomics: studying functional gene expression in microbial ecosystems. Trends Microbiol 14:92–97

Wu S, Baldwin RL VI, Li W, Li C, Connor EE, Li RW (2012a) The bacterial community composition of the bovine rumen detected using pyrosequencing of 16S rRNA genes. Metagenomics 1:e235571

Wu S, Li RW, Li W, Beshah E, Dawson HD, Urban JF Jr (2012b) Worm burden-dependent disruption of the porcine colon microbiota by Trichuris suis infection. PLoS One 7:e35470

Xia LC, Steele JA, Cram JA, Cardon ZG, Simmons SL, Vallino JJ, Fuhrman JA, Sun F (2011) Extended local similarity analysis (eLSA) of microbial community and other time series data with replicates. BMC Syst Biol 5(Suppl 2):S15

Zerbino DR, Birney E (2008) Velvet: algorithms for de novo short read assembly using de Bruijn graphs. Genome Res 18:821–829

Zhang W, Chen J, Yang Y, Tang Y, Shang J, Shen B (2011) A practical comparison of de novo genome assembly software tools for next-generation sequencing technologies. PLoS One 6:e17915

Zhang P, Gu J, He J, Gao W, Zhang W, Lindsay S, Meldrum DR (2012) Next-generation and future DNA sequencing technologies and metagenomics. In: Li RW (ed) Metagenomics and its applications in agriculture biomedicine and environmental studies. Nova Science, New York

Zhao G, Bu D, Liu C, Li J, Yang J, Liu Z, Zhao Y, Chen R (2012) CloudLCA: finding the lowest common ancestor in metagenome analysis using cloud computing. Protein Cell 3:148–152

Rumen: An Underutilised Niche for Industrially Important Enzymes

17

Gunjan Goel, Sumit Singh Dagar, Mamta Raghav, and Saurabh Bansal

Abstract

Rumen is one of the most underutilised microbial ecosystems, harbouring a diverse population of microbial species. These species thrive in this ecosystem by producing an array of enzymes for digestion and utilisation of different plant constituents. The search for novel and efficient fibrolytic cultures/enzymes will foster the development of different applications such as biofuel production from lignocellulosic biomass. Exploring and exploiting these efficient cultures/enzymes using biotechnological interventions for enhanced production are necessary before their efficient application in industries. Recent advances in molecular biology such as metagenomic studies with high-throughput screening methods are enabling the development of novel strategies for effective delivery and enhancement of these enzymes. This chapter takes a holistic review of most extensively studied enzymes produced in the rumen and their role in digestion of fibre and other associated plant cell wall polymers.

Keywords

Rumen bacteria • Rumen fungi • Enzymes • Metagenomics

17.1 Introduction

Enzymes are known to be the sustainable biocatalysts of high demand with continuously increasing industrial sector. The application of potential enzymes over the conventional chemical and thermal methods offers additional advantages such as reduction in the use of hazardous chemicals, creation of safer working environment, lowering the cost of process and product formulation, reduction in consumption of water and energy

G. Goel (✉) • M. Raghav • S. Bansal
Department of Biotechnology and Bioinformatics, Jaypee University of Information Technology, Solan 173234, India
e-mail: gunjanmicro@gmail.com

S.S. Dagar
Micorbial Science Division, Agarkhar Research Institute, Pune, India

A.K. Puniya et al. (eds.), *Rumen Microbiology: From Evolution to Revolution*, DOI 10.1007/978-81-322-2401-3_17

and enhancement of efficient use of non-renewable resources.

The rumen is an ideal habitat for the growth of anaerobic microorganisms encompassing bacteria, fungi, protozoa, archaea and bacteriophages. Out of these groups, bacteria and protozoa predominate the microbial biomass. The fungi make up only 8–20 % of microbes and occupy an important niche in the rumen because of their specific affinity towards lignin and ability to produce esterases for hydrolysing ester linkages between lignin and hemicelluloses or cellulose and thus help break down digesta particles. Further, their rhizoidal system causes physical disruption by penetrating inside recalcitrant feed stuff, which also allows bacteria to gain access to otherwise non-available sites. The bacteria being most abundant and diverse are further classified into different groups based on their enzymatic activities such as fibrolytic, amylolytic, proteolytic, etc. The diversity arises due to different microbial communities as well as multiplicity of fibrolytic enzymes produced by individual microorganisms when they encounter different plant polymers. Due to its diverse microbial population, the rumen works as a specialised fermentation vessel facilitating the microbial degradation of ingested plant materials. Therefore, the rumen has been described as microbial cell factory for biorefineries as it harbours a rich source of microbial cell and could be a potential model to study the higher organisational levels of microbial communities, finally leading to a new concept for metabolic engineering (Sauer et al. 2012). The digestion is mainly affected by the plant materials, its nature, components and structure as well as the microbial factors such as microbial load, communities, competition and others. The different characteristics of microbial population in rumen such as survival under anaerobiosis, predatory activities of rumen protozoa, recalcitrant plant components and toxic effects of plant secondary metabolites make it an ideal source for bioprospecting (Selinger et al. 1996; Wang and McAllister 2002). The majority of the enzymes discovered or studied belong to different classes that help to degrade different plant cell wall polymers (Table 17.1).

Table 17.1 Major enzyme activities required for hydrolysis of plant cell wall components

Substrate	Linkage	Enzyme required
Cellulose	β-1,4-Glucose linkage	Endo-β-1,4-glucanase
Cellulose (non-reducing ends)	β-1,4-Glucose linkage	Exo-β-1,4-glucanase
Cellobiose	β-1,4-Glucose linkage	β-1,4-Glucosidase
Soluble cello-oligomers	β-1,4-Glucose linkage	Cellulodextrinase
Cellulose/xylan	β-1,4-Glucose linkage	Xylocellulase
Xylan	β-1,4-Xylose linkage	Endo-β-1,4-xylanase
Xylobiose	β-1,4-Xylose linkage	β-1,4-Xylosidase
Arabinoxylan	α-1,3-Linkage	α-L-arabinofuranosidase
Glucuronoxylan	α-1,3 or α-1,2 linkage	α-Glucuronidase
Acetylxylan	Acetylester bond	*O*-Acetyl xylan esterase
Ferulic acid cross bridges	Feruloylester bond	Ferulic acid esterase
p-Coumaric acid cross bridge	*p*-Coumaryl ester bond or linkage	*p*-Coumaric acid esterase
Laminarin	β-1,3-Glucanase	β-1,3-hexose linkage
Lichenin	β-1,3- and β-1,4-hexose linkages	Mixed linkage β-1,3-β-1,4-glucanase
Polygalacturan	α-1,4-Galacturonide linkages	Pectate lyase
Pectin	α-1,4-Galacturonide linkages methyl ester bond	Pectin lyase Pectin methylesterase
Tannins	Depside linkage	Tannin acyl hydrolase

Modified from Wang and McAllister (2002)

17.2 Plant Cell Wall Components

Ruminal digestion of plant materials consists of numerous complex processes involving a number of microbial species and their enzymatic machinery. At the microbial level, digestion of plant materials is poorly understood. The major bacterial species associated are known; however, it is in the last decade that researchers began to isolate the pure anaerobic cultures and study the various enzymatic activities and physiological factors that influence the expression of these enzymes.

17.2.1 Cellulose

Cellulose is the most important and abundant structural part of the plants which is a homopolymer of glucose. The plant cell wall is composed of fibrils of cellulose which accounts for 20–30 % dry weight of the primary cell wall. Cellulose molecules associate with each other to form microfibrils and have crystalline formulations.

17.2.2 Hemicellulose

Hemicellulose is composed mainly of xylans with a backbone structure of β-1,4 linkage in xylose residues and attachment of various side chains (e.g. acetic acid, arabinose, coumaric acid, ferulic acid, glucuronic acid, 4-O-methylglucuronic acid). Xylan polymers may be cross-linked to other hemicellulose backbones or to lignins through ferulic acid or 4-O-methylglucuronic acid. It may also be linked to cellulose fibrils forming an extensive network of cross-links. The varying branching patterns of the surrounding structures result in different types of hemicelluloses structures.

17.2.3 Pectin

Pectic substances are prominent structural constituents of primary cell walls and middle lamella. It is one of the most complex biomolecules and can be composed of as many as 17 different monosaccharides with at least seven different polysaccharides. The predominant structure of pectin consists of homogalacturonan (HG) which is an unbranched molecule composed of polyβ -1,4- D-galacturonic acid (PGA) with α-1,4-linked residues of D-galacturonate. The galacturonic acid (GA) residues can be methyl esterified at C-6 and some of the hydroxyl groups on C2 or C3 can be acetylated. Blocks of more than 10 unesterified GA residues generally yield pectin molecules, which are sensitive to calcium cross-linking (Daas et al. 2001). The rhamnogalacturonan backbone may be interspersed with either rhamnose or galacturonic acid residues substituted with methyl ester groups or sugar side chains (Jarvis 1984; McNeil et al. 1984; Rombouts and Pilnik 1986).

17.2.4 Phytic Acid

Phytic acid (phytate) is a complex of calcium or magnesium with myo-inositol and is regarded as the primary storage form of phosphorus and inositol in almost all seeds. It is considered as an anti-nutritional constituent of plant-derived feeds. As a reactive anion, it forms a wide variety of insoluble salts with minerals including phosphorus, calcium, zinc, magnesium and copper. Phytic acid is also known to form complexes with protein and proteolytic enzymes.

17.2.5 Lignins, Polyphenols and Toxic Components

Lignin is a complex polymer of aromatic compounds which account for the most abundant polymer on earth. Instead of sugar monomers like cellulosic compounds, it is composed of up to three different phenyl propane monomers, namely, *p*-coumaryl alcohol, coniferyl alcohol and sinapyl alcohol which are methoxylated to various degrees. The lignin degradation has been reported in rumen; however, so far the lignin degradation has not been reported in pure rumen bacterial isolate.

Besides lignin, the rumen microorganisms have been reported to detoxify various plant toxic components. The exposure to these components usually results in loss of animal productivity. The most widely studied toxins include mimosine T-2 toxins, nitrotoxins, pyrrolizidine alkaloids, *trans-aconitate* and tannins. They reduce animal productivity by reducing intake, feed palatability, enzyme activity, ruminal fermentation rates, nutrient availability and wool growth or by inducing toxicosis. Tannins based on their molecular structure are classified as hydrolysable (HTs) and condensed tannins (CTs, proanthocyanidins). Hydrolysable tannins contain a carbohydrate (generally D-glucose) as a central core with hydroxyl groups esterified with phenolic groups (Haslam 1989). These HTs are metabolised to gallic acid, pyrogallol and other products by rumen microbes that are potentially toxic to the ruminants. On the other hand, CTs do not have a central carbohydrate core and are complexes of oligomers and polymers of flavonoid units linked by carbon-carbon bonds with a molecular weight of 2,000–4,000 kDa (Hagerman and Butler 1981; Foo et al. 1986). Their multiple phenolic hydroxyl groups lead to the formation of complexes primarily with proteins limiting their availability to the animal (Makkar 2003).

17.3 Microbial Enzymes

Rumen is a rich source of fibrolytic enzymes such as cellulase, xylanase and β-glucanases. A large number of anaerobic bacteria, protozoa and fungi possess very efficient cellulolytic machinery which helps in increasing the efficiency of feed conversion (Table 17.2). The fibrolytic enzymes find their application in saccharification of lignocellulosic wastes for production of biofuels, removing certain forms of polysaccharides (arabinoxylan and β-glucan) in cereals that may interfere with nutrient absorption and promote intestinal disturbances.

17.3.1 Cellulases

The researchers have gained interest in rumen bacteria and fungi for a number of biotechnological applications, mainly for the production of cellulases which are capable of hydrolysis of 1,4 β-D-glycosidic linkages in cellulose to its monomers. Based on structure and functionality, these cellulases have been categorised as:

- Endocellulase which cleaves internal bonds at amorphous sites that create new chain ends.
- Exocellulase cleaves two to four units from the non-reducing ends of the cellulose molecule produced by endocellulase.
- Cellobiase or β-glucosidase hydrolyses the exocellulase product into individual monosaccharides (Bhat and Bhat 1997; Lynd and Zhang 2002).

In some microbes, the produced cellulases may exist as free enzymes, where different enzymes act on different parts of cellulose (Fig. 17.1). In some microbes, degradation of cellulose is accomplished by large multi-enzyme complex known as cellulosome. The attachment of this complex to cellulose fibres is achieved by non-cellulolytic bacteria via cellulose-binding proteins such as scaffolding cellulosome-integrating proteins and large glycosylated proteins (Fig. 17.2). Alongside the cellulosome structure, *Ruminococcus* also uses a cellulose-binding protein type C that involves fimbrial structures that interact with cellulose. With *R. albus*, the cellulases appear to be organised into highly structured, high molecular weight and extracellular complexes whereas *B. fibrololvens* produces extracellular polysaccharides having complex sugar composition. *F. succinogenes* possesses the membrane-associated cellulases and xylanase activities. Apart from bacterial species, the cellulolytic activities have also been reported in rumen protozoal population.

17.3.2 Hemicellulases

The hemicellulases include a variety of different types of enzymes. The initial attack on hemicelluloses has been reported by celloxylanase which displays good activity either on cellulose or xylan. This enzyme is produced by cellulolytic bacteria but do not grow on xylan as substrate; therefore, it helps in the initial disruption of plant cell fibre.

Table 17.2 Rumen microbial groups possessing fibre degrading activities

	Cellulolytic	Hemicellulolytic	Pectinase
Rumen bacteria			
Fibrobacter succinogenes	+		+
Ruminococcus albus	+		+
R. flavefaciens	+		+
Butyrivibrio fibrisolvens	+		+
Eubacterium cellulosolvens	+		+
Clostridium longisporum	+	+	
Cl.locheadii		+	+
Prevotella ruminantium		+	+
Eubacterium xylanophilum		+	
Ruminobacter amylophilus		+	
Succinimonas dextrinosolvens		+	
Selenomonas ruminantium		+	
Selenomonas lactilytica		+	
Lachnospira multiparus		+	+
Streptococcus bovis		+	+
Megasphaera elsdenii			
Rumen protozoa			
Eudiplodinium maggii	+	+	+
Ostracodinium dilobum	+	+	+
Epidinium caudatum	+	+	
Metadinium affine	+	+	+
Eudiplodinium bovis	+	+	+
Ophryoscolex caudatus	+	+	+
Polyplastron multivesiculatum	+	+	+
Diplodinium pentacanthum	+		
Endoploplastron triloricatum	+		
Ophryoscolex tricoronatus	+		
Ostracodinium gracile	+		
Entodinium caudatum	+	+	
Isotricha intestinalis	+	+	+
Isotricha prostoma	+	+	+
Rumen fungi			
Neocallimastix frontalis	+	+	+
N. patriciarum	+	+	+
N. joyonii	+	+	
Caecomyces communis	+	+	
Piromyces communis	+	+	+
Orpinomyces bovis	+	+	+
Anaeromyces sp.	+	+	

Later on the activities of xylan degraders help in degradation such as mannanase, arabinofuranosidase, ferulic acid esterase and xylanase. The hemicellulase activities have been reported in cell free extracts for different protozoal species. Higher cellulolytic activity was observed in entodiniomorphid ciliates, whereas holotrich ciliates were reported to possess weak hemicellulolytic activity. The site of action of hemicellulases is shown in Fig. 17.3. These enzymes act on the various structures surrounding cellulose thereby exposing the cellulose for microbial attack.

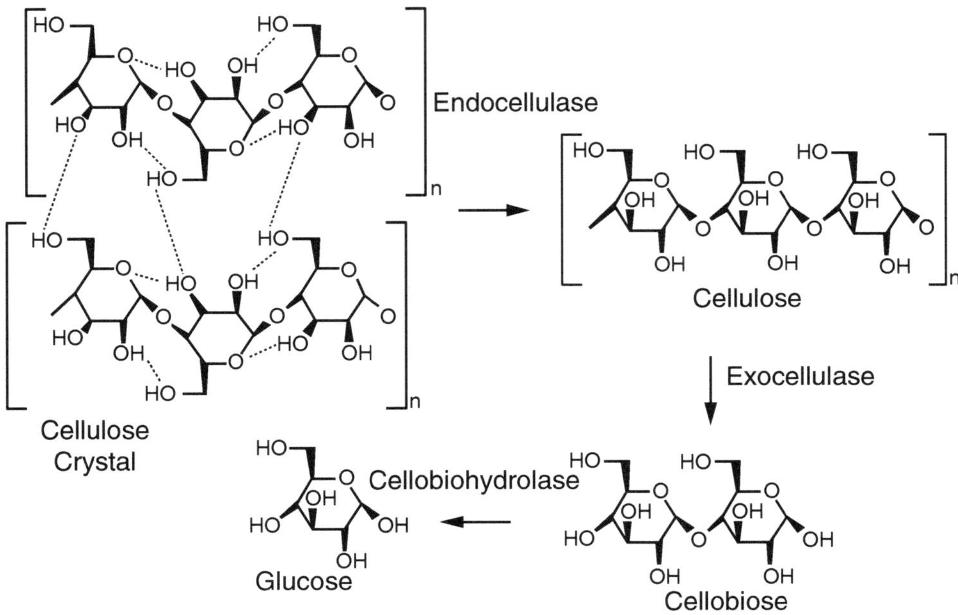

Fig. 17.1 Basic cellulose structure broken down by the three types of cellulases (Anonymous 2009)

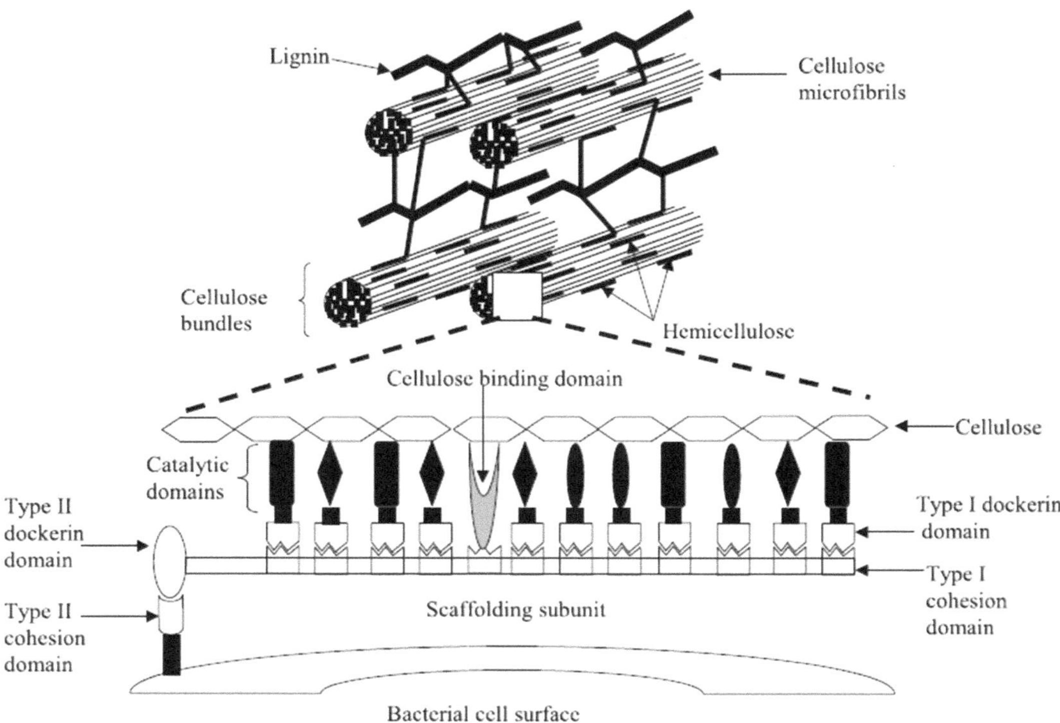

Fig. 17.2 Cellulosomal complex attached to lignocellulose (Krause et al. 2003)

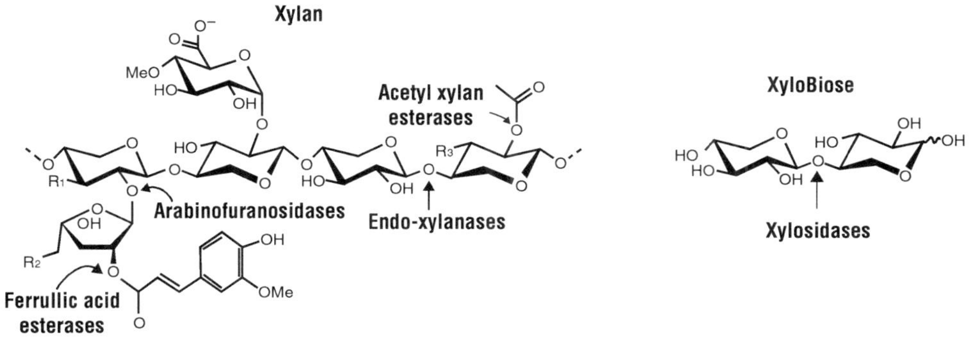

Fig. 17.3 Site of action of different hemicellulases (Shallom and Shoham 2003)

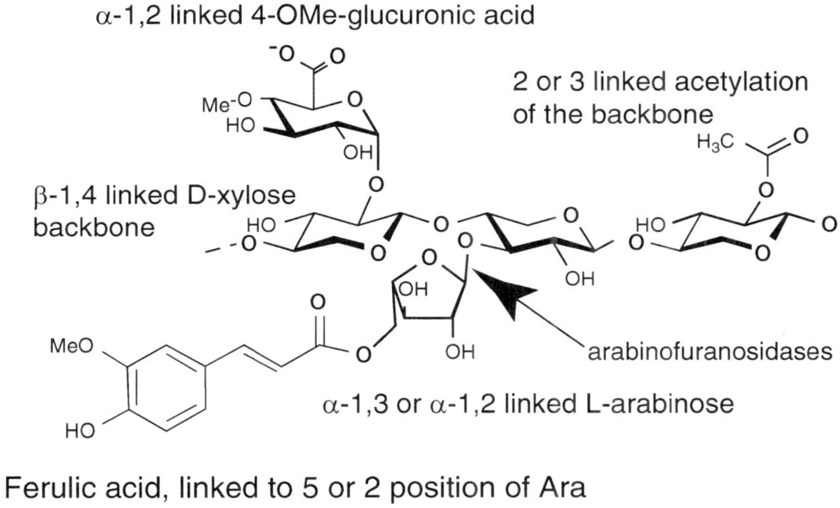

Fig. 17.4 The site of attack by FAFases on a xylan backbone (Taylor et al. 2006)

17.3.2.1 Mannanase

Mannan is a fundamental component of hemicelluloses. It consists of β-1,4 linkage between mannose monomers forming the hemicelluloses cross-linkages (Hogg et al. 2003). The mannan component is degraded to β-1,4-manno-oligomers by β-mannanases followed by action of β-mannosidases which reduce them to the monosaccharide mannose (Shallom and Shoham 2003).

17.3.2.2 Arabinofuranosidase

Arabinose is found in conjunction with xylan as hemicellulose component of plant cell wall. The arabinose units are attached to xylan via α-1,2, 1,3, 1,5 or linked to C2 or C3 position on arabinoxylan. The AFase hydrolyse the terminal, non-reducing arabinofuranosyl. The arabinose units can be cleaved off the xylose backbone by arabinofuranosidase activity of *B. fibrisolvens* and *B. ruminicola*.

17.3.2.3 Ferulic Acid Esterase

Ferulic acid esterase (FAEases) is a group of enzymes that form a subclass of carboxylic ester hydrolases. These enzymes hydrolyse the bond between hydroxycinnamates and sugars (Rashamuse et al. 2007) releasing ferulic acid (Fig. 17.4).

17.3.2.4 p-Coumaric Acid Esterase

p-Coumaric acid esterase or p-coumaroyl esterase is a very crucial enzyme for efficient degradation of lignocellulosic biomass. This enzyme helps in breakage of ester linkages that connect lignin to hemicelluloses, releasing p-coumaric acid. Very interestingly, none of the rumen bacterial group produces this and is exclusively produced by anaerobic fungi (Borneman et al. 1990), therefore further strengthening their ecological role and significance in rumen microbial ecosystem.

17.3.2.5 Xylanases

The xylan consists of β-1,4 linked xylopyranosyl residues and contains side chains with acetyl group and L-arabinofuranosyl residues. The xylanases are responsible for the hydrolysis of xylan by breaking the glycosidic linkages in xylan backbone (Shallom and Shoham 2003). Similar to cellulases, the xylanases are composed of three enzymes: endoxylanase, β-xylosidase and acetyl xylan esterase (Fig. 17.5). All the three enzymes hydrolyse the xylan molecule, rendering the D-xylan sugar usable (Kosugi et al. 2001).

17.3.3 Pectinase

There are a few rumen microorganisms possessing pectinolytic activities containing enzymes pectin lyase, polygalacturonase and pectin methylesterase (Fig. 17.6). One of the major pectinolytic bacterial species inhabiting the rumen, *Lachnospira multiparus*, produces a pectin lyase and a pectin methylesterase (Silley 1985). Apart from bacterial species, the rumen protozoa and fungi also possess the pectinolytic system (Orpin 1984; Bonhome 1990; Gordon and Philips 1992; Chesson and Forsberg 1997).

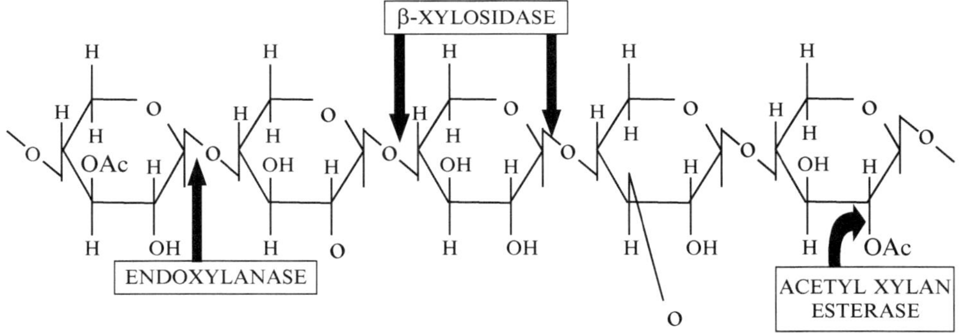

Fig. 17.5 A hypothetical xylan structure showing different sites of microbial attack (Beg et al. 2001)

Fig. 17.6 Site of action of pectin methyl esterase (Jayani et al. 2005)

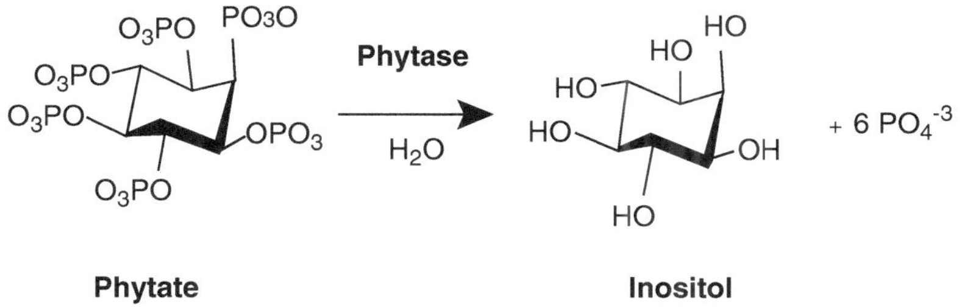

Fig. 17.7 Action of phytase on phytic acid (Mittal et al. 2011)

17.3.4 Phytase

The presence of phytase in the rumen microbes makes them able to utilise the phosphorus in phytic acid (Fig. 17.7). However, the metabolic activity of phytate degradation has been thoroughly characterised, and the study of the genetics of this process is relatively new. A recombinant phytase has been produced from *E. coli* by cloning the gene from *Selenomonas ruminantium* JY35 which possessed four to eight times higher specific activity (400 to 800 μmol phosphate released from phytate/min/mg protein) than the commercial preparation from *Aspergillus niger*. The phytase production does not depend on the coordinated activities of the enzymes such as cellulolytic enzymes; therefore, it can be genetically manipulated easily (Piddington et al. 1993; Van Hartingsveldt et al. 1993).

17.3.5 Polyphenol Degrading Enzymes

Tannin acyl hydrolase: Tannase catalyses the breakdown of hydrolysable tannins such as tannic acid, methyl gallate, ethyl gallate, n-propylgallate and isoamyl gallate. Tannase hydrolyses tannic acid completely to gallic acid and glucose through 2,3,4,6,-tetragalloyl glucose and two kinds of monogalloyl glucose. Tannase hydrolyses only those substrates that contain at least two phenolic hydroxyl groups in the acid component. The esterified carboxylic group must be on the oxidised benzene ring and must not be ortho to one of the hydroxyl groups (Fig. 17.8). Tannase activity has been reported mainly in *Streptococcus gallolyticus*, *Selenomonas ruminantium* and other inhabitants of gastrointestinal tract of ruminants such as *Enterococcus faecalis* and some other unidentified Gram-negative bacteria as reviewed by Goel et al. (2005).

17.3.6 Enzymes Involved in Biohydrogenation

The role of fatty acids in human health is very well documented. Conjugated linoleic acid (CLA) is one such polyunsaturated fatty acid that has attracted a substantial attention from the scientists from all around the world, because of its possible health effects. CLA comprises a group of positional and geometric isomers of linoleic acid and is produced as an intermediate during biohydrogenation of polyunsaturated fatty acids in the rumen of animals. Since these dietary unsaturated fatty acids are toxic to rumen microorganisms, as a defence mechanism, they secrete various enzymes to hydrolyse and hydrogenate these unsaturated fatty acids (Harfoot and Hazlewood 1988). In rumen, bacteria play the primary role in biohydrogenation (Jenkins et al. 2008). Mainly two types of bacteria, i.e. group A and group B, are involved in the biohydrogenation process (Fig. 17.9). Group A bacteria can hydrogenate linoleic acid (LA) or linolenic acid (LNA) to trans-Vaccenic acid (TVA) and are not able to hydrogenate the last step, i.e. conversion of TVA to stearic acid (SA), for example, *B. fibrisolvens* MDT5 (Fukuda et al. 2006), *Micrococcus* spp., *Ruminococcus* spp. and *Lactobacillus* spp.

Merthyl gallate = Gallic acid + Methanol Epicatechin gallate = Epicatechin + Gallic acid

Digallic acid = 2 Gallic acid Epigallocatechin gallate = Epigallocatechin − Gallic acid

Fig. 17.8 Mode of action of tannase on different polyphenols present in plant cell wall (Rodríguez et al. 2011)

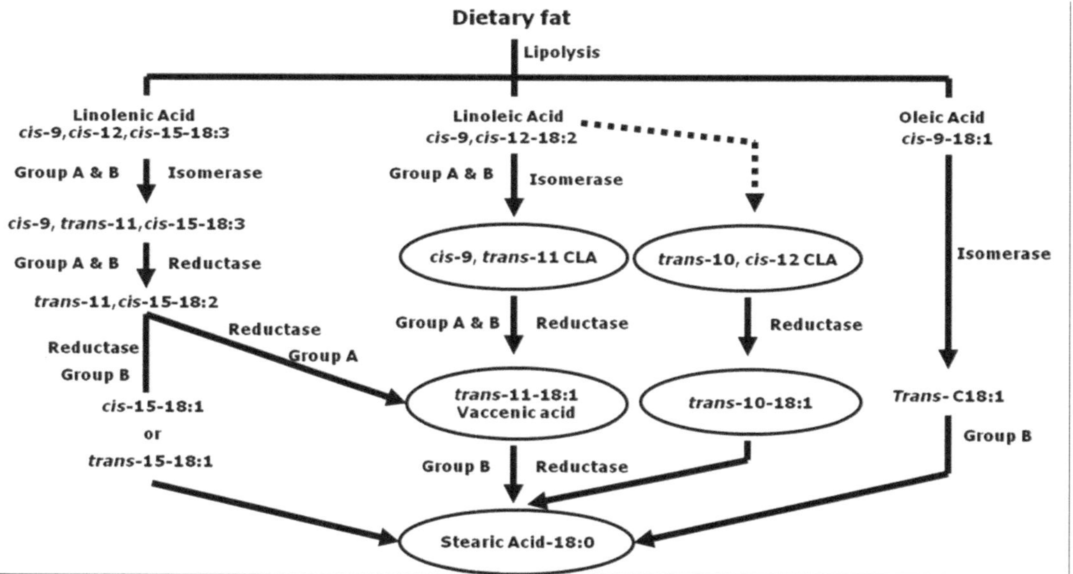

Fig. 17.9 Biohydrogenation of dietary lipids into saturated fatty acids (Griinari and Bauman 1999)

On the other hand, group B bacteria can complete all steps of biohydrogenation, i.e. conversion of LA/LNA and even oleic acid (OA) to SA, for example, *B. proteoclasticum* (Paillard et al. 2007). Many strains of *Megasphaera elsdenii* can also produce significant amounts of *trans*-10, *cis-12* CLA (Kim et al. 2002). Rumen protozoa are not involved in biohydrogenation (Boeckaert et al. 2010). The biohydrogenation activity of anaerobic fungi is reported to be very slow than rumen bacteria. Among different genera, *Orpinomyces* sp. is described as chief CLA producer (Nam and Garnsworthy 2007).

17.4 Application of Molecular Techniques in the Screening of Rumen Enzymes

From the last decade, the research has been focussed on identifying the genes encoding unique feed degradative enzymes which can be used to fortify the existing livestock production systems or to deliver novel enzymes for other industrial applications. Few ruminal microorganisms have been exploited for identification of genes and studying the expression of those genes in different microbial expression systems such as *E. coli* or *Pichia pastoris* (Table 17.3).

With the advancement in application of molecular techniques, the research has been focussed on metagenomics of the rumen microbial community. It is a method to study the DNA of entire population of microorganisms (Handelsman 2004). The metagenomics for different enzyme discovery involves creating of a metagenomic library from rumen sample and screening the library clones for specific enzymes. The advantage of metagenomics over conventional way is that it allows screening of thousands of clones in a relatively short time and enable the potential discovery of a large number of different enzymes from a sample.

Worldwide research groups are working towards the development of metagenomic approaches to characterise the structure and function of rumen microbiota in order to identify the factors that may improve the functioning of the rumen and limit undesirable environmental effects (Table 17.4). Metagenomics is the application of modern genomic techniques to the study of communities of microbes directly in their natural environments, bypassing the need for isolation and lab cultivation of individual species. Metagenomics represents a strategy for discovering diverse enzymes encoded in nature. In Table 17.4, a few studies done on metagenomics of rumen towards the identification of new enzymes which have potential biotechnological applications are listed. The metagenomic research has generated genetic information on the entire microbial community, which is important because 90 % of microbes cannot be isolated or cultured. The metagenomic method provides a global microbial gene pool without the need to culture of the microorganisms.

The major laboratories working in the area of rumen metagenomics include DOE Joint Genome Institute – Genome Technology, USA; USDA, USA; INRA, France; CSIRO, Australia; and AgResearch, New Zealand. From India, Anand Agricultural University has recently used Ion Torrent PGM next-generation sequencing technology to characterise general microbial diversity and the repertoire of microbial genes present, including genes associated with dormancy and sporulation in Mehsani buffalo rumen metagenome (Singh et al. 2014). An European project on rumenomics is underway between European partners from UK, Sweden, France, Italy, Finland and Switzerland.

17.5 Conclusions

Although rumen contains various kinds of microbes that can be used for various purposes, microbes/enzymes for lignocellulose-based biorefinery can play a crucial role in the present environment. Plant biomass being most abundant and mostly unused provides us a valuable renewable natural resource that can be exploited for various purposes ranging from production of fuels, chemicals, food or feed. However, due to

Table 17.3 Genetic characterisation of rumen microorganisms for different enzymes

Enzyme	Organism	Gene	Reference
Endoglucanase	*B. fibrisolvens* H17c	*endI*	Berger et al. (1989)
	F. succinogenes AR1	*endAFS*	Cavicchioli et al. (1991)
	F. succinogenes BL2	*endC*	B'era et al. (1996)
	F. succinogenes SD35	*end-1*	Ozcan et al. (1996)
	F. succinogenes S85	*endB*	Broussolle et al. (1994)
	P. ruminicola AR20	*celA*	Vercoe and Gregg (1992)
	P. ruminicola 23		Matsushita et al. (1991)
	R. albus F-40	*Egl*	Ohmiya et al. (1989), Duguchi et al. (1991)
	R. albus F-40	*egIV*	Karita et al. (1993)
	R. albus SY3	*celA, celB*	Poole et al. (1990)
	R. albus 8	*celA*	Attwood et al. (1996)
	R. albus AR67	*celA*	Vercoe and Gregg (1993)
	R. flavefaciens FD-1	*celE*	Wang et al. (1993)
		celB	Vercoe and Gregg. (1993)
	N. frontalis MCH3	*celA*	Fujino et al. (1995)
	N. patriciarum	*celB*	Zhou et al. (1994)
	Orpinomyces joyonii	*celA,*	Liu et al. (1996)
		celB2	Ye et al. (2001)
	F. succinogenes S85	*Cel9B, Cel5H and Cel8B*	Qi et al. (2007)
Xylanase	*B. fibrisolvens* 49	*xynA*	Mannarelli et al. (1990)
	B. fibrisolvens H17c	*xynB*	Lin and Thomson (1991)
	F. succinogenes S85	*xynC*	Paradis et al. (1993), Zhu et al. (1994)
	P. ruminicola 23	*xynA*	Whitehead (1993)
	P. ruminicola $B_1$4	*xynB*	Gasparic et al. (1995)
	R. flavefaciens 17	*xynA, xynB, xyn D*	Zhang and Flint (1992), Zhang et al. (1994), Flint et al. (1993)
	N. patriciarum	*xynA*	Gilbert et al. (1992)
		xynB	Zhou et al. (1994)
	N. patriciarum 27	*xynC*	Tamblyn et al. (1993), Selinger et al. (1995)
	N. patriciarum	*xynCDBFV*	Liu et al. (2005)
	Orpinomyces sp. PC-2	*xynA*	Chen et al. (1995)
β-Glucosidase	*R. albus* F-40	*pRA201*	Takano et al. (1992), Ohmiya et al. (1985)
	B. fibrisolvens H17c	*BglA*	Lin et al. (1990)
β-Glucanase	*F. succinogenes*	–	Liu et al. (2005)
	P. rhizinflata	*eglA*	Liu et al. (2005)
Peptidase	*Prevotella albensis M384*	*DPP-IV*	Walker et al. (2003)
Cellodextrinase	*B. fibrisolvens* H17c	*cedI*	Berger et al. (1990)

Modified from Selinger et al. (1996)

Table 17.4 Metagenome studies on rumen enzymes

Enzyme/enzyme family	Source	Screening method	Sequencing method	Reference
Cyclodextrinases	Cow	Function based	–	Ferrer et al. (2005)
	Cow	Sequential and functional screening	Shotgun sequencing	Hess et al. (2011)
Feruloyl esterase		Function based	–	Wong et al. (2013)
Endoglucanase	Cow	Function based	Pyrosequencing 454 GS FLX	Pozo et al. (2012)
	Bovine	Function based (BAC vector)	Sanger sequencing	Gong et al. (2012)
	Swamp Buffalo	Function based	–	Cheema et al. (2012)
	Buffalo	Function based	–	Rungrattanakasin et al. (2011)
	Bovine	Function based (fosmid vector)	–	Rashamuse et al. (2013)
	Buffalo	Function based (cosmid vector)	–	Liu et al. (2009)
	Cow	Function based	–	Shedova et al. (2009)
	Goat	Sequence based	Shot gun sequencing	Lim et al. (2013)
α-Glucuronidase	Cow	Function based	–	Lee et al. (2012)
Glycoside hydrolases	Bovine Ruminal Protozoan	Function based	–	Findley et al. (2011)
	Yak	Function based	–	Zhou et al. (2012)
	Yak	Function based (BAC vector)	Pyrosequencing	Dai et al. (2012)
	Bovine	Sequence based	Pyrosequencing 454 GS FLX	Brulc et al. (2009)
	Yak	Function based (cosmid vector)	–	Bao et al. (2011)
	Cow	Function based	Sanger sequencing	Zhao et al. (2010)
Carbohydrate active enzymes	Buffalo	Sequential screening	Ion torrent PGM next-generation sequencing	Patel et al. (2014)
	Cow	Function based	Pyrosequencing	Wang et al. (2013)
Mannanase-xylanase-glucanase	Cow	Function based	Sanger sequencing	Palackal et al. (2007)
Xylanase	Sheep	Function based (fosmid vector)	–	Wang et al. (2012)
Lipases	Cow	Function and sequence based	–	Liu et al. (2009)

the recalcitrant nature of biomass, its hydrolysis and further degradation are very difficult, hence also limiting industrial exploitation. Since the use of different pretreatment methods is time consuming and costly, the use of lignocellulolytic microbes of rumen can be an excellent way forward for applications in various industries including agriculture, chemicals, ethanol, animal feed, biofuel, food, paper and textiles. Furthermore, being anaerobic in nature, these microbes unlike their aerobic counterparts do not present any problems for bioprocess development.

References

Anonymous (2009) Lysing enzymes, Sigma Aldrich, vol 2009

Attwood GT, Herrera F, Weissenstein LA et al (1996) An end-β-1,4-glucanase gene (*celA*) from the rumen anaerobe *Ruminococcus albus* 8: cloning, sequencing, and transcriptional analysis. Can J Microbiol 42:267–278

B'era C, Broussolle V, Forano E et al (1996) Gene sequence analysis and properties of EGC, a family E (9) endoglucanase from Fibrobacter succinogenes BL2. FEMS Microbiol Lett 136:79–84

Bao L, Huang Q, Chang L et al (2011) Screening and characterization of a cellulase with endocellulase and exocellulase activity from yak rumen metagenome. J Mol Catal B: Enzym 73:104–110

Beg QK, Kapoor M, Mahajan L et al (2001) Microbial xylanases and their industrial applications: a review. Appl Microbiol Biotechnol 56:326–338

Berger E, Jones WA, Jones DT et al (1989) Cloning and sequencing of an endoglucanase (*end1*) gene from *Butyrivibrio fibrisolvens* H17c. Mol Gen Genet 219:183–198

Berger E, Jones WA, Jones DT et al (1990) Sequencing and expression of a cellodextrinase (*ced1*) gene from *Butyrivibrio fibrisolvens* H17c cloned in *Escherichia coli*. Mol Gen Genet 223:310–318

Bhat MK, Bhat S (1997) Cellulose degrading enzymes and their potential industrial applications. Biotechnol Adv 15:583–620

Boeckaert C, Morgavi D, Jouany JP et al (2010) Role of the protozoan *Isotricha prostoma*, liquid and solid-associated bacteria in rumen biohydrogenation of linoleic acid. Animal 3:961–971

Bonhomme A (1990) Rumen ciliates: their metabolism and relationships with bacteria and their hosts. Anim Feed Sci Technol 30:203–266

Borneman WS, Hartley RD, Morrison WH et al (1990) Feruloyl and p-coumaroyl esterase from anaerobic fungi in relation to plant cell wall degradation. Appl Microbiol Biotechnol 33(3):345–351

Broussolle V, Forano E, Gaudet G et al (1994) Gene sequence and analysis of protein domains of EGB, a novel family E endoglucanase from *Fibrobacter succinogenes* S85. FEMS Microbiol Lett 124:439–447

Brulc JM, Antonopoulos DA, Miller MEB et al (2009) Gene-centric metagenomics of the fiber-adherent bovine rumen microbiome reveals forage specific glycoside hydrolases. Proc Natl Acad Sci 106:1948–1953

Cavicchioli R, East PD, Watson K (1991) endAFS, a novel family E endoglucanase gene from *Fibrobacter succinogenes* AR1. J Bacteriol 173:3265–3268

Cheema TA, Jirajaroenrat K, Sirinarumitr T et al (2012) Isolation of a gene encoding a cellulolytic enzyme from swamp buffalo rumen metagenomes and its cloning and expression in *Escherichia Coli*. Anim Biotechnol 23:261–277

Chen H, Li XL, Liungdahl LG (1995) A cyclophilin from the polycentric anaerobic rumen fungus *Orpinomyces* sp. Strain PC-2 is highly homologous to vertebrate cyclophilin B. Proc Natl Acad Sci U S A 92:2587–2591

Chesson A, Forsberg CW (1997) Polysaccharide degradation by rumen microorganisms. In: Hobson PN, Stewart CS (eds) The rumen microbial ecosystem. Elsevier, London, pp 329–381

Daas PJH, Boxma B, Hopman AMCP et al (2001) Nonesterified galacturonic acid sequence homology of pectins. Biopolymers 58:1–8

Dai X, Zhu Y, Luo Y et al (2012) Metagenomic insights into the fibrolytic microbiome in yak rumen. PLoS One 7:e40430

Duguchi H, Watanabe Y, Sasaki T et al (1991) Structure of a *Ruminococcus albus* endo-1,4-β-glucanase gene. J Bacteriol 171:6771–6775

Ferrer M, Golyshina OV, Chernikova TN et al (2005) Novel hydrolase diversity retrieved from a metagenome library of bovine rumen microflora. Environ Microbiol 7:1996–2010

Findley SD, Mormile MR, SommerHurley A et al (2011) Activity-based metagenomic screening and biochemical characterization of bovine ruminal protozoan glycoside hydrolases. Appl Environ Microbiol 77:8106–8113

Flint HJ, Martin J, McPherson CA et al (1993) A bifunctional enzyme, with separate xylanase and β-(1,3-1,4)-glucanase domains, encoded by the *xynD* gene of *Ruminococcus flavefaciens*. J Bacteriol 175:2943–2951

Foo LY, Jones WT, Porter LJ et al (1986) Proanthocyanidin polymers of fodder legumes. Phytochemistry 21:933–935

Fujino Y, Ushida K, Miyazaki Y et al (1995) Cloning of cellulase gene from rumen chytridiomycete fungus *Neocallumastix frontalis* MCH3. Jpn Proc Jpn Soc Rumen Physiol Metabol 7:21–24

Fukuda S, Suzuki Y, Murai M et al (2006) Isolation of a novel strain of *Butyrivibrio fibrisolvens* that isomerizes linoleic acid to conjugated linoleic acid without hydrogenation, and its utilization as a probiotic for animals. J Appl Microbiol 100(4):787–794

Gasparic A, Martin J, Daniel AS et al (1995) A xylan hydrolase gene cluster in *Prevotella ruminicola* B14: sequence relationships, synergistic interactions, and oxygen sensitivity of a novel enzyme with exoxylanase and β-(1,4)-xylosidase activities. Appl Environ Microbiol 61:2958–2964

Gilbert HJ, Hazlewood GP, Laurie JI et al (1992) Homologous catalytic domains in a rumen fungal xylanase – evidence for gene duplication and prokaryotic origin. Mol Microbiol 6:2065–2072

Goel G, Puniya AK, Aguliar CN et al (2005) Interaction of gut microflora with tannins in feeds. Naturwissenschaften 92:497–503

Gong X, Gruninger RJ, Qi M et al (2012) Cloning and identification of novel hydrolase genes from a dairy cow rumen metagenomic library and characterization of a cellulase gene. BMC Res Notes 5:566

Gordon GLR, Phillips MW (1992) Extracellular pectin lyase produced by neocallimastix sp. LM1, a rumen anaerobic fungus. Lett Appl Microbiol 15:113–115

Griinari JM, Bauman DE (1999) Biosynthesis of conjugated linoleic acid and its incorporation into meat and milk in ruminants. AOCS Press, Champaign, pp 180–200

Hagerman AE, Butler LG (1981) The specificity of proanthocyanidin-protein interactions. J Biol Chem 256:4494–4497

Handelsman J (2004) Metagenomics: application of genomics to uncultured microorganisms. Microbiol Mol Biol Dev 68:669–685

Harfoot C, Hazlewood G (1988) Lipid metabolism in the rumen. In: The rumen microbial ecosystem, Elsevier, London/New York, p 285–322

Haslam E (1989) Plant polyphenols – vegetable tannins. Cambridge University Press, Cambridge, UK

Hess M, Sczyrba A, Egan R et al (2011) Metagenomic discovery of biomass-degrading genes and genomes from cow rumen. Science 331:463–467

Hogg D, Pell G, Dupree P et al (2003) The modular architecture of *Cellvibrio japonicas* mannanases in glycoside hydrolase families 5 and 26 points to differences in their role in mannan degradation. Biochem J 371:1027–1043

Jarvis MC (1984) Structure and properties of pectin gels in plant cell walls. Plant Cell Environ 7:153–164

Jayani RS, Saxena S, Gupta R (2005) Microbial pectinolytic enzymes: a review. Process Biochem 40:2931–2944

Jenkins TC, Wallace RJ, Moate PJ et al (2008) Board-invited review: recent advances in biohydrogenation of unsaturated fatty acids within the rumen microbial ecosystem. J Anim Sci 86(2):397–412

Karita S, Morioka K, Kajino T et al (1993) Cloning and sequencing of a novel endo- 1,4-â-glucanase gene from *Ruminococcus albus*. J Ferment Bioeng 76:439–444

Kim YJ, Liu RH, Rychlik J et al (2002) The enrichment of a ruminal bacterium (*Megasphaera elsdenii YJ 4*) that produces the trans 10, cis 12 isomer of conjugated linoleic acid. J Appl Microbiol 92(5):976–982

Kosugi A, Murashima K, Doi RH (2001) Characterization of xylanolytic enzymes in *Clostridium cellulovorans*: expression of xylanase activity dependent on growth substrates. J Bacteriol 183(24):7037–7043

Krause DO, Denman SE, Mackie RI et al (2003) Opportunities to improve fiber degradation in the rumen: microbiology, ecology, and genomics. FEMS Microbiol Rev 27(5):663–693

Lee CC, Kibblewhite RE, Wagschal K et al (2012) Isolation of α-glucuronidase enzyme from a rumen metagenomic library. Protein J 31:206–211

Lim SY, Seo J, Choi H et al (2013) Metagenome analysis of protein domain collocation within cellulase genes of goat rumen microbes. Asian-Aus J Anim Sci 26:1144–1151

Lin LL, Thomson JA (1991) Cloning, sequencing and expression of a gene encoding a 73 kDa xylanase enzyme from the rumen anaerobe *Butyrivibrio fibrisolvens* H17c. Mol Gen Genet 228:55–61

Lin LL, Rumbak E, Zappe H et al (1990) Cloning, sequencing and analysis of expression of a *Butyrivibrio fibrisolvens* gene encoding a β-glucosidase. J Gen Microbiol 136:1567–1576

Liu JH, Selinger LB, Hu YJ et al (1996) Cloning and characterization of endoglucanase gene from *Orpinomyces joyonii*. In: Proceedings of 96th annual meetings American society for microbiology, p 505

Liu JR, Yu B, Liu FH et al (2005) Expression of rumen microbial fibrolytic enzyme genes in probiotic *Lactobacillus reuteri*. Appl Environ Microbiol 71:6769–6775

Liu K, Wang J, Bu D et al (2009) Isolation and biochemical characterization of two lipases from a metagenomic library of China Holstein cow rumen. Biochem Ciophys Res Commun 385:605–611

Lynd LR, Zhang Y (2002) Quantitative determination of cellulase concentration as distinct from cell concentration in studies of microbial cellulose utilization: analytical framework and methodological approach. Biotechnol Bioeng 77:467–475

Makkar HPS (2003) Effects and fate of tannins in ruminant animals, adaptation to tannins, and strategies to overcome detrimental effects of feeding tannin-rich feeds. Small Rumin Res 49:241–256

Mannarelli BM, Evans S, Lee D (1990) Cloning, sequencing, and expression of a xylanase gene from the anaerobic ruminal bacterium *Butyrivibrio fibrisolvens*. J Bacteriol 172:4247–4254

Matsushita O, Russell JB, Wilson DB (1991) A *Bacteroides ruminicola* 1,4-β-D-endoglucanase is encoded in two reading frames. J Bacteriol 173:6919–6926

McNeil M, Darvill AG, Fry SC et al (1984) Structure and function of the primary cell wall of plants. Ann Rev Biochem 53:625–663

Mittal A, Singh G, Goyal V et al (2011) Isolation and biochemical characterization of acido-thermophilic extracellular phytase producing bacterial strain for potential application in poultry feed. Jushinapur J Microbiol 4:273–282

Nam IS, Garnsworthy PC (2007) Biohydrogenation of linoleic acid by rumen fungi compared with rumen bacteria. J Appl Microbiol 103(3):551–556

Ohmiya D (1989) Purification and properties of the endo-1,4-β-glucanase from *Ruminococcus albus* and its gene product in *Escherichia coli*. J Ferment Bioeng 71:221–225

Ohmiya K, Shirai M, Kurachi Y et al (1985) Isolation and properties of â-glucosidase from *Ruminococcus albus*. J Bacteriol 161:432–434

Orpin CG (1984) The role of ciliate protozoa and fungi in the rumen digestion of plant cell walls. Anim Feed Sci Technol 10:121–143

Ozcan N, Cunningham C, Harris WJ (1996) Cloning of a cellulase gene from the rumen anaerobe *Fibrobacter succinogenes* SD35 and partial characterization of the gene product. Lett Appl Bacteriol 22:85–89

Paillard D, McKain N, Chaudhary LC et al (2007) Relation between phylogenetic position, lipid metabolism and butyrate production by different *Butyrivibrio-like* bacteria from the rumen. Ant V Leeuwen 91(4):417–422

Palackal N, Lyon CS, Zaidi S et al (2007) A multifunctional hybrid glycosyl hydrolase discovered in an uncultured microbial consortium from ruminant gut. Appl Microbial Biotechnol 74:113–124

Paradis FW, Zhu H, Krell PJ et al (1993) The *xynC* gene from *Fibrobacter succinogenes* S85 codes for a xylanase with two similar catalytic domains. J Bacteriol 175:7666–7672

Patel DD, Patel AK, Parmar NR (2014) Microbial and carbohydrate active enzyme profile of buffalo rumen metagenome and their alteration in response to variation in the diet. Gene 545:88–94

Piddington CS, Houston CS, Paloheimo M et al (1993) The cloning and sequencing of the genes encoding phytase (phy) and pH 2.5-optimum acid phosphatase (aph) from *Aspergillus niger* var. awamori. Gene 133(1):55–62

Poole DH, Hazlewood GP, Laurie JI et al (1990) Nucleotide sequence of the *Ruminococcus albus* SY3 endoglucanase genes *celA* and *celB*. Mol Gen Genet 223:217–223

Pozo MV, Arrojo LF, Martinez JG et al (2012) Microbial β-glucosidases from cow rumen metagenome enhance the saccharification of lignocellulose in combination with commercial cellulase cocktail. Biotechnol Biofuel 5:73

Qi M, Jun HS, Forsberg CW (2007) Characterization and Synergistic Interactions of Fibrobacter succinogenes glycoside hydrolases. Appl Environ Microbiol 73:6098–6105

Rashamuse KJ, Burton SG, Cowan DA (2007) A novel recombinant ethyl ferulate esterase from *Burkholderia multivorans*. J Appl Microbiol 103:1610–1620

Rashamuse KJ, Visser DF, Hennessy F et al (2013) Characterisation of two bifunctional cellulase-xylanase enzymes isolated from a bovine rumen metagenome library. Curr Microbiol 66:145–151

Rodriquez LV, Valdivia B, Contreas JC et al (2011) Novel strategies for upstream and downstream processing of tannin acyl hydrolase. Enzy Res 20

Rombouts FM, Pilnik W (1986) Pectinases and other cell degrading enzymes of industrial importance. Symbiosis 2:79–90

Rungrattanakasin B, Jirajaroenrat K, Maneewan1 K et al (2011) Molecular cloning of glycoside hydrolase family 9 cellulase gene from buffalo rumen. In: International conference on bioscience, biochemistry and bioinformatics IPCBEE, vol 5

Sauer M, Marx H, Mattanovich D (2012) From rumen to industry. Microb Cell Fact 11:121

Selinger LB, Liu JH, Hu YJ et al (1995) Characterization of a xylanase gene from *Neocallimastix patriciarum*. In: Proceedings of the 23rd biennial conference on rumen function, p 39, Chicago

Selinger LB, Forsberg CW, Cheng KJ (1996) The rumen: a unique source of enzymes for enhancing livestock production. Anaerobe 2:263–284

Shallom D, Shoham Y (2003) Microbial hemicellulases. Curr Opin Microbiol 6:219–228

Shedova EN, Lunina NA, Berezina OV et al (2009) Expression of the genes *CelA* and *XylA* isolated from a fragment of metagenomic DNA in *Escherichia coli*. Mol Genet Microbiol Virol 24:76–78

Silley P (1985) A note on the pectinolytic enzymes of *Lachnospira multiparus*. J Appl Bacteriol 58:145–149

Singh KM, Reddy B, Patel AK et al (2014) Metagenomic analysis of buffalo rumen microbiome: effect of roughage diet on dormancy and sporulation genes. Metagene 2:252–268

Takano M, Moriyama R, Ohmiya K (1992) Structure of a β-glucosidase gene from *Ruminococcus albus* and properties of the translated product. J Ferment Bioeng 73:79–88

Tamblyn Lee JM, Hu Y, Zhu H et al (1993) Cloning of a xylanase gene from the ruminal fungus *Neocallimastix patriciarum* 27 and its expression in *Escherichia coli*. Can J Microbiol 39:134–139

Taylor EJ, Smith NL, Turkenburg JP et al (2006) Structural insight into the ligand specificity of a thermostable family 51 arabinofuranosidase, Araf 51, from *Clostridium thermocellum*. Biochem J 395:31–37

Van Hartingsveldt W, van Zeijl CM, Harteveld GM et al (1993) Cloning, characterization and overexpression of the phytase-encoding gene (phyA) of *Aspergillus niger*. Gene 127(1):87–94

Vercoe PE, Gregg K (1992) DNA sequence and transcription of an endoglucanase gene from *Prevotella (Bacteroides) ruminicola* AR20. Mol Gen Genet 233:284–292

Vercoe PE, Gregg K (1993) Sequence and transcriptional analysis of an endoglucanase gene from *Ruminococcus albus* AR67. GenBank Accession Number L10243

Walker ND, McEwan NR, Wallace RJ (2003) Cloning and functional expression of dipeptidyl peptidase IV from the ruminal bacterium *Prevotella albensis* M384^T. Microbiol 8:2227–2234

Wang Y, McAllister TM (2002) Rumen microbes, enzymes and feed digestion-a review. Asian-Aust J Anim Sci 15:1659–1676

Wang WY, Reid SJ, Thomson JA (1993) Transcriptional regulation of an endoglucanase and a cellodextrinase gene in *Ruminococcus flavefaciens* FD-1. J Gen Microbiol 139:1219–1226

Wang J, Sun W, Zhou Y et al (2012) Screening of a xylanase clone from a fosmid library of rumen microbiota in Hu sheep. Anim Biotechnol 23:156–173

Wang L, Hatem A, Catalyurek UV et al (2013) Metagenomic insights into the carbohydrate-active enzymes carried by the microorganisms adhering to solid digesta in the rumen of cows. PLoS One 8:e78507

Whitehead TR (1993) Analyses of the gene and amino acid sequence of the *Prevotella (Bacteroides) ruminicola* 23 xylanase reveals unexpected homology with endoglucanases from other genera of bacteria. Curr Microbiol 27:27–33

Wong DWS, Chan VJ, Liao H et al (2013) Cloning of a novel feruloyl esterase gene from rumen microbial metagenome and enzyme characterization in synergism with endoxylanases. J Ind Microbiol Biotechnol 40:287–295

Ye XY, Ng TB, Cheng KJ (2001) Purification and characterization of a cellulase from the ruminal fungus *Orpinomyces joyonii* cloned in *Escherichia coli*. Int J Biochem Cell Biol 33:87–94

Zhang JX, Flint HJ (1992) A bifunctional xylanase encoded by the xynA gene of the rumen cellulolytic bacterium *Ruminococcus flavefaciens* 17 comprises two dissimilar domains linked by an asparagines/glutamine rich sequence. Mol Microbiol 6:1013–1023

Zhang JX, Martin J, Flint HJ (1994) Identification of non-catalytic conserved regions in xylanases encoded by *xynB* and *xynD* genes of the cellulolytic rumen anaerobe *Ruminococcus flavefaciens*. Mol Gen Genet 245:260–264

Zhao SG, Wang JQ, Bu DP et al (2010) Novel glycoside hydrolases identified by screening a Chinese Holstein dairy cow rumen-derived metagenome library. Appl Environ Microbiol 76:6701–6705

Zhou L, Xue GP, Orpin CG et al (1994) Intronless *celB* from the anaerobic fungus *Neocallimastix patriciarum* encodes a modular family A endoglucanase. Biochem J 297:359–364

Zhou J, Bao L, Chang L et al (2012) Beta-xylosidase activity of a GH3 glucosidase/xylosidase from yak rumen metagenome promotes the enzymatic degradation of hemicellulosic xylans. Lett Appl Microbiol 54:79–87

Zhu H, Paradis FW, Krell PJ et al (1994) Enzymatic specificities and modes of action of the two catalytic domains of the XynC xylanase from *Fibrobacter succinogenes* S85. J Bacteriol 176:3885–3895

Ruminal Fermentations to Produce Liquid and Gaseous Fuels

18

Paul J. Weimer

Abstract

The ruminant animal has evolved to effectively utilize relatively recalcitrant cellulosic biomass to fulfill its needs for growth and reproduction. Although the animal has developed a unique biomass pretreatment process based on mastication and rumination, the key driver of biomass utilization is the microbiota residing in the rumen. The ruminal microflora, largely comprised of anaerobic bacteria, can ferment all of the non-lignin biomass components, primarily to volatile fatty acids (VFAs), methane, and carbon dioxide. Methane is a well-known and widely utilized fuel for human activities, but most of the energy in the fermentation products resides in the VFA, which serves as the primary energy source for the animal. Because VFA can be converted by chemical and electrochemical routes to hydrocarbons, alcohols, or esters, they are the central components of the "carboxylate platform" for biofuel production. The ruminal fermentation can readily be conducted in bioreactors (i.e., extraruminally) and shares several attractive features with carboxylate platform processes based on "stuck" anaerobic digestion: feedstock flexibility, high solid loading rate, non-aseptic operation, and wide substrate range. The extraruminal fermentation has the additional benefit of a short run time (days, as opposed to weeks). This chapter summarizes current knowledge in the potential for fuel production using the ruminal fermentation, along with the potential for certain members of the ruminal community to be used for production of specific chemicals in pure culture or defined mixed culture.

Keywords

Carboxylate platform • Fermentation • Fuels • Volatile fatty acids

P.J. Weimer (✉)
US Dairy Forage Research Center, US Department of Agriculture – Agricultural Research Service, Madison, WI USA

Department of Bacteriology, University of Wisconsin, Madison, WI 53706, USA
e-mail: Paul.Weimer@ars.usda.gov

A.K. Puniya et al. (eds.), *Rumen Microbiology: From Evolution to Revolution*, DOI 10.1007/978-81-322-2401-3_18

18.1 Introduction

The ruminant animal is an intricate evolutionary construct and a shining example of mutualism between a host organism and its resident microflora (Russell 2002). Ruminal microbes are essential to feed utilization by the host animal, in that they convert feedstuffs to volatile fatty acids (VFAs, sometimes called short-chain fatty acids, SCFA) that serve as the primary energy source for the host. In addition, the microbial cells themselves provide a substantial fraction of the protein needs of the host, much of it produced by conversion of nonprotein nitrogen to microbial cell protein of very high quality (i.e., having a favorable balance of essential amino acids). Although the ruminal microflora effectively ferment grains and protein concentrates that provide much of the diet of high-producing ruminant livestock (such as cattle, sheep, and goats), the true evolutionary innovation of the ruminal symbiosis was the ability of the ruminal microbes to degrade and utilize fibrous plant material that is not a significant source of nutrition to nonruminant animals. Simply put, the ruminant evolved as a grazing or browsing animal, and its physiology and microbiology are best understood within this evolutionary context.

The challenges faced by ruminal microbes in the digestion of fibrous feeds are nearly identical to those faced by the biorefinery industry in converting these same plant materials to fuels useful to human society. This observation has led to the notion that the rumen represents a natural example of consolidated bioprocessing (CBP) of cellulosic biomass (Lynd et al. 2002; Weimer et al. 2009). In this chapter, the ruminal fermentation will be evaluated from two rather different perspectives: (1) how the ruminal fermentation might inform or enlighten our efforts to convert cellulosic feedstocks to fuels in microbial systems based on the conversion of relatively recalcitrant cellulosic biomass by natural and engineered nonruminal microbes and (2) how ruminal microbes themselves might be harnessed to produce fuels or fuel precursors from this same plant material in "extraruminal" fermentations conducted in controlled bioreactors.

We can distinguish at the outset between compounds that can be used directly as fuels (ethanol, H_2, and methane) and compounds that contain high caloric value but are not directly utilizable as fuels due to issues such as low volatility or high corrosivity (viz., VFA). The development of a successful biofuel process based on the latter compounds, which we will here term "fuel precursors," is inextricably linked to the development of downstream technologies that convert these compounds to more conventional fuels (e.g., hydrocarbons and alcohols). In this chapter, we will focus the discussion of fuel precursors primarily on their fermentative production, but will briefly discuss downstream processing, its major challenges, and some newer technologies to address these challenges.

18.2 Some Basic Principles of the Ruminal Fermentation

18.2.1 The Ruminal Environment

The essential features of the feed processing in ruminants have been excellently reviewed (Hungate 1966; Van Soest 2002) and evaluated from the perspective of biofuel production (Weimer et al. 2009). The rumen is the largest of the four pre-intestinal digestive chambers in the ruminant animal; in dairy cattle, well studied for their agricultural importance and high productivity, the rumen can exceed 100 l in volume. Despite the ingestion of large amounts of oxygen during feed and water consumption by the host, the ruminal habitat is highly anaerobic ($E_h < -200$ mV) due to the uptake of oxygen by ruminal epithelial tissue, activity of facultative anaerobic bacteria, and the microbial production of various reducing agents.

On a mass basis, VFAs are the major products of the ruminal fermentation. Production of VFA at the rates observed in the rumen has the potential to reduce ruminal pH to values inhibitory to the microflora (particularly the cellulolytic bacteria), but the host animal effectively regulates pH within a range of ~5.5–7 by a combination of VFA absorption through the epithelial cells in the ruminal wall and secretion of copious amounts of bicarbonate-buffered saliva (over 100 L d^{-1} in dairy cattle). Ruminal temperature is controlled even more tightly than is ruminal pH, typically within a few degrees of 39 °C. Ruminal environmental conditions have selected a microbial community that is relatively stenothermal, relatively uncomfortable outside its natural pH range, and not well adapted to very high concentrations of fermentation end products. As will be seen below, this does provide some challenges in maximizing the productivity of extraruminal fermentations.

18.2.2 The Mechanics of the Rumen as a Bioreactor

A key feature of the rumen is its relatively high solid content, typically ~ 15 % (w/v). This solid content is similar to the targeted solid concentrations in cellulosic biorefinery designs (Benz 2008). At these concentrations, mechanical mixing of reactor contents and their movement through the multiple unit operations is challenging, and the ruminant serves as an example of how this might be most effectively accomplished in a biorefinery. The ruminant host regulates feed passage in a selective and discontinuous manner, such that the passage rate of liquids is substantially more rapid than that of solid feed materials (averaging about 8 h and 50 h, respectively). This selective retention of feed particles provides sufficient time for most of the accessible organic material to be fermented before passage of relatively indigestible fiber out of the rumen. The process of rumination (chewing of the cud) is of central importance in digestion, in that grinding and regrinding of the feed (particularly its fibrous components) in the buccal cavity facilitate continued degradation in the rumen by exposing degradable fiber surface. The exemplary ability of the ruminant to effectively utilize most of the energy in highly fibrous feeds via an energy-efficient type of physical pretreatment has been largely ignored by a biorefinery industry obsessed with chemical pretreatments to enhance product yields in biofuel fermentations. The physical process of rumination presages the fermentative conversion of up to 80 % of the potentially digestible dry matter at an energetic cost of ~1 % of the digestible energy content of the feedstock (Weimer et al. 2009). By contrast, chemical pretreatments facilitate 90 % or greater conversion of the fermentable content of the feedstock, but have inherently high costs (the chemical agents themselves, energy required for operation at high temperature and pressure, waste disposal) and often generate chemical inhibitors of downstream fermentation.

18.2.3 Ruminal Fermentation Pathways and Product Yields

Bacteria are the dominant members of the ruminal community, and the details of their physiology and biochemistry have been studied in much more detail than have those of protozoa or fungi. Fungi are a quantitatively minor component of the microflora and seem to contribute to the fermentation of fiber in large part by physically splitting plant tissue apart during the course of their extensive mycelial growth. Protozoa digest primarily starch and ingested bacterial cells, and their quantitative contribution to fiber digestion is likely very small. However, the all three major microbial groups share generally similar pathways for catabolism of soluble substrates, although the bacteria as a group are more metabolically diverse and carry out certain catabolic reactions (e.g., lactate fermentation or succinate decarboxylation) uniquely.

Polysaccharides – cellulose, hemicelluloses, starch, fructans, and pectin – provide most of the substrate for the ruminal fermentation. These polysaccharides are depolymerized primarily to oligosaccharides by various species of fibrolytic and amylolytic microbes, mostly bacteria, with the resulting soluble sugars fermented by both the fibrolytic and amylolytic species, and a host of generalist sugar-fermenting commensal microbes. The central catabolic intermediate of these sugar fermentations is pyruvic acid, which is converted mainly to acetate, succinate, and lactate. Acetate in turn is used to produce butyrate, succinate is decarboxylated to propionate, and lactate is converted by different pathways to yield both acetate and propionate. The proportions of VFA vary with diet. Rapidly fermented carbohydrates (e.g., starch and sugars) typically yield acetate and propionate in similar molar proportions, along with smaller amounts of butyrate, while cellulose and hemicelluloses usually yield acetate, propionate, and butyrate in ratios approximating 6:2:1. VFA contents in the rumen are typically in the range of 80–200 mM total VFA.

In addition to VFA, the ruminal fermentation yields substantial amounts of gas. Most fermentations yield CO_2. H_2 is produced in considerable amounts by several carbohydrate fermenters, but virtually all of this H_2 is oxidized by methanogenic archaea, using CO_2 as electron acceptor. Consequently, the ruminal gas phase is a mixture of methane and CO_2. Smaller amounts of methane are also produced from formate (which is produced from both carbohydrate and nucleic acid fermentations or by CO_2 reduction) or from methanol liberated from methoxylated pectins or phenolic acids. The oxidation of both H_2 and formate is thermodynamically very favorable, and neither accumulates to significant amounts in a normally functioning rumen.

18.2.4 Nutrient Requirements of Ruminal Microbes

In addition to the general requirements for growth (carbon and energy sources, nitrogen source, mineral nutrients such as P, S, and trace metals), some ruminal microbes have additional growth requirements. Among ruminal microbes, these requirements have been studied in detail only in bacteria, which tend to have more complex nutrient requirements than most bacteria used in industrial fermentations. In the rumen, most of these nutrient requirements are met by either feed components or by production by other members of the rumen microflora; however, these nutrients must be supplied when pure cultures are grown in the laboratory on purified substrates such as starch or cellulose. Noncellulolytic species often require particular vitamins, and many species are stimulated by peptides, amino acids, or other components found in peptones, yeast extract, or clarified rumen fluid. Ruminal cellulolytic bacteria (RCB) require several vitamins and at least one branched chain VFA from the group isobutyric, 2-methylbutyric, and isovaleric acid. Little or no incorporation of these acids was observed by Allison (1969), and more recent genomic studies (Suen et al. 2011a, b) have revealed that at least two of the predominant RCB (*Fibrobacter succinogenes* and *Ruminococcus albus*) contain the complete biosynthetic pathways for all 20 protein amino acids. It is thus likely that the branched chain VFA is required instead as a precursor for synthesis of branched long-chain fatty acids in cell membranes. One species, *R. albus*, also requires 3-phenylpropanoic acid (PPA) for growth on cellulose, but not on glucose or cellobiose; the PPA is apparently required for synthesis of extracellular polymeric substances that facilitate adherence of cells to the cellulose surface (Morrison et al. 1990).

18.3 Fuel Production by Ruminal Microbes: Ethanol and Hydrogen

Two compounds that have received great attention as biofuels are ethanol and hydrogen (H_2). The former is used in tremendous volumes for blending with gasoline (in the USA) or directly in modified vehicle engines (particularly in Brazil). Hydrogen is touted as an environmentally superior fuel due to the fact that its combustion generates only water and CO_2 generated during its production from either fossil fuels or biomass is readily captured. Both ethanol and H_2 are normal products of anaerobic fermentations by many microbial species, including several from the rumen. Ethanol production occurs via sequential reduction of acetyl-CoA and acetaldehyde, typically with NADH as electron donor. H_2 can be produced by a formic hydrogen lyase ($HCOOH \rightarrow H_2 + CO_2$) but is more commonly produced by direct reduction of protons, using hydrogenase and reduced ferredoxin as an electron carrier. The latter reaction is less thermodynamically favorable and proceeds to only a limited extent unless H_2 is continually removed (e.g., by H_2-oxidizing methanogens). In natural systems, neither H_2 nor ethanol accumulates to significant quantities, because they are attractive energy sources for other microorganisms, either within the anaerobic habitat in which they are produced or upon diffusion to nearby habitats where O_2 is available as an electron acceptor. Consequently, producing either ethanol or H_2 industrially requires pure cultures or carefully constructed defined mixed cultures.

The highly cellulolytic Gram-positive bacterium *Ruminococcus albus* is among the best-known ruminal producers of H_2 and ethanol. Phylogenetically, the genus *Ruminococcus* is within the phylum *Firmicutes*, family *Clostridiales*, and displays many fermentation characteristics of related *Clostridium* species. Individual strains display clear differences in the rate of cellulose degradation. In continuous culture, *R. albus* strain 7 has been reported (Pavlostathis et al. 1988) to ferment crystalline cellulose up to 0.91 mol ethanol and 1.58 mol H_2 per mol glucose equivalent (0.289 g ethanol and 0.020 g H_2 [g cellulose consumed]$^{-1}$). Efforts focused on maximizing H_2 production from paper waste have achieved yields of 46–280 L H_2 [kg paper]$^{-1}$ (Ntaikou et al. 2009), equivalent to a yield of 0.9–2.5% on a weight basis. These hydrogen yields are substantially lower than those produced from glucose by certain nonruminal bacteria (e.g., *Clostridium butyricum*, Masset et al. 2012), but direct comparisons are misleading because none of these high-yielding nonruminal hydrogen producers are capable of utilizing polymeric cellulose or hemicelluloses.

There are three major challenges to utilizing ruminal bacteria on an industrial scale to produce H_2 and ethanol, particularly from cellulosic biomass. First, the ruminal microbial species most proficient in converting plant cell wall material are also very strictly anaerobic and nonsporulating, limiting their ability to withstand any transient nonreducing conditions in a bioreactor. Second, these species display poor product tolerance and an inability to produce high concentrations of product. In the case of ethanol, this likely reflects the fact that these species have never had to withstand significant ethanol concentrations in their natural environment (where the compound is either not produced or produced only as an intermediate) or during their history of cultivation under axenic conditions (where cultivation has typically been conducted at low substrate concentrations for physiological experiments unrelated to biofuel production). In the case of H_2, ruminal microbes are faced with the same thermodynamic constraints, described above, that are faced by all organisms that produce H_2 by proton reduction. Third, no genetic systems are available for the most proficient of the fermentative species isolated from the rumen; this greatly limits our ability to conduct targeted metabolic engineering. Fortunately, the genome sequences of several ruminal fibrolytic bacteria have been completed (Suen et al. 2011a, b), which should stimulate attempts to heterologously express some of these organisms' more interesting metabolic features in more genetically tractable host species.

18.4 Production of Fuel Precursors

While the prospects for using ruminal microbes for industrial production of the authentic fuels ethanol and H_2 appear to be very limited, the same does not hold true for the production of organic acids that can serve as fuel precursors.

18.4.1 Succinate

Succinic acid is a four-carbon dicarboxylic acid that is widely used as a food acidulent and polymer intermediate (Zeikus et al. 1999). It will be considered here for two reasons. Firstly, it is inherently reactive via Kolbe or Hofer-Moest electrochemistry (discussed later in this chapter) and as a diacid can potentially provide additional extension to the carbon chain of the product fuel molecule. Secondly, it is a major precursor to propionate, a VFA particularly reactive in its electrochemistry.

Succinate production by anaerobic fermentation of carbohydrate proceeds through a pathway in which either phosphoenolpyruvate (PEP) or pyruvate is carboxylated to oxaloacetate (OAA); this step requires elevated levels of CO_2, which is common in the rumen but must be supplied exogenously in a bioreactor. OAA is then sequentially converted to malate, fumarate, and succinate. The terminal step is a reductive reaction typically employing $FADH_2$ as electron donor. Succinate yields are dependent on the flux of reducing equivalents to fumarate reduction versus their flux to other reductive reactions (e.g., those which form H_2 or formate).

The rumen is host to several species of bacteria that are among the most prolific succinate producers in the nature. *Actinobacillus succinogenes* $130Z^T$ (Guettler et al. 1999) is a facultative anaerobic, Gram-negative rod, first isolated from the bovine rumen, that can ferment a wide variety of sugars, including the glucose, cellobiose, xylose, and arabinose. Under a 100 % CO_2 atmosphere, the type strain produced up to 74 g of succinate L^{-1}, and some derived strains produced over 100 g L^{-1} in media additionally amended with yeast extract and corn steep liquor.

Two other ruminal bacteria have attracted interest as succinate producers, primarily due to their ability to use cellulose as a growth substrate. *Fibrobacter succinogenes* is a highly cellulolytic anaerobe that is nearly the sole representative of its bacterial phylum (Suen et al. 2011b). It typically comprises 1–2 % or more of the bacterial community in the bovine rumen. As its name implies, *F. succinogenes* produces succinate as a major product. It also produces smaller amounts of acetate and formate, but produces no other reduced compounds (H_2, ethanol, butyrate, etc.). The proportion of its fermentation products does not change significantly with growth rate (Weimer 1993). *F. succinogenes* can grow reasonably rapidly on highly crystalline cellulose and can hydrolyze a wide variety of polysaccharides, although it cannot ferment the hydrolytic products. This last fact suggests that the bacterium can be a major component of a mixed culture system containing other species that utilize these various saccharides. *Ruminococcus flavefaciens* is another highly cellulolytic ruminal bacterium that hydrolyzes fewer polysaccharides but is less restricted in its use of hydrolytic products. Although acetate is its major fermentation product, reducing equivalents generated during carbohydrate fermentation are used to produce succinate, H_2, and formate (the latter from CO_2). Disposal of reducing equivalents to the latter two products decreases the overall yield and productivity of succinate. Because of its lower molar yield of succinate, *R. flavefaciens* is likely inferior to *F. succinogenes* for industrial production of succinate. For example, fermentation of pulped paper (a highly reactive form of cellulose) resulted in succinate productivities of 0.0597 g L^{-1} h^{-1} for *F. succinogenes* and 0.0431 g L^{-1} h^{-1} for *R. flavefaciens* FD-1 (Gokarn et al. 1997). While the ability to produce succinate from cellulose is attractive on the surface, these productivities are far lower than those achieved either by *Actinobacillus* (Lee et al. 2000) or by recently engineered strains of *Escherichia coli* (Lee et al. 2005), and thus succinate production by these cellulolytic species is unlikely to have commercial promises.

18.4.2 VFA

As noted above, acetate is the most common and, on a mass basis, the most abundant product of carbohydrate fermentation by most carbohydrate fermenters in pure culture. Propionate is produced from two major routes: from lactate via dehydration to acrylate followed by a reduction to propionate or by direct decarboxylation of succinate. In principle, VFA production can be attained using either pure culture fermentations or defined cocultures, for example, the fermentation of cellulose by *Fibrobacter succinogenes* to succinate, with decarboxylation to propionate by *Selenomonas ruminantium* (Scheiffinger and Wolin 1973). This latter interaction represents a classic case of bacterial mutualism, but has not been exploited for commercial propionate production. On the other hand, production of propionate from lactate has been examined using lactilytic strains of *S. ruminantium*, which have been shown to produce up to 24 g propionate L^{-1} (Eaton and Gabelman 1992).

Butyrate is produced via a condensation of two acetyl-CoA units, followed by a reduction, dehydration, and another reduction; this pathway is sometimes referred to as reverse β-oxidation, because the VFA chain is lengthened by incorporating the two-carbon unit of acetic acid, the compound produced during sequential degradation of long-chain fatty acids (Fig. 18.1). The best-known ruminal producer of butyrate is *Butyrivibrio*, a genus of generalist bacteria that can ferment a wide range of carbohydrates, a few polysaccharides (particularly hemicelluloses), as well as some proteins and amino acids. The best-known reverse β-oxidizer, *Clostridium kluyveri*, is easily enriched from ruminal contents in ethanol/acetate media amended with inhibitors of methanogenesis (the latter can be removed from the media upon pure culture isolation). Pure cultures of one ruminal isolate, strain 3231B (= NRRL B-59667), produced up to 110 mM (13.2 g L^{-1}) caproate, near its solubility limit (Weimer and Stevenson 2012). This organism thus has promise as a caproate producer in pure culture. However, the organism is present in only trace numbers in the rumen (and is likely a transient entering the rumen with feeds such as alfalfa silage), and its requirement for ethanol as an electron donor would complicate mixed culture fermentation of cellulosic biomass. A more authentic member of the ruminal community, *Megasphaera elsdenii*, can ferment lactate to propionate (C_3) and then to valerate (C_5) and can ferment glucose successively to acetate (C_2), butyrate (C_4), and caproate (C_6). H_2 is produced on both substrates. From DL-lactate, the maximum reported concentrations of acetate, propionate, butyrate, and valerate in the same fermentation broth were 2.9, 3.9, 0.9, and 1.1 g L^{-1}, respectively (Weimer and Moen 2013). Moreover, cocultures of *M. elsdenii* and the lactate producer, *Streptococcus bovis*, rapidly convert sugars in whole plant biomass to valerate and propionate in high yield (Weimer and Digman 2013).

18.5 Mixed Culture Fermentations

18.5.1 The Carboxylate Platform

From a practical standpoint, use of either pure cultures or defined mixed cultures of ruminal microbes for producing fuels and fuel precursors is complicated by the same issues of culture contamination that challenge nonruminal production of these same compounds. However, these issues can largely be circumvented by exploiting the robustness and efficiency characteristic of ruminal fermentations conducted by the entire microbial community in undefined mixed culture. The concept of using mixed cultures to convert complex biomass materials to a mixture of useful products was originally proposed over 30 years ago by Levy et al. (1981), who coined the term "biorefinery" to describe the process. This term is now generally applied to almost any collection of processes for converting biomass to fuels. In their concept of a biorefinery, Levy et al. (1981) proposed operating a conventional anaerobic digestion with a sewage sludge inoculum under conditions where methanogenesis, the terminal step of the process, was intentionally inhibited (Fig. 18.2). Inhibition of methanogens via either low operating pH or addition of specific methanogenic inhibitors, such as iodoform (CHI_3), results in a "stuck" anaerobic digestion in which both H_2 and acetate accumulate. This in turn inhibits the proton-reducing acetogenic bacteria that normally assist the methanogens by converting propionate and butyrate to acetate and H_2. When methanogenesis is inhibited, these acids can accumulate and are used as substrates by other bacteria for reductive synthesis of longer-chain (C_5-C_8) fatty acids by reverse β-oxidation. As discussed later in this chapter, these longer-chain VFA can be extracted using solvents or resins and converted to a variety of fuel compounds.

The biorefinery concept was further extended by Holtzapple and coworkers (Granda et al. 2009; Holtzapple and Granda 2009), who carried the mixed culture fermentation to pilot plant scale, including downstream processing of the fatty acids; the resulting process was named the "car-

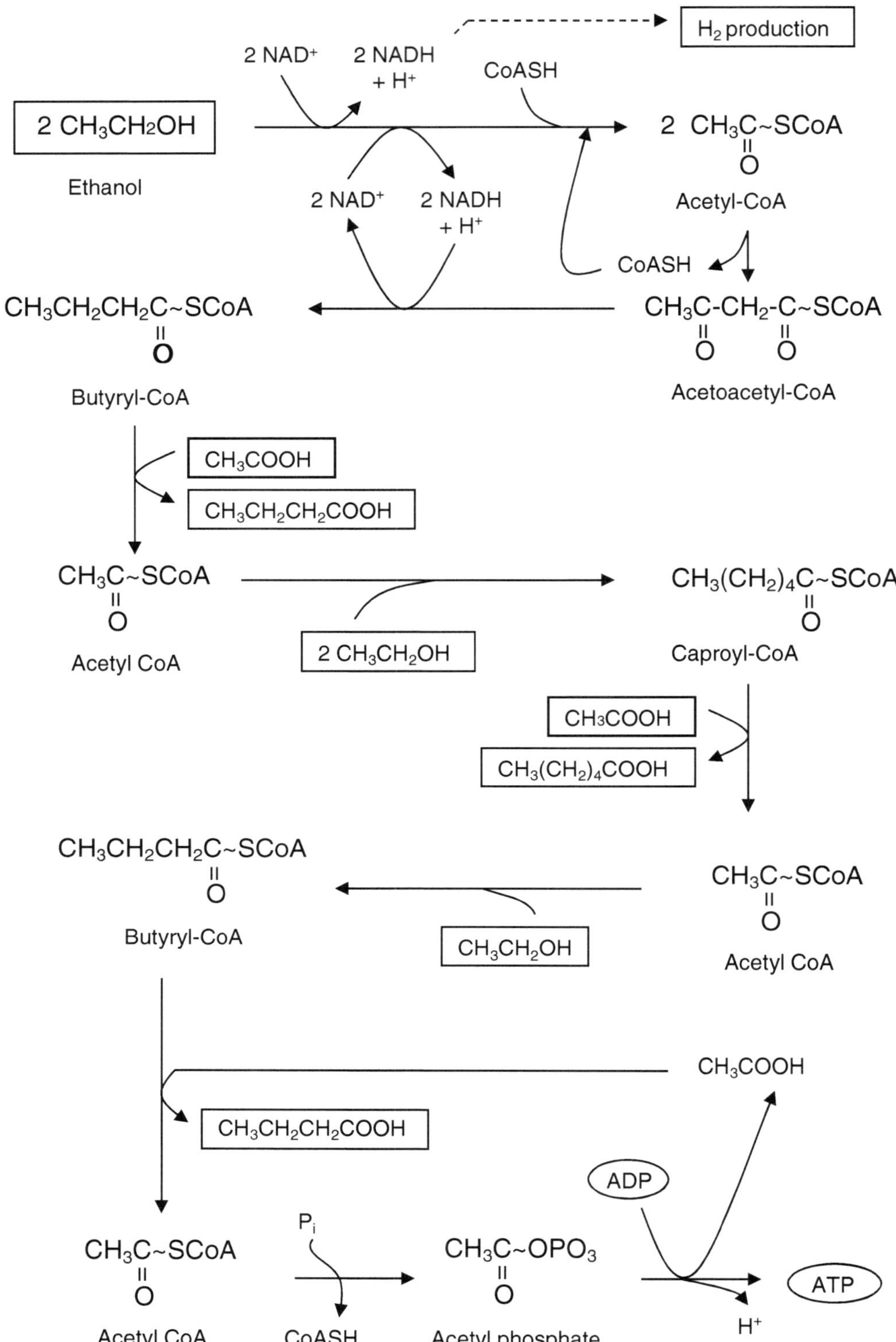

Fig. 18.1 Fermentation pathway of *Clostridium kluyveri*. This anaerobic bacterium oxidizes ethanol and reduces acetate successively to butyrate and caproate (and likely to the C_8 acid, caprylate; not shown). The pathway, sometimes termed "reverse β-oxidation," is used by another ruminal bacterium, *Megasphaera elsdenii*, to convert acetate (C_2) and reducing equivalents generated by glucose fermentation to butyrate (C_4) and caproate (C_6) and to convert propionate (C_3, generated by fermentation of lactate) to valerate (C_5)

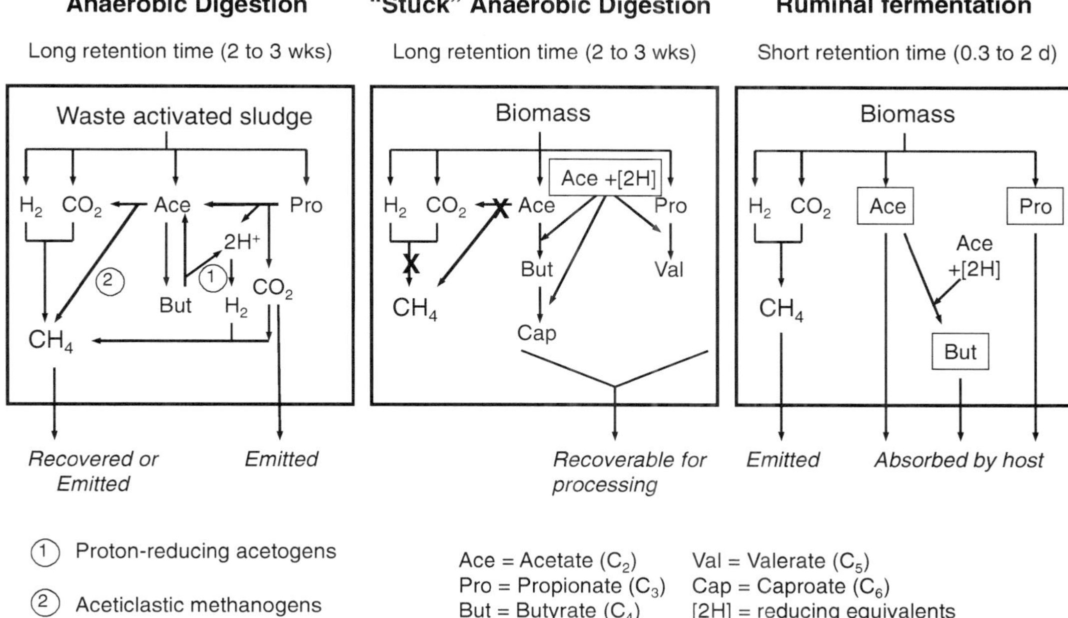

Fig. 18.2 Comparison of conventional anaerobic digestion, "stuck" anaerobic digestion, and ruminal fermentation. Conventional anaerobic digestion results in nearly complete conversion of the digestible components of biomass to methane and CO_2. C_3-C_6 VFAs produced during the fermentation are ultimately converted to acetate, H_2, and CO_2 by proton-reducing acetogens, and acetate is converted to methane and CO_2 by aceticlastic methanogens. Long retention times in the bioreactor are required to accommodate the very slow growth rates of these two microbial groups. Some methane is also produced via microbial reduction of CO_2 with H_2. "Stuck" anaerobic digestion is achieved by inhibition (usually via chemical inhibitors) of both H_2-oxidizing and aceticlastic methanogens, which results in accumulation of acetate and other VFAs, to the point where excess reducing equivalents drive the synthesis of longer-chain (primarily C_5-C_8) VFA, which can be recovered for chemical or electrochemical conversion to fuels. Long retention times are required to accommodate the slow growth of these chain-extending bacteria. In the rumen, the rapid dilution rate of ruminal contents exceeds the growth rates of both proton-reducing acetogens and aceticlastic methanogens. As a result, the VFA (primarily C_2-C_4, with relatively little C_5-C_6)) accumulates to relatively high concentrations (>100 mM total), which provide most of the energy for the ruminant host. Methane is produced almost exclusively via CO_2 reduction

boxylate platform" of fuel production. Over the last decade, several other groups (Agler et al. 2011; Chang et al. 2010; Kleerebezem and van Loosdrecht 2007) have advocated for the process, reiterating its many advantages: feedstock flexibility, nearly complete fermentation of non-lignin organic components of the feedstock, culture robustness and stability, non-aseptic operation, and low cost.

The ruminal fermentation operated in bioreactors (i.e., extraruminally) can be viewed as an alternative configuration of the carboxylate platform, but with some interesting and useful differences. Because digesta (including the microflora) have a relatively short retention time in the rumen, several key functional groups of microbes present in the anaerobic digester – namely, the proton-reducing acetogens and aceticlastic methanogens – do not establish themselves in the rumen and would not become established or persistent in an extraruminal fermentation conducted at run times of 2–3 days. As a result, extraruminal fermentations will convert biomass materials to VFA (predominantly acetate, propionate, and butyrate, totaling up to 200 mM) with the primary benefits of the carboxylate platform (feedstock flexibility and non-aseptic operation) and at higher productivity due to a shorter run time and without the necessity of inhibiting methanogenesis.

18.5.2 Recruitment of Additional Pathways Via Coculture

Operation of an extraruminal fermentation without inhibition of methanogenesis should reduce costs and obviate some waste disposal issues, but it does have one disadvantage. Unless methanogenesis is inhibited, the fermentation would appear to have more limited prospects for chain elongation of the VFA than would a stuck anaerobic digestion, because methane draws off substantial fraction of the reducing equivalents that could otherwise be used for the reductive steps of chain elongation. However, it is important to note that in the nonruminal configurations of the carboxylate platform, inhibition of methanogenesis to create a "stuck" anaerobic digestion is only successful at prolonged incubation times. Both Levy et al. (1981) and Holtzapple et al. (1999) routinely used incubation times of 2–3 weeks, and even these long incubation times may be insufficient to produce high concentrations of the longer-chain VFA. For example, Hollister et al. (2010), in a 30-day stuck anaerobic digestion of alkaline-pretreated sorghum, demonstrated a ratio of acetate/propionate/butyrate of ~4:1:1, with less than 6 % of the total acids produced as valerate and caproate. These ratios are fairly similar to those of many ruminal fermentations conducted at 2–3-day incubation times. In reactors fed a mixture of ethanol and acetate, Steinbusch et al. (2011) demonstrated that the microbial community was eventually dominated by uncharacterized bacterial strains resembling *C. kluyveri*, but they did not observe production of caproic or caprylic acid until 30 days and 60 days of fermentation, respectively. Thus, for both the conventional and ruminal configurations of the carboxylate platform, a major research objective is to extend the VFA products to chain lengths sufficient for facile recovery by extraction and for conversion to liquid fuels while minimizing fermentation run time. The prospects for such rapid chain elongation have not been systematically explored. Producing longer-chain VFA by reverse β-oxidation will require the proper electron donor, available carbon skeletons (viz., acetate and possibly other VFAs), and the presence of one or more specific strains of microbe with the required enzymatic capabilities. The strategy of bioaugmentation, in which the microbial community is amended with specific microbes with enhanced capacity for reverse β-oxidation, may be useful to achieve this end.

18.5.3 Product Recovery

A major challenge in developing commercially viable biofuel processes is the economic recovery and separation of the fuels and fuel precursors from the fermentation broth (i.e., from dilute aqueous solution). Ethanol is most conventionally recovered by distillation, but distillation cost increases dramatically and nonlinearly as concentration in solution decreases, and the concentration typically regarded as the minimum for cost-effective recovery, 52 g L^{-1} (Varga et al. 2004), exceeds the highest reported ethanol concentration in any ruminal microbial culture. Even H_2, which is virtually insoluble in water, would have to be separated from CO_2 and other gases present in the bioreactor headspace in order for it to be useful as a fuel. For fuel precursors (VFA), the most useful recovery strategy is highly dependent on the chain length of the VFA. The extractability of VFA into organic solvents is estimable by their partition coefficients, normally measured with *n*-octanol as solvent. The octanol/water partition coefficients for VFA in their free acid forms increase exponentially with chain length (Fig. 18.3). Consequently, C_5 and C_6 VFAs are easily recovered by extraction into organic solvents, providing a strong incentive to pushing fermentations to maximize VFA chain length and to minimize final culture pH. The more polar C_2-C_4 VFAs are not effectively extracted by organic solvents, and their recovery by distillation is energy intensive; this provides incentive to develop cost-effective membrane separations and/or electrophoretic concentration.

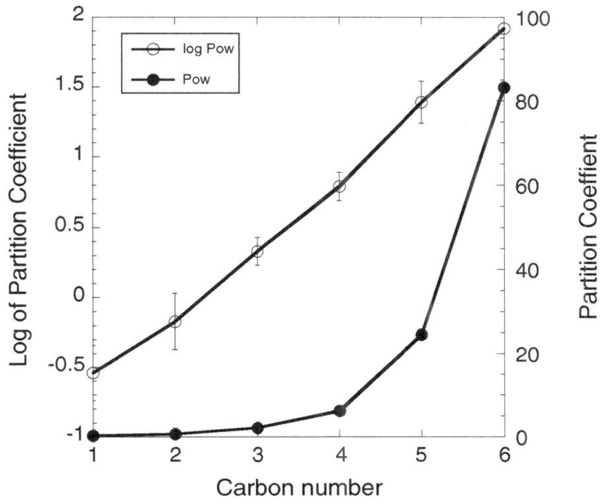

2
1.5
1
0.5
0
-0.5
-1
Log of Partition Coefficient
log Pow
Pow
100
80
60
40
20
0
Partition Coeffient
1 2 3 4 5 6
Carbon number

Fig. 18.3 n-Octanol/water partition coefficients for the free acid forms of the volatile fatty acid series C_1-C_6 (formic through caproic). The recommended values and associated uncertainty values (represented by error bars) are from the literature compilation of Sangster (1989). The partition coefficient continues to rise dramatically with chain length for the medium-chain fatty acids; for example, the partition coefficient for caprylic acid (C_8) is 1,122, about 14 times higher than that of caproic acid (C_6)

18.6 Conversion of Fuel Precursor to Fuels

Economically viable production and recovery of VFA at high concentrations are only half of the battle for producing fuels by the carboxylate platform. These VFAs must then be efficiently and economically converted to fuel compounds having adequate fuel properties (high energy content, high volatility, non-corrosivity, and compatibility with current fuel infrastructure). Conversion of VFA to fuel compounds can be accomplished by several reaction schemes, including chemical conversion to esters and alcohols and electrochemical conversion to hydrocarbons, alcohols, and esters. Purification of individual VFA prior to chemical conversion is not economically viable. However, such purifications are unnecessary because all of these conversions can utilize mixtures of VFA to form a mixture of chemical reaction products, which provides the opportunity to tune the composition of the mixture to desired fuel application.

18.6.1 Chemical Conversion to Esters

Mixtures of fatty acid alkyl esters (FAAE) are the centerpiece of the biodiesel industry and are produced in amounts exceeding 12 million metric tons annually (Röttig et al. 2010). The compounds are the methyl, ethyl, and propyl esters of C_{10}-C_{12} monocarboxylic acids (sometimes called "medium-chain fatty acids," MCFA). These esterification reactions can be carried out chemically, enzymatically, or microbially. Until recently, there have been few attempts to extend esterification technology to lower chain (C_5-C_6) VFA, heretofore due to a lack of an inexpensive source of concentrated VFA. Lange et al. (2010) have proposed chemical conversion of lignocellulosic feedstocks to levulinic acid, followed by reduction to valerate via a valerolactone intermediate and conversion of valerate to several esters. These workers further demonstrated acceptable vehicle performance using 15 % blends of valerate esters in gasoline. Given the broad spectrum of esterification reactions, there is little reason to believe that esterification cannot be applied to VFA mixtures upon successful application of the carboxylate platform for their production.

18.6.2 Chemical Conversion to Secondary Alcohols

It has long been known (Young 1922) that Ca salts of VFA are readily converted to ketones by a condensation reaction (e.g., one mol acetone produced from two mol of Ca acetate) at high temperature (e.g., 380–430 °C). Care must be taken to remove the ketone products rapidly to prevent degradative reactions or the formation of undesirable tar compounds; this can be accomplished by carrying out the reaction in the presence of an inert sweep gas or under moderately high vacuum (e.g., 20 mmHg). $CaCO_3$ is formed as a product of the reaction and must be continuously removed, but can be partially recycled for use as a pH control agent in the bioreactors. Once recovered, the ketones can be hydrogenated to secondary alcohols, some of which are potential liquid fuels. The reaction conditions are somewhat milder than for the ketonization (e.g., 130 °C, 1.5 MPa H_2 partial pressure), but the reaction requires high concentrations of catalyst (200 g Raney nickel per liter) and vigorous agitation.

18.6.3 Electrochemical Conversions

The biorefinery system originally described by Levy et al. (1981) incorporated an electrochemical conversion of extracted VFA. In the presence of an electrical current, VFA can be decarboxylated on an anodic surface to yield an alkyl radical that can condense tail to tail with other alkyl radicals (Fig. 18.4) to form alkanes. This electrolysis converts to alcohols, alkenes, or esters. The electrons reduce protons at the cathode to produce H_2. This H_2 can be recovered for direct use as a fuel, converted to methane by methanogenic archaea, or combusted to generate electricity to drive the anodic reaction

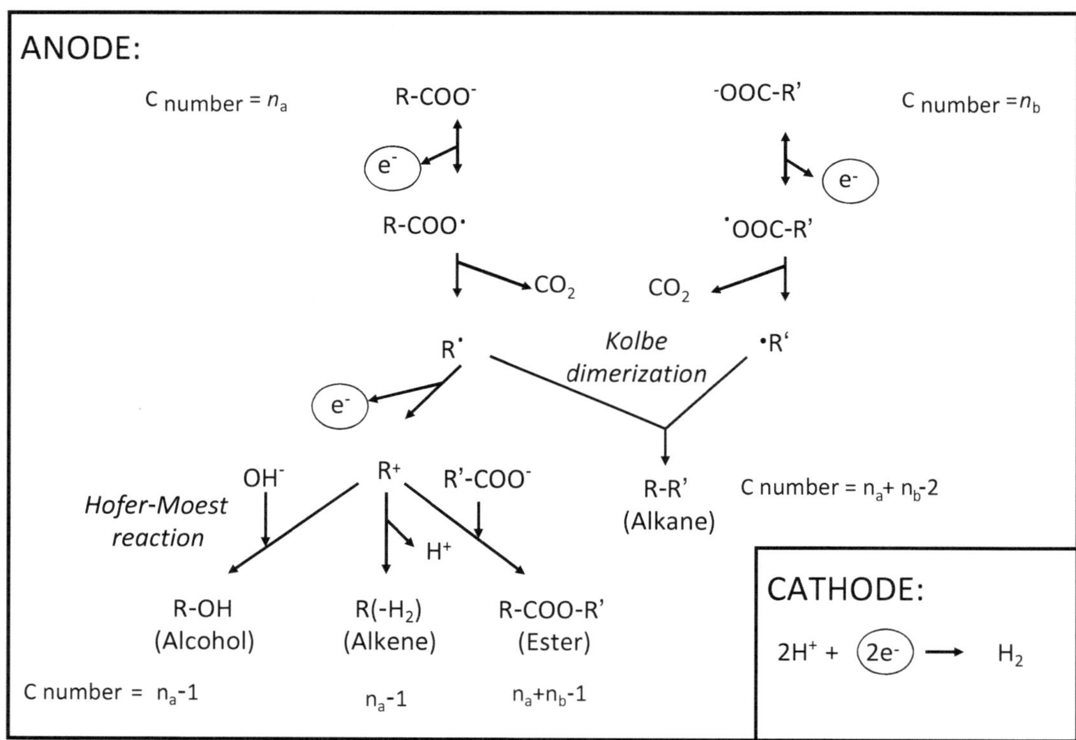

Fig. 18.4 Reactions for electrochemical conversion of monocarboxylic acids to fuel compounds. Decarboxylation at the anode releases an electron and generates a reactive alkyl radical that may either combine to form alkanes (Kolbe reaction) or oxidize further to a carbonium ion that

reaction, first described by Kolbe (1849), is one of the oldest known reactions in organic chemistry. By virtue of its chain extension capability, the Kolbe synthesis provides the opportunity of converting fermentative VFA to liquid hydrocarbons, the most desirable form of biofuels. Pioneering work by Sanderson et al. (1983) demonstrated high yields of C_6-C_{10} alkanes from C_5 and C_6 VFA solutions under specialized electrochemical reaction conditions. However, the high reactivity of the alkyl radicals can result in a number of side reactions to produce less desirable, shorter chain products (Fig. 18.4). The relative proportion of these products is highly dependent on a host of reaction conditions, including VFA chain length, VFA concentration, solvent type, electrode composition, current density, pH, temperature, and pressure (Torii and Tanaka 2001). In an electrochemical cell, decarboxylation of VFA at the anode is balanced by reduction of protons at the cathode to yield H_2. This H_2 could potentially be recovered as an additional fuel product or be used to generate the modest amount of electricity needed to operate the electrochemical cell.

One of the attractive features of electrochemical conversion of VFA is that it can be conducted at low voltages (typically less than 3V), and in fact operation at low voltages is essential to prevent unproductive evolution of O_2 at the anode. On the other hand, the major challenge associated with the electrochemical conversion of VFA to hydrocarbons is overcoming the requirement for high current densities at the anodic surface, which limits electrode surface area (and thus reaction rate) and requires close (submillimeter) separation of the anode and cathode. Overcoming these limitations is a worthy goal for future research, as it is perhaps the main stumbling block to applying the carboxylate platform for electrochemical production of drop-in hydrocarbon fuels.

18.6.4 Use of Ruminal Microbes in Microbial Fuel Cells

Although they generate electrical power rather than liquid or gaseous fuels, microbial fuel cells merit consideration here owing to the considerable interest in economically and sustainably powering instrumentation and other devices in remote locations. A microbial fuel cell (MFC) is an electrochemical device in which chemical energy (in the form of oxidizable substrates) is converted to electricity via the catalytic action of microorganisms. MFCs are typically divided cells in which microbial oxidation of substrate occurs in the anode compartment, yielding protons, electrons, and CO_2. The electrons are delivered to the cathode by an electrical circuit, where they react with an electron acceptor. One of the most common configurations of an MFC uses a proton-permeable membrane between the cells. The protons diffusing through this membrane react with electrons and O_2 (the electron acceptor) at the cathode to produce water, with the net generation of an electrical current. MFCs can use pure microbial cultures (the nonruminal *Geobacter* and *Shewanella* are two of the more popular examples) but can also use natural consortia from soils, sediments, or the rumen. Because of their substantial ability to degrade cellulose, MFCs employing ruminal inocula have been most widely studied with cellulosic feedstocks added to the anodic compartment (Rismani-Yadzi et al. 2007; Wang et al. 2012). MFC technology is still very immature, and it is expected that improvements in culture composition, electrode materials, and cell configurations can substantially exceed power yields well beyond the low values currently observed (~0.003 mW m^{-2} electrode surface; Wang et al. 2012).

18.7 Fermentation Coproducts

Production of biofuels has been historically hampered by the small price differential between the feedstock and the fuel product. It is thus imperative that biofuel production systems include additional sources of revenue in the form of vendable coproducts. The optimal coproduct scenario would involve a moderately high-value product offering unique end use properties, properly scaled to its intended market. For most biofuel processes, the intended market is typically animal feeds. However, there are several challenges with targeting animal feeds as a coproduct of biofuel production. Firstly, animal feed is a highly competitive market in that it is the conventional dumping ground for by-products of the human food chain. Indeed, animal feed is the major use of the millions of metric tons of food by-products (e.g., citrus pulp, sugar beet pulp, soy hulls, etc.) produced in the USA annually, along with existing biofuel residues – particularly distiller's dried grains and solubles (DDGS), whose US production now exceeds 9.6 million metric tons annually. Successful by-products thus must feature availability and pricing that are matched to seasonal and geographical demands of the livestock industry within a highly elastic market that includes a wide array of competing feed additives. Secondly, the coproducts are not stand-alone feeds but must be included in rations in a way that provides nutritional balance with other ration components.

By these standards, the residues from an extraruminal fermentation to VFA are relatively attractive. Ruminal microbes are regarded as the ideal protein source for ruminants (Russell 2002), with an excellent amino acid balance, particularly high in methionine and lysine, the two main limiting amino acids in milk protein production (National Research Council 2001).

A second potential use of extraruminal fermentation residues is as a wood adhesive. For example, residues from fermentation of alfalfa stems by *Ruminococcus* or *Clostridium* species have some utility as a plywood adhesive, although for optimum performance, the residues would still need to be combined with a traditional adhesive such as phenol-formaldehyde, PF (Weimer et al. 2003). Nevertheless, the prospect for decreasing by two-thirds the industrial use of PF, a toxic, petroleum-derived chemical mixture, hits every checkbox on a green chemical technology list.

18.8 Concluding Remarks

The ruminant animal and its symbiotic ruminal microbiota have evolved over eons into a remarkably efficient processing system for cellulosic biomass. The microbiota has been shown to effectively convert a wide array of biomass feedstocks to useful products, particularly VFA, in bioreactors under conditions that should attract the attention of a biorefinery industry: high solid loading, minimal feedstock pretreatment, the potential for nearly complete fermentation of organic components other than lignin, non-aseptic operation, continuous or semicontinuous operation, and scalability. Until recently, the many benefits of the ruminal fermentation have been ignored by biorefiners, perhaps due to an underappreciation of the flexibility of the process and a lack of obvious ways to convert the VFA to fuels. Additional research in VFA recovery and conversion, whether by revitalization and improvement of established chemical routes or by development of new routes, will be key to translating the potential of extraruminal fermentations to new technologies for liquid and gaseous fuel production.

References

Agler MT, Wrenn BA, Zinder SH et al (2011) Waste to bioproduct conversion with undefined mixed cultures: the carboxylate platform. Trends Biotechnol 29:70–78

Allison MJ (1969) Biosynthesis of amino acids by ruminal microorganisms. J Anim Sci 29:797–807

Benz GT (2008) Agitation challenges in cellulosic ethanol production. Ethanol Producer, 5 Feb 2008, http://www.ethanolproducer.com/articles/3731/agitation-challenges-in-cellulosic-ethanol-production

Chang HN, Kim NJ, Kang J et al (2010) Biomass-derived volatile fatty acid platform for fuels and chemicals. Biotechnol Bioprocess Eng 15:1–10

Eaton DC, Gabelman A (1992) Production of propionic acid salts by fermentation of lactates using *Selenomonas ruminantium*. US Patent 5,137,736

Gokarn RR, Eiteman MA, Martin SA et al (1997) Production of succinate from glucose, cellobiose, and various cellulosic materials by the ruminal anaerobic bacteria *Fibrobacter succinogenes* and *Ruminococcus flavefaciens*. Appl Biochem Biotechnol 68:69–80

Granda C, Holtzapple M, Luce G et al (2009) Carboxylate platform: The MixAlco process part 2: Process economics. Appl Biochem Biotechnol 156:537–554

Guettler MV, Rumler D, Jain MK (1999) *Actinobacillus succinogenes* sp nov, a novel succinic-acid-producing strain from the bovine rumen. Int J Syst Bacteriol 49:207–216

Hollister EB, Forrest AK, Wilkinson HH et al (2010) Structure and dynamics of the microbial communities underlying the carboxylate platform for biofuel production. Appl Microbiol Biotechnol 88:388–389

Holtzapple MT, Granda C (2009) Carboxylate platform: The MixAlco process part 1: Comparison of three biomass conversion platforms. Appl Biochem Biotechnol 156:525–536

Holtzapple MT, Davison RR, Ross MK et al (1999) Biomass conversion to mixed alcohol fuels using the MixAlco process. Appl Biochem Biotechnol 77–79:609–631

Hungate RE (1966) The rumen and its microbes. Academic, New York

Kleerebezem R, van Loosdrecht MCM (2007) Mixed culture biotechnology for bioenergy production. Curr Opin Biotechnol 18:207–212

Kolbe H (1849) Untersuchungen über die Elektrolyse organischer Verbindungen. Justus Liebigs Ann Chem 69:257–294

Lange JP, Price R, Ayoub PM et al (2010) Valeric biofuels: A platform of cellulosic transportation fuels. Angew Chem 49:4479–4483

Lee PC, Lee WG, Kwon S et al (2000) Batch and continuous cultivation of *Anaerobiospirillum succiniciproducens* for the production of succinic acid from whey. Appl Microbiol Biotechnol 54:23–27

Lee SJ, Lee D et al (2005) Metabolic engineering of *Escherichia coli* for enhanced production of succinic acid, based on genome comparison and in silico gene knockout simulation. Appl Environ Microbiol 71:7880–7887

Levy PF, Sanderson JE, Kispert RG et al (1981) Biorefining of biomass to liquid fuels and organic-chemicals. Enzyme Microb Technol 3(3):207–215

Lynd LR, Weimer PJ, Van Zyl WH et al (2002) Microbial cellulose utilization: fundamentals and biotechnology. Microbiol Mol Biol Rev 66:506–577

Masset J, Calusinska M, Hamilton C et al (2012) Fermentative hydrogen production from glucose and starch using pure strains and artificial co-cultures of *Clostridium* spp. Biotechnol Biofuels 5:35

Morrison M, Mackie RI, Kistner A (1990) 3-Phenylpropanoic acid improves the affinity of *Ruminococcus albus* for cellulose in continuous culture. Appl Environ Microbiol 56:3220–3222

National Research Council (2001) Nutrient requirements for dairy cattle, 7th edn. National Academy of Sciences, Washington, DC

Ntaikou I, Koutros E, Komaros M (2009) Valorisation of wastepaper using the fibrolytic/hydrogen producing bacterium *Ruminococcus albus*. Bioresour Technol 100:5928–5933

Pavlostathis SG, Miller TL, Wolin MJ (1988) Fermentation of insoluble cellulose by continuous cultures of *Ruminococcus albus*. Appl Environ Microbiol 54:2655–2659

Rismani-Yadzi HR, Christy AD, Dehority BA et al (2007) Electricity generation from cellulose by rumen microorganisms in microbial fuel cells. Biotechnol Bioeng 97:1398–1407

Röttig A, Wenning L, Bröker D et al (2010) Fatty acid alkyl esters: perspectives for production of alternative biofuels. Appl Microbiol Biotechnol 85:1713–1733

Russell JB (2002) Rumen microbiology and its role in ruminant nutrition. James B.Russell, Ithaca, Available at: http://www.ars.usda.gov/services/software/download.htm?softwareid=409

Sanderson JE, Levy PF, Cheng LK et al (1983) The effect of pressure on the product distribution in Kolbe electrolysis. J Electrochem Soc 130:1844–1848

Sangster J (1989) Octanol-water partition coefficients of simple organic compounds. J Phys Chem Ref Data 18:1111–1227

Scheiffinger CC, Wolin MJ (1973) Propionate formation from cellulose and soluble sugars by combined cultures of *Bacteroides succinogenes* and *Selenomonas ruminantium*. Appl Microbiol 26:789–795

Steinbusch KJJ, Hamelers HVM, Plugge CM et al (2011) Biological formation of caproate and caprylate from acetate: fuel and chemical production from low grade biomass. Energy Environ Sci 4:216–224

Suen G, Stevenson DM, Bruce DC et al (2011a) Complete genome of the cellulolytic ruminal bacterium *Ruminococcus albus* 7. J Bacteriol 193:5574–5575

Suen G, Weimer PJ, Stevenson DM et al (2011b) The complete genome sequence of *Fibrobacter succinogenes* S85 reveals a cellulolytic and metabolic specialist. PLoS One 6:e0018814

Torii S, Tanaka H (2001) Carboxylic acids. In: Lund H, Hammerich O (eds) Organic electrochemistry, 4th edn. Marcel Dekker, New York, pp 499–535

Van Soest PJ (2002) Nutritional ecology of the ruminant, 2nd edn. Comstock Press, Ithaca

Varga E, Klinke HB, Reczey K et al (2004) High solids simultaneous saccharification and fermentation of wet oxidized corn stover to ethanol. Biotechnol Bioeng 88:567–574

Wang C-T, Yang C-MJ, Chen Z-S (2012) Rumen microbial volatile fatty acids in relation to oxidation reduction potential and electricity generation from straw in microbial fuel cells. Biomass Bioener 37:318–329

Weimer PJ (1993) Effect of dilution rate and pH on the ruminal cellulolytic bacterium *Fibrobacter succinogenes* S85 in cellulose-fed continuous culture. Arch Microbiol 160:288–294

Weimer PJ, Digman MF (2013) Fermentation of alfalfa wet-fractionation liquids to volatile fatty acids by *Streptococcus bovis* and *Megasphaera elsdenii*. Bioresour Technol 142:88–94

Weimer PJ, Moen GN (2013) Quantitative analysis of growth and volatile fatty acid production by the anaerobic ruminal bacterium *Megasphaera elsdenii* T81. Appl Microbiol Biotechnol 97:4075–4081

Weimer PJ, Conner AH, Lorenz LF (2003) Solid residues from *Ruminococcus* cellulose fermentations as components of wood adhesive formulations. Appl Microbiol Biotechnol 63:29–34

Weimer PJ, Russell JB, Muck RE (2009) Lessons from the cow: what the ruminant animal can teach us about consolidated bioprocessing of cellulosic biomass. Bioresour Technol 100:5323–5331

Weimer PJ, Stevenson DM (2012) Isolation, characterization and quantification of *Clostridium kluyveri* from the bovine rumen. Appl Microbiol Biotechnol 94:561–566

Young S (1922) Distillation principles and processes. Macmillan, London

Zeikus JG, Jain MK, Elankovan P (1999) Biotechnology of succinic acid production and markets for derived industrial products. Appl Microbiol Biotechnol 51:545–552

19 Commercial Application of Rumen Microbial Enzymes

Saurabh Bansal and Gunjan Goel

Abstract

The enzyme as a biocatalyst is an important tool in various industries where these enzymes have been used in the development of various products and services in the eco-friendly manner. The use of enzymes commercially not only reduces the cost but also lowers the environmental concerns particularly related to the water and energy consumption. Their applications also improve the quality of the bioproducts. Therefore, today, the enzymes provide a sustainable alternative to the conventional chemical processes.

While searching of new enzymes and new enzyme sources, various researchers find rumen as an attractive source of various enzymes. The rumen is the first chamber of the digestive system of the ruminants capable in the bioconversion of various lignocellulosic plant material into easily metabolizable compounds by virtue of the presence of various microbial populations: bacteria, fungi, and protozoa. The functioning of rumen is quite similar to various biorefineries which aim at the bioconversion of lignocellulosic material into various useful bioproducts such as fuels and chemicals. The current chapter will discuss about the various industrial/ commercial applications of rumen enzymes along with their desired properties.

Keywords

Rumen enzymes • Biofuel • Ligno-cellulosic biomass • Ruminants • Feed additives • Lignocellulosic substrates

S. Bansal (✉) • G. Goel
Department of Biotechnology and Bioinformatics,
Jaypee University of Information Technology,
Solan 173234, India
e-mail: saurab.bansal02@gmail.com

A.K. Puniya et al. (eds.), *Rumen Microbiology: From Evolution to Revolution*, DOI 10.1007/978-81-322-2401-3_19

19.1 Introduction

The enzymes are biocatalysts known to be commercially important because of their catalytic properties, specificities and efficiencies. They can be obtained from various forms of life such as animals, plants, bacteria, human, fungi, etc. These enzymes are involved in various anabolic and catabolic reactions to maintain the various metabolic functions and activities of the living beings. All the enzymes fasten the reaction rate without altering equilibrium of the reaction. Most of the enzymes are protein in nature. Recently some catalytic activities are also found in RNA (ribozymes) and DNA (DNA enzymes) (Breaker and Joyce 1994).

In the current scenario, enzymes become important tools for the various industries as their use in various industrial processes not only improve the quality and quantity of the product but also fasten the processes. Also the uses of enzyme at industrial scale help in reducing the water and energy consumption and so the pollution. Therefore, they are now continuously playing important role in the various areas of life, e.g., in medicine and in detergent, pulp, leather, and food industries (Godfrey and West 1996). Various such industries recognize the importance of enzymes and exploit their potential for:

(i) Reduction in product cost
(ii) Improvement in quality and quantity of products
(iii) Eco-friendliness
(iv) Safety to their workers

Therefore, the demands of such enzymes are continuously increasing in the global market. The BCC research showed in 2011 that the global market of enzymes will be worth more than $ 4.5 billion by year 2015 from the worth of $ 3.3 billion per annum as was in 2010 (Fig. 19.1) (BCC Research 2011). The most common source for obtaining such commercially important enzymes is microbes in particular bacteria and fungi. However, few of them are also obtained from other sources like plants and animals. The development of fermentation processes, recombinant DNA technology, protein engineering, directed evolution and advanced bioinformatics tools makes possible manufacturing of enzymes as purified and well-characterized form at large scale with the improved characteristics (Kirk et al. 2002). These recent advancement in biotechnology not only allows the production of commercial enzymes at large scale but also helps in the development of new enzymes displaying new and/or improved properties.

Although microbes are preferred choice for the enzyme production, animals are also a source of several enzymes having industrial and medicinal importance. However, they are not an attractive source of enzymes due to difficulties in isolation, enzyme purification, and the cost factor. Also there is always a risk of contamination of various animal diseases such as bovine spongiform

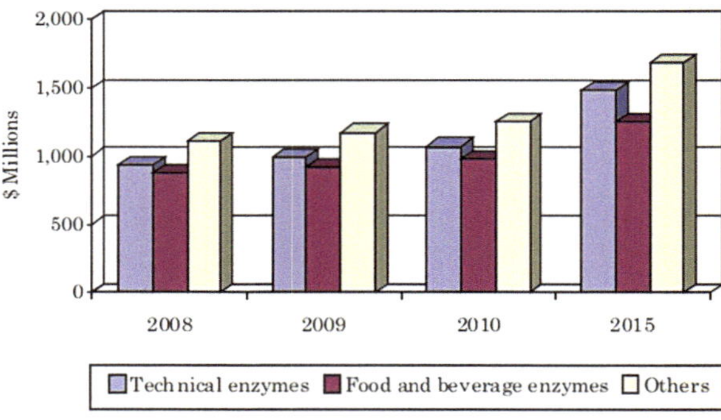

Fig. 19.1 Yearwise trend of industrial enzymes demand in the global market (BCC Research 2011)

Table 19.1 Enzymes obtained from various animals and their commercial applications

Enzymes	Sources	Applications
Aldolases	Liver and muscle	Fructose digestion
Alkaline phosphatase	Calf intestine/kidney	Diagnostic (indicator in ELISA)
Ancrod	Snake venom	Anticoagulant
Catalase	Liver	Food industry
Chymotrypsin	Pancreas	Leather industry
Lipase	Pancreas	Food industry
Pepsin	Porcine stomach	Body fortifying agents
Proteases	Bovine and porcine pancreas	Digestive enzymes, anti-inflammatory agents, health food additives
Rennet (chymosin)	Abomasum (fourth compartment of ruminant's stomach)	Cheese manufacture
Trypsin	Pancreas	Leather industry
Urokinase	Urine	Thrombolytic agent

encephalopathy (BSE) or mad cow disease, a prion-induced disease caused by ingestion of abnormal proteins. Despite this, animals are known to be very good sources of enzymes such as lipases, proteases, and esterase. The choice of animal tissue for the purification of a particular enzyme depends on the extent of expression of that particular enzyme in a tissue, for example, the pancreas for the production of digestive enzymes and the liver for the catalase and aldolase production. Few of enzymes obtained from various animals and their application has been listed in Table 19.1.

In lieu of the identification of new sources for the enzyme production, researchers in recent years found rumen as an interesting source of enzymes due to the presence of a wide variety of microbial population. Furthermore, the advent of new advanced technology and very high throughput systems biology tools makes possible for understanding this highly complex natural ecosystem. It is well established that the rumen livestock are prominent for digesting fibric food materials over the non-rumen livestock due to the presence of microbial system which produce various enzymes in rumen for the fiber digestion. The current chapter will be focused on the commercial applications of various ruminal enzymes.

19.2 Microbial Diversity of Rumen

The presence of a variety of microbial population in the rumen makes the rumen one of the most complicated and fascinating microbial ecosystems in nature. The rumen is the first chamber of the digestive system (foremost stomach) of the ruminant livestock such as cows. The environment of the rumen is anaerobic, weakly acidic (pH 5.3–6.7). The rumen also contains 12–18 % solids content. All this make rumen environment favorable to the ruminal microbial community (Sauer et al. 2012). The rumen of the ruminants is usually populated by a diverse, symbiotic population of obligate anaerobic microorganisms including fungi, protozoa, and bacteria (Selinger et al. 1996). These microbial populations have evolved the capacity for the efficient utilization of complex and recalcitrant plant polymers such as cellulose and hemicellulose. They make the ruminants able to degrade the various carbohydrates usually polysaccharides and structural polysaccharides constituting the major part of the ruminants diets and the primary source of the energy in forage-based diets (Wang and McAllister 2002). These various rumen microbial species digest the plant materials to produce a spectrum of microbial metabolites which are

Table 19.2 Fermentation characteristics of rumen microorganisms

Rumen microorganisms	Substrates digested	Fermentation products
Bacteroides succinogenes	Cellulose, xylan	Acetate, succinate, formate
Ruminococcus flavefaciens	Cellulose, xylan	Acetate, succinate, formate, H_2
Ruminococcus albus	Cellulose, xylan	Acetate, formate, hydrogen, CO_2
Butyrivibrio fibrisolvens	Cellulose, xylan	Acetate, butyrate, formate, H_2, CO_2
Bacteroides ruminicola	Xylan, soluble carbohydrates	Acetate, succinate, propionate, formate
Bacteroides amylophilus	Soluble carbohydrates, starch	Acetate, succinate, formate
Selenomonas ruminantium	Lactate, succinate	Acetate, propionate, H_2, CO_2
Megasphaera elsdenii	Lactate, succinate	Acetate, propionate, butyrate, valerate, caproate, H_2, CO_2
Anaerovibrio lipolytica	Soluble carbohydrates, lipid	Acetate, propionate, succinate
Spirochete species	Soluble carbohydrates	Acetate, succinate, formate
Lachnospira multiparus	Pectins, soluble carbohydrates	Acetate, formate, H_2, CO_2
Succinivibrio dextrinosolvens	Soluble carbohydrates	Acetate, succinate, formate
Succinomonas amylolytica	Soluble carbohydrates	Acetate, succinate
Streptococcus bovis	Soluble carbohydrates	Lactate, acetate, formate, CO_2
Lactobacillus vitulinus	Soluble carbohydrates	Lactate
Methanobacterium ruminantium	H_2, CO_2, formate	Methane
Methanobacterium mobilis	H_2, CO_2, formate	Methane

Adapted from Davies et al. (2007)

then further utilized by the other microbial species present in the rumen ecosystem (Table 19.2). Thus, the digestion and metabolism of structural polysaccharides and other feed ingredients are accomplished by the action of an extensive array of microbial species or by the microbial enzymes produced in the rumen (Davies et al. 2007). These enzymes play an important role in the ruminant digestive process. The rumen microbial enzymes involved in the digestion of fibers and other associated plant cell wall polymers have extensively been studied. Bacteria and fungi contribute approximately 80 % of the total degradative activity whereas protozoa about 20 %. Among them, ruminant fungi are reported as potent fibrolytic enzyme producers which have the ability to degrade the most recalcitrant plant cell wall polymers (Forsberg and Cheng 1992; Forsberg et al. 1993; Wubah et al. 1993; Trinci et al. 1994). However, bacteria like *Fibrobacter succinogenes*, *Ruminococcus albus*, and *Ruminococcus flavefaciens* are primarily involved in the degradation of plant cell walls in the rumen (Forsberg and Cheng 1992).

19.2.1 Enzymes Involved in Degradation of Plant Cell Wall in Rumen

The plant cell wall mainly consists of the cellulose, hemicelluloses, and pectins. Cellulose is one of the most prominently occurring components in the plant cell wall which usually accounts for about 20–30 % of the dry weight of most plant primary cell walls. Cellulose is formed from linear chains of β-1, 4-linked glucose units (Chafe 1970; McNeil et al. 1984).

Hemicellulose is another major component of the cell wall of both stems and leaves of grasses and legumes. In the cell wall, the ratio of cellulose to hemicellulose ranges from 0.8:1 to 1.6:1 (Wilkie 1979). Hemicellulose is composed mainly of xylans with a backbone structure of β-1, 4-linked xylose residues attached with various side chains like acetic acid, coumaric acid, glucuronic acid, ferulic acid, and arabinose (Chesson et al. 1983; McNeil et al. 1984). Xylan polymers in hemicelluloses may further be cross-linked to other hemicelluloses backbones or to lignin through ferulic acid or 4-O-methyl-α-D-glucuronic acid residues (Hartley and Ford 1989; Lam et al. 1990). The linear xylan backbone also interacts to cellulose at one surface and interlocks with other xylan polymers, forming an extensive network of cross-links between cellulose microfibrils (Carpita and Gibeaut 1993).

Pectins, made up of a chain of α-1, 4-D-galacturonate, are also a part of primary cell wall. Besides these, rhamnogalacturonan, extensions (structural proteins) are also found as a part of cell walls of some plants. Pectin, a minor component of the cell wall, is digested either by the strictly pectinolytic species or by those species possessing the combined activity of pectinases (pectin lyase, polygalacturonase, pectin methylesterase) and xylanases (Orpin 1984; Cheng et al. 1991; Gordon and Phillips 1992). For example, rumen inhabitant *Lachnospira multiparus* produces a pectin lyase and a pectin methylesterase (Silley 1985). Ruminal bacteria such as *Streptococcus bovis* and protozoa like *Treponema saccharophilum* are also identified as the pectinolytic microbes (Wojciechowicz and Ziolecki 1984; Paster and Canale-Parola 1985).

Although the composition of plant cell wall is well known, the organization of these individual compounds in plant cell walls is not still clear. However, it is well established that a three-dimensional complex matrix formed by polysaccharides, hydroxycinnamic acids, lignin, protein, and ions is intermolecularly cross-linked through various ionic, hydrogen, and covalent (glycosidic, ester, and ether) bonds that entrap the polysaccharide within the cell wall. Thus, overall composition of cell wall shows that the plant cell wall is an interwoven matrix of polymers. Therefore, the degradation of plant cell wall need not only hydrolytic enzymes but also those able to cleave other kinds of linkages occurred in the cell wall (Wang and McAllister 2002). Till date, ligninolytic enzymes which usually function by oxidative processes are found to be good in degrading the plant cell wall but poorly active in rumen (Selinger et al. 1996).

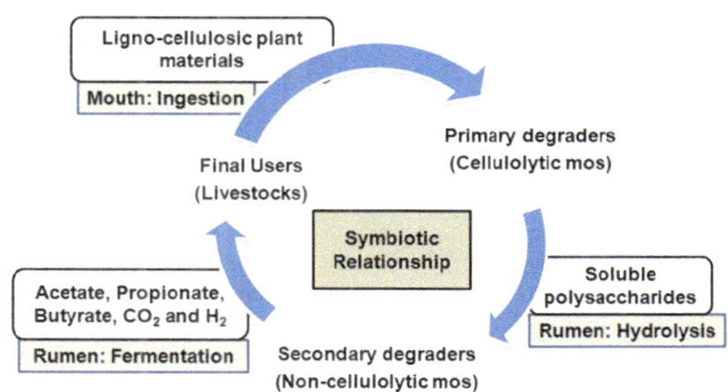

Fig. 19.2 Symbiotic relationship between livestocks and rumen microorganisms

In the rumen, a fine interaction between lingo-cellulolytic (lignin, cellulose, and hemicelluloses degrading) and non-cellulolytic bacteria occurs (Flint et al. 2008). Non-cellulolytic bacteria usually use the soluble polysaccharides released by the primary degraders and convert them through fermentation into products such as acetate, propionate, butyrate, carbon dioxide, and hydrogen which are then used as nutrients by the livestock (Weimar et al. 2009) (Fig. 19.2). Thus, the microbial population in the rumen converts the complex food stuffs (lingo-cellulosic plant materials) into the simpler compounds which can be easily taken up and metabolized by the ruminant. Therefore, the direct use of microbial population from the rumen or the enzymes obtained from them may provide the solution to a current industrial problem, i.e., the bioconversion of lignocellulosic material into fuels, chemicals, and other valuable bioproducts.

The outermost layer of plant cell wall which is usually composed of epicuticular waxes, cuticle, and pectin, functions as a plant's first line of defense against dehydration and penetration of phytopathogens (Selinger et al. 1996). This outermost layer of the plant material particularly cuticle layer is also a potent barrier to penetration by ruminal microbes (Forsberg and Cheng 1992). Therefore, mastication of forages and pretreatment of cereal grains and legumes are carried out to disrupt the cuticular layer to minimize its deleterious effect on digestion (Akin 1989).

19.3 Enzymes from Ruminants

Because of the presence of diversified microbial population in rumen, activities of enzymes confirmed in rumen are diverse. The most commonly occurring enzyme activity in the rumen is related to the fibrolytic (plant cell wall polymer degrading) activities which include cellulases, xylanases, β-glucanases, and pectinases. Apart from fibrolytic activities, other enzyme activities in rumen are also identified which include amylases, proteases, phytases, and specific plant toxin-degrading enzymes (tannases) (Selinger et al. 1996; Cheng et al. 1999; Wang and McAllister 2002). All these enzymes have different valuable commercial applications when they get isolated from the rumen microorganisms. However till date, no report of commercial application of rumen enzyme is available. Many research groups are trying to establish the importance of rumen enzymes in various industries in particular feed industries and biorefineries (biofuel production industries). All major enzyme activities occurred in the rumen, and their prospective commercial applications are listed in Table 19.3.

19.3.1 Rumen Enzymes as Feed Additives

The use of various rumen enzymes (cellulases, xylanases, pectinases, glucanases, and phytases) in the feed of non-ruminants enhances the food stuffs utilization by enhancing the degradation of various plant cell wall polymers and so increases the growth of the livestock (Cheng et al. 1999).

The non-ruminants are not able to digest the various plant materials due to their physical and

Table 19.3 Major rumen enzymes along with substrates and their prospective applications

Substrate	Enzyme	Applications
Lignocellulosic substrates		
Cellulose	Endo-β-1, 4-glucanase	Ligno-cellulose based industries such as paper, biofuel, sugar industries and feed industries
Cellulose (Non-reducing end)	Exo-β-1, 4-glucanase	
Cellobiose	β-1, 4-glucosidase	
Soluble cellooligomers	Cellulodextrinase	
Xylan	Endo-β-1, 4-xylanase	Few of them can be used in detergent industries
Xylobiose	β-1, 4-xylosidase	
Arabinoxylan	α-L-arabinofuranosidase	
Glucuronoxylan	α-glucuronidase	
Acetylxylan	O-acetyl xylan esterase	
Ferulic acid linkage or cross bridge	Ferulic acid esterase	
p-coumaric acid cross bridge	p-coumaric acid esterase	
Laminarin	β-1, 3-glucanase	
Lichenin	β-1, 3 and β-1, 4-glucanase	
Polygalacturan	Pactate lyase	
Pectin	Pectin lyase	
	Pectin methylesterase	
Non-lignocellulosic substrates		
Starch	Amylases	Paper and sugar industries
Dietary proteins	Proteases and peptidases (e.g., aminopeptidases, carboxypeptidases, leucine aminopeptidases, dipeptidyl-aminopeptidases	Animal feed supplements, detergent industries
Phytate phosphate	Phytase	Animal feed supplements
Tanins	Tannase	Animal feed supplements, juice industries

chemical complexities. Many other feed ingredients are also not fully digestible by livestock. The addition of exogenous enzymes in their feed enhances the digestibility of the feed ingredient and so as to improve the efficiency in digestion of the raw material.

Enzymes are usually added in the granulated form. The coating of enzymes plays an important role in their protection from heat treatments as well as from the deactivation by other feed additives such as choline chloride.

The ruminants (livestock with rumen) are able to degrade the various plant materials due to the presence of various microbial populations in their rumen. In contrast to this, non-ruminants are unable to digest various plant cell wall polymers. The addition of enzymes like cellulases, xylanases, pectinases, and glucanases in foodstuff of the non-rumen livestock increases the foodstuff utilization by enhancing the degradation of various plant cell wall polymers (cellulose, xylan, pectin, and glucans) into simple sugars. Such enzyme utilization allows the availability of carbon to the non-rumen livestock as their intestinal enzymes are not able to make these sources available to the livestock (Cheng et al. 1999).

Cereal grains are the largest component of animal feed. However, the type of cereal grains available to the livestock depends on the local availability like viscous cereal (wheat, barley, or rye) or non-viscous cereal (maize or sorghum). The animal feed that usually contains a large proportion of wheat, barley, and rye has high fiber content as a main anti-nutritional factor (ANF). This high fiber content mainly consists of non-starch polysaccharides (NSP), e.g., arabinoxylan in wheat and rye and β-glucan in barley and oats. The presence of anti-nutritional compounds reduces the nutrient absorption efficiency and promotes intestinal disturbances by pathogenic enteric microorganisms (Kumar and Singh 1984; Bae et al. 1993) whereas NSP usually interfere with the digestion by increasing the viscosity of the digesta in the animals' small intestine (Singh et al. 2001). Therefore, utilization of a wide range of enzymes such as cellulases, glucanases, and pectinases decreases the anti-nutritional compounds from the foodstuffs. The addition of other enzymes such as xylanase and β-glucanase hydrolyze the NSP to reduce the digesta viscosity and also release the nutrients (proteins and starch) (Pack et al. 1998). Thus, the use of exogenous enzymes in animal feed improves the feed utilization and makes the healthier digestive system of the animals.

Another important component of animal diet is phosphorous for their proper growth as it plays an important role in the formation and maintenance of skeletal structure and also in the various metabolic processes. An important source of phosphorous is phytic acid which is found in many of the cereals and legumes fed to the animals. But the phytate-bound phosphorous cannot be directly used by the monogastric animals (non-ruminants); consequently they face the problems of the phosphorous deficiency which cause loss in their weight gain, leg disorders, and reduced fertility. Therefore, the enzyme phytase (a kind of phosphatase) releases the inorganic phosphorous by cleaving phytic acid into inositol phosphate intermediates, myo-inositol, and orthophosphates and make it available to the animals (Cheng et al. 1999). The major advantages of the use of enzyme phytase in feed are:

- The enzyme releases the inorganic phosphorous and makes available to the livestock and so reduces the need of orthophosphate supplementation to the feed. As a result, environmental pollution due to excessive manure phosphorous runoffs in the form of animal excretion is also reduced.
- The enzyme also decreases the chelating properties of phytic acid so as to release the various nutrient elements like calcium, magnesium, iron, and zinc from the phytic acid to make them available to be absorbed by the livestock.

Thus, the use of exogenous enzymes in the animal diet has a number of advantages as listed below:

- Releases the extra nutrients from the breakdown of cereals to increase the feed value of the cereal

- Breaks down the anti-nutritional factors (ANF) in the feedstuff to improve the weight gain and better feed utilization
- Increases the availability to the animal of non-accessible proteins, starches, and minerals from feed
- Supplements the endogenous enzymes of the young animals
- Reduces the variability inherent in feedstuffs to increase the uniformity of the animals and so increases the profitability.
- Reduces the environmental pollution caused by the excretion of undigested feed

As a whole, the use of enzyme as a feed additive improves the growth of animals, livestock production, feed utilization, and health status. It also improves the environment for chickens due to reductions in "sticky droppings."

19.3.2 Rumen Enzymes for the Lignocellulose-Based Industries

The use of enzymes or sometimes direct microorganisms in the conversion of biomass for the production of valuable products provides a cleaner, energy-efficient, and less toxic process compared to the conventional chemical processes. The major goal of bioindustries (biorefineries) is to maximize the value derived from the feedstock either by using cheaper raw materials or by minimizing the waste (Cherubini 2010; Yang and Yu 2013). The use of food crops or edible oil crops as a feed in the industries for the production of various bioproducts such as biofuel, beer, and wine is not sustainable in the consideration of growing global population and their demands. Therefore, there is always a need of such raw materials which are easily available and a sustainable source for the industries. Therefore, lignocellulosic plant materials can be the good carbon source fuel and chemical production due to their easy availability and their lower cost. However, the deconstruction of lignocellulosic plant cell wall through enzymatic method is a significant challenge due to its physical and chemical complexity. Therefore, a continuous focused research has been going on for the development of an effective and cheaper method for the deconstruction of the lignocellulosic plant material which is a critical point for the establishment of economic lignocelluloses-based industries. In most of the lignocellulose-based industries, plant material needs to be subjected to a pretreatment (mechanical and/or chemical) followed by enzymatic hydrolysis to obtain a microbially accessible carbon source. This phenomenon in bioindustries is quite similar to a phenomenon that happened in the ruminants (Sauer et al. 2012). In ruminants, the plant materials first undergo mastication, salivation, and rumination to render it enzymatically accessible which is then converted into simpler compounds by the action of rumen microbial digestive enzymes. This procedure supplies the carbon, energy, amino acids, and vitamins to the ruminant hosts (Selinger et al. 1996). Thus, the enzymes obtained from rumen have the tremendous potential for being used in various lignocellulose-based industries such as biofuel and paper industries.

19.3.2.1 Rumen Enzymes in Biofuel Production

In the recent scenario, biofuels come as the alternative to the rapidly depleting non-renewable fossil fuels for fulfilling the energy demand of all the sectors. They look like as the tomorrow's fuels with a sustainable solution to the fossil fuels which also deal with the environmental concerns caused by limiting fossil fuels (Yue et al. 2013). Thus, the biofuel industry is one of the rapidly growing industries in the world. However, the biofuels are currently economically uncompetitive. The cost of production of biofuels (biodiesel, bioethanol) in a sustainable manner is the major challenge to the researchers. The higher cost of biofuel production is usually because of the use of costlier raw materials and the higher enzyme loading during the lignocellulose saccharification (Gardner et al. 2012). The production cost can be reduced by using easily available and cheaper raw materials and by optimizing the production procedure and by the use of catalytic efficient enzymes in a cost-effective manner. Also the method can be made sustainable by using raw materials which should qualify the following points:

- Should be cheaper
- Should be evergreen and naturally available in very large quantities
- Shouldn't be edible crops
- Shouldn't be grown over the agricultural lands

Therefore, the use of agricultural and agro-industrial by-products or wastes in the biofuel production would maximize the resource use, reduce waste management, and also deal with the cost reduction of the added-value products. It also helps in the dealing with the various environmental concerns such as reduction in greenhouse gas emissions, diminishing of potential eutrophication of water bodies near the production plants (Cherubini and Ulgiati 2010). For example, wheat straw is such a kind of a lignocellulosic agricultural residue which is abundant in nature and has the lower cost. It also does not compete with the food/feed chain. Thus, wheat straw can be a good feedstock to the biorefineries for the biofuel production.

The increasing demand of biofuels enforces the researchers for the development of the sustainable advanced biochemical bioprocesses. In the recent years, various works have been done in the area of biofuel production. However, there is a lot of scope in the development of the processes. Among them, the use of enzyme technology is more fascinating to overcome the limitations of conventional methods of the biofuel production. Such technology makes the process eco-friendly and less energy intensive as compared to its conventional alkaline-catalyzed processes. The rumen enzymes such as proteases, lipases cellulases, and their various different forms allow the researchers to work upon the various agro-wastes and lignocellulosic plant materials for the biofuel production. For example, in the biodiesel production, rumen lipase can be used as one of the major enzymes due to the following advantages/properties:

- Their ability to work in biphasic or monophasic system (in the presence of hydrophilic and hydrophobic solvents).
- Robustness and versatility in nature and their bulk production are easier.
- Able to catalyze transesterification reaction with long or branched chain alcohols, which can hardly be converted to fatty acid esters through conventional alkaline processes.
- Make the downstream processing easier.
- More convenient for the biodiesel production because of the availability of thermostable and short-chain alcohol-tolerant lipase.

However, lipase application in the biodiesel production is restricted due to:

- Their significant cost.
- The by-product of lipase reaction, i.e., glycerol, may cause the risk of inhibiting lipase activity.

Other rumen enzymes such as cellulases, cellulobiase, hemicellulases, amylases, etc., can also be useful in biofuel industries as they are involved in the conversion of various lignocellulosic plant materials and agro-wastes into fermentable sugars for the bioethanol production. However, an extensive research is needed for the development of catalytically efficient enzymes and the bioprocess for making the biofuel production more efficient and cost-effective so that these biofuels can replace the utilization of fossil fuels.

19.3.3 Rumen Enzymes in Bio-Based Chemical (Organic Acids) Production

Various organic acids and other chemicals are widely used as key gradients for the production of various derivatives, fragrance, and additives in food for giving good texture, essence, etc. For example, succinic acid and its salt derivatives are being used as a platform chemical for the synthesis of various commercially important products. They are used in the food and pharmaceutical industries. Also they are involved in the manufacture of plastics, detergents, biodegradable solvents, cosmetics, etc. (Davies et al. 2007). Therefore, with time, the demands of such organic acids are continuously growing. Nowadays most of them are synthesized from non-renewable feedstock. However, few of them are obtained from fermentation of various renewable feedstocks. Organic acid producers are the best studied ruminal cell factories so far. One classical example of such organic acid production is the bioproduction of succinic acid which can be produced through fermentation carried out by various ruminal bacteria which include *Bacteroides succinogenes*, *Ruminococcus flavefaciens*, and *Succinimonas amylolytica.* Such bacterial species give the freedom of usage of various plant materials like grass or maize depending on the area of production, making the production of succinic acid more sustainable and cost-effective (Davies et al. 2007).

Other classical examples of bioproduction of commercially important organic acid are the production of short-chain aliphatic organic acids such as butyric, valeric, or caproic acids using ruminal bacteria such as *Megasphaera elsdenii*, under appropriate conditions (Elsden et al. 1951). Particularly, the production of caproic (hexanoic) acid using biomethods is important as it may open the sustainable route for the production of nylon. The ruminal bacteria *M. elsdenii* can be used as an effective microbial cell factory as they are resistant to the short-chain fatty acids (Sauer et al. 2012). However, core enzymatic bioprocess method for such organic acids production has not yet been developed.

19.3.4 Other Applications

The enzymes isolated from the ruminal microorganism can also be used in the various other industries. Various fibrolytic enzymes obtained from ruminal microorganisms have potential applications in the diverse industries such as food processing and manufacturing of chemicals, clothing, paper, and detergents. For example, pectin esterases, endopectin lyases, and endopolygalacturonases from the ruminants can be used in fruit processing, paper, textile, detergent, and pharmaceutical industries (Singh et al. 2001). Fibrolytic enzymes can be used in the paper industries, cellulases in the textile and detergent industries, and proteases in the detergent and food industries. The enzyme tannase can be used in the commercial production of gallic acid from tannic acid and in the preparation of instant tea. Tannase is extremely useful in food, feed, beverages, and pharmaceutical industries (Singh et al. 2001). Enzymes from various ruminal fungi may be more advantageous for the industrial applications as they are stable, highly active, and extracellular and require milder reaction conditions (Cheng et al. 1999).

19.3.5 Future Prospective

Till date, the various studies suggest that the ruminal microorganisms are the good source of various hydrolytic and acidogenic enzymes in the presence of lignocellulosic biomass. However, there is a need of extensive exploration of the ruminal enzymes for the various applications. The further characterization of such ruminal microorganisms and the purification of enzymes from rumen may further allow us for their complete exploitation so as to make their applications in the various industries.

References

Akin DE (1989) Histological and physical factors affecting digestibility of forages. Agron J 81:17–25

Bae HD, McAllister TA, Yanke LJ et al (1993) Effects of condensed tannins on endoglucanase activity and filter paper digestion by *Fibrobacter succinogenes S85.* Appl Environ Microbiol 59:2132–2138

BCC Research (2011) Enzymes in industrial applications: global markets. Jan 2011 BIO030F

Breaker RR, Joyce GF (1994) A DNA enzyme that cleaves RNA. Chem Biol 1:223–229

Carpita NC, Gibeaut DM (1993) Structural models of primary cell walls in flowering plants: consistency of molecular structure with the physical properties of the walls during growth. Plant J 3:1–30

Chafe SC (1970) The fine structure of the collenchyma cell wall. Planta 90:12–21

Cheng KJ, Forsberg CW, Minato H et al (1991) Microbial ecology and physiology of feed degradation within the rumen. In: Tsuda T, Sasaki Y, Kawahima R (eds) Physiological aspects of digestion and metabolism in ruminant. Academic, New York, pp 595–624

Cheng KJ, Lee SS, Bae HD et al (1999) Industrial application of rumen microbes. Asian-Aust J Anim Sci 12:84–92

Cherubini F (2010) The biorefinery concept: using biomass instead of oil for producing energy and chemicals. Energ Convers Manag 51(7):11412–11421

Cherubini F, Ulgiati S (2010) Crop residues as raw materials for biorefinery systems – a LCA case study. Appl Energ 87(1):47–57

Chesson A, Gordon AH, Lomax JA (1983) Substituent groups linked by alkali-labile bonds to arabinose and xylose residues of legume, grass and cereal straw cell walls and their fate during digestion by rumen microorganisms. J Sci Food Agric 34:1330–1340

Davies D, Theodorou M, Newbold J (2007) Biotransformation and fermentation – exploiting the rumen and silo. IGER Innov 11:37–41

Elsden SR, Gilchrist FM, Lewis D et al (1951) The formation of fatty acids by a Gram-negative coccus. Biochem J 49:lxix–lxx

Flint HJ, Bayer EA, Rincon MT et al (2008) Polysaccharide utilization by gut bacteria: potential for new insights from genomic analysis. Nat Rev Microbiol 6(2):121–131

Forsberg CW, Cheng KJ (1992) Molecular strategies to optimize forage and cereal digestion by ruminants. In: Bills D, Kung S-D (eds) Biotechnology and nutrition. Butterworth Heinmann, Stoneham, pp 107–147

Forsberg CW, Cheng KJ, Krell PJ et al (1993) Establishment of rumen microbial gene pools and their manipulation to benefit fibre digestion by domestic animals. In: Proceedings VII world conference on animal production, World Association for Animal Production, Edmonton, pp 281–316

Gardner JG, Zeitler LA, Wigstrom W, Jo S et al (2012) A high-throughput solid phase screening method for identification of lignocellulose-degrading bacteria from environmental isolates. Biotechnol Lett 34:81–89

Godfrey T, West SI (1996) Introduction to industrial enzymology. In: Godfrey T, West S (eds) Industrial enzymology, 2nd edn. Macmillan Press, London, pp 1–8

Gordon GLR, Phillips MW (1992) Extracellular pectin lyase produced by *Neocallimastix* sp LM1, a rumen anaerobic fungus. Lett Appl Microbiol 15:113–115

Hartley RD, Ford CW (1989) Phenolic constituents of plant cell wall and biodegradability. In: Lewis NG, Paice MG (eds) Plant cell wall polymers: biogenesis and biodegradation. American Chemical Society, Washington, DC, pp 137–147

Kirk O, Borchert TV, Fuglsang CC (2002) Industrial enzyme applications. Curr Opin Biotechnol 13(4):345–351

Kumar R, Singh M (1984) Tannins: their adverse role in ruminant nutrition. J Agric Food Chem 32:447–453

Lam TBT, Iiyama K, Stone BA (1990) Primary and secondary walls of grasses and other forage plants: taxonomic and structural considerations. In: Akin DE, Ljungdahl LG, Wilson JR, Harris PJ (eds) Microbial and plant opportunities to improve lignocellulose utilization by ruminants. Elsevier, London, pp 43–69

McNeil M, Darvill AG, Fry SC et al (1984) Structure and function of the primary cell wall of plants. Ann Rev Biochem 53:625–663

Orpin CG (1984) The role of ciliate protozoa and fungi in the rumen digestion of plant cell walls. Anim Feed Sci Technol 10:121–143

Pack M, Bedford M, Wyatt C (1998) Feed enzymes may improve corn, sorghum diets. Feedstuffs 70:18–19

Paster BJ, Canale-Parola E (1985) *Treponema saccharophilum sp* Nov, a large pectinolytic spirochete from the bovine rumen. Appl Environ Microbiol 50:212–219

Sauer M, Marx H, Mattanovich D (2012) From rumen to industry. Microb Cell Fact 11(121):1–3

Selinger LB, Forsberg CW, Cheng KJ (1996) The rumen: a unique source of enzymes for enhancing livestock production. Anaerobe 2:263–284

Silley P (1985) A note on the pectinolytic enzymes of *Lachnospira multiparus*. J Appl Bacteriol 58:145–149

Singh B, Bhat TK, Singh B (2001) Exploiting gastrointestinal microbes for livestock and industrial development-Review. Asian-Aust J Anim Sci 14(4):567–586

Trinci APJ, Davies DR, Gull K et al (1994) Anaerobic fungi in herbivorous animals. Mycol Res 98:129–152

Wang Y, McAllister TA (2002) Rumen microbes, enzymes and feed digestion – a review. Asian-Aust J Anim Sci 15(11):1659–1676

Weimer PJ, Russell JB, Muck RE (2009) Lessons from the cow: what the ruminant animal can teach us about consolidated bioprocessing of cellulosic biomass. Bioresour Technol 100(21):5323–5331

Wilkie KCB (1979) The hemicelluloses of grasses and cereals. Adv Carbohyd Chem Biochem 36:215–264

Wojciechowicz M, Ziolecki A (1984) A note on pectinolytic enzymes of *Streptococcus bovis*. J Appl Bacteriol 56:515–518

Wubah DA, Akin DE, Borneman WS (1993) Biology, fiber- degradation, and enzymology of anaerobic zoosporic fungi. Crit Rev Microbiol 19:99–115

Yang ST, Yu M (2013) Integrated biorefinery for sustainable production of fuels, chemicals, and polymers bioprocessing technologies: In: Biorefinery for sustainable production of fuels, chemicals, and polymers. Wiley, Hoboken, pp 1–26

Yue Z, Li WW, Yu HQ (2013) Application of rumen microorganisms for anaerobic bioconversion of lignocellulosic biomass. Bioresour Technol 128:738–744

20 Molecular Characterisation of Euryarchaeotal Community Within an Anaerobic Digester

K. Arunasri, S. Shivaji, Phil J. Hobbs, Mamatha Potu, S. Kishore, and Sreenivas Rao Ravella

Abstract

Two euryarchaeota-specific 16S rRNA gene clone libraries (MT1 and MT2) and two bacterial 16S rRNA gene clone libraries (HW1 and HW2) were constructed using DNA isolated from two sludge samples (sample 1 and sample 2) collected from two separate tanks of commercial biogas plant operating at the methanogenesis stage, each receiving the same feedstock, after heat pretreatment for an hour at 70 °C, located near Holsworthy, Devon, UK. In the euryarchaeotal-based libraries, both samples revealed clone sequenses assigned to Methanomicrobiales and Methanosarcinales of the phylum Euryarchaeota. Although not much difference was observed in the Archaeal diversity, phylogenetic tree analysis demonstrates that 25 % of clone sequences of sample 1 and 42 % of clone sequences of sample 2 formed separate clades from cultured diversity. 16S rRNA gene-based bacterial libraries also revealed the ubiquitous presence of phyla *Firmicutes* and *Proteobacteria*. Cellulolytic bacterial clone sequences were abundant in both the samples. Fluorescent in situ hybridisation (FISH) results also revealed the presence of Methanomicrobiales and Methanosarcinales as a dominant group.

Keywords

Biogas • Methanogens • Molecular characterisation • 16S rRNA • Fluorescent in situ hybridisation

K. Arunasri • S. Shivaji
Centre for Cellular and Molecular Biology (CCMB), Uppal Road, Hyderabad 500007, India

P.J. Hobbs
Anaerobic Analytics Ltd., 2 East Street, Okehampton EX20 1AS, UK

M. Potu
Institute of Biological and Environmental and Rural Sciences (IBERS), Aberystwyth University, Gogerddan, Aberystwyth, Ceredigion SY23 3EE, UK

S. Kishore
Rothamsted Research, North Wyke, Okehampton, Devon EX20 2SB, UK

S.R. Ravella (✉)
Institute of Biological and Environmental and Rural Sciences (IBERS), Aberystwyth University, Gogerddan, Aberystwyth, Ceredigion SY23 3EE, UK

Rothamsted Research, North Wyke, Okehampton, Devon EX20 2SB, UK
e-mail: rsr@aber.ac.uk

A.K. Puniya et al. (eds.), *Rumen Microbiology: From Evolution to Revolution*, DOI 10.1007/978-81-322-2401-3_20

20.1 Introduction

Biological methane production or methanogenesis is an important process in the global carbon cycle. About 74 % of the emitted methane is derived from biological methanogenesis. Anaerobic digestion is a process of biological means of methane production from biological waste. Anaerobic digestion involves a series of biochemical processes involving hydrolysis, acidogenesis, acetogenesis and methanogenesis (Yu et al. 2010). Methanogens are broadly classified into acetoclastic or hydrogenotrophic Archaea based on the methanogenic precursors such as H_2/CO_2, acetate, formate or methanol to methane (Zinder 1993), used by these organisms to convert to methane. Methanogens have been isolated from a wide variety of anaerobic environments. The common methanogenic habitats include marine sediments, freshwater sediments, flooded soils, human and animal gastrointestinal tracts, termites, anaerobic digesters, landfill, geothermal systems, and hardwood of trees. Cultured methanogens are grouped into five orders, namely, Methanobacteriales, Methanococcales, Methanomicrobiales, Methanosarcinales and Methanopyrales based upon their phylogeny and phenotypic properties (Liu and Whitman 2008). Methane from an anaerobic digestion of biological wastes can be utilised to produce green energy which leaves little or a negative carbon footprint. This process is advantageous over the conventional aerobic treatment of the biological wastes as there is a lower level of sludge production, low space requirement and the production of valuable biogas (McHugh et al. 2003). The biogas generated comprises principally methane (55–75 %) and carbon dioxide (45–25 %) with traces of water vapour, ammonia and sulphide (Edelmann et al. 1999).

A full scale stable operation of anaerobic digester for commercial production of methane depends mainly on the feedstock used, physiochemical characteristics of the operational conditions and on the microbial communities. Different biological wastewaters are treated by anaerobic digestion and successfully produced the biogas. Traditionally biogas is produced from the organic wastes like sludge or wastewaters. The production of biogas from the plants or organic waste as a renewable source of energy is of increasing economic and social importance. Despite the widespread use of anaerobic digesters for the production of biogas, knowledge of the structure, function and biological properties of the microbial communities involved is incomplete. Knowledge of the microbial diversity will provide valuable information to help design and provide process control. There is good evidence from sewage treatment works that during biogas production organisms accumulate in small aggregations. These are bound by extra-polymeric substances (EPS). Different organisms may also be bound together in a symbiotic relationship to produce biogas more effectively. It is evident from several studies that in general, bacterial EPS help in the formation of bioflocs in activated sludge and contribute to its structural, surface charge, settling properties in biogas reactors and wastewater treatment plants (Houghton et al. 2001). Bioflocs are formed with interaction among microbial aggregates, filamentous bacterial strains and organic and inorganic particles, which are held together by EPS (Bala Subramanian et al. 2008, 2010). EPS have been identified as one of the major components in bioflocs/biofilms (Urbain et al. 1993). They mainly consist of polysaccharides, proteins, nucleic acids and other cellular components (Higgins and Novak 1997). The availability of trace metals as micronutrients plays a very significant role on the performance and stability of agricultural biogas digesters (Feng et al. 2010; Demirel and Scherer 2011; Schattauer et al. 2011). Thus, in the present study chemical composition of the sludge samples was also studied.

The microbial communities involved in the anaerobic digestion are responsible for the methane formation and are often difficult to cultivate and classify systematically into taxonomy. Thus, alternatively culture-independent molecular approaches are more informative on microbial diversity and can provide evolutionary relationships (Woese et al. 1990; Handelsman 2004; Riesenfield et al. 2004; Tolvanen and Karp 2011). The culture-independent characterisation of microbial communities includes analysis of microbial 16S rDNA sequence (Hofman-Bang et al. 2003) which is the most widely used target for gene surveys (Traversi et al. 2011). Molecular and physiological characterisation of methanogens is important to understand the diversity of methanogens in methane production (Kumar et al. 2009, 2012, 2013; Sirohi et al. 2010, 2012, 2013; Dagar et al. 2011). The methanogenic microbial composition in anerobic digesters for methane production was demonstrated in several laboratory scale based experiments. However studies on the commercial scale anaerobic digesters operating for long periods of time are scanty (Deborah et al. 2011). Further the methanogenic microbial composition varies based on the physiochemical operational conditions and the feedstock used for methane production. Hence in the present work the methanogenic diversity from the two commercial biogas digesters that received same feedstock and operated under similar conditions were compared. This was accomplished by 16S rDNA clone library construction of amplified euryarchaeotal gene sequences from the collected sludge samples of the methanogenic stage of the digesters.

We also investigated microbial community especially methanogenic Archaea present in the anaerobic digesters using FISH technique. A molecular approach that enables the culture-independent characterisation of microorganisms, FISH with rRNA-targeted oligonucleotide probes has been one of the most powerful and widely used techniques. The main advantage of FISH, particularly, is it can detect, identify and quantify microbes at a wide range of phylogenetic levels, ranging from whole domain to particular genera and species (Kumar et al. 2011).

20.2 Materials and Methods

20.2.1 Source of Sludge Sample

For the analysis of the methanogenic diversity, slurry samples, sample 1 and sample 2, were collected each from methanogenesis stage operating anaerobic digester unit of two different commercial biogas plants located in Devon, UK (50° 48′N 4°21′W). The digetsers was fed with cattle manure, chicken manure, pig manure as well as blood and different sources of food waste, was operated at mesophilic temperature (approx. 37 ± 2 °C) in wet fermentation conditions and at pH 7.8.

20.2.2 Sludge Chemical Characteristics

The total and volatile solids (TS and VS) of the samples were determined using standard methods (APHA 1992), and their pH was determined with a microcomputer pH metre (Hanna Instruments, H19025). The alkalinities of the reactors were monitored using the discrete photometric analysis method (STM 234) with an Aquatek 250 analyser. Digestate samples from the digesters were also analysed for their volatile fatty acid (VFA) content with a Thermo Electron High-Performance Liquid Chromatography (HPLC) using a Bio-Rad column (125–0115) and H_2SO_4 (1 mM) as a mobile phase with a flow rate of 0.5 ml/min. The detection was carried out at ambient temperatures with a diode array UV detector at 220 nm (Nicholas et al. 2012).

EPS from sludge are extracted based on formaldehyde and NaOH extraction protocol adopted from Liu and Fang (2002). The separation of extracted EPS was carried out with HPLC equipped with a Zorbax Bio Series column (GF-250, 25 cm 9.4 mm, Agilent Technologies, France). Five protein standards of molecular mass 13,700, 45,000, 67,000, 200,000 and 670,000 Da (ribonuclease, amylase, chicken albumin and bovine serum albumin), provided by Sigma-Aldrich, were injected for column and size calibration.

20.2.3 Extraction of Total DNA from Sludge

Total DNA was isolated in triplicates essentially according to the methods described earlier (Yeates et al. 1998; Tsai and Olson 1991, 1992;

Shivaji et al. 2004). To 1 g of the sludge sample, 2 ml of the lysis buffer (prepared using 100 mM Tris HCl- pH 8.5; 50 mM EDTA and 50 mM NaCl) was added. The suspension was centrifuged at 4,000 rpm for 5 min. To the supernatant obtained, 50 µl of lysozyme was added and kept for incubation at 37 °C for 1 h. Then 40 µl of proteinase *K* (10 mg/ml) was added and incubated again under the same conditions for 1 h. Later 1 ml of SDS (20 % w/v) was added and incubated at 65 °C for 90 min. At this stage, five cycles of freeze and thaw are necessary. The samples were centrifuged at 6,000 rpm for 10 min and the supernatant was collected. Half volume (inrelation to the supernatant obtained) of polyethylene glycol (30 % w/v) and sodium chloride (1.6 M) was added to the supernatant and incubated at room temperature for 2 h. Samples were centrifuged at 10,000 rpm for 20 min and the pellet was suspended in TE buffer (pH 8.0). The aqueous phase was extracted with phenol, chloroform and isoamyl alcohol mixture, and DNA was precipitated by adding isopropanol. After 2 h of incubation at room temperature, DNA was pelleted by centrifugation at 12,000 rpm for 30 min, and the obtained pellet was suspended in sterile distilled water. Extracted DNA was visualised after gel electrophoresis on a 1 % agarose TAE (for 1,000 ml 4.84 gm Tris base; 1.14 ml acetic acid; 2 ml of 0.5 M EDTA; adjust pH to 8.5 using KOH) gel containing 1 µg/ml ethidium bromide.

20.2.4 Generation of 16S rDNA Clone Library

Euryarchaeota-specific 16S rDNA primers 21 F (5′–TTCCGGTTGATCCYGCCG–3′) and Eury 498R (5′–CTTGCCCRGCCCTT–3′) were used to amplify about 480-bp fragment of the total 1.5-kb 16S rRNA gene. The amplification was done by polymerase chain reaction (PCR) under the following conditions: denaturation at 95 °C for 1 min, annealing of primers at 45 °C for 1 min and extension at 72 °C for 1 min for a total of 30 cycles, followed by 10 min of extension at 72 °C. Controls containing no DNA were also used to determine whether contaminants were being amplified. Each PCR product was visualised on 1 % agarose TAE gel containing 1 µg/ml ethidium bromide.

20.2.5 Bacterial Universal 16S rDNA Primers

Amplification was done as described earlier (Reddy et al. 2000; Shivaji et al. 2004, 2011). Primers PA-F8 (5′–AGAGTTTGATCCTGGCTCAG–3′) and PH-1546R (5′–AAGGAGGTGATCCAG CCGCA–3′) were used to amplify about total 1.5-kb fragment of the 16S rRNA gene. The PCR product was purified with the QIAquick PCR Purification Kit (Qiagen Inc.) according to the manufacturer's instructions.

20.2.6 Cloning and Library Construction of the Amplified Gene Sequences

The purified PCR products were ligated into the plasmid vector pMOS (Amersham Biosciences, New Jersey, USA), and the hybrid vectors were used to transform *Escherichia coli* DH5α ultracompetent cells by following manufacturer's instructions. Transformants were screened using blue-white selection on Luria-Bertani agar plates (82 mm) containing 20 µg/ml X-gal (5-bromo-4-chloro-3-indolyl-β-galactopyranoside) and 20 µl of 100 mm IPTG (isopropyl B-D-thiogalactopyranoside) and 60 µg/ml ampicillin. Clone libraries were constructed by randomly picking up white-coloured colonies derived from each slurry sample and maintained on LB agar plate containing X-gal and ampicillin.

20.2.7 Partial rRNA Gene Sequencing and Phylogenetic Analysis

The 16S rRNA gene from both euryarchaeotal- and bacterial-specific libraries was amplified from the transformants by colony PCR using the vector-targeted M13 forward (5′–GTAAAACGACGGCCAGT–3′) and M13 reverse (5′–GGAAACAGCTATGACCATG–3′) primers, respectively, and sequenced using the primers M13 forward and M13 reverse for only euryarchaeota-specific clones, and bacterial clone libraries were sequenced using the primers M13 forward, M13 reverse, pD (5′–CAG CAG CCG CGG TAA TAC–3′) and pF (5′–ACG AGC TGA CGA CAG CCA TG–3′) (Pradhan et al. 2010). Vector and chimeric sequences were eliminated using Gene Tool Version 2 (www.biotools.com). Sequences were then subjected to BLAST to identify the nearest taxa and aligned with sequences belonging to the nearest taxa (downloaded from NCBI databasehttp://www.ncbi.nlm.nih.gov/) using Clustal X. Phylogenetic trees were constructed using neighbour-joining and maximum likelihood methods (Pradhan et al. 2010; Shivaji et al. 2011). Bootstrap analysis, based on 1,000 replicate data sets, was performed to assess stability among the clades.

20.2.8 Statistical Analyses of the Cloned Libraries

To compare the bacterial diversity within the two samples, euryarchaeota-specific 16S rRNA gene sequences and that of bacterial universal 16S rRNA gene sequences of the clones from their respective libraries showing ≥97 % sequence similarity were grouped into the same OTU (phylotype). Shannon-Wiener Diversity Index (http://www.changbioscience.com/genetics/shannon.html) was used to calculate Shannon index (H'), evenness (Lefauconnier et al. 1994). Rarefaction analysis was done using the site Online Calculation (http://biome.sdsu.edu/fastgroup/cal-tools.htm). Rarefaction curves were generated to compare the relative diversity and coverage of each library.

20.2.9 Fluorescent In Situ Hybridisation (FISH)

16S rRNA-targeted oligonucleotide probe set (*Metabion international AG, Germany*) used to identify methanogenic Archaea present in anaerobic bioreactors. The fluorescent in situ hybridisation (FISH) slides were analysed with confocal scanning laser microscopy (CSLM) Zeiss LSM 780 system with Axio Observer inverted microscope, using LSM software ZEN 2010 to see the spatial distribution of specially labelled target cells present in the anaerobic bioreactors.

The 16S rRNA-targeted oligonucleotide probe set targeted bacteria and most methanogenic Archaea that were used in this research:

EUB338	Most *Bacteria*	5′–GCT GCC TCC CGT AGG AGT–3′
ARC915	*Archaea*	5′–GTG CTC CCC CGC CAA TTC CT–3′
MG1200	Methanomicrobiales	5′–CGG ATA ATT CGG GGC ATG CTG–3′
MS821	Methanosarcina	5′–CGC CAT GCC TGA CAC CTA GCG AGC–3′
MX825	Methanosaetaceae	5′–TCG CAC CGT GGC CGA CAC CTA GC–3′
MB311	Methanobacteriales	5′–ACC TTG TCT CAG GTT CCA TCT CC–3′
MB1175	Most Methanobacteriales	5′–TAC CGT CCA CTC CTT CCTC–3′

In order to visualise microbial community in the activated sludge samples, sample preparation, fixation, hybridisation, washing and visualisation are required. Fresh sludge samples were treated with enzymes followed by incubation for an overnight with 200 rpm. Later cells were fixed with 4 % paraformaldehyde (PFA) in a 3:1 PFA to sample ratio with incubation at 4 °C. Fixed samples was centrifuged for 5 min at 4 °C with 15,000 rpm; after centrifugation, the cells were resuspended with one volume of ice-cold phosphate-buffered saline (PBS), and finally the cells were subjected to the same centrifugation and resuspended with one volume of ice-cold PBS and one volume of 96 % ice-cold ethanol. 15 μl of sample was placed onto the centre of the slide and allowed to dry for 15 min at 46 °C; once dry, the sample was then dehydrated for 3 min in 50 %, 80 % and 96 % ethanol. 10 μl of 35 % hybridisation buffer and 1 μl of desired probe are mixed and placed onto the slide and incubated overnight. Overnight incubated samples with respective probes were washed with 35 % hybridisation washing buffer (48 °C). Finally the slides were washed with ice-cold water and visualised with confocal scanning laser microscopy (CSLM) (Fig. 20.1).

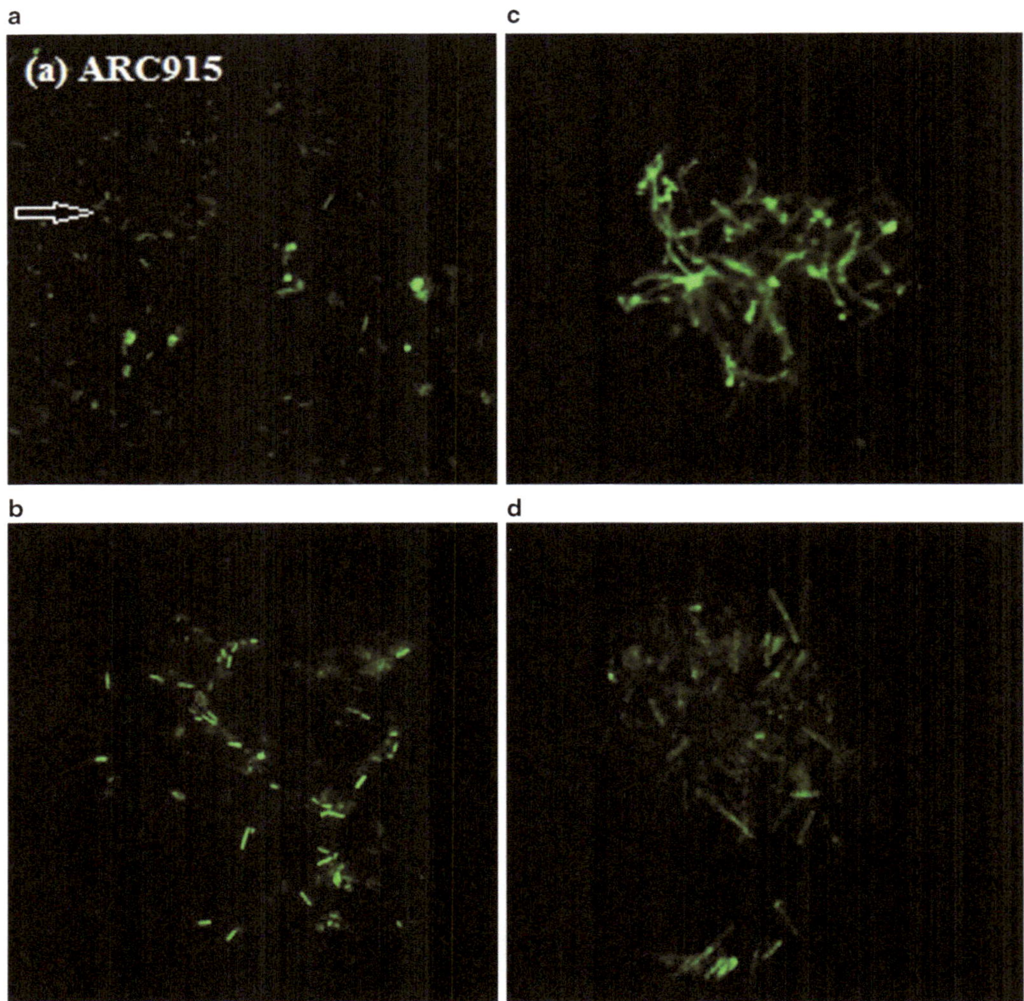

Fig. 20.1 Hybridisation of activated sludge samples treated with different oligonucleotide probes (Cy3 – labelled) to target their respective species. CSLM micrographs of (**a**) Archaea spp., when sludge sample is treated with ARC915. (**b** and **c**) Fluorescence micrographs of Methanosarcina spp., when sludge samples are treated with MX821 probe. (**d**) Fluorescence micrographs of Methanomicrobiales spp. when sludge sample is treated with MG1200 probe

20.3 Results

Two sludge samples (sample 1 and sample 2) were collected from two separate biogas tanks (tank 1 and tank 2) operated at methanogenesis stages each receiving the same feedstock, located near Holsworthy, Devon, UK. The pH of the two tanks 1 and 2 are 8.3 and 8.2 respectively and the C:N ratio is 2.3:1 and 2.6:1, respectively. In the other physicochemical characteristics like total solids, volatile solids, conductivity, total nitrogen, ammonium nitrogen, total carbon, COD, total phosphorus and total potassium, no major differences were observed in the two tanks (Table 20.1). Among the trace elements, Fe, Mg, Mn, Zn, Ni and Cr showed high concentrations in tank 1 compared to tank 2. The concentration of sulphur is high in tank 2. In other elements such as Ca, Mo, Cu, Co, Se and B, concentrations were similar in both tanks (Table 20.2).

Sugars from the two sludge samples were analysed; pentoses were 32 μg/mg of the slurry sample in tank 1 and 40 μg/mg of the slurry sample in tank 2, whereas hexoses were 65 μg/mg of the slurry sample in tank 1 and 90 μg/mg of the slurry sample in tank 2. The protein concentration was 0.45 μg/mg in tank 1 and 0.56 μg/mg in tank 2. Typical molecular chromatogram profile of EPS extracted from Holsworthy biogas plant sludge is shown in Fig. 20.2. Polysaccharides with a molecular weight between 14 and 44 kDa were dominant in the extracted EPS.

The two slurry samples from the biogas reactor yielded about 30 and 32 μg/g of DNA from the sludge samples 1 and 2, respectively. About 200 ng of the DNA from sample 1 and sample 2 was used for the construction of euryarchaeota-specific and bacterial universal 16S rRNA gene-specific libraries. The number of clones in sample 1 and sample 2 of euryarchaeota-specific gene libraries is 107 and 130 clones, respectively, with an insertion of approximately 480 bp, and for bacterial universal 16S rRNA gene-specific libraries, it is 135 and 144, respectively, with an insertion of approximately 1.5 kb.

Table 20.1 Physicochemical characteristics of the two sludge samples collected from two separate commercial biogas plants (tank 1 and tank 2)

Sludge parameter	Tank 1	Tank 2
Total solids (%)	3.71	3.48
Volatile solids (% of TS)	69.54	70.40
Gross energy (dry) (MJ/kg)	17.8	19.1
Conductivity 1:6 (μS/cm)	7,130	6,840
Ammonium nitrogen (mg/kg)	6,664	6,991
Total nitrogen (%w/w)	0.735	0.750
Total carbon (%w/w)	1.68	1.96
C:N ratio	23:1	26:1
pH	8.3	8.2
COD	5,530	5,410
Total (P)	286	270
Total (K)	738	755
Euryarchaeaotal-specific 16S rRNA gene clone library	MT1	MT2
Bacterial 16S rRNA gene clone library	HW1	HW2

Table 20.2 Trace element composition of the two sludge samples collected from two separate commercial biogas plants (tank 1 and tank 2)

	mg/kg [fresh matter]	
Trace elements	Tank 1	Tank 2
Fe	1,134	783
Mg	72.6	66.3
S	67.32	119.35
Ca	896	886
Mn	7.98	5.58
Mo	0.19	0.16
Zn	14.20	8.63
Cu	1.1	0.9
Co	0.06	0.06
Ni	0.50	0.34
Se	0.02	0.02
Cr	1.73	1.33
B	1.10	1.08

20.3.1 Archaeal Diversity Analysis

Phylogenetic analysis of the Archaea indicated that sample 1 clone libraries representing 107 sequences and sample 2 represented by 130 sequences were affiliated to two classes

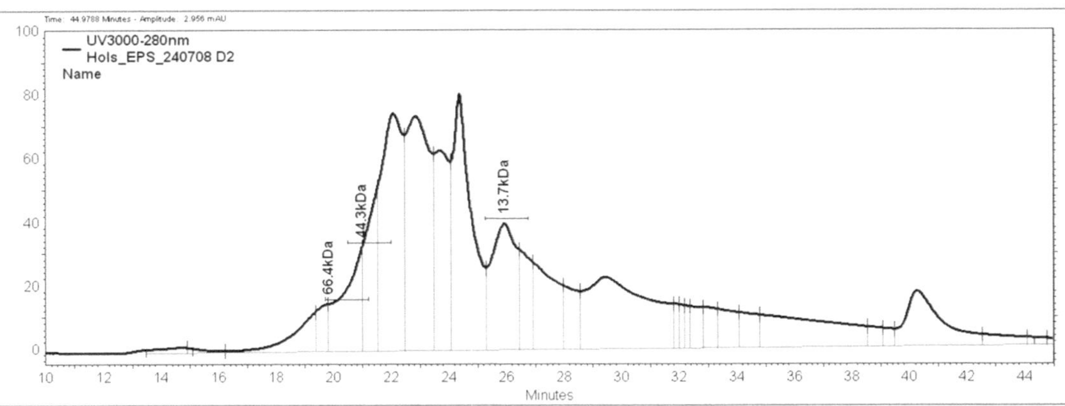

Fig. 20.2 Typical molecular chromatogram profile of EPS extracted from commercial biogas plant sludge. Polysaccharides with a molecular weight between 14 and 44 kDa were dominant in the extracted EPS

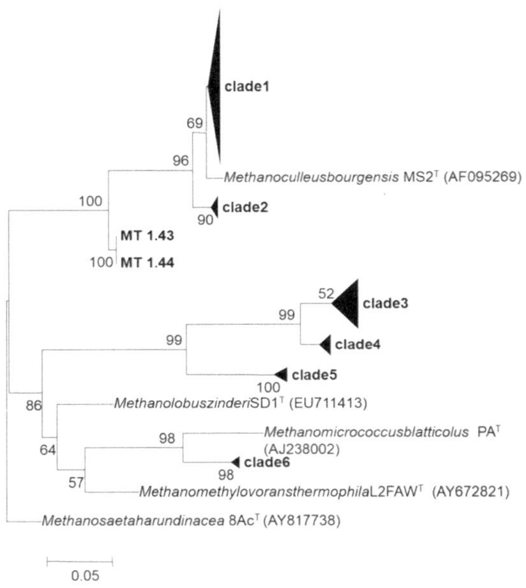

Fig. 20.3 Neighbour-joining phylogenetic trees of euryarchae al-specific 16S rRNA gene clones from MT1 library. Numbers at nodes are bootstrap values. The bar represents 0.05 substitutions per alignment position

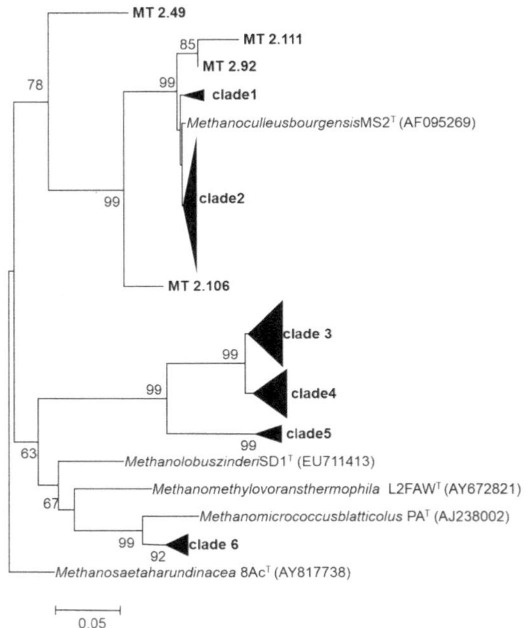

Fig. 20.4 Neighbour-joining phylogenetic trees of euryarchae al-specific 16S rRNA gene clones from MT2 library. Numbers at nodes are bootstrap values. The bar represents 0.05 substitutions per alignment position

Methanomicrobiales and Methanosarcinales of phylum Euryarchaeota. These sequences were grouped and assigned operational taxonomic unit (OTU) for which a threshold of 97 % similarity was used (Goebel and Stackebrandt 1994). Sample 1 encompasses 6 OTUs, 2 sequences, and that of sample 2 comprises 6 OTUs, 4 sequences (Figs. 20.3 and 20.4). In sample 1, 55 % of the sequences showed more than 97 % sequence similarity with *Methanoculleus bourgensis* MS2^{T}. Similarly in sample 2, the majority of sequences (48.5 %) were clustered with *Methanoculleus bourgensis* MS2^{T}. In sample 1 clone library, sequences of clade 1 and clade 2 formed a separate cluster with *Methanoculleus bourgensis* MS2^{T} that shows 98–99 % sequence similarity with the uncultured archaeon gene for

16S rRNA, partial sequence, clone: 3MP-A-3ME-76. This archaeon gene was obtained from methane production process which was operated using food waste as organic source. Twenty-five percent of the sequences representing clades 3, 4 and 5 of sample 1 and 42.5 % of the sequences representing clades 3, 4 and 5 of sample 2 formed a separate cluster from the cultured archaeal sequences that signifies novel uncultured representatives of Euryarchaeota. However, BLAST analysis showed that sequences of clade 3 of sample 1 and sample 2 showed sequence similarity (ranging from 96 % to 99 %) with uncultured archaeon PY-2 16S ribosomal RNA gene, partial sequence obtained from permafrost samples from the northwest and southwest of China. Sequences within the clade 4 of sample 1 and sample 2 showed sequence similarity (ranging from 95 % to 99 %) with uncultured archaeon clone BGA22C 16S ribosomal RNA gene, partial sequence obtained from anaerobic digesters receiving food processing wastes. Sequences of the clade 5 of sample 1 and sample 2 showed a sequence similarity (ranging from 98 % to 99 %) with uncultured archaeon clone ATB-EN-8289-R062 16S ribosomal RNA gene, partial sequence obtained from agricultural biogas plants. Clone library sequences of clade 6 of sample 1 and sample 2 are distantly clustered with *Methanimicrococcus blatticola* PA^T that showed similarity with uncultured archaeon clone BA8-EN5190N-LBR27 16S ribosomal RNA gene, partial sequence archived from two-phase biogas reactor systems operated with plant biomass. Sequences MT 1.43 and MT1.44 of sample 1 showed 93.8 % sequence similarity with *Methanoculleus bourgensis* $MS2^T$ and 94 % similarity with uncultured archaeon clone 5.5 ft C30A 16S ribosomal RNA gene, partial sequence attained from landfill environment. Sequences MT 2.92, MT2.111 and MT 2.106 of sample 2 showed similarity of 98.3 %, 96.4 % and 94.0 %, respectively, with *Methanoculleus bourgensis* $MS2^T$. Sequence MT 2.49 showed similarity of 88.3 % with uncultured archaeon clone *Methanosarcina thermophila* 5.5 ft C17A obtained from landfill environment.

20.3.2 Bacterial Diversity Analysis

BLAST analysis of 16S rDNA bacterial clone library of the two samples identified that clones in sample 1 and sample 2 were affiliated to six phyla (Table 20.3). Clones affiliated to phyla *Firmicutes* and *Proteobacteria* were commonly present in both the libraries. Clone sequences that are affiliated to phyla *Bacteroidetes* and *Thermotoga* were present only in MT1 clone library, and clone sequences belonging to *Actinobacteria* and *Chloroflexi* are present only in MT2 clone library. Based on the BLAST analysis, the 135 clones of MT1 belonging to 22 bacterial taxa and the 144 clones of MT2 belong to 18 bacterial taxa. In both libraries of sample 1 and sample 2, clones belonging to phylum *Firmicutes* were the most predominant representing 94.8 % and 97.9 %, respectively, of the total clones.

20.3.3 *Firmicutes*

Clones affiliated to different classes of the phylum *Firmicutes*, namely, *Bacilli, Carboxydocella, Clostridia, Erysipelotrichi* and *Syntrophomonadaceae*, were present in both sample 1 and sample 2 (Figs. 20.5 and 20.6). Clones affiliated to class *Natranaerobiales* are present only in sample 1. BLAST analysis shows that clones affiliated with class *Clostridia* were predominant representing 114 clones out of 128 in sample 1 and 133 out of 141 in sample 2. Clones HW1.4, HW1.82, HW1.117, HW1.46, HW1.76, HW1.81, HW1.96, HW2.8, HW2.77, HW2.18 and HW2.71 showed greater than 97 % sequence similarity with their nearest phylotype *Syntrophaceticus schinkii* $SP3^T$ and clustered with the same. All the sequences that demonstrated more than 97 % similarity with in the same cluster with their neighbour sequences were grouped and given the clade numbers. Clone sequences of sample 1 representing clade 1 were affiliated to *Saccharofermentans acetigenes* and also showed similarity with uncultured diversity obtained from mesophilic anaerobic digester representing uncultured bacterium clone

Table 20.3 Affiliation of 16S rRNA gene sequences of two 16S rRNA gene libraries (HW1 and HW2) constructed using two sludge samples collected from methanogenesis stage of two operating bioreactor units of commercial biogas plants located in Devon, UK

Phylogenetic neighbour and nucleotide accession number	Clone number and 16S rRNA gene sequence similarity (%)	
	Sample 1 library	Sample 2 library
Phylum – *Actinobacteria*	0	2
Class – Actinobacteria	0	2
Propionibacterium acnes KPA171202^T (AE017283)		53, 54 [99]
Phylum – *Bacteroidetes*	3	0
Class – Bacteroidia	3	0
Proteiniphilum acetatigenes TB107^T (AY742226)	43, 54, 72 [90.2]	
Phylum – *Chloroflexi*		1
Class – Anaerolineae		1
Levilinea saccharolytica KIBI-1^T (AB109439)		117a [88.5]
Phylum – *Firmicutes*	4	1
Class – *Bacilli*	4	
Bacillus vedderi JaH^T (Z48306)	111 [83.4]	
Paenibacillus chitinolyticus IFO 15660^T (AB021183)		46 [93.3]
Planifilum fulgidum 500275^T (AB088362)	21, 22 [85.2]	
Class – *Clostridia*	121	135
Carboxydocella manganica SLM 61^T (GU584133)	47, 62, 59, 60, 27, 17, 36 [86.7]	43, 143 [86.5]
Clostridium acetireducens 30A^T (X79862)	95 [83.6]	
Clostridium alkalicellulosi Z-7026^T (AY959944)		119, 116, 131, 123 [84]
Clostridium jejuense HY-35-12T (AY494606)		68, 127 [92.8]
Clostridium straminisolvens CSK1^T (AB125279)	33, 35, 8, 49, 19, 31, 53, 70, 18, 132, 10, 15, 20, 30, 50, 93, 99, 24, 102, 83, 89, 75, 79, 106, 23, 26, 91, 1, 29, 2, 28, 109, 26, 21, 110, 112, 108, 92, 103, 65, 68, 48, 107, 37, 88, 94 [86.9] (clade 1) 100, 114, 32, 6, 133, 113a, 128, 113, 127, 124, 119, 129, 130, 16 [87.3] (clade 3) 40 [86]	61 [87.4], 102 [83.5]
Clostridium thermocellum ATCC 27405^T (CP000568)	9, 58, 86, 34, 51, 63, 25, 77, 85 [86] (clade 2) 105 [87.9], 42 [87.6], 73 [85.7], 98 [86.9]	101, 125, 121, 109, 129, 100, 115, 118, 132, 128, 114, 133, 124, 110, 130, 105, 134, 111, 117 [86.15] (clade 2a), 137, 65, 74, 136 [84.3] (clade 1a), 41, 76, 97, 72, 60, 92, 87, 73, 96, 67, 86, 79, 94, 85, 50, 82, 83, 84, 69, 93, 70, 78, 88, 91, 36, 17, 42, 16, 24, 15, 35, 37, 39, 38, 31, 40, 5, 9, 4, 20, 10, 11, 25, 26, 27, 19, 14, 6, 29, 28, 22, 48, 21, 13, 47, 62, 56, 113, 99, 95, 140, 108, 49, 64, 89, 107, 75, 81, 104, 58, 80, 66, 112, 120, 122, 141, 23, 1 [86.12] (clade 1)

(continued)

Table 20.3 (continued)

Phylogenetic neighbour and nucleotide accession number	Clone number and 16S rRNA gene sequence similarity (%)	
	Sample 1 library	Sample 2 library
Clostridium stercorarium NCIMB 11754^T (L09174)	38, 41, 14 [87.2]	126, 142, 12
Clostridium saccharogumia SDG-Mt85-3Db^T (DQ100445)	135 [84.3]	
Clostridium ultunense BS^T (Z69293)	101, 80, 97, 104 [93.2]	63, 139, 90 [93.2] (clade 3) 135, 3, 45, 44, 103, 98 [93.5] (clade 4)
Desulfotomaculum solfataricum V21^T (U33455)	122	
Desulfotomaculum thermocisternum ST90^T (NR 025979)		2
Desulfosporosinus acidiphilus SJ4^T (NC_018068)	71 [89]	
Pelotomaculum propionicicum MGP^T (NR_041000)	66, 69 [84.6]	
Peptoniphilus methioninivorax NRRLB-23883^T (GU440754)		138
Papillibacter cinnamivorans CIN1^T (NR_025025)	13, 118 [90]	
Saccharofermentans acetigenes P6^T (AY949857)	3, 64, 78, 125, 134, 123, 131, 89, 52, 61, 57, 87, 56, 7, 12 [86.5]	7 [86.6]
Syntrophaceticus schinkii Sp3^T (EU386162)	4, 82, 117 [100], 46 [96], 76 [96.6], 81 [92.7], 96 [93.2]	8, 77 [100], 18, 71 [96.6]
Tepidanaerobacter syntrophicus JL^T (AB106353)	90[95]	
Tissierella creatinophila KRE4^T (X8022)	44, 45, 120 [88]	51
Class – Erysipelotrichi	3	1
Eubacterium tortuosum ATCC 25548^T (L34683)	11, 116, 67 [84.5]	
Holdemania filiformis ATCC51649^T (Y11466)		106 [86]
Class – *Natranaerobiales*	1	0
Natranaerobius truperi JW/NM-WN-LU^T (EU338490)	5 [88.2]	
Class – *Syntrophomonadaceae*		4
Syntrophomonas curvata GB8-1^T (NR_025752)		32, 33, 34, 57 [91.3] (clade 5)
Phylum – *Proteobacteria*	1	1
Class – *Gammaproteobacteria*	1	1
Erwinia toletana A37^T (AF130910)		5 [55]
Pseudomonas caeni HY-14^T (EU620679)	74 [91.9]	
Phylum – *Thermotogae*	2	0
Class – *Thermotogae*	2	0
Defluviitoga tunisiensis SulfLac1^T (FR850164)	115, 55 [99.46]	

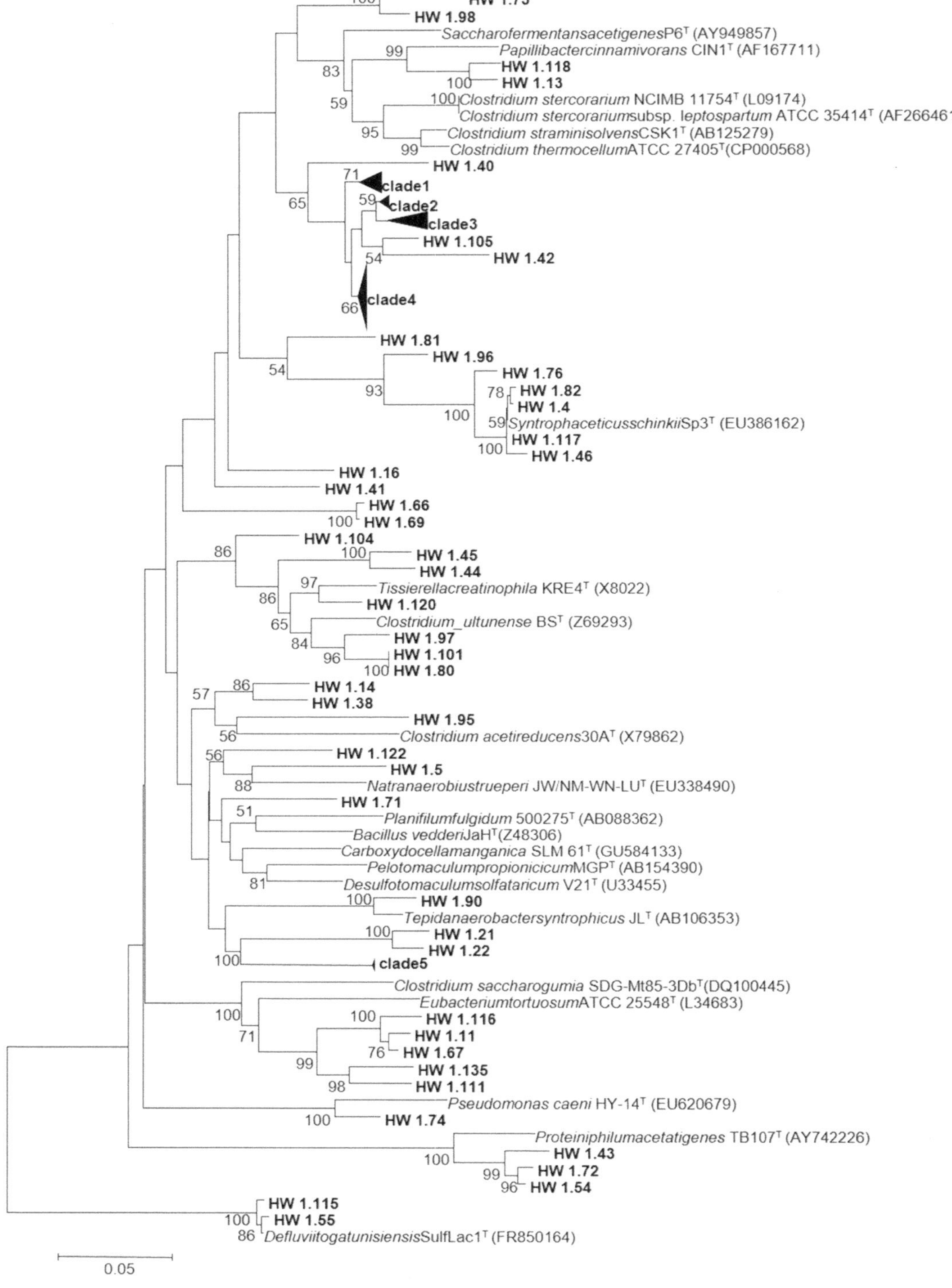

Fig. 20.5 Neighbour-joining phylogenetic trees of bacterial 16S rRNA gene clones from HW1 library. Numbers at nodes are bootstrap values. The bar represents 0.05 substitutions per alignment position

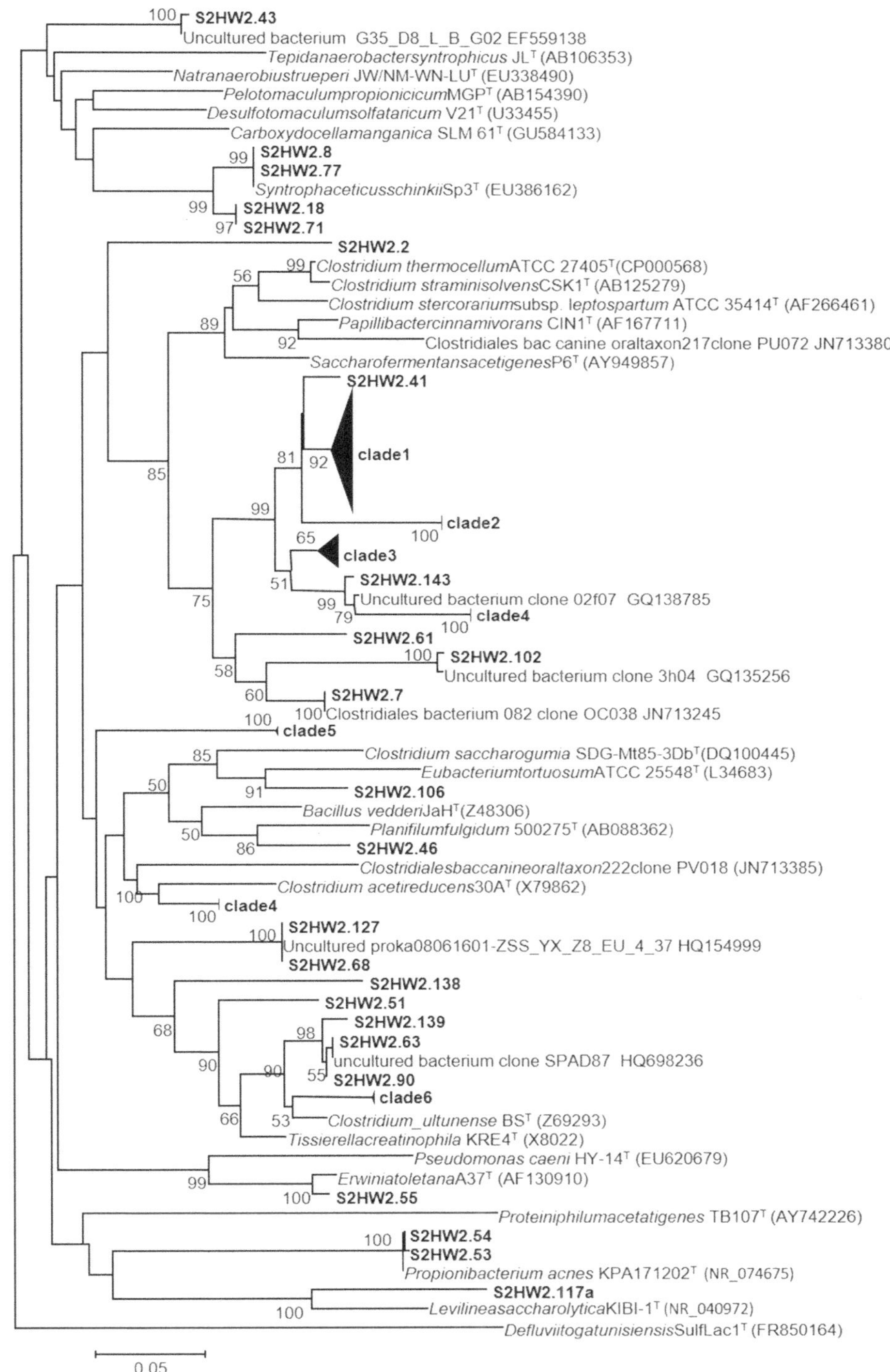

Fig. 20.6 Neighbour-joining phylogenetic trees of bacterial 16S rRNA gene clones from HW2 library. Numbers at nodes are bootstrap values. The bar represents 0.05 substitutions per alignment position

G35_D8_L_B_G02 partial sequences. Clone sequences of sample 1 representing clade 2 were affiliated to *Clostridium thermocellum*, and clone sequences of clades 3 and 4 were affiliated to *Clostridium straminisolvens*. However, sequences of clones belonging to clade 2 were more closely related to uncultured bacterium clone 12Bac (95–99 %), clade 3 with uncultured bacterium clone 02f07 (98–99 %) and clade 4 with uncultured bacterium clone 4Bac R8 (94–99 %), which were obtained from an anaerobic biogas reactor utilising food waste. Similarly clones representing clade 5 were affiliated with *Carboxydocella manganica* and showed 93–99 % sequence similarity with uncultured bacterium clone 4Bac R8.

The cluster of clone sequences that showed more than 97 % sequences imilarity with their neighbouring sequences were grouped and given the clade numbers. Clone sequences of sample 2 that formed clades 1, 2 and 3 showed sequence similarity with *Clostridium thermocellum* ranging from 84 to 87. However, sequences of clade 1 shared a sequence similarity with uncultured bacterium clone 4Bac R8 16S ribosomal RNA gene, partial sequence (ranging from 93 % to 96 %), sequences of clade 2 shared similarity with uncultured bacterium clone 12Bac R8 16S ribosomal RNA gene, partial sequence (ranging from 94 % to 98 %), both of these sequences were obtained from biogas plant utilising predominantly food waste. Sequences of clade 3 showed its sequenc similarity with uncultured bacterium clone G35_D8_L_B_G02 (99 %) 16S ribosomal RNA gene, partial sequence obtained from mesophilic anaerobic digester. Sequences of clade 4 showed 92.8 % sequence similarity with *Clostridium alkalicellulosi* Z-1026^T and 99 % with uncultured bacterium clone 01a06 16S ribosomal RNA gene, partial sequence obtained from Anaerobic Sequencing Batch Reactor (ASBR) reactor treating swine waste. Clone sequences of clade 5 showed only 91.3 % similarity with *Syntrophomonas curvata* and showed 99 % similarity with uncultured bacterium partial 16S rRNA gene, clone 3wk_1LB11 that was obtained from anaerobically digested sludge. Clone sequences of clade 6 showed a sequence similarity with *Clostridium stercorarium* subsp. *leptospartum* ATCC 35414^T (87.2 %) and 99 % with uncultured bacterium clone A35_D28_L_B_B05 16S ribosomal RNA gene, partial sequence. Clone sequences that formed clade 7 showed sequence similarity (93–93.5 %) with *Clostridium ultunense* and 98 % with uncultured bacterium clone SPAD87 obtained from swine manure anaerobic digester and 99 % with uncultured prokaryote clone 07122804-ZSS_Z7_EU_4_2_27 16S ribosomal RNA gene, partial sequence obtained from pig ordure-based biogas digester.

20.3.4 *Proteobacteria*

One clone sequence each from sample 1 and sample 2 represented class *Gammaproteobacteria* of phylum *Proteobacteria.* Clone sequence of HW1.74 showed 91.9 % similarity with *Pseudomonas caeni* and that of HW2.5 showed 95 % similarity with *Erwiniatoletana*.

20.3.4.1 Other Phyla

Representatives of clone sequences of phylum *Bacteroidetes* and *Thermotoga* are present only in sample 1 (Fig. 20.5), while representatives of clone sequences of *Actinobacteria* and *Chloroflexi* are present only in sample 1 (Fig. 20.5).

20.3.5 Statistical Analysis

The rarefaction curves indicated that the archaeal population in samples S1 and S2 is plateaued while the bacterial population in these samples is not (Fig. 20.7). The rarefaction curve analysis implied that these are likely to be minimal estimates of diversity for the bacterial population. The Shannon index of the euryarchaeotal clones from samples 1 and 2 is 3.9 and 4.2, respectively, while for bacterial libraries, it is 4.5 and 3.8, respectively. The evenness of species richness showed 1.9 and 2.0 for euryarchaeotal clone samples and 2.1 and 1.8 for bacterial samples.

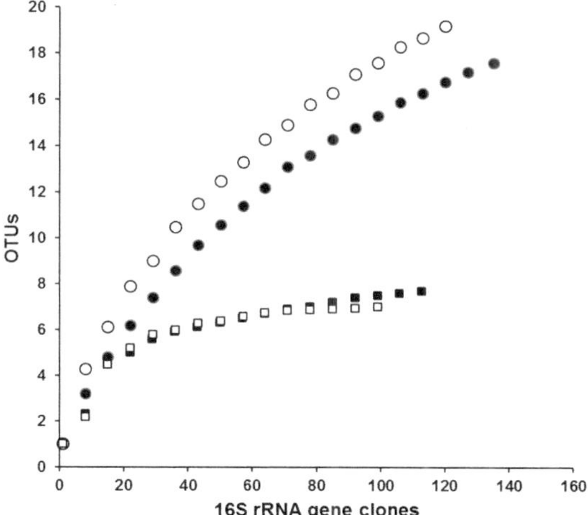

Fig. 20.7 Rarefaction curves of observed OTUs in the two sludge samples collected from methanogenesis stage operating bioreactor unit of two different commercial biogas plants located in Devon, UK (sample 1 and sample 2). Rarefaction curves generated for methanogenic diversity of the study for sample 1 (□) and sample 2 (■). Rarefaction curves generated for bacterial diversity of the study for sample 1 (O) and sample 2 (●)

20.4 Discussion

Animal manure represents an important source of energy for production of biogas through anaerobic digestion. Like many European countries, Britain has also focused on utilising manure from farms and waste from abattoirs and food processors to create gas and electricity. Methanogenic archaeal diversity has been studied from various anaerobic digesters that are fed wastewaters (McHugh et al. 2003), a permanently frozen Lake Fryxellin, Antarctica (Karr et al. 2006), and reactors fed with plant biomass (Klocke et al. 2007), from agricultural biogas plant (Krause et al. 2008; Nettmann et al. 2010). Diversity of archaeal communities within 44 bioreactors fed with different organic materials was demonstrated by LeClerc et al. (2004). Archaeal diversity has also been determined from a commercial biogas plant located in Germany, which is being operated using herbal biomass (Nettmann et al. 2008). The present study gives an outlines about the archaeal and bacterial diversity within two commercial anaerobic digesters located in Devon, UK, that are both fed cattle manure and food waste. The comparative analysis of the diversity within these two commercial anaerobic reactors helps in understanding the stability of the bioreactor. We have used Fluorescent in Situ Hybridisation (FISH) to identify methanogenic Archaea which are dominant in the production of biomethane in the anaerobic digesters. The availability of trace metals as micronutrients plays a very significant role on the performance and stability of agricultural biogas digesters (Feng et al. 2010; Demirel and Scherer 2011; Schattauer et al. 2011). The trace elements as shown in Table 20.2 are abundant in both digesters and play a key role in enzymatic pathways of methanogens. The other chemical characteristics of samples are shown in Table 20.1.

In this study, sludge samples collected from two anaerobic digesters operated under similar conditions were compared. Previous studies demontrated the microbial diversity characterisation from anaerobic digesters fed with cattle manure were operated at thermophilic temperatures. These studies reported that two archaeal species *Methanoculleus thermophilus* and the acetate utilising *Methanosarcina thermophila* were predominant (Chachkiani et al. 2004). However, in our study around 50 % of the sequences from the archaeal clone libraries of the two samples are dominated by one species,

Methanoculleus bourgensis. It is interesting to note that the population of our investigation of a mesophilic digester and the predominant species from previous studies from the thermophilic anaerobic digester belonging to Methanomicrobiales of Archaea. Thus, our study corroborates with previous studies for digesters that are fed with cattle manure (Chachkiani et al. 2004). However, in our study clone sequence representatives belonging to Methanosarcinales were also present in both samples 1 and 2. This is in agreement with previous studies on characterisation of methanogens from municipal solid waste digesters and sewage sludge digesters, where they reported that Methanosarcinales are prevalent under the mesophilic conditions but not found in thermophilic conditions (Chouari et al. 2005). At the same time, these sequences based on the BLAST analysis results showed similarity with uncultured archaeon clone sequences of the anaerobic digester fed with different organic sources like food waste, food processing waste and plant waste and from other methanogenic environments like permafrost and landfill environments.

Phylogenetic analysis revealed that around 25 % of the clone sequences of sample 1 and 42 % of sample 2 formed a separate cluster and may be representing novel uncultured diversity within the anaerobic digester fed with cattle manure operated at mesophilic temperature. The studies on methanogenic diversity in anaerobic digesters reveal that the diversity is influenced by the operational and environmental conditions, which leads to digester stability. Thus, the identification of a considerable fraction of novel uncultured methanogenic diversity in the present study signifies a complex role leading to stability of the anaerobic digester at mesophilic temperature.

Previous studies revealed that members belonging to the genus *Methanoculleus* are predominantly present in various anaerobic digesters. Representative of this genus, *Methanoculleus bourgensis* was identified in various anaerobic digesters that are operated using maize silage and manure (Feng et al. 2010; Jaenicke et al. 2011; Krober et al. 2009). Studies of Schnurer et al. (1999) and Weiss et al. (2009) reported that *Methanoculleus bourgensis* is well adapted to some of the adverse conditions like high salt and ammonium concentrations that usually hinder the functioning of the anaerobic digester. *Methanoculleus bourgensis* is a hydrogenotrophic methanogen utilising hydrogen and carbon dioxide or formate as the substrates for methane synthesis. Hydrogenotrophic methanogens predominate in the anaerobic digester system by growing on the simple substrates like acetate, ethanol, methanol, glucose and propionate (Demirel and Scherer 2008). The characterisation of the Eurychaeota bacterial diversity from both anaerobic digesters has shown no variation.

The comparative bacterial analysis in both samples 1 and 2 shows that members belonging to the phylum *Firmicutes* were abundant. Further clone sequences that shared similarity with *Clostridium straminisolvens* were abundant in sample 1, while the clone sequences that shared similarity with *Clostridium thermocellum* were predominantly present in sample 2. Previous report on bacterial community dynamics in liquid swine manure reported that *Clostridium* sp. are active in the stored liquid swine manure and survived by fermenting lipids, sugars and amino acids to simultaneously producing malodorous volatile organic acids, hydrogen sulphide and amines (Leung and Topp 2001). Hence, from the results of the present study, it appears that predominance of *Clostridium* sp. in the digester has favoured the production of high concentrations of ammonium compounds and thus simultaneously favoured the growth of *Methanoculleus bourgensis*. *Clostridium thermocellum* that predominated in sample 2 has the ability to hydrolyze cellulose efficiently by means of its extracellular cellulases, which are organised into cellulosomes (Wirth et al. 2012). *Clostridium straminisolvens* having cellulolytic properties was abundant in sample 1. However, it is interesting to note that both *Clostridium straminisolvens and Clostridium thermocellum* are moderately thermophilic organisms, and the former does not occur at or below 45 °C (Kato et al. 2004). Another phylogenetically close relative of the above-discussed two species of genus *Clostridium*, *Saccharofermentans acetigenes* P6^T

that ferments several hexoses, polysaccharides and alcohols (Chen et al. 2010) was also identified in both samples. Identification of clone sequences belonging to *Syntrophaceticus schinkii*, *Clostridium ultunense* and *Tepidanaerobacter syntrophicus* indicates the prevalence of syntrophic acetate oxidation (Westerholm et al. 2010).

20.5 Conclusions

The results of this study imply that no major differences are found in the microbial diversity of the two anaerobic digesters however varied in the relative abundance of the species richness. The results also reveal that cellulolytic bacteria were predominantly present in the reactors more particularly *Clostridium thermocellum* were predominant in tank 2. Majority of the archaeal clone sequences belonged to *Methanoculleus bourgensis*. A huge portion of the sequences of methanogenic clone libraries has showed similarity with the uncultured archaeal clones identified from the anaerobic digesters. It is also revealed that hydrogenotrophic methanogens are abundant in the reactors. This may be because of the cellulolytic bacteria that are present in the reactors favoured the production of substrates that enhanced sympathetic conditions for hydrogenotrophic methanogens. We conclude that Archaea, especially *Methanomicrobiales* and *Methanosarcinales* are the dominant groups in the digesters.

Acknowledgements We thank the Biotechnology and Biological Sciences Research Council (BBSRC), UK, for the financial support. Sreenivas would like to thank Welsh European Funding Office (WEFO) for financial support.

References

APHA (1992) Standard methods for the examination of water and wastewater, 18th edn. American Public Health Association, American Water Works Association and Water Environment Federation, Washington, DC

Bala Subramanian S, Yan S, Tyagi RD et al (2010) Extracellular polymeric substances (EPS) producing bacterial strains of municipal wastewater sludge: isolation, molecular identification, EPS characterization and performance for sludge settling and dewatering. Water Res 44:2256–2266

Chachkiani M, Dabert P, Abzianidze T et al (2004) 16S rDNA characterization of bacterial and archaeal communities during start-up of anaerobic thermophilic digestion of cattle manure. Bioresour Technol 93:227–232

Chen S, Niu L, Zhang Y (2010) *Saccharofermentans acetigenes* gen. nov., sp. nov., an anaerobic bacterium isolated from sludge treating brewery wastewater. Int J Syst Evol Microbiol 60:2735–2738

Chouari R, Paslier DL, Daegelen P et al (2005) Novel predominant archaeal and bacterial groups revealed by molecular analysis of an anaerobic sludge digester. Environ Microbiol 7:1104–1115

Dagar SS, Kumar S, Mudgil P et al (2011) D1/D2 domain of large-subunit ribosomal DNA for differentiation of *Orpinomyces* spp. Appl Environ Microbiol 77:6722–6725

Deborah T, Silvia V, Marco A et al (2011) The role of different methanogen groups evaluated by Real-Time qPCR as high-efficiency bioindicators of wet anaerobic co-digestion of organic waste. AMB Express 1:28

Demirel B, Scherer P (2008) The roles of acetotrophic and hydrogenotrophic methanogens during anaerobic conversion of biomass to methane: a review. Rev Environ Sci Biotech 7:173–190

Demirel B, Scherer P (2011) Trace element requirements of agricultural biogas digesters during biological conversion of renewable biomass to methane. Biomass Bioenerg 35:992–998

Edelmann W, Joss A, Engeli H (1999) Two step anaerobic digestion of organic solid wastes. In: Mata Alvarez J, Tilehe A, Cecchi J (eds) 11 international symposium on anaerobic digestion of solid wastes. International Association of Water Quality, Barcelona, pp 150–153

Feng XM, Karlsson A, Svensson BH et al (2010) Impact of trace element addition on biogas production from food industrial waste-linking process to microbial communities. FEMS Microbiol Ecol 74:226–240

Goebel BM, Stackebrandt E (1994) Cultural and phylogenetic analysis of mixed microbial populations found in natural and commercial bioleaching environments. Appl Environ Microbiol 60:1614–1621

Handelsman J (2004) Metagenomics: application of genomics to uncultured microorganisms. Microbiol Mol Biol Rev 68:669–685

Higgins MJ, Novak JT (1997) Characterization of exocellular protein and its role in bioflocculation. J Environ Eng 123:479–485

Hofman-Bang J, Zhang D, Westermann P et al (2003) Molecular ecology of anaerobic reactor systems. Adv Biochem Eng Biotechnol 81:151–203

Houghton JI, Quarmby J, Stephenson T (2001) Municipal wastewater sludge dewaterability and the presence of microbial extracellular polymer. Water Sci Technol 44:373–379

Jaenicke S, Ander C, Bekel T et al (2011) Comparative and joint analysis of two metagenomic datasets from a biogas fermenter obtained by 454-pyrosequencing. PLoS One 6:e14519

Karr EA, Ng JM, Belchik SM, Sattley WM, Madigan MT, Achenbach LA (2006) Biodiversity of methanogenic and other archaea in the permanently frozen Lake Fryxell, Antarctica. Appl Environ Microbiol 72:1663–1666

Kato S, Haruta S, Cui ZJ et al (2004) *Clostridium straminisolvens* sp. nov., a moderately thermophilic, aerotolerant and cellulolytic bacterium isolated from a cellulose-degrading bacterial community. Int J Syst Evol Microbiol 54:2043–2047

Klocke M, Mähnert P, Mundt K et al (2007) Microbial community analysis of a biogas producing completely stirred tank reactor fed continuously with fodder beet silage as mono-substrate. Syst Appl Microbiol 30:139–151

Krause L, Diaz NN, Goesmann A, Kelley S, Nattkemper TW, Rohwer F, Edwards RA, Stoye J (2008) Phylogenetic classification of short environmental DNA fragments. Nucleic Acids Res 36:2230–2239

Kröber M, Bekel T, Diaz NN et al (2009) Phylogenetic characterization of biogas plant microbial community integrating clone library 16S-rDNA sequences and methagenome sequence data obtained by 454-pyrosequencing. J Biotechnol 142:38–49

Kumar S, Puniya AK, Puniya M et al (2009) Factors affecting rumen methanogens and methane mitigation strategies. World J Microbiol Biotechnol 25:1557–1566

Kumar S, Sumit SD, Mohanty AK et al (2011) Enumeration of methanogens with a focus on fluorescence in situ hybridization. Naturwissenschaften 98:457–472

Kumar S, Sumit SD, Puniya AK (2012) Isolation and characterization of methanogens from rumen of Murrah buffalo. Ann Microbiol 62:345–350

Kumar S, Sumit SD, Puniya AK et al (2013) Changes in methane emission, rumen fermentation in response to diet and microbial interactions. Res Vet Sci 94:263–268

Leclerc M, Delgènes JP, Godon JJ (2004) Diversity of the archaeal community in 44 anaerobic digesters as determined by single strand conformation polymorphism analysis and 16S rDNA sequencing. Environ Microbiol 6:809–819

Lefauconnier B, Hagen JO, Rudant JP (1994) Flow speed and calving rate of Kongsbreen glacier, 70°N Spitsbergen, Svalbard, using SPOT images. Polar Res 13:59–66

Leung K, Topp E (2001) Bacterial community dynamics in liquid swine manure during storage: molecular analysis using DGGE/PCR of 16S rDNA. FEMS Microbiol Ecol 38:169–177

Liu H, Fang HH (2002) Extraction of extracellular polymeric substances (EPS) of sludges. J Biotechnol 95:249–256

Liu Y, Whitman WB (2008) Metabolic, phylogenetic, and ecological diversity of the methanogenic archaea. Ann N Y Acad Sci 1125:171–189

McHugh S, Carton M, Mahony T et al (2003) Methanogenic population structure in a variety of anaerobic bioreactors. FEMS Microbiol Lett 219:297–304

Nettmann E, Bergmann I, Mundt K et al (2008) Archaea diversity within a commercial biogas plant utilizing herbal biomass determined by 16S rDNA and mcrA analysis. J Appl Microbiol 105:1835–1850

Nettmann E, Bergmann I, Pramschufer S et al (2010) Polyphasic analyses of methanogenic archaeal communities in agricultural biogas plants. Appl Environ Microbiol 76:2540–2548

Nicholas A, Ravella SR, Hobbs PJ (2012) Optimizing anaerobic digestion by selection of the immobilizing surface for enhanced methane production. Bioresour Technol 120:248–255

Pradhan S, Srinivas TNR, Pindi PK et al (2010) Bacterial diversity from Roopkund Glacier, Himalayan Mountain ranges, India. Extremophiles 14:377–395

Reddy GSN, Aggarwal RK, Matsumoto GI, Shivaji S (2000) *Arthrobacter flavus* sp. nov., a psychrophilic bacterium isolated from a pond in McMurdo Dry Valley, Antarctica. Int J Syst Evol Microbiol 50:1553–1561

Riesenfield SC, Goodman RM, Handelsman J (2004) Uncultural soil bacteria are a reservoir of new antibiotic resistance genes. Environ Microbiol 6:981–989

Schattauer A, Abdoun E, Weiland P et al (2011) Abundance of trace elements in demonstration biogas plants. Biosyst Eng 108:57–65

Schnürer A, Zellner G, Svensson BH (1999) Mesophilic syntrophic acetate oxidation during methane formation in biogas reactors. FEMS Microbiol Ecol 29:249–261

Shivaji S, Reddy GSN, Aduri RP et al (2004) Bacterial diversity of a soil sample from Schirmacher Oasis, Antarctica. Cell Mol Biol 50:525–536

Shivaji S, Kumari K, Kishore KH et al (2011) Vertical distribution of bacteria in a lake sediment from Antarctica by culture-independent and culture-dependent approaches. Res Microbiol 162:191–203

Sirohi SK, Pandey N, Singh B et al (2010) Rumen methanogens: a review. Indian J Microbiol 50:253–262

Sirohi SK, Singh N, Dagar SS et al (2012) Molecular tools for deciphering the microbial community structure and diversity in rumen ecosystem. Appl Microbiol Biotechnol 95:1135–1154

Sirohi SK, Chaudhary PP, Singh N et al (2013) The 16S rRNA and mcrA gene based comparative diversity of methanogens in cattle fed on high fibre based diet. Gene 523:161–166

Subramanian B, Yan S, Tyagi RD et al (2008) Isolation and molecular identification of extracellular polymeric substances (EPS) producing bacterial strains for sludge settling and dewatering. J Environ Sci Heal-A 43:1495–1503

Tolvanen KES, Karp MT (2011) Molecular methods for characterizing mixed microbial communities in hydrogen-fermenting systems. Int J Hydrog Energy. doi:10.1016/j.ijhydene.2011.01.029

Traversi D, Villa S, Acri M, Pietrangeli B, Degan R, Gilli G (2011) The role of different methnogen groups evaluated by real-time qPCR as high-efficiency bioindicators of wet anaerobic co- digestion of organic waste. AMB Express 1:28

Tsai YL, Olson BH (1991) Rapid method for direct extraction of DNA from soil and sediments. Appl Environ Microbiol 57:1070–1074

Tsai YL, Olson BH (1992) Rapid method for separation of bacterial DNA from humic substances in sediments for polymerase chain reaction. Appl Environ Microbiol 58:2292–2295

Urbain V, Block JC, Manem J (1993) Bioflocculation in activated sludge: an analytical approach. Water Res 27:829–838

Weiss A, Jerome V, Burghardt D et al (2009) Investigation of factors influencing biogas production in a large-scale thermophilic municipal biogas plant. Appl Microbiol Biotechnol 84:987–1001

Westerholm M, Roos S, Schnürer A (2010) *Syntrophaceticus schinkii* gen. nov., sp. nov., an anaerobic, syntrophic acetate-oxidizing bacterium isolated from a mesophilic anaerobic filter. FEMS Microbiol Lett 309:100–104

Wirth R, Kovács E, Maróti G et al (2012) Characterization of a biogas-producing microbial community by short-read next generation DNA sequencing. Biotechnol Biofuels 5:41

Woese CR, Kandler O, Wheelis ML (1990) Towards a natural system of organisms: proposal for the domains of Archaea, Bacteria, and Eucarya. Proc Natl Acad Sci U S A 87:4576–4579

Yeates C, Gillings MR, Davison AD et al (1998) Methods for microbial DNA extraction from soil for PCR amplification. Biol Proced Online 1:40–47

Yu Z, Morrison M, Schanbacher FL (2010) Production and utilization of methane biogas as renewable fuel. In: Vertes AA, Qureshi N, Yukawa H, Blaschek H (eds) Biomass to biofuels: strategies for global industries. Wiley, New York

Zinder SH (1993) Physiological ecology of methanogens. In: Ferry JG (ed) Methanogenesis: ecology, physiology, biochemistry, and genetics. Chapman & Hall, New York, pp 128–206

Part V

Intestinal Disorders and Rumen Microbes

Acidosis in Cattle

21

Rodrigo Dias Lauritano Pacheco
and Gustavo Durante Cruz

Abstract

The ruminants do not directly compete with human beings for food resources, unlike simple-stomach animals, such as poultry or swine. However, if high levels of animal productivity are necessary, forage alone cannot sustain it. Thus, it is necessary to feed ruminants with grains and/or coproduct agricultural production units. Digestive disorders are the second most commonly reported health issues in North American and Brazilian feedlot operations, while respiratory issues tend to rank toward the top of the list of health problems. Ruminal fermentation is a result of fine-tuned cooperation between the host animal and the rumen microorganisms. Rumen provides an optimum environment for microbial population cultivation while fermentation end products serve as an energy source and microbes supply high-quality protein for the host animal. The production of end products differs according to the diet consumed by the animal, mainly due to microorganism carbohydrate preferences and affinity. Besides being the most studied digestive disorder in cattle, ruminal acidosis still presents blind spots. The first intriguing fact is individual susceptibility. If one challenges animals from same herd and similar background with a high-concentrate diet protocol, animals will present a range of acidosis symptoms, ranging from no symptoms to moderate and even severe ruminal acidosis within the same feeding period. Further studies should be conducted to determine whether the differences in acidosis symptoms in one herd are based on behavior, physiological, microbial, or multifactorial effects.

R.D.L. Pacheco
Universidade Federal do Mato Grosso,
Cuiabá, MT, Brazil

G.D. Cruz (✉)
Cargill Animal Nutrition, Elk River, MN, USA
e-mail: Gustavo_Cruz@cargill.com

A.K. Puniya et al. (eds.), *Rumen Microbiology: From Evolution to Revolution*, DOI 10.1007/978-81-322-2401-3_21

Keywords
Ruminants • Acidosis • Sustainability

21.1 Overview of Ruminal Acidosis in Intensive Cattle Production

For millions of years, as a result of natural selection, ruminants developed the capability to obtain the greatest part of its energy, necessary to reproduce and maintain vital function, by ruminal microbial fermentation. Ruminal microorganisms produce enzymes that degrade plant complex carbohydrates. Such biological strategy enables ruminants to eat fibrous plant materials, which cannot sustain vital functions for most single-stomach animals. Consequently, ruminants present competitive advantages during the natural selection process, either in natural conditions or in production systems.

The ruminants do not directly compete with human being for food resources, unlike simple-stomach animals, such as poultry or swine. However, if high levels of animal productivity are necessary, forage alone cannot sustain it. Thus, it is necessary to feed ruminants with grains and/or coproducts.

"An Essay on Principle of Population" (Malthus 1798) states that population growth occurs in geometrical progression, while natural resources, such as food sources, grow linearly. According to Malthus, at some point in the future, all of mankind will perish from starvation, illness, and/or war, due to the lack of food production.

Even after two centuries, Malthus' theory is still widely accepted by scientists in the biological disciplines. For example, acidosis, the central topic in this current chapter, can be used as a great example of Malthus' theory. Unbridled bacteria growth leads to elimination of many types of acid-intolerant bacteria (war) and to faster depletion of ruminal carbohydrates (the lack of food) when compared to normal fermentation. Nevertheless, Malthus' theory does not apply to human population because scientists cannot predict the direct impact technology will have on food production. As an example the Green Revolution, after World War II, was able to eliminate or at least postpone Malthus' apocalyptical prediction.

Nonetheless, according to FAO (2002) reports, it will be necessary for a second Green Revolution to go until 2030 to end world hunger. Interestingly, FAO's study reported that world population growth is decelerating, due mainly to the influence of modern lifestyle globalization. The most intriguing facts roused by the reports are related to the physical condition of the land. The most productive areas of food production are affected by erosion. In other words, performance of these areas does not match their highest production potential. Moreover, they are not managed under perspectives of sustainable concepts.

Based on the world conditions, higher yield production on the available arable lands and maintenance on the currently protected areas are necessary for complete hunger eradication around the world. These land and yield issues can often be observed in high food production areas, especially in tropical land areas. As an example, Meyer et al. (2011) reported that the amount of specified for use as pasture in Brazil increased 0.07 % per year, from 1970 to 2006. During the same period, the number of cattle in the country increased from 78.6 million to 171.6 million. Thus, by improving pasture management and supplementing the actual land used to feed the cattle, the cattle-to-area ratio almost doubled, ranging from 0.51 cattle per hectare in 1970 to 1.08 cattle per hectare in 2006.

Moreover, countries with tropical climates, such as Brazil, are investing in feedlot production, which will result in an increased grain inclusion on cattle diet.Temperate countries have been investing in intensive cattle production since the 1950s to better supply food for the post-war society. The metabolic consequences for cattle fed with high-grain diets are well known. For decades the animal science community has debated whether or not feeding ruminants with more energetic diets would achieve higher production levels.

Some of the debates are centered on energetic dense diets being on the verge of either maximum efficiency or almost a nutritional disorder. Additionally, individual susceptibility may contribute to acidosis occurrence in commercial cattle operations that feed animals with such diets, even when provided a minimum level of fiber. Thus, differently from past decades and accordingly with the discussion of the following topics, acidosis may turn to a cosmopolitan digestive disorder, not present only in intensive cattle operations of temperate countries but as a reality of previously pasture cattle production countries.

Research topics focusing on genetic influence (*Bos indicus* vs. *Bos taurus*), individual susceptibility causes, influence of coproducts on traditional grain diets, alternative feed additives, and feeding behavior may elucidate current acidosis blind spots.

The aim of this chapter is to suggest combining current knowledge on acidosis and research topics with the objective of preparing for a future of intensive worldwide cattle production for hunger eradication without deforesting new agricultural frontiers.

21.1.1 Effect on Production Costs

Measurements of costs regarding decreased performance related to acidosis-affected animals are often difficult because of the differences in the possible diagnoses. As it will be more deeply discussed in the sections that follow, a diagnosis of acidosis is classified as either being acute or subacute according to its intensity. Acute acidosis may result in severe diarrhea, "downers" and poor performance, and often death. Thus, it is easier to diagnose and few animals instead of a great proportion of the herd are affected. Production costs are related directly with animal death or its very poor performance in conjunction with treatment.

On the other hand, a subacute acidosis symptom is not easily detected and often leads to a larger portion of a herd being infected. The major response by the animal to subacute acidosis is reduced feed intake with an accompanying reduction in performance (Stock and Britton 1996). Additionally, identification of acidotic animals in commercial herds, which normally allots animals in groups ranging from 100 to 200 heads, is extremely difficult.

Nearly every animal will experience subacute acidosis at least one time during feeding period if they are part of an intensive beef production operation. Reduction in feed intake due to subacute acidosis can vary from 0.11 to 0.45 kg, resulting in profit loss ranging from US$ 1.71 to $7.21 (Stock and Britton 1996). Additionally, Schwartzkopf-Genswein et al. (2003) reported that North American feedlot nutritionists and managers attribute subacute acidosis and reduced performance to erratic feeding behavior and intake by cattle, which is believed to result in deficits between US$15 and $20 per animal due to the animals' deficiency. Similarly, in a simulation with 2012's feed and operational costs for feedlot operation in Midwest Brazil and 10 % decreased feed efficiency due to subacute acidosis, the present authors calculated that acidosis cost resulted in a deficit of R$ 41.23. Based on historical currency exchange rates between US dollars and the Brazilian real, acidosis costs are similar for the temperate country of North America, when compared to tropical South America, for emerging intensive feeding operations.

Digestive disorders are the second most commonly reported health issue in North American feedlot operations, while respiratory issues tend to rank toward the top of the list of health problems (Vogel and Parrot 1994). Similarly, in Brazil, acidosis is the second most reported health issue for feedlots and respiratory difficulties also classified as being the biggest health concern (Millen et al. 2009). In addition, acidosis may lead to a vast range of nutritional disorders. Loads of acid in a rumen can cause ruminal epithelial damage, which decreases short-chain fatty acid absorption and consequently resulting in less energy being available for the animal to obtain nourishment and restricting growth. Moreover, acidosis often leads to a disruption of the rumen epithelium resulting in bacteria gaining access to blood circulation and ultimately causing liver abscesses. Stock and Britton (1996) calculated that severe liver abscess may decrease daily gain by 11 % and feed efficiency by 9 %, require extra carcass trimming, and lead to lost value of condemned livers. An extra cost of US$ 3.00 per animal occurs in pens with 15 % prevalence of liver abscess.

With respect to dairy production, Oetzel (2007) concluded that ruminal acidosis is not only a concern of economic reasons but also for welfare reasons. According to the Oetzel, lameness is one of most valued animal welfare topics in today's dairy herds in which a great proportion is secondary to high-grain feeding. Lameness along with reproductive failure and low milk production are the main causes of involuntary culling and unexplained cow death in dairy herds, all of which can be caused by acidosis.

21.2 Etiology

Ruminal fermentation is a result of fine-tune cooperation between the host animal and rumen microorganisms. Rumina provide an optimum environment for microbial population cultivation while fermentation end products serve as an energy source and microbes supply high-quality protein for host animal. Fermentation of end products differs according to the diet consumed by the animal, mainly due to microorganism carbohydrate preferences and affinity.

Roughage diets are rich in cellulose, intermediary in soluble sugars, and low in starch (Kauffmann et al. 1980). Thus, cellulolytic and saccharolytic bacteria are more active and, consequently, cellulose digestion and fermentation of the soluble sugars by the microbial community result in greater acetate production (Owens and Goetesch 1988). Oppositely, in high-starch diets, the majority of the ruminal microbe population is composed of amylolytic bacteria which compete for substrates (soluble carbohydrates, starch and sugar fermentation products) in lower pH (the opposite occurs in roughage diets); as a result, more propionic acid is produced (Kauffmann et al. 1980).

Steady and health ruminal fermentation is only possible if microbial population is adapted to respective dietary carbohydrates. Thereby, abrupt dietary changes, mainly from roughage to high-concentrate diets, result in a mutualism break between rumen microbial population and host animal. This is the initial trigger for the most studied nutritional disorder present in intensive cattle production: the ruminal acidosis.

Acidosis can be defined as decreased alkali (base excess) in body fluids related to acid (hydrogen ions) content (Stedman 1982). This process is initiated when a non-adapted animal consumes large amounts of rapidly fermentable carbohydrates, in most cases, starch. Hydrolysis of the latter carbohydrate increases glucose in ruminal fluid which is typically low and in other cases is higher than in blood concentration (Galyean and Rivera 2003). Free glucose in ruminal fluid promotes an optimum environmental condition for a fast increase of lactate-producing bacteria populations, mainly *Streptococcus bovis* (Dawson et al. 1997), along with an osmolality increment which decreases the rate of absorption of volatile fatty acids (**VFAs**). Consequently, VFA accumulation leads to decreased ruminal pH.

Lactate (an organic acid ten times stronger than VFAs) accumulation arises from acid-tolerant microorganisms' proliferation which ferments glucose to lactic acid. Additionally, lactate is an intermediary metabolite of propionate production in grain-fed animals. Normally during a steady ruminal fermentation process, a low amount of lactate is observed in rumen fluid. Interestingly, the group of bacteria (lactate utilizers) that converts lactate to propionate or other VFAs is sensitive to low pH (Owens et al. 1998). Moreover, it is known that a pH of 5.5 is optimum for lactate-producing bacteria growth (Slyter 1976; Nagaraja 2003) such as *S. bovis* and *Lactobacillus*. In cases where the pH is 5.5 or lower (normally caused by VFA accumulation), lactate-utilizer bacteria population is injured resulting in lactate accumulation where the expected in pH is similar or lower than the superscripted (DiLorenzo 2004).

When animals are slowly fed with an adapted diet high in grain, lactate accumulation is avoided or at least ameliorated. However in transition or growing diets in which starch content is lower than a finishing diet or in high-intake dairy cows which receive intermediary concentrate rations, acidosis is often observed. This occurs because VFA production overcomes the rate of ruminal absorption (Nagaraja 2003).

Lactate-producing bacteria such as *S. bovis* are facultative anaerobes and can be found in the rumen, cecum, and colon of ruminants. In grass-fed animals, the number of bacteria present in rumen fluids varies from 10^4 to 10^7 per gram, and it can increase for up to 10^{11} per gram when there is an excess of readily fermentable carbohydrates (Nagaraja and Titgemeyer 2007). In a rumen, there are more efficient bacteria in starch utilization; however, *S. bovis* presents fast replication (double its population in 12 min) and utilization of carbohydrates (McCallister et al. 1990), which explains the boom of its population during the initial stage of acidosis. Normally, this microorganism produces acetate and ethanol from glucose, but when pH drops to 5.6 or below, a shift in the fermentation end-product production occurs and L-lactate is produced (Russel and Hino 1985; Finlayson 1986).

The reason for fermentation shift is due to enzymatic activity in pyruvate utilization by *S. bovis*. Inside these bacteria, pyruvate is either transformed to lactate by lactic dehydrogenase or converted in acetyl CoA by pyruvate formate lyase and then to acetate and ethanol. Lactate production mainly occurs because *S. bovis* allows intracellular drop in pH to 5.5, which is the optimum pH for lactic dehydrogenase activity (Russel and Hino 1985).

Based on abovementioned observations, one possible strategy that can be used to prevent acidosis is to prevent the fast growth population of *S. bovis* during the initial stage of acidosis. Nevertheless, unbridle growth of these bacteria happens as a consequence of fast degradable carbohydrate ingestion mostly when non-adapted animals receive high-concentrate diets. Surprisingly, the same concentrations of *S. bovis* are observed in both forage-fed animals and in high-concentrate-fed animals adapted to these diets (Wells et al. 1997).

Acidosis is classified in either acute (clinic) or subacute (subclinic) acidosis. In acute acidosis, the pH drops considerably and lactate accumulates in ruminal fluid. The concentration of 40 mM of lactate (either D or L isomers) in ruminal fluid is considered acute, whereas a pH of 5.2 is considered the threshold between acute and subacute acidoses (Owens et al. 1998; Galyean and Rivera 2003).

In animals affected by acute acidosis, due to a great osmolality difference between blood and rumen fluid, large quantity of acid flows to bloodstream and bicarbonate base excess responsible for blood buffering is suppressed. Osmolality differences cause leg swelling and lameness (Nocek 1997). In critical cases, animals affected die if blood homeostasis is not reestablished.

Related to subacute acidosis, external signs are difficult to detect and symptoms include lower feed intake and performance (DiLorenzo 2004). An average daily pH of 5.6 is considered the threshold between acidotic and a healthy rumen (Owens et al. 1998; Galyean and Rivera 2003). After the development of indwelling measurement systems for continuously measuring rumen pH, subacute acidosis intensity is determined by how long pH drops from threshold, for example, below 5.6 for more than 12 h (Owens et al. 1998). Recent research suggests that the length of time the pH remains lower than a set pH is a better indication than a daily mean pH value when describing treatment effect on rumen health. Recent studies have also shown the benefits in studying the daily fluctuation of rumen pH.

A comparison between the two types of acidosis and their respective implication in animal performance has been presented by Nagaraja and Titgemeyer (2007) and it has been summarized in Table 21.1.

Certain types of feedstuffs provide greater chances to induce acidosis compared to others. Rate, site, and extent of starch digestion are the main considerations in classifying acidosis

Table 21.1 Comparison between acute and subacute acidosis in beef cattle

Items	Acute	Subacute
Clinical signs	Present	Missing
Death	Yes	No
Rumen changes		
pH	<5.2	5.2–5.5
Total organic acids	Increased	Increased
Lactic acid	Increased (50–120 mM)	Normal (0–5 mM)
SCFA	Below normal (100 mM)	High (150–225 mM)
Microorganisms		
Gram-negative bacteria	Decreased	No change
Gram-positive bacteria	Increased	No change
Streptococcus bovis	Initial increased	No change
Lactobacillus spp.	High	High
Lactate producers	High	High
Lactate utilizers	Decreased	High
Bloodstream changes		
pH	Decreased (<7.35)	No change
Lactate		Normal
Bicarbonate	Below normal (<20)	Normal
Outcome		
Ruminitis	Yes	Yes
Laminitis	Yes	Yes
Polioencephalomalacia	Yes	Yes
Liver abscess	Yes	Yes

potential with certain diet ingredients. For example, grain type affects starch degradation. Wheat and barley are more degradable than corn and sorghum due to starch granule protection by protein or starch granule conformation. Additionally, intensive grain processing improves starch ruminal degradation and consequently increases the incidence of acidosis. Likely due to higher moisture content, corn presents greater frequency in potential acidosis cases than dry-rolled corn. On the other hand, coproducts with low starch content, such as pectin present in citrus pulp, are rapidly fermentable but do not produce lactate during its fermentation.

In order to understand the rumen pH oscillation throughout the day and the duration under a certain pH value, Marino et al. (2011) fed crossbred milking cows with a diet with roughage to concentrate ratio of 70:30 and tested different sources of concentrate: high-moisture corn, dry-grounded corn, and citrus pulp. The lowest pH values were observed 4 h after feeding and the recorded values differentiated between the diets with high-moisture corn, the dry-grounded corn, and the citrus pulp. The pH value for the animals fed with dry-grounded corn or high-moisture corn was 5.71, whereas the animals fed with citrus pulp had slightly a higher pH value of 6.1. The length of time the rumen pH was below 6.0 was greater for animals fed with dry-grounded corn (61.5 % during 12 h of measurement), 20.1 % for animals fed with citric pulp, and 33 % for animals fed with high-moisture corn. Due to the rapid fermentation pattern, high-moisture corn presented lower pH values during first postprandial hours and higher pH values later, which helps explain the intermediary time below pH 6.0. Thus, the addition of coproducts and the combination of rapidly and slower fermentable grains may improve performance of high-concentrate-fed cattle.

21.3 Adaptation of Beef Cattle to High-Concentrate Diets

The process of adaptation of newly received cattle is a multifactorial complex, which compresses stress, behavioral, nutritional, microbiological, and physiological effects along with management, feedstuff costs, and logistics. The adaptation process includes recovery of body water, establishing or improving immunity to common virus pathogens, establishing social structure in the pen, and adapting microbes in the rumen to new feeds (Brown and Millen 2009).

Counette and Prins (1981) proposed a practical definition to determine when a ruminant could be considered fully adapted to high-concentrate diets, suggesting that an animal is considered adapted when it ingests concentrate without adverse effects in such levels that would cause ruminal acidosis in non-adapted animals. Moreover, consequences of high-concentrate abuse may persist for several months (Krehbiel et al. 1995).

Rumen microbial population undergoes various changes either in population size or type of most active microorganisms. During the grazing period of life for ruminants, cellulolytic bacteria are the most active. The ruminal fermentation rate of grass-fed animals is steady and slower when compared to high-concentrate-fed animals. Cole and Hutcheson (1981) reported that a 48 h feed and water deprivation resulted in 75 % reduction of ruminal fermentative capacity and it continued to decrease until the fifth day after restriction. Protozoa numbers also decreased as the length of fasting time increased (Loerch and Fluharty 1999). Protozoa are related to starch engulfment, which leads to its slower ruminal degradation by bacteria and results in a more stable fermentation (Nagaraja and Towne 1990). Based on that, one may infer that reduced protozoa numbers in rumen fluid caused by fasting transportation and decreased dry matter intake during the first days after arrival lead to an increase in animal's susceptibility to acidosis.

Brown et al. (2006) conducted a literature review comprising bacterial changes during dietary roughage to concentrate ratio transitions. In the consulted literature, as more concentrate was fed, a greater number of amylolytic bacteria were observed in ruminal fluid while lactate-utilizing bacteria increased dramatically when more than 60 % of concentrate was fed. Complementary research was done by Tajima et al. (2000, 2001), in which they investigated bacterial population changes in cattle in which their feed was switched from 21 % to 82 % of grain. Cellulolytic bacteria were immensely reduced 3 days after the diet switch and continued to decrease for 28 days. Transient population peaks were observed for *S. bovis* and *Prevotella* on the third day although both approached the initial numbers on 28th day. Lactate-utilizer *Selenomonas ruminantium* peaked on the third day and was twice as much in terms of number on the end of 82 % concentrate diet feeding period compared to initial 21 % concentrate diet.

Based on the literature review and research studies cited above, the most critical period for acidosis development is during dietary transitions while cattle adapt to finishing or growing rations. Nutritionists have developed several feed adaptation protocols ranging from duration to methods. The most common approach is the step-up (stepwise) in which the concentrate in the feed increases in predetermined number of diets and days within diets. Another approach is limiting the amount of feed in the final ration and slowly increasing it a few grams per day in the diet according to bunk score.

According to Brown et al. (2006), cattle that adapt from 55 % to 90 % of concentrate in 14 days or less are more likely to reduce performance during the adaptation period and possibly persisting for the entire feeding period. Similarly, Bevans et al. (2005) reported that rapidly adapting cattle for 5 days results in greater rumen pH fluctuation and acidosis occurrence compared to gradual adaptation for 17 days.

Choat et al. (2002) adapted steers to a 90 % concentrate final diet comparing a step-up protocol (ad libitum initial diet composed by 70 % of concentrate, increasing 5 % of concentrate every 5 days), with two restricted adaptation protocols. The first restricted treatment received 1.5 % of body weight in a final diet, with daily increases of

0.45 kg per animal, while the other treatment received 1.25 % of body weight in the final diet with 0.25 kg daily increases of diet offer. Restricted steers consumed 22 % less ration during the adaptation period and presented a decreased average daily gain in the same period compared to ad libitum step-up adapted animals. Nevertheless, similar performance and results were observed in all treatments for the entire feeding period.

Weichenthal et al. (1999) also executed a research study where adapted cattle were fed with a 95 % concentrate diet using two different protocols. In the first treatment, cattle were ad libitum fed with 65 %, 75 %, 82 %, and 90 % of concentrate, fed with each diet for 6 days, until the final diet had 95 % of concentrate. Restricted animals were fed initially with the final diet of 1.77 % of body weight which was increased from 0.23 to 0.45 kg of dry matter per animal over 24 days. Restricted animals consumed 8 % less ration and feed efficiency increased 8 %.

In addition to the number of diets utilized during the adaptation period, management strategies, such as restricting feed offered, could be a potential option to minimize metabolic issues. The objective of restricting feed intake is to avoid excess acid accumulation in the rumen, consequently causing acidosis. Restricting feed intake can cause a reverse effect, causing subacute acidosis, because of the increase in the frequency of bunk visits which would lead to greater intake per meal (Zinn 1995).

Restricting feed offered could also lead to competition among animals and this can be a limiting factor in feedlot that does not have the proper proportion of feed bunk space per animal. Also this strategy could prolong compensatory growth of animals that were in poor-quality pasture during dry periods as is often observed in tropical countries.

Adaptation strategies studied by Brazilian feedlot nutritionist, Millen et al. (2009), included the observation that 48.8 % of the consultants used step-up protocols in commercial feed yards, 2.9 diets to achieve final diet, and feeding each diet for 5.9 days. They found that the average for a whole adaptation period in feedlot cattle in Brazil is 17.1 days, around 4–7 days shorter than the results obtained in North American research trials and commercial operations (Brown et al. 2006; Vasconcelos and Galyean 2007). One possible explanation for the difference in the length of adaption period is the higher inclusion of forage and fibrous coproducts in Brazilian feedlots.

However, it is observed that after high demand of grains for ethanol production in North American countries, more coproducts are included in feedlot diets which results in diets higher in fiber. Conversely, Brazilian feedlots are beginning to invest in intensive grain processing, so it may be possible that in the future acidosis will be of a greater concern in tropical countries using intensive feeding operations.

In summary, adaptation protocols have been widely studied in the United States, but research studies are just starting in tropical countries. Restriction protocols have the tendency to increase performance of feedlot cattle in research trials; however, it is difficult to adopt this management in commercial conditions due to bunk space and cattle competition in newly formed large pens. The increase of coproducts in feedlot diets has increased drastically in the last few years, leading to new study opportunities in the future for learning how coproducts will affect the management and nutritional strategies during the adaptation period.

21.4 Acidosis Versus Roughage

The basic strategy to reduce acidosis is to provide a minimum amount of fiber (physically effective NDF) in order to stimulate rumination and saliva production for ruminal pH and buffering capacity. However, it is possible to induce subacute ruminal acidosis (SARA) by replacing alfalfa hay by alfalfa pellets in moderate concentrate rations of dairy cows (Khafipour et al. 2009a, b; Li et al. 2011). This may happen due to an increase in eating rates and portion sizes (Li et al. 2011). Thus, the physical structure of roughage also plays an important role in the acidosis potential of diets.

As described in the previous section, currently scientific and practical efforts are being made to better understand the roughage potential value of coproducts. Despite its importance in cattle feeding, the role of fiber that is derived from coproducts for feedlot cattle is not well defined (Galyean 2012). By replacing starch with high fermentable fiber present in coproducts such as pectin in beet and citrus pulp, one would expect a more stable rumen pH, due to the fact that pectin does not produce lactate as an intermediary fermentation factor. As more fiber derived from coproducts is added to feedlot diets, a faster rate of passage could be promoted leading to a potential to shift in fiber digestion to lower tract increases (Galyean 2012), unlike traditional roughage sources like alfalfa and corn silage. Moreover, depending on fiber content in feedlot diets (if higher levels are used more than the usual), lower ruminal pH may decrease overall fiber digestion. Lower pH decreases bacterial attachment to cellulose fraction of fiber. In the cases where pH drops below 6.2, cellulolytic bacteria are affected; thus, less fiber digestion is observed, whereas the optimum pH for fiber digestion ranges from 6.2 to 7.1 (Ørskov 1988).

On pasture-fed animals, acidic rumen pH is also a critical issue. Pasture supplementation may be misused if ruminal pH value drops below the suggested pH (cited above), although SARA may not be present in supplemented herds. Malafaia et al. (2003) reviewed Brazilian literature and reported that for tropical grazing beef cattle, supplementation levels under 2 g per kg of live body weight did not negatively affect ruminal pH and remain in the range for normal cellulolytic bacteria activity.

In addition, pasture feeding, the program entitled "semi-feedlot," in which animals are kept in pasture and receive supplementation levels ranging from 1.2 % to 1.7 % of body weight, may present conditions for the establishment of nutritional disorders. On those managements, SARA may be a problem if the following occur: paddocks do not supply enough pasture availability and bunk linear space is not sufficient in conjunction with animals that present high live weight variability within paddocks (leading to competition for supplement).

A whole shelled corn diet is another noteworthy nutritional program currently being used in tropical beef feeding programs. This type of diet has been revealed as one of the most challenging feeding management strategies. In such programs, cattle normally are shifted from a low-quality pasture to a diet composed basically of a whole shelled flint-type corn plus pellets containing protein, urea, and mineral and feed additives. The advantage of this program is to reduce operational investment with feed mixers, particularly appropriate for small- and medium-herd cattle producers. In this diet, whole corn provides the necessary physically effective fiber for rumination. In most cases, restricted dry matter intake is the only option for diet adaptation. Anecdotal field observations have been reported with high incidences of acidosis, lameness, and bloat in herds with adaptation management problems. Currently, studies are being conducted to determine if low inclusions of fibrous ingredients will improve cattle performance and health, as well as feed additive interactions. An example of a change in this type of diet would be combining ionophores with probiotics, essential oils, and non-ionophore antibiotics.

Temperate climate beef production and coproducts found in nontropical locations also play an important role in grazing cattle supplementation in both backgrounding and cow-calf operations. The past severe droughts have led to roughage shortage in North American cattle production, especially in United States, forcing researchers and cattlemen to look for alternative feeding programs. Studies related to roughage substitution for fibrous coproducts and intermittent roughage feeding delivery are being conducted and may raise important information about cattle physiology and management strategies during unusual climate changes.

21.5 Feed Additives

For more than three decades, cattle nutritionists have been studying and recommending several types of feed additives. Currently a variety of additives are present in commercial cattle diets.

Acidosis-preventative feed additives include active compounds with the capability to modify rumen fermentation, like buffers, ionophore antibiotics, non-ionophore antibiotics, direct-fed microbial (**DFM**), essential oils, and antibodies.

Dietary buffers promote productivity and health in ruminants by neutralizing excessive acids in the rumen in situations where natural buffering systems, primarily salivary flow, may be inadequate (Nagaraja 2007). Currently, the most used buffers in ruminant nutrition are sodium bicarbonate, calcium carbonate, and magnesium oxide. Buffers have a direct effect on rumen pH fluid because they neutralize acidity through H^+ sequestration and increase the buffering capacity (Kohn and Dunlap 1998; González et al. 2011) of the fluids. However, research findings may be contradictory, especially for beef cattle diets. According to González et al. (2008a, b), the concentration of 12.5 g/kg of dry matter of sodium bicarbonate promotes higher intake and ruminal pH and also diminishes meal size and increases intake distribution throughout the day for heifers.

Direct-fed microbials (DFM) have been referred to as "probiotics," a term defined by Fuller (1989) as "a live microbial feed supplement, which beneficially affects the host animal by improving its intestinal microbial balance (Krehbiel 2007). Beneficial modification of ruminal fermentation might occur when DFM are fed to ruminants by modification of ruminal microbial population, for example, an increase in lactate-utilizing bacteria may possibly reduce the risk of lactate accumulation (Elam et al. 2003). In several experiments, supplementing feedlot cattle with lactate-utilizing or lactate-producing bacteria has been shown to improve feed efficiency and daily gain (approximately 2.5 %) with little change in dry matter intake and also results in improved milk yield (0.75–2.0 kg/d) with little to no variation in milk compositions in dairy cows (Krehbiel et al. 2003). Similarly with buffer feed additives, research results are contradictory. The use of strains nonnative to the rumen as DFM and the process to maintain such microorganisms viable in challenging production systems are the main critical points of this technology.

Ionophore antibiotics are one of the most recommended classes of acidosis-preventative feed additives. These molecules present the capability to form lipid-soluble complexes with cations and mediate their transport across the lipid barrier of microbial cells. Ionophores are generally bacteriostatic and not bactericidal. The mechanism of bacteriostatic activity of ionophores is related to their ability to alter the flow of cations across the cell membrane (Nagaraja 2007). Target bacteria are mainly Gram-positive such as lactate-producing *S. bovis* and *Lactobacillus ssp.* In addition, ionophores, specifically monensin, cause significant reduction on bloat incidence due to decreased slime and microbial gas production by rumen microbes in a lower pH environment (Bartley et al. 1983; Katz et al. 1986; Nagaraja 2007). Erickson et al. (2003) and Mariani (2010) observed that animals supplemented with monensin presented more frequent feeding behavior and smaller meals, decreasing chances of acidosis development caused by overload intake of rapidly fermentable carbohydrates.

Non-ionophore antibiotics, second to tylosin and virginiamycin (used in combination with ionophores with the objective to reduce liver abscess incidence), is the most potent inhibitor of lactic acid-producing bacteria and lactic acid production in the rumen (Nagaraja et al. 1987; Coe et al. 1999; Nagaraja 2007). It is likely that tylosin and virginiamycin help to control liver abscess in feedlot cattle by providing steady ruminal fermentation and/or by reducing *Fusobacterium necrophorum* numbers in rumen fluid (Coe et al. 1999; Nagaraja 2007).

Recently, consumers' concern about the effects of the antibiotic cross-resistance of human pathogens derived from animals fed with growth-promoting antibiotics has increased. Recently, the use of monensin and virginiamycin as growth-promoting compounds was banned in Europe. The actual risks implicated in the adoption of these compounds are still unknown. The general public's safety concerns, increase in health-related questions and banning, seem to be correlated more with consumer perception and political issues than scientific knowledge of the effects of the growth-promoting hormones in animals. Nevertheless, it opens new frontiers for "natural" feed additive research development. Essential oils derived from spice plants like cinnamon, pepper, and oregano are receiving special attention due to the antimicrobial capacity and may result in altered rumen fermentation processes. In addition, innovative approaches like avian-derived polyclonal antibodies which target lactate-producing and/or liver abscess bacteria may present a greater potential to be used as feed additive in the future for high-concentrate-fed cattle. Polyclonal antibody preparations reduced *S. bovis* numbers in ruminal fluid of steers and increased rumen pH in acidosis-induced heifers (DiLorenzo et al. 2006; Blanch et al. 2009). Additionally, Pacheco et al. (2012) observed similar performance of feedlot bullocks supplemented with polyclonal antibodies preparation or monensin.

21.6 Acidosis Blind Spots

Besides being the most studied digestive disorder in cattle, ruminal acidosis still presents blind spots. The first intriguing fact is individual susceptibility. If one challenges animals from same herd and similar background with a high-concentrate diet protocol, animals will present a range of acidosis symptoms, ranging from no symptoms to moderate and even severe ruminal acidosis within the same feeding period. Further studies should be conducted to determine whether the differences in acidosis symptoms in one herd are based on behavior, physiological, microbial, or multifactorial effects.

We know that genetic differences also play an important role in acidosis blind spots. Pacheco et al. (2012) reported notorious higher incidence of rumenitis in pure *Bos indicus* feedlot bullocks compared to crossbred animals (3/8 Nellore, 5/8 Angus). Distinct evolution between *Bos indicus* and *Bos taurus* cattle may explain the origin of susceptibility. However, the specific biological causes are currently unknown but will be important for designing better feeding programs in the future for *Bos indicus* cattle fed with high-concentrate diets.

In conclusion, better knowledge of the advantages and disadvantages of roughage substitution by coproducts in high-concentrate diets may complete the research topics for complete comprehension of ruminal acidosis. The union of the above-cited research may turn in one of the masterpieces for cattle production issues arising in the future. If the current conditions of mismanaged land continue, the world population will no longer tolerate food production in protected areas, consumers will choose not to eat meat derived from animals raised with antibiotics as growth-promoting compounds, sustainability will be present in every child's life story, and most importantly millions of people will die of hunger in a worldwide growing population. Are we still able to delay Malthus' predictions based on our current scientific and political blind spots?

References

Bartley EE, Nagaraja TG, Pressman ES et al (1983) Effects of lasalocid or monensin on legume or grain (feedlot) bloat. J Anim Sci 56:1400–1406

Bevans DW, Beauchemin KA, Schwartzkopf-Genswein KS et al (2005) Effect of rapid or gradual grain adaptation on subacute acidosis and feed intake by feedlot cattle. J Anim Sci 83:1116–1132

Blanch M, Calsamiglia S, DiLorenzo N et al (2009) Physiological changes in rumen fermentation during acidosis induction and its control using a multivalent polyclonal antibody preparation in heifers. J Anim Sci 87:1722–1730

Brown MS, Millen DD (2009) Practices for adapting feedlot cattle to high concentrate rations. In: Arrigoni MDB, Martins CL, Millen DD, Pacheco RDL (eds) 2nd Brazilian ruminant nutrition conference-Recent advances in feedlot cattle nutrition, 1st edn. Ruminant Nutrition Group (NUTRIR). Botucatu/São Paulo

Brown MS, Ponce CH, Pulikani R (2006) Adaptation of beef cattle to high concentrate diets: performance and ruminal metabolism. J Anim Sci 84:E25–E33

Choat WT, Krehbiel CR, Brown MS et al (2002) Effects of restricted versus conventional dietary adaptation on feedlot performance, carcass characteristics, site and extent of digestion, digesta kinetics, and ruminal metabolism. J Anim Sci 80:2726–2739

Coe ML, Nagaraja TG, Sun YD, Wallace N et al (1999) Effect of virginiamycin on ruminal fermentation in cattle during adaptation to a high concentrate diet and during an induced acidosis. J Anim Sci 77:2259–2268

Cole NA, Hutcheson DP (1981) Influence on beef steers of two sequential short periods of feed and water deprivation. J Anim Sci 53:907–915

Counette GHM, Prins RA (1981) Regulation of lactate metabolism in the rumen. Vet Res Commun 5:101–115

Dawson KA, Rasmussen MA, Allison MJ (1997) Digestive disorders and nutritional toxicity. In: Hobson PN, Stewart CS (eds) The rumen microbial ecosystem, 1st edn. Blackie Academic, New York

DiLorenzo N (2004) Effects of feeding polyclonal antibody preparations against rumen starch and lactic-fermenting bacteria on target bacteria populations and steer performance. Dissertation, University of Minnesota

DiLorenzo N, Diez-Gonzalez F, DiCostanzo A (2006) Effects of feeding polyclonal antibody preparations on ruminal bacterial populations and ruminal pH of steers fed high-grain diets. J Anim Sci 84:2178–2185

Elam NA, Gleghorn JF, Rivera JD, Galyean ML et al (2003) Effects of live cultures of *Lactobacillus acidophilus* (strains NP45 and NP51) and *Propionibacterium freudenreichii* on performance, carcass, and intestinal characteristics, and *Escherichia coli* strain O157 shedding of finishing beef steers. J Anim Sci 81:2686–2698

Erickson GE, Milton CT, Fanning KC et al (2003) Interaction between bunk management and monensin concentration on finishing performance, feeding behaviour, and ruminal metabolism during an acidosis challenge with feedlot cattle. J Anim Sci 81:2869–2879

FAO (2002) World agriculture: towards 2015/30. Summary reports, Rome

Finlayson HJ (1986) The effect of pH on the growth and metabolism of Streptococcus bovis in continuous culture. J Appl Bacteriol 61:201–208

Fuller R (1989) A review: probiotics in man and animals. J Appl Bacteriol 66:365–378

Galyean ML (2012) Traditional and alternative sources of fiber – Roughage values, effectiveness, levels in today's starting and finishing diets. 2012 Plains nutrition council spring conference. http://amarillo.tamu.edu/files/2010/11/2012PNC-Proceedings.pdf. Accessed 27 May 2013

Galyean ML, Rivera JD (2003) Nutritionally related disorders affecting feedlot cattle. Can J Anim Sci 83:13–20

González LA, Ferret A, Manteca X, Calsamiglia S (2008a) Increasing sodium bicarbonate level in high-concentrate diets for heifers. I. Effects on intake, water consumption and ruminal fermentation. Animal 2:705–712

González LA, Ferret A, Manteca X, Calsamiglia S (2008b) Increasing sodium bicarbonate level in high-concentrate diets for heifers. II. Effects on chewing and feeding behaviors. Animal 2:713–722

González LA, Manteca X, Calsamiglia S et al (2011) Ruminal acidosis in feedlot cattle: interplay between feed ingredients, rumen function and feeding behavior (a review). Anim Feed Sci Techno 172(2012):66–79. doi:10.1016/j.anifeedsci.2011.12.009

Katz MP, Nagaraja TG, Fina LR (1986) Ruminal changes in monensin- and lasalocid-fed cattle grazing bloat-provocative alfalfa pasture. J Anim Sci 63:1246–1257

Kauffmann W, Hagesmeister H, Dirksen G (1980) Adaptations to changes in dietary composition, level and frequency of feeding. In: Ruckebush Y, Thivend P (eds) Digestive physiology and metabolism in ruminants, 1st edn. MTP Press, Lancaster

Khafipour E, Krause DO, Plaizier JC (2009a) A grain-based subacute ruminal acidosis challenge causes translocation of lipopolysaccharide and triggers inflammation. J Dairy Sci 92:1060–1070

Khafipour E, Krause DO, Plaizier JC (2009b) Alfalfa pellet-induced subacute ruminal acidosis in dairy cows increases bacterial endotoxin in the rumen without causing inflammation. J Dairy Sci 92:1712–1724

Kohn RA, Dunlap TF (1998) Calculation of the buffering capacity of bicarbonate in the rumen and in vitro. J Anim Sci 76:1702–1709

Krehbiel CR (2007) Effects of direct-fed microbials on rumen function. In: Arrigoni MDB, Martins CL, Millen DD, Pacheco RDL (eds) 1st Brazilian ruminant nutrition conference-Rumen health, 1st edn. Ruminant Nutrition Group (NUTRIR). Botucatu/São Paulo

Krehbiel CR, Britton RA, Harmon DL et al (1995) The effects of ruminal acidosis on volatile fatty acid absorption and plasma activities of pancreatic enzymes in lambs. J Anim Sci 73:3111–3121

Krehbiel CR, Rust SR, Zhang G et al (2003) Bacterial direct fed microbials in ruminant diets: performance response and mode of action. J Anim Sci 81:E120–E132

Li S, Khafipour E, Krause DO et al (2011) Effects of grain-pellet and alfalfa-pellet subacute ruminal acidosis (SARA) challenges on feeding behaviour of lactating dairy cows. Can J Anim Sci 91:323–330. doi:10.4141/CJAS10090

Loerch SC, Fluharty FL (1999) Physiological changes and digestive capabilities of newly received feedlot cattle. J Anim Sci 77:1113–1119

Malafaia P, Cabral LS, Vieira RAM et al (2003) Suplementação protéico-energética para bovinos criados em pastagens: Aspectos teóricos e principais resultados publicados no Brasil. Livest Res Rural Dev 15 (12). Retrieved 7 Aug, 113, from http://www.lrrd.org/lrrd15/12/mala1512.htm

Malthus TR (1798) An essay on the principle of population. J. Johnson, London

Mariani TM (2010) Suplementação de Anticorpos Policlonais ou Monensina Sódica sobre o Comportamento Ingestivo e Desempenho de Bovinos Brangus e Nelore Confinados. Dissertation, Sao Paulo State University

Marino CT, Otero WG, Rodrigues PHM et al (2011) Effects of adding polyclonal antibody preparations on ruminal fermentation patterns and digestibility of cows fed different energy sources. J Anim Sci 89:3228–3235. doi:10.2527/jas. 2010-3062

McCallister TA, Cheng KJ, Lyle M et al (1990) Digestion of barley, maize and wheat by selected species of ruminal bacteria. Appl Environ Microbiol 56:3146–3153

Meyer PM, Barros TA, Cassiano EC et al (2011) Evolution of the land use efficiency by Brazilian bovine husbandry during the last three decades. In: Proceedings of 3rd international conference on sustainable animal agriculture for developing countries, 1st edn. Nakhon Ratchasima

Millen DD, Pacheco RDL, Arrigoni MD et al (2009) A snapshot of management practices and nutritional recommendations used by feedlot nutritionists in Brazil. J Anim Sci 87:3427–3439. doi:10.2527/jas.2009-1880

Nagaraja TG (2003) Response of the gut and microbial populations to feedstuffs: the ruminant story. In: Minnesota nutrition conference, proceedings. Saint Paul

Nagaraja TG, Titgemeyer EC (2007) Ruminal acidosis in beef cattle: the current microbiological and nutritional outlook. J Dairy Sci 90:E17–E38

Nagaraja TG (2007) Manipulation of rumen to minimize problems. In: Arrigoni MDB, Martins CL, Millen DD, Pacheco RDL (eds) 1st Brazilian ruminant nutrition conference-Rumen health, 1st edn. Botucatu

Nagaraja TG, Taylor MB, Harmon DL et al (1987) In vitro lactic acid inhibition and alterations in volatile fatty acid production by antimicrobial feed additives. J Anim Sci 65:1064–1076

Nagaraja TG, Towne G (1990) Ciliated protozoa in relation to ruminal acidosis and lactic acid metabolism. In: Hoshino S, Onodera R, Minato H, Itabashi H (eds) The rumen ecosystem: the microbial metabolism and its regulation. Japanese Scientific Societies Press, Springer, Tokyo

Nocek JE (1997) Bovine acidosis: implications on laminitis. J Dairy Sci 80:1005–1028

Oetzel GR (2007) Subacute ruminal acidosis in dairy herds: pathophysiology, milk fat responses, and nutritional management. http://www.vetmed.wisc.edu/dms/fapm/fapmtools/2nutr/sara1aabp.pdf. Accessed 25 Jun 2013

Ørskov ER (1988) Nutrición proteica de los ruminantes. Acribia, Zaragoza

Owens FN, Goetesch AL (1988) Ruminal fermentation. In: Church DC (ed) The ruminant animal: digestive physiology and nutrition. Prentice-Hall, Englewood Cliffs

Owens FN, Secrist DS, Hill WJ et al (1998) Acidosis in cattle: a review. J Anim Sci 76:275–286

Pacheco RDL, Millen DD, DiLorenzo N et al (2012) Effects of feeding a multivalent polyclonal antibody preparation on feedlot performance, carcass characteristics, rumenitis, and blood gas profile in *Bos indicus* biotype yearling bulls. J Anim Sci 90:1898–1909. doi:10.2527/jas. 2010-3521

Russel JB, Hino T (1985) Regulation of lactate producing in *Streptococcus bovis*: a spiraling effect that contributes to rumen acidosis. J Dairy Sci 68:1712–1721

Schwartzkopf-Genswein KS, Beauchemin KA, Gibb DJ et al (2003) Effect of bunk management on feeding behavior, ruminal acidosis and performance of feedlot cattle: a review. J Anim Sci 81(E. Suppl. 2): E149–E158

Slyter LL (1976) Influence of acidosis on rumen function. J Anim Sci 43:910–929

Stedman TL (1982) Stedmans's medical dictionary. Lippincott Williams & Wilkins. Baltimore/Maryland

Stock R, Britton R (1996) Acidosis. Beef cattle hand book, University of Nebraska, Lincoln. http://ianrpubs.unl.edu/animaldisease/g1047.htm. Accessed 22 May 2013

Tajima K, Aminov RI, Nagamine T et al (2001) Diet-dependent shifts in the bacterial population of the rumen revealed with real-time PCR. Appl Environ Microbiol 67:2766–2774

Tajima K, Arai S, Ogata K, Nagamine T et al (2000) Rumen bacterial community transition during adaptation to a high-grain diet. Anaerobe 6:273–284

Vasconcelos JT, Galyean ML (2007) Nutritional recommendations of feedlot consulting nutritionists: the 2007 Texas Tech University survey. J Anim Sci 85:2772–2781

Vogel GL, Parrot JC (1994) Mortality survey in feedyards: the incidence of death from digestive, respiratory and other causes in feedyards in the great plains. Comp Cont Ed Pract Vet 16:227–234

Weichenthal B, Rush I, Van Pelt B (1999) Dietary management for starting finishing yearlings steers on feed. Beef Cattle Report, Lincoln

Wells JE, Krause DO, Callaway TR et al (1997) A bacteriocin-mediated antagonism by ruminal lactobacilli against *Streptococcus bovis*. FEMS Microbiol Ecol 22:237–243

Zinn RA (1995) Effects of levels and patterns of intake on digestive function in feedlot steers. Proceedings symposium Intake by feedlot Cattle. http://beefextension.com/proceedings/feed_intake95/feed_intake95_23.pdf. Accessed 25 May 2013

22 Urea/Ammonia Metabolism in the Rumen and Toxicity in Ruminants

Amlan Kumar Patra

Abstract

Urea and other nonprotein nitrogen compounds in the ration of ruminants as an economical replacement of vegetable and animal proteins have been investigated for more than 100 years. Initially, the death of animals from urea toxicity owing to insufficient knowledge of urea feeding impeded the widespread use of urea in the diets of ruminants. However, following comprehensive research demonstrating its safety and usefulness in many feeding conditions, urea has commonly become an accepted ingredient for the diets of ruminants. A large body of information and understandings related to the mechanisms of urea and other nonprotein nitrogen compound utilization by ruminal microorganisms has been documented. This chapter discusses urea recycling in the gut, urea and ammonia metabolism by rumen microorganisms, ammonia absorption and mechanisms, and symptoms and treatments of urea/ammonia toxicity in ruminants. The ammonia/urea toxicity problems could easily be prevented through proper employment of scientific knowledge of urea feeding to ruminants discussed in this chapter. Opportunities exist for enhancing anabolic use of urea-N by the microorganisms through modulating urea-N recycling into the rumen by dietary and feeding management, which could decrease N wastage into environment and improve efficiency of N utilization in ruminants.

Keywords

Ruminants • Urea metabolism • Urea recycling • Urea metabolising microorganisms • Ammonia toxicity • Rumen

A.K. Patra (✉)
Department of Animal Nutrition, West Bengal University of Animal and Fishery Sciences, 37 K. B. Sarani, Belgachia, Kolkata 700037, India
e-mail: patra_amlan@yahoo.com

A.K. Puniya et al. (eds.), *Rumen Microbiology: From Evolution to Revolution*, DOI 10.1007/978-81-322-2401-3_22

22.1 Introduction

It has been recognized for more than a century that urea and nonprotein nitrogen compounds may be incorporated in the diets of ruminants. As early as 1891, Ehrenberg et al. (1891) and Zuntz (1891) suggested that nonprotein nitrogen (NPN) compounds could be converted to true protein by ruminal microflora and may replace a portion of protein in the diets of ruminants (cited by Huntington and Archibeque 2000; Kertz 2010). Thereafter, few studies were conducted on the use of NPN in ruminant diets until 1937; it was, however, not widely recognized that urea can be converted to proteins in substantial amounts in ruminants (Reid 1953). In reviewing the earlier research, Krebs (1937) concluded that there remained considerable doubt whether NPN compounds were converted to protein in amounts significant to ruminants.

Research on urea feeding took an impetus after Bartlett and Cotton (1938) reported that satisfactory growth of young cattle resulted when urea supplemented protein in the diet. Hart et al. (1939) showed that replacing vegetable protein with urea or ammonium bicarbonate resulted in normal growth in growing cattle and inclusion of soluble sugars to the diet improved the utilization. A number of studies, as reviewed by Reid (1953), subsequently demonstrated that urea-nitrogen fed to ruminants was definitely retained in the body and that the tissues of the growing animals were of normal composition. Digestion studies showed that urea supplements sometimes increased the digestibility of cellulose and crude fiber of low-protein rations. Balance studies provided evidence of increased nitrogen retention in ruminants that gained body weight with supplemental urea. In vitro fermentation techniques and analyses of rumen ingesta showed that concentrations of urea or ammonia decreased, while true protein content increased in the fermentation medium. Chemical and microbiological analyses and the use of tracers proved that urea-nitrogen was, indeed, converted into true protein in the rumen, which subsequently appeared as tissue and milk proteins.

Most extensive research had been conducted on urea or NPN feeding to ruminants throughout the world in animal nutrition. Initially, the death of animals from urea toxicity owing to insufficient knowledge of urea feeding impeded the widespread use of urea in the diets of ruminants. However, following comprehensive research demonstrating its safety and usefulness in many feeding conditions, urea has commonly become an accepted ingredient for the diets of ruminants. This chapter focuses on urea and ammonia metabolism in the rumen in relation to ammonia toxicity in ruminants.

22.2 Ruminal Urea/Ammonia Metabolism

22.2.1 Urea/Ammonia Pool in the Rumen

Ammonia and urea arise in the rumen through diet, urea recycling via ruminal wall, and salivary secretion. Transfer of urea into rumen occurs by simple diffusion down the concentration gradient, which is greatly variable depending upon dietary and rumen environmental factors (Owens and Bergen 1983). The transport across the rumen epithelium is also mediated via urea transporters in the luminal and basolateral membrane of the epithelium (Abdoun et al. 2006). Urea transporters in the rumen wall are differentially expressed depending on dietary N concentration (Marini and Amburgh 2003). In ruminants, the amount of urea-N recycled to the gut (as a proportion of total hepatic urea-N output) varies between 29 % and 99 % and N transfer across the gut can be much greater than N intake (Kiran and Mutsvangwa 2007). This mechanism is proposed to have constant supply of nitrogen to preserve rumen microbial population when the nitrogen supply in feed is limited. Urea-N recycling to the gut is influenced by several dietary and ruminal factors. Amount of N intake (Marini et al. 2004), total dry matter intake (Sarraseca et al. 1998), type and frequency of feeding dietary N that is degraded in the rumen (Wickersham et al. 2008; Rémond et al. 2009; Kiran and Mutsvangwa

2010), ruminal fermentable carbohydrate intake, and organic matter digestibility (Kennedy and Milligan 1980) are the major dietary factors regulating the proportion of hepatic urea-N output returning to the gut. Concentrations of ammonia-N and volatile fatty acids, particularly butyrate, urease activity, ruminal CO_2 and pH, and plasma urea concentration also exert an important role in trans-epithelial movement of blood urea-N into the rumen (Kennedy and Milligan 1980; Kiran and Mutsvangwa 2010). Ruminal ammonia-N concentration is negatively correlated with urea-N transfer to the rumen (Kennedy and Milligan 1980), while greater urease activity, butyrate, and CO_2 increase urea transport through rumen wall (Huntington and Archibeque 2000; Abdoun et al. 2010). Defaunation may enhance the urea cycling as a proportion of endogenous urea synthesis through the rumen wall (Kiran and Mutsvangwa 2010).

As outlined in the review by Lapierre and Lobley (2001), contributions from salivary flow to urea-N entry to the rumen can vary between 15 % and 100 % depending on the type of the diet. Depending on the dietary and ruminal factors, 19–96 % of endogenous urea production may be recycled to the gut, 15–94 % of the recycling may be transferred in saliva, and 25–90 % of urea degraded in the gut may be degraded in the postruminal digestive tract (Huntington and Archibeque 2000). Salivary flow of urea-N into the rumen as a percent of total urea-N entry to the gut was 36 % in forage-fed (Taniguchi et al. 1995) and 16 % in concentrate-fed (Guerino et al. 1991) ruminants. Urea excreted in the urine represents from 25 % to 60 % of endogenous urea production in ruminants (Huntington and Archibeque 2000).

Ammonia is also derived from degradation of dietary protein depending upon source. Amino groups are also split from amino acids and from intact proteins and used by bacteria in the same manner. The solubility of natural proteins varies highly, and thus the rate at which they are hydrolyzed by bacteria differs appreciably. For the less soluble proteins, the process of ammonia liberation is much less rapid and fairly large proportions of the protein may pass through the rumen to the abomasum without being broken down. There are several major known protein-degrading rumen bacteria, i.e., *Prevotella ruminicola, Ruminobacter amylophilus, Butyrivibrio fibrisolvens, Selenomonas ruminantium, Prevotella bryantii*, and *Streptococcus bovis*, and amino acid-fermenting bacteria, i.e., *Clostridium aminophilum, Clostridium sticklandii, Megasphaera elsdenii, Prevotella ruminicola, Butyrivibrio fibrisolvens, Selenomonas ruminantium*, and *Streptococcus bovis* (Russell 1998; Wallace 1996). Degradation of feed protein and predation of bacteria by protozoa might play a major role in protein degradation in the rumen influencing the ammonia concentration in the rumen (Wallace et al. 1987). The inhibition of the growth of the protein-degrading bacteria and protozoa may decrease protein degradation and metabolism in the rumen, consequently ruminal ammonia concentration (Patra and Saxena 2009; Patra and Yu 2014).

22.2.2 Hydrolysis of Urea

Upon entering into the rumen, urea is rapidly hydrolyzed to ammonia within 30 min to 2 h (Rekib and Sadhu 1986) by the urease enzyme produced by the ruminal microorganisms.

22.2.2.1 Bacterial Urease Enzyme

The urease activity depends upon the concentration of urea entering into the rumen. It has been noted that the magnitude of the urease activity was a direct function of the amounts of urea infused. The urease activity may be repressed with high ammonia concentrations and complex organic nitrogen sources in the rumen fluid. Wozny et al. (1977) noted that the activity of the urease in two anaerobes (*Selenomonas ruminantium* and *Peptostreptococcus productus*) was depressed two- to six-folds by the inclusion of 50 mM urea as compared to inclusion of none. There is evidence that glutamine synthatase can regulate the production of urease in *Selenomonas ruminantium* (Smith and Bryant 1979) and the activities of both these enzymes could be increased manyfolds when ammonia was limiting. In general, urease activity increased 2–6 h after feeding and the increases were greater with roughage diets. The purified urease enzyme was inhibited by divalent cations (Mn^{2+}, Mg^{2+}, Ba^{2+}, Hg^{2+}, Cu^{2+}, Zn^{2+}, Cd^{2+}, Ni^{2+}, and Co^{2+}) (Mahadevan et al. 1976), while Spears and Hatfield (1978) using incubated ruminal fluid reported that ruminal urease was stimulated by a number of inorganic ions including Mn^{2+}, Ni^{2+}, and Ba^{2+}, but was inhibited by Cu^{2+}, Zn^{2+}, and Cd^{2+}.

Urease activity is mostly localized on rumen wall epithelium with adherent bacteria, as well as in the rumen fluid (Rybosová et al. 1984). The high ureolytic activity of bacteria adhering to the rumen wall in the presence of a low nitrogen intake is assumed to be one of the partial mechanisms of the hydrolysis of blood urea entering the rumen across the rumen wall and its reutilization in the rumen-liver nitrogen cycle in ruminal mucosa. In *in vivo* studies with sheep fed with a low-protein diet (3.7 g N/day), urease activity was found to be the highest in the bacteria adhering to the rumen wall, lower in the rumen fluid bacteria, and lowest in the bacteria adhering to feed particles in the rumen (Javorský et al. 1987). However, the urease activities of bacteria adhering to the rumen wall and the rumen fluid bacteria in sheep on a high-protein diet (21 g N/day) were similar, but significantly lower than in sheep with a low N intake; it was lowest in bacteria adhering to feed particles in the rumen (Javorský et al. 1987). In rusitec fermenter study (Czerkawski and Breckenridge 1982), the ureolytic activity per unit volume was usually higher in compartment 2 (space occupied by microorganisms that are loosely associated with the solids) than in compartment 1 (strained rumen contents) or compartment 3 (space occupied by microorganisms that are tightly associated with solid matrix). Specific urease activity (per unit weight of protein or deoxyribonucleic acid) was much higher in compartment 1 than in compartment 2, which decreased markedly with depth of compartment (Czerkawski and Breckenridge 1982).

22.2.2.2 Ureolytic Bacteria in the Rumen

Following extensive research on the utilization of urea as a replacement of protein and its supplementation in ruminant diet, interests were raised to isolate urea-hydrolyzing microbes for a better understanding of urea metabolism in the rumen. Appleby (1955) and Blackburn and Hobson (1962) reported only a few facultative anaerobes (predominantly staphylococci or micrococci) as being ureolytic. Among the other reported experiments on ureolytic bacteria, Gibbons and Doetsch (1959) isolated an ureolytic bacterium, assigned to the species *Lactobacillus bifidus* from normally fed cattle. Slyter et al. (1968) reported presumptively identified ureolytic isolates of *Propionibacterium* sp., *Bacteroides* sp., *Ruminococcus* sp., *Streptococcus bovis*, and an anaerobic *Lactobacillus* sp. However, these bacteria isolated from cattle fed with semisynthetic, purified diets and the levels of urease activity in the bacteria were not determined. John et al. (1974) isolated an ureolytic strain of *Selenomonas ruminantium* from the rumen of a steer. Screening of over 1,000 strains of rumen bacteria isolated from sheep rumen on different media, Cook (1976) showed that urease activity was apparently confined to species of *Staphylococcus*, *Lactobacillus casei* var. *casei*, and *Klebsiella aerogenes*. The ureolytic strain of *Streptococcus faecium* had a higher urease activity than the other bacteria occurring in higher numbers and possessed urease activity sufficient to account for most of the ureolytic activity of the rumen of roughage-fed sheep. Van Wyk and Steyn (1975) noted that all urease positive isolates were facultative anaerobic, Gram-positive, catalase-positive cocci. Out of ten isolates, nine were identified as *Staphylococcus saprophyticus* and one as *Micrococcus varians*. The facultative anaerobic Gram-positive cocci were probably responsible for a large proportion of the urease activity of the rumen fluid.

Ureolytic activity of bacteria differs among the strains. Lauková and Koniarová (1995) tested several strains of bacteria, including *Selenomonas ruminantium*, *Lactobacillus* sp., *Enterococcus* sp., and *Staphylococcus* sp., from the rumen of domesticated and wild ruminants for their urease activity. It was noted that 56.7 % of *Selenomonas ruminantium* strains and 18.5 % of lactobacilli manifested medium urease activity. Most of the *Enterococcus faecium* (62.2 %) and all of the *Enterococcus faecalis* isolates expressed low urease activity. *Streptococcus bovis* and *Streptococcus uberis* did not produce any urease. All the staphylococci screened were urease-producing strains, mostly with medium or low urease activity.

Urease activity was also detected in many strains of nonselectively isolated rumen species, which include *Succinivibrio dextrinosolvens*, *Treponema* sp., *Ruminococcus bromii*, *Butyrivibrio* sp., *Bifidobacterium* sp., *Bacteroides ruminicola*, and *Peptostreptococcus productus* (Wozny et al. 1977). Most *Peptostreptococcus productus* strains contain urease; however, the uniformity of this feature in the other species noted above is not known. *Propionibacterium, Veillonella*, and *Megasphaera* did not possess urease activity.

22.2.3 Ammonia Utilization

The ammonia can be utilized by the bacteria for synthesis of amino acids required for their growth. Protein synthesis in the rumen by microorganisms is very closely associated with the activity of these same organisms in breaking down cellulose and other carbohydrate materials and in the formation of organic acids as by-products of this fermentation process. There is evidence, however, that a fairly high proportion of the more soluble proteins such as casein will be utilized by bacteria in about the same way as the ammonia from urea. Through the years many attempts were made to determine the optimum concentrations of ammonia in the rumen and to relate it to synthesis of microbial protein. The *in vitro* studies suggest that ammonia concentration required for maximal microbial protein synthesis is approximately 50–60 mg/L (Satter and Slyter 1974; Mehrez et al. 1977), while with *in vivo* studies, Hume et al. (1970) and Pisulewski et al. (1981) obtained best results with 88–133 and 27–100 and mg/L, respectively, which varied depending upon diet. Mehrez et al. (1977) obtained maximal rate of fermentation with 235 mg/L. In all these studies, the concentration of ammonia in the rumen meant the concentration in compartment 1 only. Since the concentration of carbohydrate in this compartment is low, it is unlikely to be the most active metabolic site (Czerkawski and Breckenridge 1982). Utilization of ruminal ammonia-N into the bacterial protein in the rumen is energy dependent, and hence, providing adequate ruminally available energy is associated with lower ruminal ammonia-N concentration and, consequently, increased microbial protein synthesis. Defaunation may also increase urea-N recycling to the GIT and microbial non-ammonia-N supply, thus improving efficiency of N utilization (Kiran and Mutsvangwa 2011).

Utilization of ammonia for synthesis of amino acids by rumen microorganisms involves amination and transamination reactions. When the activities of those involved in ammonia production (urease and NAD and NADP-linked glutamate dehydrogenases) exceed the activities of enzymes those concerned with ammonia utilization (glutamine synthetase, carbamyl phosphokinase, and NADH- and NADPH-linked glutamate dehydrogenases), ammonia concentration in the rumen increases. Ruminal ammonia utilization is also controlled by other factors such as availability of energy and cofactors, substrate levels, and bacterial cell permeability (Allison 1969). Since branched-chain volatile organic acids derive in the rumen mainly from degradation of dietary protein and the deamination of branched-chain amino acids, the feeding of protein free diets can cause depressions in these acids (Oltjen 1969). Concentrations of isobutyrate and isovalerate usually decrease and other volatile fatty acids may also be affected due to urea feeding. Rumen fluid from urea-added treatment contained more total volatile fatty acids, but decreased molar percentages of isobutyrate, isovalerate, and caproate (Czerkawski and Breckenridge 1982). Rumen microorganisms have pathways similar to those of the tricarboxylic acid cycle, which will provide carbon skeletons for amino acid biosynthesis (Allison 1969). However, for synthesis of some amino acids, specific carbon sources are required. For example, isobutyrate, phenylacetate, indole-3-acetate, isovalerate, and 2-methyl-butyrate are precursors of valine, phenylalanine, tryptophan, leucine, and isoleucine, respectively (Allison 1969). When urea is the primary dietary nitrogen source, energy, carbon skeletons, and cofactors required for amino acid biosynthesis may be needed in greater quantities.

22.2.4 Ammonia Absorption

When ammonia is produced too rapidly in the rumen compared to the utilization by the rumen microorganisms, or if the concentration becomes too high, appreciable amounts are absorbed directly into the bloodstream, reconverted to urea in the liver, excreted through the kidneys in the urine, or reenter through the saliva and the ruminal wall into the rumen. The absorption of ammonia occurs in the lipophilic form as ammonia (NH_3) via simple diffusion, the magnitude of which is linearly related to the pH in the ruminal fluid at pH values above 7, while at normal rumen pH of 6–7, ammonia is predominantly absorbed as NH_4^+ via putative potassium channels in the apical membrane (Abdoun et al. 2006). At normal ruminal pH, NH_4^+ is converted to ammonia at the entry site in the rumen epithelium before being absorbed into portal blood. In addition, Abdoun et al. (2007) also suggested that the absorption of NH_4^+ may occur through some transport proteins and the movement of NH_4^+ across the ruminal epithelium is probably regulated by both chemical and electrical gradients. Absorption of both forms of ammonia across the ruminal wall increases with the increase in ruminal pH and total ammonia-N concentrations. Hence, the relative transport rates of ammonia or NH_4^+ are determined by the ruminal pH according to the Henderson-Hasselbalch equation (Abdoun et al. 2006). At ruminal pH of 6.5 and low, which is normally observed in most feeding conditions, most of the ammonia is absorbed in the form of NH_4^+ (Abdoun et al. 2007).

The amount of N absorbed as ammonia ranged from 16 % to 73 % of N intake, which can be several times the amount of N absorbed as amino acids (Parker et al. 1995). Ammonia-N is absorbed across all the sections of the gut and on average 67 % of ammonia-N is absorbed from the reticulo-rumen, while the lower gut, including the small and large intestines, and cecum can account for 33 % (Reynolds and Huntington 1988). However, these proportions vary considerably depending upon dietary chemical composition (Parker et al. 1995). Ammonia-N absorbed across the ruminal wall into the portal blood accounts for up to 50 % of total ammonia-N flow to the liver (Parker et al. 1995). The quantity of ammonia-N absorbed across the ruminal wall is mainly determined by dietary as well as ruminal factors, with the most important factors being dietary protein that is degraded in the rumen, contributions of endogenous urea to the ruminal ammonia-N pool, and dietary ruminally available energy (Reynolds and Kristensen 2008). Under a wide variety of dietary and physiological conditions in growing and lactating cattle, Firkins and Reynolds (2005) concluded that ammonia-N absorption across the gut accounts for about 42 % of dietary N intake. Supply of adequate amount of ruminally available energy is associated with lower ruminal ammonia-N concentration and, consequently, reduced ammonia-N absorption into portal blood.

22.3 Toxicity of Urea

Urea is not normally toxic, but hydrolysis of urea produces ammonia that is toxic to all mammals. Inability of the liver to convert absorbed ammonia from the rumen to nontoxic urea results in an increased concentration of ammonia in the blood, which is responsible for urea toxicity. Any compounds forming ammonia in the rumen exert toxic effects if the rumen ammonia level is highly increased so that the rate of ammonia absorption from the rumen exceeds the capacity of the liver to remove ammonia from blood and thus leading to a rise in peripheral blood ammonia to a level of more than 10 mg/L (McDonald 1958). Toxicity of ammonia in ruminants may be attributed to a direct toxic effect of ammonium ion in body systems (Lewis 1960). Besides, a disturbance of the acid–base balance and a change in electrolyte balances due to high ammonia concentrations in blood may modify the signs of toxicity (Lewis 1960). Indicators of ammonia toxicity include ruminal ammonia concentration above 1,000 mg/L, rumen pH above 8, and blood ammonia concentration above 20 mg/L (Owens and Bergen 1983). Death of animals may usually occur when blood ammonia concentrations exceed 40 mg/L. Several contributing factors e.g., rumen pH, diets, and adaptation to the animals, modify the degree of ammonia/urea toxicity. Rumen pH, rather than rumen ammonia concentrations, is one of most important contributing factors triggering the symptoms of urea toxicity, since rumen pH determines how quickly and how much ammonia is absorbed into the blood. The classical urea toxicity studies of Bartley et al. (1976) showed that rumen pH had a greater correlation ($r=0.317$) with toxicity than rumen ammonia-N concentration ($r=0.039$) and blood ammonia-N had the highest correlation ($r=0.707$) with toxicity. Kertz et al. (1983) demonstrated this finding in a study in which ammonium chloride was dosed into the rumen of cows at urea-equivalent levels. Because ammonium chloride simply dissociates in the rumen, it would not have the rumen pH-elevating effect of urea hydrolysis. Although ammonium chloride considerably increased rumen ammonia, because rumen pH was not elevated, as occurs with urea hydrolysis, the rumen ammonia was essentially trapped, resulting in no significant increases in blood urea-N and ammonia and no decreases in feed intake.

Urea can be tolerated in greater amounts in the presence of readily available carbohydrate feeds such as cereal grains. Animals fed with balanced or high-carbohydrate feeds and those adjusted to urea-containing feeds can handle larger amounts of urea than those subsisting on low-protein roughage (Kertz 2010). Urea may produce harmful effects under unusual conditions such as for starved or fasted animals due to rapid consumption of urea-containing feeds. However, urea toxicity would not be expected in animals that are fed properly with mixed rations containing urea in the recommended amounts. Coombe and Tribe (1960) reported that 75 g of urea per day was not toxic for sheep when it was carefully mixed with hay. Clark et al. (1951) showed that putting urea into the rumen of sheep (10 g or in aqueous solution) when feed was withheld caused an acute intoxication. Toxicity was shown to be associated with increased alkalinity of the rumen contents. Toxicity of urea was reduced when sheep were fed with poor-quality hay and was nonexistent when lucerne hay or casein plus low-quality hay was fed. It was concluded that the toxicity of urea depends on the activity of the rumen flora in utilizing the ammonia and the presence of available carbohydrate. They noted that sheep fed with poor-quality diets were more susceptible than those well-fed sheep.

22.4 Signs and Symptoms

Toxicity of urea feeding to ruminants has been investigated since early 1950s. Ammonia exerts toxic effects on the central nervous system, kidney, and heart, producing symptom of urea toxicity. Subclinical toxicity may alter carbohydrate metabolism in the liver and energetic efficiency of lactation (Owens and Bergen 1983). Acute toxicity of ammonia is immediate threat to the life of animals depending upon the dose of urea feeding. Hart et al. (1939) found that cattle fed for a year with a ration containing 4.3 % urea showed hypertrophy of the kidneys upon slaughter, but there was no evidence of toxicity. Those fed with a ration having 2.8 % urea were normal. Harris and Mitchell (1941) could not detect any evidence of injury to withers fed with a ration having 3.2 % urea. When larger doses were given, urea caused toxic effects and even death of ruminants. Dinning et al. (1948) observed that a single dose of 116 g of urea caused ataxia, severe tetany, a retarded respiration rate, and excessive salivation in cattle. However, feeding 200 g properly mixed in the daily feed did not result in unfavorable effects. Clark et al. (1951) showed that putting urea into the rumen of sheep (10 g or more in aqueous solution) after withholding the feed caused an acute intoxication characterized by atony of the rumen, muscular spasms, and sudden death caused by circulatory failure. Toxicity was shown to be associated with increased alkalinity of the rumen contents. Pathological changes may also be found in the liver, lungs, kidneys, and brain (Hart et al. 1939; Antonelli et al. 2004; Srinivasan et al. 2008). There are no characteristic lesions on postmortem examination, but most cases show generalized exfoliation of mucosa, edema and degeneration of rumen mucosa, necrotic enteritis, congestion and hemorrhages in the brain, and lymphocytic infiltrations of the heart, lungs, and kidney (Randhawa et al. 1989).

Davis and Roberts (1959) described the progressive symptoms of urea toxicity. Following administration of a toxic dose of urea, the animals showed uneasiness, muscle and skin tremors, excess salivation, respiratory difficulties, incoordination or ataxia, bloat, tetany, and death. When blood ammonia concentrations exceeded 40 mg/L, all animals died. The toxic dose was about 0.30 g/kg of body weight given as a drench, whereas an 18 g dose was not fatal. In buffaloes, a urea dose of 1.25 g/kg body weight caused similar symptoms and death after 60–150 min (Randhawa et al. 1989). Coombe and Tribe (1958) and Coombe et al. (1960) showed that sheep could ingest very large amounts of urea (100 g/day) provided that the intake was spread over several hours each day; this was achieved by mixing a solution of urea with the roughage; under these circumstances, rumen pH and ammonia concentration remained low. They noted that dosing with urea resulted in a high ammonia concentration and an elevated pH in the rumen, and when the pH value increased to 7.0, rumination time declined, and at pH 7.3, there was complete rumen stasis. This effect was not due to high ammonia concentration as a similar concentration produced by administration of ammonium chloride did not cause stasis at normal rumen pH. It appeared that a rise in rumen pH could enhance any tendency to ammonia intoxication due to increased rate of absorption from the rumen. Coombe and Tribe (1962) noted that feeding of molasses had the effect of reducing rumen pH and ammonia concentration, which could be of significance in reducing the risk of toxicity from urea.

Lewis (1960) reported experiments on ammonia intoxication with different forms of ammonia feeding to sheep. The metabolic observations showed that there were distinct differences following the intraruminal administration of ammonium chloride, ammonium acetate, or urea. Ammonium chloride produced acidosis, with low blood pH, respiratory hyperventilation, and a consequent low blood bicarbonate level; however, it was considered that these changes were not adequate to explain the toxic signs and that toxicity was probably directly related to the blood ammonia level, with a critical concentration about 8 mg/L. Ammonium acetate had no effect on blood pH, but caused a marked fall in blood bicarbonate, which was considered to be due to a respiratory alkalosis produced by the respiratory stimulation from the blood ammonium ion. The critical blood ammonia concentration for toxicity was about 6 mg/L. Addition of urea led to transitory increases in blood pH and bicarbonate, and symptoms of toxicity occurred when blood ammonia attained a level of about 5 mg/L. Lewis (1960) concludes that the toxicity of urea is almost certainly due to a direct effect of the circulating ammonium ion. Clark et al. (1963) confirmed that a 30 g dose of urea, given as a drench, was fatal to the sheep. However, biuret proved nontoxic to sheep in very large doses (250 g) even after the sheep had become adapted by feeding biuret for 9 weeks. Some 20–30 % of a dose of biuret given per rumen fistula was excreted unchanged in the urine (Gray and Clark 1964).

22.5 Treatments

Acetic acid is an effective antidote, whereas the addition of sodium bicarbonate increased the severity of the symptoms. A 5 % solution of acetic acid or vinegar is an effective cure in many cases if administered orally before severe tetany develops. A case of acute urea toxicity in a buffalo heifer was successfully treated after administering with 2.5 L of 5 % acetic acid followed by 1 L of acetic acid after 30 min (Kulkarni and Kulkarni 2002). Intravenous calcium and magnesium solutions can be effective in decreasing tetany. Using rumen pH and ability to neutralize ammonia as indicators, Oltjen et al. (1964) suggested that acetic acid is several times more effective than glutamic acid in preventing signs of urea toxicity. Though drenching of acetic acid decreases ammonia absorption, it does not reduce ammonia concentration in the blood quickly, and thus, death of animals may occur once tetany develops. However, emptying the rumen contents with surgery may result in rapid decrease of blood ammonia concentration, and possibility of animal survivability increases (Bartley et al. 1976). Animals that exhibit severe toxicity could be treated intravenously with 1 mL/kg body weight of commercial solution of urea-cycle amino acids (improve ammonia metabolism in the liver), 1 mg/kg body weight of furosemide (diuretic), and 20 mL/kg body weight isotonic saline solution (Antonelli et al. 2004). However, acute ammonia toxicity sometimes causes death of animals so rapidly that there is normally not adequate time for many of these treatments.

22.6 Factors Influencing Urea Utilization and Toxicity

22.6.1 Readily Available Carbohydrates

Supply of readily available digestible energy is the important factor influencing the efficiency of urea utilization by ruminants. A low level of protein and high level of starch in the ration favor urea utilization. Rations high in digestible energy (high grains) result in better urea utilization; conversely those that are low in digestible energy (high forage) result in a lowered utilization of urea. Utilization of urea by animals fed with high-forage rations may be improved by the addition of grain or molasses. Soluble sugars and cellulose are inferior to starch as sources of energy for ruminal microorganisms because sugars are rapidly fermented by the rumen microorganisms and cellulose is too slowly degradable (Reid 1953). Urea-molasses block and uromol preparations containing urea and readily available carbohydrate and minerals have proved to be good protein supplements for different ruminants (Dass et al. 1996; Mehra et al. 1998; Toppo et al. 1997).

22.6.2 Frequency of Feeding Urea

A continuous intake of urea will improve its utilization over abrupt or periodic intake (Kertz 2010). Constant supply of urea-N decreases ammonia concentrations in the rumen fluid, which is advantageous for rumen microbial growth and nutrient utilization (Alvarez Almora et al. 2012).

22.6.3 Level of Urea Fed

Low levels of urea are utilized more efficiently with less problems than high levels. Urea may provide up to 3 % of the concentrate ration or up to 1 % of the total ration for milking cows (Reid 1953). However, urea should not provide more than 25 % of the N in rations containing 12 % protein equivalent for fattening lambs and for pregnant or lactating ewes (Reid 1953).

22.6.4 Thorough Mixing of Urea-Containing Supplements into the Daily Feed

If urea-containing supplements are mixed with the entire daily ration, the intake of urea at any one time will not likely be great, and the ability of the microbes to synthesize protein will not be exceeded. However, small quantities of urea undiluted by feed (116 g in cattle and 10 g in sheep) and introduced suddenly into the rumen resulted in rapid onset of toxicosis, whereas 180–272 g urea was consumed daily by beef calves or cows without toxicosis when fed along with hay or corn silage (Reid 1953).

22.6.5 Adequate Supply of Phosphorus, Sulfur, and Trace Minerals

Substitution of urea for natural protein sharply changes the quality and quantity of minerals available for ruminal bacteria and animals. Although needed only in small quantities, these elements are necessary building blocks for microbial protein synthesis. Addition of methionine or S has improved the retention of N by ruminants fed with urea-containing rations (Reid 1953). These often are found in many urea-containing supplements.

22.6.6 Amount and Solubility of Proteins

Urea is fairly inferior for dairy and beef calves fed with rations containing 12 % or more of protein equivalent, of which three-fourths of the N is supplied by conventional protein sources (Reid 1953). Different vegetable and animal protein sources have different solubilities or rates of hydrolysis in the rumen. Bacteria may prefer highly soluble and readily hydrolyzable protein rather than urea in the ration; thus, dietary proteins may be competitive with urea. Supplementation of tannins or tannin-containing feeds may reduce the degradability of protein (Patra and Saxena 2011), which might be beneficial to decrease ammonia concentrations in the rumen due to urea feeding.

22.6.7 Urease Inhibitors or Slow-Release Urea Products

Slowing the urea hydrolysis may minimize ammonia wastage and improve utilization. Several inhibitors of ruminal urease activity substantially reduce urease activity; but, inhibitors usually provide a short-term effect because remaining urease capacity is still great enough to hydrolyze ruminal urea and possibly because of microbial adaptation to the inhibitors (Whitelaw et al. 1991; Ludden et al. 2000). Coated urea or slow-release urea products as N supplements have shown equal performance to urea supplements without the potential hazards associated with feed-grade urea (Taylor-Edwards et al. 2009; Kertz 2010). Slow-release urea products provide constant supply of ammonia to rumen microorganism for their growth, which also improves nutrition utilization for low-quality forages and reduces plasma ammonia concentrations (Ribeiro et al. 2011; Huntington et al. 2009).

22.7 Conclusions

Urea and other NPN sources in the ration of ruminants as an economical replacement of vegetable and animal proteins have been investigated for more than 100 years. A large body of information and understandings related to the mechanisms of urea and other NPN utilization by ruminal microorganisms has been documented. Conventionally, it has been recommended that urea should not contribute more than 25 % of total dietary protein, not more than 3 % of concentrate, and not more than 1 % of the total ration (Reid 1953). On the safer side, Kertz (2010), however, suggests that a more reasonable recommendation for feeding urea to adult cattle is 1 % of the concentrate, 135 g/animal daily, and not more than 20 % of total protein, counting other added NPN sources, which would be appropriate under the most adverse conditions for maintaining normal feed intake. Depending on other dietary situations, the urea source and ration balancing programs, these levels may be safely exceeded. The ammonia/urea toxicity problems could easily be prevented through proper employment of knowledge of urea feeding to ruminants. As most of intake N is excreted through urine as urea, opportunities exist for enhancing anabolic use of urea-N by the microorganisms through modulating urea-N recycling into the rumen, which could decrease N wastage into environment, and improve efficiency of N utilization in ruminants.

References

Abdoun K, Stumpff F, Martens H (2006) Ammonia and urea transport across the rumen epithelium: a review. Anim Health Res Rev 7:43–59

Abdoun K, Stumpff F, Martens H (2007) Ammonia and urea transport across the rumen epithelium: a review. Anim Health Res Rev 7:43–59

Abdoun K, Stumpff F, Rabbani I et al (2010) Modulation of urea transport across sheep rumen epithelium *in vitro* by SCFA and CO_2. Am J Physiol Gastrointest Liver Physiol 298:G190–G202

Allison MJ (1969) Biosynthesis of amino acids by ruminal microorganisms. J Anim Sci 29:797–807

Alvarez Almora EG, Huntington GB, Burns JC (2012) Effects of supplemental urea sources and feeding frequency on ruminal fermentation, fiber digestion, and nitrogen balance in beef steers. Anim Feed Sci Technol 171:136–145

Antonelli AC, Mori CS, Soares PC et al (2004) Experimental ammonia poisoning in cattle fed extruded or prilled urea: clinical findings. Braz J Vet Res Anim Sci 41:67–74

Appleby C (1955) The isolation and classification of proteolytic bacteria from the rumen of the sheep. J Gen Microbiol 12:526–533

Bartlett S, Cotton AG (1938) Urea as a protein substitute in the diet of young cattle. J Dairy Res 9:263–272

Bartley EE, Davidovich AD, Barr GW et al (1976) Ammonia toxicity in cattle. I. Rumen and blood changes associated with toxicity and treatment methods. J Anim Sci 53:835–841

Blackburn TH, Hobson PN (1962) Further studies on the isolation of proteolytic bacteria from the sheep rumen. J Gen Microbiol 29:69–81

Clark R, Oyaert W, Quin JI (1951) Studies on the alimentary tract of the Merino sheep in South Africa. XXI. The toxicity of urea to sheep under different conditions. Onderstepoort J Vet Res 25:73–78

Clark R, Barrett EL, Kellerman JH (1963) A comparison between nitrogen retention from biuret and urea by sheep on a low – protein roughage diet. J S Afr Vet Med Assoc 34:419–423

Cook AR (1976) The elimination of urease activity in *Streptococcus faecium* as evidence for plasmid coded urease. J Gen Microbiol 9:49–58

Coombe JB, Tribe DE (1958) Toxicity of urea to sheep. Nature (London) 182:116–117

Coombe JB, Tribe DE (1960) The effect of urea on the utilization of low-quality roughage by the ruminant. Proc Aust Soc Anim Prod 3:83–85

Coombe JB, Tribe DE (1962) The feeding of urea supplements to sheep and cattle: the results of penned feeding and grazing experiments. J Agric Sci 59:125–141

Coombe JB, Tribe DE, Morrison JWC (1960) Some experimental observations on the toxicity of urea to sheep. Aust J Agric Res 11:247–256

Czerkawski JW, Breckenridge W (1982) Distribution and changes in urease *(EC* 3.5.1.5) activity in rumen simulation technique (Rusitec). Br J Nutr 47:331–348

Dass RS, Verma AK, Mehra UR (1996) Effect of feeding urea molasses liquid diet on nutrient utilization, rumen fermentation pattern and blood profile in adult male buffaloes. Buffalo J 12:1–2

Davis GK, Roberts HF (1959) Urea toxicity in cattle, Bulletin No. 611. Florida Agricultural Experiment Station, Gainesville

Dinning JS, Briggs HM, Gallup WD et al (1948) Effect of orally administered urea on the ammonia and urea concentration in the blood of cattle and sheep, with observations on blood ammonia levels associated with symptoms of alkalosis. Am J Physiol 153:41–46

Ehrenberg P, Nitsche H, Muller J (1891) Report on protein substitutions in feeding experiments at Bettlerm (translated title). Z Tierernahr Futtermittlek 1:33–40

Firkins JL, Reynolds CK (2005) Whole animal nitrogen balance in cattle. In: Pfeffer E, Hristov A (eds) Nitrogen and phosphorus nutrition of cattle and environment. CAB International, Cambridge, MA, pp 167–185

Gibbons RJ, Doetsch RN (1959) Physiological study of an obligately anaerobic ureolytic bacterium. J Bacteriol 77:417–428

Gray RS, Clark R (1964) The excretion of biuret in the urine of sheep fed biuret. Onderstepoort J Vet Res 31:91–96

Guerino F, Huntington GB, Erdman RA (1991) The net hepatic flux of metabolites and oxygen consumption in growing beef steers given postruminal casein. J Anim Sci 69:387–395

Harris LE, Mitchell HH (1941) The value of urea in the synthesis of protein in the paunch of the ruminant. 1. In maintenance. J Nutr 22:167–182, 2. In growth J Nutr 22:183–196

Hart EB, Bohstedt G, Deobald HJ et al (1939) The utilization of simple nitrogenous compounds such as urea and ammonium bicarbonate by growing calves. J Dairy Sci 22:785–798

Hume ID, Moir RJ, Somers M (1970) Synthesis of microbial protein in the rumen. I. Influence of the level of nitrogen intake. Aust J Agric Res 21:283–296

Huntington GB, Archibeque SL (2000) Practical aspects of urea and ammonia metabolism in ruminants. J Anim Sci 78:742–749

Huntington GB, Magee K, Matthews A et al (2009) Urea metabolism in beef steers fed tall fescue, orchardgrass, or gamagrass hays. J Anim Sci 87:1346–1353

Javorský P, Rybosová E, Havassy I et al (1987) Urease activity of adherent bacteria and rumen fluid bacteria. Physiol Bohemoslov 36:75–81

John A, Isaacson HR, Bryant MP (1974) Isolation and characteristics of an ureolytic strain of *Selenomonas ruminantium*. J Dairy Sci 57:1003–1014

Kennedy PM, Milligan LP (1980) The degradation and utilization of endogenous urea in the gastrointestinal tract of ruminants: a review. Can J Anim Sci 60:205–221

Kertz AF (2010) Review: urea feeding to dairy cattle: a historical perspective and review. Prof Anim Sci 26:257–272

Kertz AF, Davidson LE, Cords BR et al (1983) Ruminal infusion of ammonium chloride in lactating cows to determine effect of pH on ammonia trapping. J Dairy Sci 66:2597–2601

Kiran D, Mutsvangwa T (2007) Effect of barley grain processing and dietary ruminally degradable protein on urea nitrogen recycling and nitrogen metabolism in growing lambs. J Anim Sci 85:3391–3399

Kiran D, Mutsvangwa T (2010) Effects of partial ruminal defaunation on urea-nitrogen recycling, nitrogen metabolism, and microbial nitrogen supply in growing lambs fed low or high dietary crude protein concentrations. J Anim Sci 88:1034–1047

Kiran D, Mutsvangwa T (2011) Feeding sunflower oil to partially defaunate the rumen increases nitrogen retention, urea-nitrogen recycling to the gastrointestinal tract and the anabolic use of recycled urea-nitrogen in growing lambs. Br J Nutr 105:1453–1464

Krebs K (1937) Der Wirt der Amide bei der Fütterung des Rindes. Biedermanns Z Tierernähr 9:394–507

Kulkarni S, Kulkarni S (2002) Urea poisoning in a buffalo heifer. Buffalo Bull 2:27–28

Lapierre H, Lobley GE (2001) Nitrogen recycling in the ruminant: a review. J Dairy Sci 84:E223–E236

Lauková A, Koniarová I (1995) Survey of urease activity in ruminal bacteria isolated from domestic and wild ruminants. Microbios 84:7–11

Lewis D (1960) Ammonia toxicity in the ruminant. J Agric Sci 55:111–117

Ludden PA, Harmon DL, Huntington GB et al (2000) Influence of the novel urease inhibitor N-(n-butyl) thiophosphoric triamide on ruminant nitrogen metabolism: II. Ruminal nitrogen metabolism, diet digestibility, and nitrogen balance in lambs. J Anim Sci 78:188–198

Mahadevan S, Sauer F, Erfle JD (1976) Studies on bovine rumen bacterial urease. J Anim Sci 42:745–753

Marini JC, Van Amburgh ME (2003) Nitrogen metabolism and recycling in Holstein heifers. J Anim Sci 81:545–552

Marini JC, Klein JM, Sands JM et al (2004) Effect of nitrogen intake on nitrogen recycling and urea transporter abundance in lambs. J Anim Sci 82:1157–1164

McDonald IW (1958) The utilization of ammonia-nitrogen by the sheep. Proc Aust Soc Anim Prod 2:46–51

Mehra UR, Verma AK, Dass RS (1998) Effect of restricted and ad libitum feeding of urea molasses liquid diet (UMLD) on the performance of adult crossbred cattle. Asian Austral J Anim Sci 1:30–34

Mehrez AZ, Orskov ER, McDonald I (1977) Rates of rumen fermentation in relation to ammonia concentration. Br J Nutr 38:437–443

Oltjen RR (1969) Effects of feeding ruminants non-protein nitrogen as the only nitrogen source. J Anim Sci 28:673–681

Oltjen RR, Robbins JD, Davis RE (1964) Studies involving the use of glutamic acid in ruminant nutrition. J Anim Sci 23:767–770

Owens FN, Bergen WG (1983) Nitrogen metabolism of ruminant animals: historical perspective, current understanding and future implications. J Anim Sci 57:498–518

Parker DS, Lomax MA, Seal CJ et al (1995) Metabolic implications of ammonia production in the ruminant. Proc Nutr Soc 54:549–563

Patra AK, Saxena J (2009) The effect and mode of action of saponins on the microbial populations and fermentation in the rumen and ruminant production. Nutr Res Rev 22:204–219

Patra AK, Saxena J (2011) Exploitation of dietary tannins to improve rumen metabolism and ruminant nutrition. J Sci Food Agric 91:24–37

Patra AK, Yu Z (2014) Effects of vanillin, quillaja saponin, and essential oils on *in vitro* fermentation and protein-degrading microorganisms of the rumen. Appl Microbiol Biotechnol 98:897–905

Pisulewski PM, Okorie AIU, Buttery PJ et al (1981) Ammonia concentration and protein synthesis in the rumen. J Sci Food Agric 32:759–766

Randhawa SS, Dhaliwal PS, Gupta PP et al (1989) Studies on clinico-biochemical and pathological changes in the urea-induced acute rumen alkalosis in buffalo calves. Acta Vet Brno 58:225–243

Ribeiro SS, Vasconcelos JT, Morais MG et al (2011) Effects of ruminal infusion of a slow-release polymer-coated urea or conventional urea on apparent nutrient digestibility, *in situ* degradability, and rumen parameters in cattle fed low-quality hay. Anim Feed Sci Technol 164:53–61

Reid JT (1953) Urea as a protein replacement for ruminants: a review. J Dairy Sci 36:955–996

Rekib A, Sadhu DP (1986) Effect of feeding higher doses of urea on the rumen metabolism in goat. Ind Vet J 45:735–739

Rémond D, Bernard L, Savary-Auzeloux I et al (2009) Partitioning of nutrient net fluxes across the portal-drained viscera in sheep fed twice daily: effect of dietary protein degradability. Br J Nutr 102:370–381

Reynolds CK, Huntington GB (1988) Partition of portal-drained visceral net flux in beef steers. 1. Blood flow and net flux of oxygen, glucose and nitrogenous compounds across stomach and post-stomach tissues. Br J Nutr 60:539–551

Reynolds CK, Kristensen NB (2008) Nitrogen recycling through the gut and the nitrogen economy of ruminants: an asynchronous symbiosis. J Anim Sci 86(E Suppl):E293–E305

Russell JB (1998) Strategies that ruminal bacteria use to handle excess carbohydrate. J Anim Sci 76:1955–1963

Rybosová E, Javorský P, Havassy I et al (1984) Urease activity of adherent bacteria in the sheep rumen. Physiol Bohemoslov 33:411–416

Sarraseca A, Milne E, Metcalf MJ et al (1998) Urea recycling in sheep: effects of intake. Br J Nutr 79:79–88

Satter LD, Slyter LL (1974) Effect of ammonia concentration on rumen microbial protein production *in vitro*. Br J Nutr 32:199–208

Slyter LL, Oltgen RR, Kern DL et al (1968) Microbial species including ureolytic bacteria from the rumen of cattle fed purified diets. J Nutr 94:185–192

Smith CJ, Bryant MP (1979) Introduction to metabolic activities of intestinal bacteria. Am J Clin Nutr 32:149–157

Spears JW, Hatfield EE (1978) Nickel for ruminants i. Influence of dietary nickel on ruminal urease activity. J Anim Sci 47:1345–1350

Srinivasan P, Balasubramaniam GA, Sivaseelan S et al (2008) Accidental acute and sub acute urea poisoning in dairy cows. Indian Vet J 85:243–245

Taniguchi K, Huntington GB, Glenn BP (1995) Net nutrient flux by visceral tissues of beef steers given abomasal and ruminal infusions of casein and starch. J Anim Sci 73:236–249

Taylor-Edwards CC, Hibbard G, Kitts SE et al (2009) Effects of slow-release urea on ruminal digesta characteristics and growth performance in beef steers. J Anim Sci 87:200–208

Toppo S, Verma AK, Dass RS et al (1997) Nutrient utilization and rumen fermentation pattern in crossbred cattle fed different planes of nutrition supplemented with urea molasses mineral block. Anim Feed Sci Technol 64:101–112

Van Wyk L, Steyn PL (1975) Ureolytic bacteria in sheep rumen. J Gen Microbiol 91:225–232

Wallace RJ (1996) Ruminal microbial metabolism of peptides and amino acids. J Nutr 126:1326S–1334S

Wallace RJ, Broderick GA, Brammall ML (1987) Microbial protein and peptide metabolism in rumen fluid from faunated and ciliate-free sheep. Br J Nutr 58:87–93

Whitelaw FG, Milnde JS, Wright SA (1991) Urease (EC 3.5.1.5) inhibition in the sheep rumen and its effect on urea and nitrogen metabolism. Br J Nutr 66:209–225

Wickersham TA, Titgemeyer EC, Cochran RC et al (2008) Effect of rumen degradable intake protein supplementation on urea kinetics and microbial use of recycled urea in steers consuming low-quality forage. J Anim Sci 86:3079–3088

Wozny MA, Bryant MP, Holdeman LV et al (1977) Urease assay and urease-producing species of anaerobes in the bovine rumen and human feces. Appl Microbiol 33:1097–1104

Zuntz N (1891) Observations on the digestion and nutritive value of cellulose (translated title). Pflugers Archiv – Eur J Physiol 49:477–483

23 Nitrate/Nitrite Toxicity and Possibilities of Their Use in Ruminant Diet

Devki Nandan Kamra, Neeta Agarwal, and L.C. Chaudhary

Abstract

Nitrate is not a normal component of livestock feed due to toxicity of nitrate/nitrite. Sometimes nitrate might enter into the rumen either accidentally or intentionally incorporated in ration as a source of nitrogen, but the nitrate-reducing bacteria present in the rumen *(Selenomonas ruminantium, Veillonella parvula* and *Wolinella succinogenes)* take care of it by reducing nitrate to nitrite and further to ammonia, which can be used for synthesis of amino acids in the rumen. The number of such bacteria is not plenty enough to take care of excess nitrate in diet, but slow increase in the level of nitrate might help in the adaptation of the rumen microbes, which give capability of tolerating even toxic levels of nitrate in livestock ration. The incorporation of nitrate in the diet of adapted animals can be as high as 30 % protein requirement of the animals without affecting adversely the rumen fermentation and health of the animals. There was no accumulation of nitrite in the rumen liquor, and methaemoglobin levels in blood were also within safe limits, when the animals were adapted to higher levels of nitrate (Unpublished report from this laboratory). Simultaneously, nitrate inclusion in diet also results in decreased emission of methane, making the process of livestock production eco-friendly. Long-term feeding trials are essential before the technique can be recommended to the farmers for practical application.

Keywords

Ruminants • Nitrate/nitrite toxicity • Methane inhibition • Methaemoglobin • *Selenomonas ruminantium* • *Veillonella parvula* • *Wolinella succinogenes*

D.N. Kamra (✉) • N. Agarwal • L.C. Chaudhary
Animal Nutrition Division, ICAR-Indian Veterinary Research Institute, Izatnagar 243122, India
e-mail: dnkamra@rediffmail.com

A.K. Puniya et al. (eds.), *Rumen Microbiology: From Evolution to Revolution*, DOI 10.1007/978-81-322-2401-3_23

23.1 Introduction

The presence of nitrate/nitrite in animal feed is of concern to the well-being of the livestock. The levels of nitrates and nitrites are low in the environment and are formed during the nitrogen cycle. The atmosphere is full of nitrogen, having about 79 % N in the form of N_2 which cannot be used by the living things due to its inertness. Nitrogen cycle includes transformation of nitrogen into inorganic forms (nitrate and ammonia) and then into organic nitrogen. The first step is the fixation of atmospheric nitrogen by nitrogen-fixing bacteria, which convert atmospheric nitrogen into ammonia. The term 'fixation' depicts transformation of atmospheric nitrogen into plant nitrogen. By the action of nitrifying bacteria, the ammonia nitrogen transforms into nitrite and then to nitrate which is the major nutrient assimilated by the plants. Nitrate is an essential nutrient for the growth of the plants, as it is a precursor for protein synthesis. Nitrate is converted to NH_3 by the action of plant nitrate reductase enzymes which are highly active in light but are inhibited in dark and in cloudy or foggy weather. Plants convert most of their nitrate to amino acids through NH_3, and plant proteins are formed from these amino acids. Nitrate levels are generally higher in roots as compared to tops including stem, leaves and tops; nitrate level is highest in the lower stems and leaves of plants with least in the seeds (Walker 1990). The level of nitrite/nitrate in the plants is governed by many factors like soil status, environment, plant type, growth of plants, etc. which in turn affects animal health.

On one hand, nitrate feeding has deleterious effects on the animal, but it is also beneficial as it improves feed conversion efficiency of the animals through reducing methane synthesis and finally improving the feed conversion efficiency of the animal. In the rumen under anaerobic conditions, glucose is fermented, and reduced cofactors like NADH and NADPH are produced through EMP pathway. For reuse in the anaerobic ecosystem, these reduced cofactors have to be oxidized to NAD^+ and $NADP^+$ by electron transfer to the acceptors other than oxygen like CO_2, SO_4, NO_3, fumarate, etc., as a result of which these electron acceptors are reduced to methane, hydrogen sulphide and ammonia. If this reducing power concentrated in the reduced cofactors is not utilized for the production of methane (the major reduced product in rumen under normal conditions), the further fermentation of carbohydrates will be stopped and no release of energy from feed will be possible. Therefore, to continue the release of energy from the feed taken by the animals, methane has to be generated in the rumen. But in this process about 5–15 % of gross energy intake by the animals is wasted in the form of methane (Johnson and Johnson 1995). On oxidation of reduced cofactors, molecular hydrogen is released which is utilized by different electron sinks resulting in negligible hydrogen available in the rumen gases, in spite of the fact that more than 600 l of hydrogen is generated in the rumen of a cow. Various electron or hydrogen sinks available in the rumen are:

- Growth of microbial biomass
- Production of propionate
- Conversion of carbon dioxide to acetate by reductive acetogens
- Hydrogenation of unsaturated fatty acids
- Generation of methane from carbon dioxide
- Reduction of nitrate to nitrite and ammonia
- Reduction of sulphate to sulphide

All the above electron sinks, except biosynthesis of methane, are useful for the rumen microbes and the animal. Thermodynamically, the reduction of nitrate, nitrite or sulphate is more favourable than the conversion of carbon dioxide either to acetate or methane. The utilization of these alternate electron sinks has not attracted attention of researchers for their inclusion in the ration of animals due to their toxic effects on the rumen microbes or the host animal. The inclusion of nitrate in the ration of animals has an added advantage of being a good source of nonprotein nitrogen which can be incorporated into good quality microbial protein.

23.2 Where Do Nitrates Come from in the Rumen?

Nitrates usually are neither a natural ingredient of ruminant diets nor are these molecules added in the diet. Plants are the natural source of nitrate/nitrite. During cultivation of green fodders nitrogen, fertilizers are applied as a source of nitrogen especially in cereal crops like maize, sorghum, oats, etc. Therefore, the level of unutilized nitrates on the surface of plant material (the fodder) serves as a rich source of nitrate in ruminant diet. The level of nitrate in green forages also depends upon the time gap between fertilizer application and forage harvest. Nitrates are not evenly distributed in different parts of the plant. There is a maximum level of nitrate found in stems, which is lower in leaves and the least in seeds (Fjell et al. 1991). The level of nitrates is more in young plants than that in mature plants.

Second source might be a sudden and accidental approach of nitrogenous fertilizer being stored in the field near the vicinity of ruminants. Nitrates and other nitrogenous compounds are sometimes used as adulterants in manufactured complete feeds to increase the level of crude protein.

23.2.1 Nitrate Accumulation in Plants

In normal conditions, nitrate is not accumulated in the plants because nitrates are converted to plant proteins which are not hazardous to animal or environment. But due to some reasons, if the growth of the plant is retarded, there is an accumulation of nitrate in the plants. The activities of plant nitrate reductases are suppressed in cloudy weather or low temperature leading to accumulation of nitrate. Sidhu et al. (2011) assessed nitrate contents in different crops of Punjab in different seasons and found that in hot and humid climate, there is an accumulation of nitrate in the plants as compared to normal climate. Damage caused by hail, frosty or any other physical reasons hampers photosynthesis in the plants, and nitrates accumulate. Strong dry winds also cause nitrate accumulation in the plants due to water scarcity. During heavy rains, just after drought, there is very fast nitrate uptake, but conversion of nitrate into amino acids takes several days; therefore, nitrate accumulates in the plants. But when there is normal growth after heavy rains, the conversion of nitrate to amino acids becomes faster due to more requirements of amino acids, but this might take a week's time. The parts of the plants nearer to ground have lower activity of these enzymes; hence, nitrate levels are higher in these parts as compared to upper parts of the plants. The young plants are more prone to nitrate accumulation because at initial growing stage, nitrate is required for the growth of root and shoot, and at this stage, roots are capable to take up nitrate more than the requirement. Once the plant is grown, its leaves can convert nitrate more efficiently, and there is no accumulation of nitrate. The tendency of plants to accumulate nitrogen is variable in different species, but in extreme conditions, the low potential plants can also accumulate nitrate at high rate. Members of Brassicaceae and Cruciferae like cabbage, sprouts, rape, freshly pulled turnips, etc. contain nitrate ranging from 0.9 to 12 % of DM (Rogers 1999). Annual forage crops are planted in well-fertilized and freshly ploughed lands; therefore, they can accumulate more nitrate than the perennial forages. Also, the annual crops are harvested at an early stage when there was the highest level of nitrate. The nitrate level will remain high in the plant until a new growing tissue utilizes it that is why when the plants die or harvested, its nitrate content remain as such in the plant. Use of herbicides is also one of the reasons for nitrate accumulation because herbicides also interfere in photosynthesis. Cash et al. (2007) reported 1,800 and 3,200 mg nitrate/kg DM in the first crop and the second crop alfalfa (vegetative), respectively, which indicates the impact of harvesting regimen.

Soil is the main source for nitrate accumulation in the plants. Soil status changes due to use of fertilizers, water pollutants, environment, etc. By increased use of nitrogen fertilizers like urea, there is an increase in nitrogen contents in the soil which finally results in high contents of nitrate/nitrite in the plants growing on that soil. The excessive use of animal waste as organic fertilizer beyond the soil holding capacity, results in leaching of soil nitrate to groundwater in a faster rate, which is ultimately taken by the animals. When the soil nitrate increases due to use of nitrogen-rich fertilizers or animal wastes, the plant enzymes responsible for nitrate breakdown are overburdened, not able to break them so speedily, and nitrate accumulates in the plants. Cash et al. (2007) estimated nitrate contents (mg/kg DM) in common feeds grown on soil with different levels of fertilization (67 and 134 kg/ha) and found increased nitrate concentration from 2,149 to 5,613 ppm in oat hay and from 868 to 2,627 ppm in barley hay. Mineral deficiency and imbalances also lead to nitrate accumulation. Fertilizing low sulphur soil with sulphur can check this nitrate accumulation up to some extent.

Presence of nitrate in the plants does not mean that they are always problematic. These plants can be very well used as animal feed; only some management practices are to be followed.

23.3 Management Practices to Minimize Nitrate Toxicity

To avoid nitrate toxicity, some precautions have to be followed. The best and proper way is to analyse all the diet components and water for their nitrate contents, and then with the advice of an expert, formulate diet accordingly to avoid toxicity.

- The N-rich pond waters, N-rich fertilizers and manure should be used in limited quantities. The soil saturated with water or fertilized immediately after rain should not be flooded with pond water.
- Check nitrate level of the plants frequently, and as the nitrate level increases above the acceptable level, reduce the N fertilization of the soil.
- The nitrate level of the crop can be reduced by ensiling. The reduction in nitrate level depends on the days of ensiling and fermentable carbohydrates present in the forages. Ensiling for longer period with high fermentable carbohydrates efficiently reduces nitrate accumulation in the plants. Through ensiling, the nitrate contents of the forages can be reduced by about 30–60 %. For example, lucerne hay contains nitrate (% DM) ranging from 0.06 to 0.6 and silage contains 0.0–0.36, Sudangrass green chop contained 0.12–2.88 and silage contains 0.12–0.24 and Sorghum stalks contain 0.0–1.62 and silage contains 0–0.90 (Faulkner and Hutjens 1989).
- During cool and cloudy conditions, the nitrate accumulation is more; therefore, during this period, N fertilization of the soil should be kept to the minimum possible depending upon the minimal requirement.
- Try to avoid harvesting in cool and cloudy climate. Let the plant be exposed at least for 2 days in bright sunshine, as during sunshine nitrate level falls drastically.
- Selection of crop is also one way to avoid nitrate toxicity. The crops with high nitrate accumulation potential should be avoided. Agricultural crops like barley green feed, wheat green feed, oat green feed, rye green feed, canola plants and sugar beet tops accumulate high levels of nitrate.
- Do not feed animals suddenly on nitrate-rich feeds. First give a little amount of nitrate feed and then increase it gradually. Generally within a week, animal gets adapted to high nitrate feeds.
- A hungry animal should not be allowed to graze on fields of immature plants because it will consume feed very fast which might lead to high nitrate intake. First, feed the animal normal diet in small amounts, then allow the animal to graze only for an hour, and thereafter gradually increase grazing hours every day. Studies have shown that rumen microbes easily get adapted to nitrate, and by continuously feeding of nitrate-rich feed, they efficiently degrade nitrate to ammonia.
- Extra care should be taken when immature forages are used as green chop for animal feeding. The green chop feed should be fed to the animal immediately. Never practise chopping of green in the evening and feed it in the

morning, because it is 10 times dangerous as the heating pile causes conversion of nitrate into nitrite.

- Try to harvest crops at maturity stage. For example, nitrate concentration may be 20–25 % lower in corn silage harvested at a height of 6′ to 8′ as compared to when harvested at 2′ to 4′ height.
- The high nitrate forages should be fed along with grain, grain by-products and other feedstuffs that are high in energy to enhance the conversion of nitrite into ammonia.
- Do not feed animal on high level of fermentable carbohydrate diet. On such diet, due to excessive production of lactic acid, the pH of the rumen goes down which affects protozoa, one of the major nitrate degraders in the rumen.
- Feeding of sulphur can also ameliorate nitrate toxicity. Due to high nitrate (0.4–0.8 %) in the diet, cellulose digestion is affected adversely, but inclusion of sulphate-sulphur at the rate of 0.1–0.4 % maintains the activity of cellulolytic bacteria (Spears et al. 1977).
- It is the rumen microbial ecosystem which is responsible for nitrate degradation; therefore, the animal should be offered well-balanced diet for optimum rumen microbial activity.

23.3.1 Nitrate Toxicity

Nitrate toxicity is a problem for ruminants. In monogastric animals, nitrate is degraded in the intestine where there is hardly any scope of absorption and the major amount is excreted out of the body. In ruminants, the rumen nitrate is converted to nitrite by the enzyme nitrate reductase which then converts into ammonia by the activity of nitrite reductase. There is a group of microbes which are responsible for nitrate degradation. The microbes include *Selenomonas ruminantium, Veillonella parvula* and *Wolinella succinogenes* as the major components. There is an equilibrium between these two enzymes, so there is no accumulation of nitrite in the rumen. But, when there is excess of nitrate because of the feeding of high nitrate feed, the rumen enzymes cannot cope up with this level of nitrate and nitrite gets accumulated, or also if the activity of nitrite reductase is not matching with the nitrate reductase in that case, there is an accumulation of nitrite. This nitrite does not stay in the rumen but enter into the blood stream and converts oxyhaemoglobin into methaemoglobin (met-Hb) which is dangerous. Unlike haemoglobin, met-Hb cannot transport oxygen to the tissues, and the animal develops anaemic anoxia. In the rumen, there is an enzyme named met-Hb reductase which converts back met-Hb into oxyhaemoglobin, but the activity is very low. With the increasing level of met-Hb, the colour of the blood changes. At about 20 % conversion of Hb to met-Hb, the colour of the blood may be light brown; at 50 %, it may be of dark brown, and at 80 %, it is usually dark brown to black. If about 80 % haemoglobin is converted in to methaemoglobin, the animal can die. According to Guerink et al. (1982), the conversion of Hb to met-Hb is regulated by the level of nitrite in the blood. The conversion increases from 24 to 70 % with the increase in nitrite from 50 to 150 mg/L, and the level of nitrite reaches to 250 mg/L; the conversion rate is 100 %. Within 1–2 h of feeding, met-Hb reaches to the peak.

The toxicity of nitrate also depends upon the type of animal (sheep is the most tolerant to nitrate toxicity, O'Hara and Fraser 1975), level of nitrate consumed, period of adaptation, etc. The toxic levels of nitrate are three times higher when fed in mixed feed than when it is given in solution to cattle (Bradley et al. 1940; Crawford et al. 1966).

23.3.2 Symptoms of Nitrate Toxicity

Acute nitrate poisoning occurs due to fast release of nitrate from the plant and its conversion into nitrite in the rumen and usually occurs when there is 80–90 % conversion of Hb into met-Hb. Generally, in acute poisoning, animal dies before showing any symptom; this condition is termed as peracute. An animal dies within a few hours of feeding (Burrows et al. 1987). Anorexia is the main reason for the death of the animal. Clinical signs of acute poisoning include increased heart rate, blue-grey mucous membranes, lacrimation, grinding of the teeth, excess salivation, weakness, respiratory distress, staggered gait, frequent urination, vomiting, muscle tremors, low body temperature, disorientation and an inability to get up.

Chronic nitrate toxicity occurs when the animal is fed on feed containing moderate levels (0.5–1.0 % DM) of nitrate. During chronic nitrate poisoning, generally production-related problems occur and animal did not show any clinical symptom. The production problems include reduced weight gain, lower milk yield (Jainudeen et al. 1964) and reduced feed intake (Farra and Satter 1971). The animal suffering from chronic nitrate toxicity becomes more susceptible to the infections (Johnson et al. 1983). Reproductive problems like abortion and stillbirth due to hormonal imbalances may also take place, or if calf is born live, then nitrate toxicity can be transmitted to the newly born calf, and it may suffer from convulsions and seizures. A stage of an animal, where only 10–20 % ofHb is converted to met-Hb, it is considered as subclinical nitrate poisoning with no symptoms except darkening of mucosae.

23.3.3 Safe Level of Nitrate in Animal Feeds

It is difficult to assign a definite safe level of nitrate that can be fed to the animals, because the tolerance level of nitrate also depends on the frequency of feeding, type of feed given to the animal, etc. As has already been mentioned that when there is a slow and steady increase of nitrate level in the diet, the animal can gradually adapt to it but a sudden high dose of nitrate in the diet may be lethal and the animal can die within few hours of feeding.

In the author's laboratory, buffalo calves of average body weight of 83.5 kg were fed on concentrate mixture (18 % CP and 70 % TDN) and wheat straw in 1:1 ratio. The nitrate (sodium nitrate and potassium nitrate in equal proportion) feeding was started at the rate of 0.5 % of the concentrate mixture and then increased weekly by 0.5 % up to 3 % level in the final diet. The nitrate feeding (3 %) was continued for 6 months. The met-Hb concentration was checked throughout the experiment to know the status of the nitrate degradation. Up to 6 months of feeding, the level of met-Hb did not exceed the safe level, rather it was observed far below the safe level (Table 23.1) (unpublished data). Commonly, to be on the safe side, a feed having 0.3 % nitrate on dry matter basis is considered as safe and a level of 0.3–0.6 % nitrate is moderately safe, and if the animal is in any kind of stress, such type of feed should not constitute more than 50 % of total diet; 0.6–0.9 may be toxic and should not be used as sole feed source, and level above 0.9 % might be fatal for the animal.

Table 23.1 Methaemoglobin (% of blood) level in growing buffaloes fed on nitrate-supplemented diet

Period	Control	3 % nitrate
0 day feeding	ND	0.0145
3 month feeding	ND	0.2245
6 month feeding	ND	0.2836

ND not detectable

23.4 Nitrate-/Nitrite-Reducing Bacteria

As discussed above, reduction of nitrate can serve as an alternate hydrogen sink in the rumen as nitrate-reducing bacteria have higher affinity for hydrogen than reductive acetogenic bacteria or methanogenic archaea. Nitrate-reducing bacteria in the rumen of goat are *Selenomonas ruminantium, Veillonella parvula* and *Wolinella succinogenes* (Stewart et al. 1997). Not all the types of these bacteria are able to reduce nitrate, but only a small fraction of the total number of these three genera has been found to be able to reduce nitrate to ammonia. *S. ruminantium* is the most important of these three genera to accomplish this job. The mixed ruminal microbes from the rumen of goats fed on high roughage diet reduced nitrate and nitrite to a greater extent than the rumen microbes from an animal fed on high concentrate diet (Iwamoto et al. 2001).

Feeding of potassium nitrate at the rate of 6 g/day to goats weighing 25–30 kg resulted in an increase in number of *Veillonella parvula* and *W. succinogenes*. High roughage diet also stimulates these bacteria except *S. ruminantium* which was reported to be higher on high concentrate diet (Asanuma et al. 2002). The already known nitrate-reducing bacteria *V. parvula*, *W. succinogenes*, *S. ruminantium* and *Streptococcus bovis* were more tolerant to nitrite toxicity than other rumen bacteria. Nitrate addition in the medium stimulates these bacteria (Iwamoto et al. 2002) and helps in the adaptation of the ruminants for increased intake of nitrate in the feed. These bacteria might act as alternate hydrogen sinks in the rumen by converting nitrate to ammonia, with nitrite as an intermediate. But the reduction of nitrate is 2–3 times faster than that of nitrite, resulting in an accumulation of nitrite in the rumen. Therefore, reduction of nitrite in the rumen should be enhanced to avoid intoxication of other rumen microbes.

The change in free energy (ΔG) of the reactions plays an important role in reduction reactions going on in the rumen. The conversion of nitrite to ammonia has the minimum ΔG (−371 kJ), conversion of carbon dioxide to acetate has the highest (−9 kJ) and conversion to carbon dioxide to methane (−67 kJ) and nitrate to nitrite (−130 kD) being the intermediate (Iwamoto et al. 1999; Ungerfeld and Kohn 2006). Nitrate has higher affinity with H_2 than CO_2, but has lower affinity in comparison to nitrate. Therefore, when the rumen ecosystem has option for reduction of CO_2, nitrate and nitrite, it prefers to reduce nitrate to nitrite, then nitrite to ammonia and finally CO_2 to methane. Energetically, nitrate reduction to ammonia is more favourable than methanogenesis and propionogenesis as it requires lower threshold energy to be converted to the end products in the latter cases (Ungerfeld and Kohn 2006), and this might be the reason for reduced propionate and increased acetate when nitrate is converted to ammonia.

Sakthivel (2010) studied the effect of three levels of sodium nitrate (0, 5, 10 mM) on fermentation of three diets varying in their wheat straw to concentrate ratio (700:300, low concentrate, LC; 500:500, medium concentrate, MC; and 300:700, high concentrate, HC) in vitro using buffalo rumen liquor as inoculum. Inclusion of sodium nitrate at 5 or 10 mM reduced ($P<0.01$) methane production (9.56, 7.93 vs. 21.76 mL/g DM; 12.20, 10.42 vs. 25.76 mL/g DM; 15.49, 12.33 vs. 26.86 mL/g DM) in LC, MC and HC diets, respectively. Inclusion of nitrate at both 5 and 10 mM reduced ($P<0.05$) gas production in all diets, although IVTD was reduced ($P<0.05$) in only the LC and MC diets. At 10 mM sodium nitrate, 75–118 g/kg of residual nitrate remained in the fluid, and there was an accumulation of nitrite ($P<0.01$) in the medium. In an attempt to eliminate residual nitrate and nitrite in the medium, a number of isolates of nitrate-reducing bacteria were isolated from buffalo adapted to nitrate and introduced individually (3 mL containing 1.2 to 2.3×10^6 cfu/ml) into in vitro incubations containing the MC diet with 10 mM sodium nitrate. Addition of live culture of NRBB 57 resulted in complete removal of nitrate and nitrite from the medium with a further reduction in methane and no effect on IVTD compared to the control treatments containing nitrate with autoclaved cultures or nitrate without any culture. Data indicated that nitrate-reducing bacteria could be used to prevent accumulation of nitrite when sodium nitrate is used to reduce methane emissions in vitro (Sakthivel 2010).

Sar et al. (2005) reported that nitrate inhibits methanogenesis significantly in sheep at the rate of 1.3 g $NaNO_3$/kg $W^{0.75}$ of the animals, but it was accompanied with higher levels of plasma nitrite and blood methaemoglobin. These negative effects of nitrate feeding could be reduced or completely removed by feeding *Escherichia coli* culture at the rate of 150 ml (2×10^{10} cells/ml), which had a capability to reduce nitrite with nitrite reductase activity. An in vivo study conducted at the authors' laboratory on buffaloes indicated that 20 % of protein requirement could be replaced with nitrate (at the rate of 2 % of dry matter intake, the animals were adapted in 8 weeks of time by a gradual increase of nitrate in diet) if an inoculum of nitrate-reducing bacteria was used as probiotic. There was no adverse effect on nutrient utilization and growth performance of animals (Sakthivel et al. 2012). This replacement of protein caused a significant reduction in methane emission and caused no accumulation of nitrite in the rumen liquor of buffaloes. Almost similar results have been reported by Hulshof et al. (2010) in beef cattle fed a sugarcane-based diet including 2.2 % nitrate in diet dry matter (animals adapted in 4 weeks to the highest level used for feeding). This caused a 27 % methane reduction per kg of DMI on nitrate feeding (13.6 vs. 18.6 g/kg DMI) and tended to increase acetate and decrease propionate molar proportions in rumen fluid. These results indicated that nitrate could serve as a terminal electron acceptor and could be used as a methane inhibitor if the animal was either adapted for nitrate consumption or nitrate-reducing bacteria were used as probiotic.

23.5 Adaptation of Ruminants to High Nitrate Diets

Animals can be adapted to high concentration of nitrate in the diet. Allison and Reddy (1984) reported that nitrate and nitrite concentrations in the rumen liquor of sheep-fed Lucerne and ground corn-based diet were undetectable. Infusion of nitrate at the rate 0.17 g/kg BW at 1 h post-feeding resulted in 1.7 mM nitrate and 0.26 mM nitrite, but on continuous feeding for 6 days, the levels of nitrate and nitrite in the rumen fluid were again undetectable, indicating adaptation of the rumen microbes to added nitrate in the diet. In a similar study with buffaloes in the author's laboratory, 20 % of the nitrogen requirement of animal was met through a mixture of sodium nitrate and potassium nitrate by adaptation in 8 weeks by increasing the dose of nitrate every week. This resulted in no accumulation of nitrate or nitrite in the rumen liquor, and no adverse effects on nutrient utilization (unpublished data) were observed.

When nitrate is drenched at high concentration, there is no accumulation of nitrate in the rumen, it is absorbed by the rumen, a part of it is excreted and a part is recycled through saliva (Wang et al. 1961). This absorption of nitrate through the rumen will be drastically reduced if the animals are adapted to higher concentration of nitrate in the diet (Alaboudi and Jones 1985). The adaptation of the animals increases its ability to use nitrate and nitrite more efficiently in the rumen, and these are metabolized efficiently into the body tissue. The in vitro results at IVRI, Izatnagar, indicate that an inoculum of nitrate-/nitrite-reducing bacteria can be an alternative to periodic and slow adaptation for obtaining similar results with nitrate utilizing capacity of the animals. It is very interesting to note that adapted animals return to normal level of its capability to use nitrate in diet within 3 weeks once the nitrate is withdrawn from the ration of sheep (Alaboudi and Jones 1985).

The initial rapid decline in nitrate concentration (about 80 % given to sheep, 560 out of 700 mM) in rumen liquor of animals loaded with high nitrate content without adversely affecting rumen fermentation (Tillman et al. 1965) can be justified if the rumen bacteria are able to sequester nitrate immediately as it enters the rumen and release it during the day depending upon ammonia concentration in the rumen liquor. The accumulation of ammonia in the rumen once the nitrate load has been cleared can also be explained by this hypothesis.

To achieve a significant decrease in methane emission by ruminants, the use of nitrate as a major source of fermentable nitrogen is one of the economic methods for practical application. The only condition for formulating a suitable ration is that it should be nitrogen deficit with a portion of by-pass protein and energy sufficient with supplementation of additional sulphur to make it balanced for different nutrients.

23.5.1 Effect on Nitrate/Nitrite Feeding on Methanogenesis

There have been many studies about the use of nitrates (sodium nitrate, potassium nitrate, calcium nitrate or ammonium nitrate) as the terminal electron acceptors, which compete with CO_2 and hamper the process of methanogenesis. Of course there are some difficulties due to toxicity of nitrite, formed as an intermediate in the reduction of nitrate to ammonia. But these hurdles are overcome

Table 23.2 Methane production by different types of animals fed various diets

Animal	Diet	Methane production (g/kg DMI)	Reference
Beef cattle	90:10, roughage: concentrate	21.9	Boadi and Wittenberg (2002)
Dairy cattle		22.4	
Beef cattle	20:60, roughage: concentrate	26.2	Foley et al. (2009)
Cattle	High-grain diet	22.0	Kurihara et al. (1999)
Beef cattle	Sugarcane-based diet	18.2	Hulshof et al. (2012)
	Sugarcane-based diet + nitrate supplement	13.3	
Beef cattle	Pasture grown	25.6	Hart et al. (2009)
Murrah buffalo (male)	1:1, roughage: concentrate	22.0	Sakthivel (2010)
	1:1, roughage: concentrate (20 % nitrogen replaced with nitrate)	18.6	
	1:1, roughage : concentrate (20 % nitrogen replaced with nitrate) + nitrate-reducing bacterial isolate as probiotic	17.9	

by providing some adaptation period for tolerating high concentration of nitrate in the diet. As discussed above in the chapter, the nitrate-/nitrite-reducing bacteria are present in the rumen, but their number is not sufficient to take care of reduction of nitrite into ammonia as soon as it is formed. Therefore, if we plan to replace 20 % of feed nitrogen with nitrate, it should be increased in graded levels in diet by enhancing one fourth of the target nitrate concentration every 5th day. In a fortnight, the animals will get adapted to the highest concentration of nitrate in the ration, and there will be no accumulation of nitrite in the rumen. In an experiment with buffaloes, adapted to 20 % replacement of nitrogen in the ration with nitrate, there was a significant ($P=0.04$) reduction in methane emission (169.6 vs. 132.1 l/day). There was a significant ($P=0.01$) increase in carboxymethyl cellulose and xylanase activities in the rumen, the key enzymes required for fibre degradation in the rumen. The buffaloes in the control group emitted 22 g/kg DMI, while on supplementation of nitrate, the methane emission was reduced to 17.9–18.6 g/kg DMI (Sakthivel 2010). Hulshof et al. (2012) reported that there was 32 %, 27 % and 28 % reduction in methane productions in terms of total methane production, per kg of dry matter intake and gross energy intake, respectively, in beef cattle fed sugarcane-based diet supplemented with nitrate at the rate of 22 g/kg DM. There was 6 % lower feed intake by the animals on nitrate feeding. There was no effect of nitrate feeding on VFA production, but the proportion of individual fatty acids was changed in favour of higher acetate production. Some of the experiments conducted in different laboratories are listed in Table 23.2.

In the authors' laboratory, feeding of nitrate at the rate of 3 % of diet to the adapted buffalo calves (average BW 83.5 kg) for 6 months resulted in 34.12 % decrease in methane production (l/kg DDMI), improvement in body weight gain and feed conversion efficiency by 15 and 9.65 % (unpublished data).

23.6 Future Research Needs

Keeping in view the work done in this area of research, special attention of animal nutritionists/rumen microbiologists is required in the following areas:

- Exhaustive studies are needed on nitrate feeding to the animals to control methanogenesis and to make livestock production more economic and eco-friendly.
- There is a need to study the microbial diversity of microbes responsible for nitrate

reduction in the rumen. Meta-transcriptomic studies will be helpful in identifying microbes and enzymes involved in efficient nitrate reduction and control methane emission by the ruminants.

References

Alaboudi R, Jones GA (1985) Effects of acclimation to high nitrate intake on some rumen fermentation parameters in sheep. Can J Anim Sci 65:841–849

Allison MJ, Reddy CA (1984) Adaptations of gastrointestinal bacteria in response to changes in dietary oxalate and nitrate. In: Current perspectives in microbial ecology, proceedings of the 3rd international symposium on microbial ecology, 7–12 Aug 1983. Klug MJ, Reddy CA (eds) Amer Soc Microbiol Washington, DC and feeding management. Cooperative Extension Service, Kansas State University, Manhattan

Asanuma N, Iwamoto M, Kawato M et al (2002) Numbers of nitrate-reducing bacteria in the rumen as estimated by competitive polymerase chain reaction. Anim Sci J 73:199–205

Boadi DA, Wittenberg KM (2002) Methane production from dairy and beef heifers fed forages differing in nutrient density using the sulphur hexafluoride (SF6) tracer gas technique. Can J Anim Sci 82:201–206

Bradley LH, Eppson HF, Beath OA (1940) Livestock poisoning by oat hay and other plants containing nitrate. Univ Wyo Agric Exp Station Bull 241:3–20

Burrows GE, Horn GW, McNew RW et al (1987) The Prophylactic effect of corn supplementation on experimental nitrate intoxication in Cattle. J Anim Sci 64:1682–1689

Cash D, Funston R, King M et al (2007) Nitrate Toxicity of Montana Forages. http://animalrangeextension.montana.edu/articles/Forage/General/Nitratetox.htm

Crawford RF, Kennedy WK, Davison KL (1966) Factors influencing the toxicity of forages that contain nitrate when fed to cattle. Cornell Vet Med 56:1–17

Farra PA, Satter LD (1971) Manipulation of the ruminal fermentation. III. Effect of nitrate on ruminal volatile fatty acid production and milk composition. J Dairy Sci 54:1018

Faulkner DB, Hutjens MF (1989) Nitrates in livestock feeding. www.inar.unl.edu/pubs/Beef/g170.htm#causes

Fjell D, Blasi D, Towne G (1991) Nitrate and prussic acid toxicity in forage. Causes, prevention, feeding, management. Cooperative Extension Service, Kansas State University, Manhattan

Foley PA, Kenny DA, Callan JJ et al (2009) Effect of DL-malic acid supplementation on feed intake, methane emission, and rumen fermentation in beef cattle. J Anim Sci 87:1048–1057

Guerink JH, Malestein A, Kemp A et al (1982) Nitrate poisoning in cattle. 7. Prevention. Neth J Agric Sci 301:105–113

Hart KJ, Martin PG, Foley PA et al (2009) Effect of sward dry matter digestibility on methane production, ruminal fermentation, and microbial populations of zero-grazed beef cattle. J Anim Sci 87:3342–3350

Hulshof RBA, Berndt A, Demarchi JJAA et al (2010) Dietary nitrate supplementation reduces methane emission in beef cattle fed sugarcane based diets. In: Proceedings of 4th greenhouse gases and animal agriculture conference, Banff, p 110

Hulshof RBA, Berndt A, Gerrits WJJ et al (2012) Dietary nitrate supplementation reduces methane emission in beef cattle fed sugarcane-based diets. J Anim Sci 90:2317–2323

Iwamoto M, Asanuma N, Hino T (1999) Effects of nitrate combined with fumarate on methanogenesis, fermentation and cellulose digestion by mixed ruminal microbes in vitro. Anim Sci J 70:471–478

Iwamoto M, Asanuma N, Hino T (2001) Effects of energy substrates on nitrate reduction and nitrate reductase activity in a ruminal bacterium, *Selenomonas ruminantium*. Anarobe 7:315–321

Iwamoto M, Asanuma N, Hino T (2002) Ability of *Selenomonas ruminantium, Veillonella parvula*, and *Wolinella succinogenes* to reduce nitrate and nitrite with special reference to the suppression of ruminal methanogenesis. Anaerobe 8:209–215

Jainudeen MR, Hansel W, Davidson KL (1964) Nitrate toxicity in dairy heifers. II Enthropoietic responses to nitrate ingestion during pregnancy. J Dairy Sci 47:1382–1387

Johnson KA, Johnson DE (1995) Methane emissions from cattle. J Anim Sci 73:2483–2492

Johnson JL, Schneiderm NR, Kelling CL et al (1983) Nitrate exposure in perinatal beef calves. Am Assoc Vet Lab Diagn. 30th proceedings, pp 167–180

Kurihara MT, Magner RA, McCrabb GJ (1999) Methane production and energy partition of cattle in the tropics. Br J Nutr 81:227–234

O'Hara PJ, Fraser AJ (1975) Nitrate poisoning in cattle grazing crops. N Z Vet J 23:45–53

Rogers P (1999) Poisoning in cattle and sheep by urea, nitrates and nitrites: Sources, toxic doses, treatment and prevention. philrogers@eircom.net

Sakthivel PC (2010) Studies on the effect of nitrate on methanogenesis, microbial eco-system and nutrient utilization in buffaloes. Ph.D. thesis, IVRI Deemed University, Izatnagar

Sakthivel PC, Kamra DN, Agarwal N et al (2012) Effect of sodium nitrate and nitrate reducing bacteria on in vitro methane production and feed fermentation with buffalo rumen liquor. Asian-Aust J Anim Sci 25:812–817

Sar C, Mwenya B, Pen B et al (2005) Effect of ruminal administration of *Escherichia coli* wild type or a genetically modified with enhanced nitrite reductase activity on methane emission and nitrate toxicity in nitrate infused sheep. Br J Nutr 94:691–697

Sidhu PK, Bedi GK, Meenakshi et al (2011) Evaluation of factors contributing to excessive nitrate accumulation in fodder crops leading to ill health in dairy animals. Toxicol Int 18:22–16

Spears JW, Bush LP, Ely DG (1977) Influence of nitrate and molybdenum on sulphur utilization by rumen microorganisms. J Dairy Sci 60:1889–1893

Stewart CS, Flint HJ, Bryant MP (1997) The rumen bacteria. In: Hobson PN, Stewart CS (eds) The rumen microbial ecosystem, 2nd edn. Blackie Academic and Professional, London, pp 10–72

Tillman AD, Sheriha GM, Sirny RJ (1965) Nitrate reduction studies with sheep. J Anim Sci 24:1140–1146

Ungerfeld EM, Kohn RA (2006) The role of thermodynamics in the control of ruminal fermentation. In: Sejrsen K, Hvelplund T, Nielsen MO (eds) Ruminant physiology: digestion, metabolism and impact of nutrition on gene expression, immunology and stress. Wageningen Academic, Wageningen, pp 55–64

Walker R (1990) Nitrates, nitrites and N-nitrosocompounds: a review of the occurrence in food and diet and the toxicological implications. Food Addit Contam 7:717–768

Wang LC, Garcia-Rivera J, Burris RH (1961) Metabolism of nitrate by cattle. Biochem J 81:237–242

Part VI

Future Prospects of Rumen Microbiology

Revolution in Rumen Microbiology 24

Sanjay Kumar and Dipti W. Pitta

Abstract

Meat and milk derived from ruminants comprise a significant portion of worldwide human nutrition. Ruminants and their gut microbes have coevolved over millions of years to harvest energy from indigestible plant cellulosic biomass. This symbiotic host–microbe interaction is fundamental to the ruminant's health and production. Research on rumen microbes dates back to the 1950s with the majority of work involving cultivation of several microbes in the laboratory under strict anaerobic conditions. However, culture-independent molecular methods have demonstrated that <1 % of what is known of the rumen microbiome has been accounted for with culture methodology. Molecular techniques have provided a great potential to define and describe dynamics in the rumen microbiome in response to the animal's changing diet and physiologic state. With the advent of metagenomics and next-generation sequencers, the characterization of community microbial populations has become less cumbersome. The first few reports on the use of metagenomics to study the rumen microbiome have portrayed the dominance and the phylogenetic distribution of bacteria in the rumen and their potential involvement in the conversion of dietary plant cellulosic material to microbial protein, volatile fatty acids, and other byproducts of ruminal fermentation. The initial colonization and gradual establishment of microbial communities in the rumen from birth to maturity (2 years) was recently demonstrated using metagenomics. Knowledge of not only the composition of the rumen microbiome but also the functional role of certain microbes and their genes has provided greater insights on the contribution of the rumen microbiome in host metabolism. Such advances have led to polysaccharide-hydrolyzing gene

S. Kumar • D.W. Pitta (✉)
Agriculture Systems and Microbial Genomics Laboratory, Department of Clinical Studies, School of Veterinary Medicine, University of Pennsylvania, Philadelphia, PA 19348, USA
e-mail: dpitta@vet.upenn.edu

A.K. Puniya et al. (eds.), *Rumen Microbiology: From Evolution to Revolution*, DOI 10.1007/978-81-322-2401-3_24

cataloguing which has become a useful tool. Application of advanced "omic" technologies such as transcriptomics, proteomics, and metabolomics are other examples of current technologies delving into host–microbe interactions and how the rumen microbiome can be altered to improve animal performance. Although several inconsistencies continue to pose challenges to rumen microbiology research, optimizing the rumen microbiome is an attainable goal in this post-genomics era. Overall, advances in genomic technology indicate that research contributions to the rumen microbiome physiology have the potential to not only improve ruminant production but also address global issues such as advanced biofuel production, reduction of greenhouse gases, enhancement of food safety, and improvement of the food supply.

Keywords
Gene catalogue • Metagenomics • Microbial community • Next-generation sequencers • Plasmid • Rumen

24.1 Introduction

Ruminants have a specialized forestomach called "the rumen" that allows colonization of countless numbers of microbes (Stevens and Hume 1998), comprised of bacteria, archaea, protozoa, and fungi, collectively called the rumen microbiome (Hobson and Stewart 1997). Bacteria predominate in the rumen (Hespell et al. 1997) and primarily convert energy stored in plant biomass to microbial protein and short-chain fatty acids required to manufacture food products (Hobson and Stewart 1997; Godoy-Vitorino et al. 2012). Therefore, a better understanding of microbial fermentation processes is essential to enhance feed efficiency and improve ruminant production. Since the 1950s, efforts were made to enhance our knowledge of the composition of the rumen microbiome, its functional role, and how the rumen microbiome can be manipulated to enhance animal production. We know the rumen microbiome is dynamic and changes with diet, animal parameters, physiology, and the environment; therefore, optimization of rumen parameters followed by the manipulation of the rumen microbiome is promising for the enhancement of ruminant production but still faces daunting challenges. Anaerobic cultivation procedures initiated by Hungate in the 1950s have set the golden standard for rumen microbiology research. Information on rumen microbes gradually increased with isolation and characterization of the rumen microbes from different ruminant species fed different diets using these cultivation methods. However, it became evident that cultured microbes account for less than 1 % of what have been identified using molecular methods (c.f. the "Great plate count anomaly"; Staley and Konopka 1985). In the past decade, tremendous improvements have occurred in DNA technology which substantially changed our approach to understanding the rumen microbiome. The dawn of next-generation sequencing platforms (454 Roche, Illumina/Solexa, and PGM), complemented by whole-genome sequencing (Morrison et al. 2010) and single-molecule real-time (SMRT) sequencing (Roesch et al. 2007; Liu et al. 2012), has led to a better characterization and construction of a metagenome of the rumen microbial environment in different ruminant species (Brulc et al. 2009; Liu et al. 2009; Kong et al. 2010; Pitta et al. 2010, 2014a, b; Wang et al. 2011; Hess et al. 2011; Mohammed et al. 2012). In addition, different "omic" technologies such as transcriptomics, proteomics, and metabolomics provide capabilities to explore the functional aspects of the rumen microbiome.

The increase of genetic information on the rumen microbiome cataloguing of genes has been a boon for collection and future reference for microbial ecologists, ruminant nutritionists, geneticists, and molecular biologists. There are existing databases which provide enormous amounts of information on rumen microbes such as the FibRumBa (http://jcvi.org/rumenomics/), Hungate 1000 (http://www.hungate1000.org.nz/) and Fungi 1000, and the CAZyme databases (http://www.cazy.org). These databases demonstrate that the rumen is filled with specialized organisms and enzymes (polysaccharide and toxin degrading organisms/enzymes, etc.) that not only are adapted to fill their ecological niches in the rumen but also provide unique genetic resources for investigation of lignocellulose degradation, organic acid synthesis, and bioremediation. Studies that have applied "omic" technologies extensively to the rumen microbiome have revealed that the rumen microbiome is a great genetic resource with the potential capability to not only improve animal production but also address global concerns such as production of next-generation biofuels, reduction of greenhouse gases, enhancement of food safety, and increasing global food supply (Ransom-Jones et al. 2012; Krause et al. 2013; Pitta et al. 2014a). More recently, studies on the rumen plasmidome (the collection of plasmids) have demonstrated the capabilities for gene transfer and the evolution of shared functions among heterogeneous microbial populations in the rumen, which is of great significance to animal physiology and productivity (Kav et al. 2012; Mizrahi 2012).

In the present chapter, we discuss the recent trends and evolution in techniques that have revolutionized rumen microbiology research, particularly our understanding of rumen microbes, their genetic elements, and their interactions with the host metabolism. We attempt to describe the historic developments that link the rumen microbiome to animal production and beyond.

24.1.1 16S/18S rDNA Gene Versus Genomic DNA

The 16S/18S rRNA (rrs) gene sequences are widely adopted for the identification of bacterial, archaeal, fungal, and protozoal species. The rrs gene is evolutionarily conserved and characteristics of the rrs gene make it possible to quantify the target organisms (Zoetendal et al. 2004; Deng et al. 2008) and also aid in inferring the natural phylogenetic relationships between microbes within a microbial community (Tajima et al. 2001; Larue et al. 2005). Construction of 16S/18S rRNA gene clone libraries of rumen microbes using molecular tools has revealed a several-fold increase in diversity among ruminal microbes when compared to traditional cultural techniques. The cloning-based diversity analysis was derived from ruminants fed different diets and also across different ruminant species (Tajima et al. 2001; Larue et al. 2005). With increasing information on bacterial populations based on the 16S rrs gene, it became obvious that cultivated cellulolytic bacterial genera such as *Ruminococcus* and *Fibrobacter* were not among the most abundant members of the community and identified the presence of various other fiber-degrading genera (Brulc et al. 2009; Pitta et al. 2010). Among ruminal archaeal community, Whitford et al. (2001) analyzed 41 methanogen-related rDNA sequences from the bovine rumen. The most abundant clones recovered were from *Methanobrevibacter ruminantium*, followed by *M. stadtmanae*, not previously described in the rumen. The molecular diversity of rumen methanogens in sheep was also shown on the basis of 16S rRNA, when sheep were fed with two different diets, i.e., pasture-grazed sheep or fed with oat or Lucerne hay. On the basis of sequence similarity, 65 phylotypes were found to be from order *Methanobacteriales* and greater diversity was found to be present in pasture-grazed sheep (Wright et al. 2004). In a study that involved reindeers, methanogens in the rumen were from families *Methanobacteriaceae* and *Methanosarcinaceae*, based on 16s rRNA sequence similarity (Sundset et al. 2009). In-depth rrs sequencing proved that there are considerable shifts in microbial community structure of the rumen with changes in dietary regime, thus allowing researchers to recognize the predominant core microbial species of the community and also rare community members that could be associated with feeding practices

(Pitta et al. 2010; Callaway et al. 2010; Shanks et al. 2011).

Quantification of the 16S/18S rRNA gene of a particular genus or species has been assessed either by hybridization techniques or by qPCR (Chaucheyras-Durand and Ossa 2014). Blot hybridization techniques revealed that the main cellulolytic species (*Fibrobacter succinogenes, Ruminococcus flavefaciens*, and *Ruminococcus albus*) represent 10 % of the total bacteria in the rumen (Stevenson and Weimer 2007). Further, fluorescence in situ hybridization (FISH) detection suggests that these bacteria account for about 50 % of the total active cellulolytic bacteria (Soliva et al. 2004; Kong et al. 2012). Studies using FISH-based enumeration have established the predominant specific microorganisms in the rumen. Soliva et al. (2004) employed the technique for in vitro rumen methanogen determination, and similarly Yanagita et al. (2000) estimated that 54 % of total sheep methanogens (10^8 per mL rumen sample) were found to be *Methanomicrobium mobile*. FISH is, however, less sensitive than PCR-based techniques as it could only detect 10^3 cells per mL and is not amenable to high sample throughput. Moreover, an extensive knowledge of the community is required before designing specific probes. Therefore, to measure microbial species in a very sensitive manner, quantitative PCR is being used by researchers. QC-PCR (quantitative competitive PCR) is a rapid enumeration technique, which involves the co-amplification of a target DNA sample with known amounts of a competitor DNA having the nucleotide sequence similar to the target (Sirohi et al. 2012). Precision of this method exceeds that of other quantitative PCR procedures including RT-PCR (real-time PCR). Reilly and Attwood (1998) measured the *Clostridium proteoclasticum* population in the cow rumen. Subsequently, Koike and Kobayashi (2001) used this approach to enumerate the major cellulolytic bacterial populations in the sheep rumen. Competitive PCR also has been proven useful in monitoring bacterial populations at different time periods in diet-based animal studies (Mrazek et al. 2006; Sekhavati et al. 2009)

Real-time PCR targeting the 16S rRNA gene also overcame the hybridization disadvantages and has been used successfully on nucleic acids extracted from rumen contents to monitor microbial populations in the rumen. Tajima et al. (2001) monitored bacterial shifts in the rumen during diet transition, whereas Ouwerkerk et al. (2002) used this technique to quantify the population of *Megasphaera elsdenii*. Similarly, Klieve et al. (2003) adopted this method to determine the populations of *M. elsdenii* and *B. fibrisolvens* in the rumen of cattle. With the availability of many different primer pairs for exploring other rumen bacterial populations (*Anaerovibrio lipolytica, Butyrivibrio fibrisolvens, Eubacterium ruminantium, Prevotella albensis, P. brevis, P. bryantii, P. ruminicola, Ruminobacter amylophilus*, genus *Prevotella*), qPCR assays can provide useful information for evaluating dietary effects on the rumen microbiome, such as changes associated with acidogenic diets in dairy cows (Khafipour et al. 2009).

Researchers use many techniques for community fingerprinting, as there is as yet no "gold standard" technique in the anaerobic field. Commonly used techniques consist of terminal restriction fragment length polymorphisms (T-RFLP), denaturing gradient gel electrophoresis (DGGE), and temperature gradient gel electrophoresis (TGGE) (Zoetendal et al. 2004). Additionally, single-strand conformation polymorphism (Lee et al. 1996), length heterogeneity PCR, and ribosomal intergenic spacer analysis are employed. The amplicons generated from amplified nucleic acids, using specific primers, are separated according to their intrinsic properties and the resulting pattern is reflective of the community diversity (Kumar et al. 2011). These techniques have been developed to assess shifts in the microbial composition from a given environment. Studies using 16S rDNA DGGE/TGGE found that the bacterial population was more diverse and abundant in corn-fed ruminants compared to hay-fed ruminants (Kocherginskaya et al. 2001).

DGGE was used by Mackie et al. (2003) to monitor the uncultured bacterium *Oscillospira* spp. in different ruminants. The results indicated that the presence as well as abundance of *Oscillospira* spp. in various rumen ecosystems is diet dependent and the highest counts were found in cattle and sheep fed with fresh-forage diets. Profiling of protozoal communities in the rumen was performed by Sylvester et al. (2004) and Regensbogenova et al. (2004) and diet was found to influence the protozoal diversity. Major protozoal species identified were *Epidinium caudatum*, *Entodinium caudatum*, and *Isotricha prostoma*. Mosoni et al. (2011) studied the effect of defaunation on methanogen diversity employing DGGE based on the *mcrA* gene. Although defaunation reduced methane emission considerably, neither the diversity nor the abundance of the dominant methanogen population was affected by defaunation. Therefore, it is clear that the application of combined molecular approaches will enable researchers to gain insight into a complete description of the genetic diversity of the rumen microbiome. In addition, with the advent of different "omic" technologies – genomics, transcriptomics (for functional analysis), metabolomics (for metabolic profile), chemogenomics/proteomics (for key proteins and enzymes in cellular functions), and nutrigenomics (study on the effect of nutrition on gene expression) – fundamental biological processes can also be examined (Fig. 24.1). In fact, the collective use of high-throughput DNA sequencing and "omic" disciplines offers two advantages: (a) potential to understand the lifestyle of a specific microbe and (b) potential to evaluate its genetic characteristics for a functional and comparative aspect.

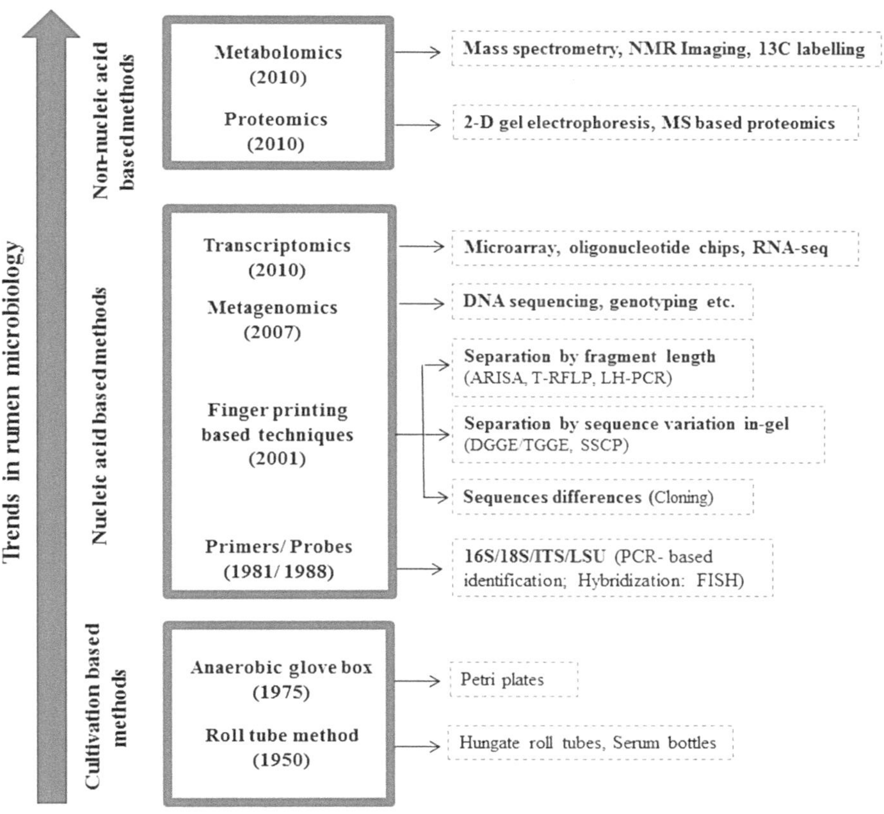

Fig. 24.1 Advances in rumen microbiological technologies

24.2 Next-Generation Sequencing

De novo assemblies of microbial genomes have been accomplished with the new high-throughput sequencing approaches, which are referred to as next-generation sequencing (NGS). With the wide applications of NGS technologies, genome sequence-based information is now within reach to aid understanding of the rumen microbial ecosystems and their associated functions (Liu et al. 2012). Next-generation sequencing is a second-generation sequencing technique, after Sanger sequencing, whose instrumentation includes Roche/454, Illumina/Solexa HiSeq/MiSeq or PGM (Personal Genome Machine) system, Applied Biosystems SOLiD, and Helicos Bioscience (Table 24.1). Though these methods allow rapid characterization of targeted sequences and are cost-effective, they suffer limitations such as a short read assembly with SOLiD and HiSeq and an error rate with polybase more than 6, high cost, and low throughput with the 454 Roche platform. On the contrary, third-generation sequencing such as single-molecule real-time (SMRT) and nanopore sequencing are cost-effective as well and require shorter DNA preparation time (do not require PCR), and the signal is captured in real time. Single-molecule real-time sequencing, developed by Pacific Bioscience, USA, employs fluorescence-based analysis (Timp et al. 2010; Liu et al. 2012). The technique is useful not only in genome sequencing but also in the prediction of structural variance in the sequences, particularly in epigenetic studies such as DNA methylation (Branton et al. 2008). The NGS has dramatically changed our approach to microbial ecology and taxonomy.

24.2.1 Next-Generation Sequencing: Advances in the Knowledge of the Rumen Microbiome

24.2.1.1 Rumen Bacteria

The abundance and diversity of rumen bacteria and archaea have been investigated through meta-analysis of all the deposited 16S rRNA sequences in the Ribosomal Database Project (RDP). The RDP 11.1, released in March 2014 (http://rdp.cme.msu.edu/), contains 2,929,433 aligned and annotated bacterial and archaeal small subunit (SSU) rRNA gene sequences. Newly released RDP data represent 27 existing bacterial phyla. Members of the phylum *Bacteroidetes*, *Proteobacteria*, and *Firmicutes* constituted a major fraction of the total rumen bacterial population, irrespective of animal age; however, with dietary changes, host, and different environmental conditions, a substantial shift in the composition of these bacterial phyla has been reported by several researchers (Brulc et al. 2009; Pitta et al. 2010, 2014a; Uyeno et al. 2010; Hess et al. 2011). The lineages within the phylum *Bacteroidetes* are predominately represented by *Bacteroidales* (*Prevotellaceae*, *Porphyromonadaceae*), *Flavobacteriales*, and *Sphingobacteriales*, while phylum *Firmicutes* showed an abundance of *Clostridiales* (*Lachnospiraceae*, *Ruminococcaceae*, *Veillonellaceae*, and *Clostridiales* Family XIII) representatives. The predominant bacterial genera in phylum *Firmicutes* include *Butyrivibrio*, *Acetivibrio*, *Ruminococcus*, and *Succiniclasticum*. The RDP of rumen sequences represents all the five classes of phylum *Proteobacteria* (alpha-, beta-, gamma-, delta-, and epsilon-*Proteobacteria*). Other phyla

Table 24.1 A comparison of different platforms available for next-generation sequencing

Platforms	454 Roche	Ion torrent (PGM)	Illumina
Principle	Pyrosequencing	Proton detection	Reversible terminator
Read length	700	400	(2×150 bases)
Run time	10 h	2 h	2 d
Gb/run	0.7	150	120 (rapid mode)
Cost/run	$1,100	$225–625	$750
Advantage	Long read length	Short run times	High throughput/cost
Disadvantage	High error rate in homopolymer	Short reads	New

that contribute to less than 3–5 % of the rumen bacterial populations are comprised of *Tenericutes, Verrucomicrobia, Cyanobacteria, Actinobacteria*, and *Fusobacteria* (Kong et al. 2010; Jami et al. 2013; Kittelmann et al. 2013; Pitta et al. 2014a, b).

The rumen microbiome continues to adapt, mature, and establish with age, a new finding that is revealed with NGS technology. In a recent study exploring bacterial diversity of five different age groups, from 1-day-old calves to 2-year-old cows, functionally essential rumen bacteria were detected from day 1 after birth irrespective of rumen activity and even before ingestion of plant material. Parallel to findings reported for human gut microbiota, a core bacterial community exists across most dairy cows. This core microbial community was composed of 35 genera that were present across 90–100 % of the cow studied (Jami et al. 2013). More recently, employing a metagenomics approach, a direct relation between the rumen microbiome and animal's physiological parameters and productivity was established (Jami et al. 2014). Further, Pitta et al. (2014b) used 16S pyrotag sequencing to explain differences in the rumen microbiome of primiparous and multiparous cows during the transition period. Among 17 bacterial phyla analyzed, the most abundant phyla were *Bacteroidetes* and *Firmicutes*. As the cows transitioned into lactation, the ratio of *Bacteroidetes* to *Firmicutes* increased from 6:1 to 12:1 and this ratio was higher in primiparous cows compared to multiparous cows. Few studies on lactating dairy cows have been published, but they were limited to small numbers of animals and displayed large inter-animal microbiome variation (Weimer et al. 2010a; Wang et al. 2012; Mohammed et al. 2012) Therefore, NGS has dramatically changed our approach as well as an understanding of rumen bacterial community composition under different conditions.

24.2.1.2 Rumen Methanogens

Methanogenic archaea comprises a diverse and composite group that utilizes hydrogen in the rumen and reduces carbon dioxide to methane (Kumar et al. 2012). Until recently, the rumen methanogens were assigned to a few genera of the orders *Methanobacteriales, Methanomicrobiales*, and *Methanosarcinales* within the phylum *Euryarchaeota* (Kumar et al. 2014a). However, based on meta-analysis, a novel group distantly related to the *Thermoplasmatales* (named as rumen cluster C; previously described as rice cluster C *Thermoplasmata*) are estimated to have approximately 16 % abundance in ruminants (Poulsen et al. 2013; Janssen and Kirs 2008). Analysis of rumen methanogen diversity has previously relied on the 16S rRNA gene, but this gene occurs as multiple heterogeneous copies within a genome (Sirohi et al. 2012). These intra-genomic copies can vary in sequence, leading to the identification of multiple ribotypes for a single organism, and hence may fail to provide an accurate depiction of microbial community structure (Dahllöf 2002; Crosby and Criddle 2003). Further, the 16S rRNA-based method can have limitations when analyzing multiple samples in response to dietary treatments over time. As an alternative, the most efficient and reliable method to infer phylogeny and to estimate abundance of rumen methanogens is targeting the methyl-coenzyme-reductase (*mcrA*) gene (Luton et al. 2002; Tatsuoka et al. 2004; Denman et al. 2007; Evans et al. 2009; Kumar et al. 2009, 2011, 2012). Compared to 16S rRNA quantification showing 10^8 gene copies/gram of rumen contents, *mcrA* gene-based qPCR showed 10^{11} gene copies of methanogens/gram of rumen contents (Mosoni et al. 2011). In a recent study on the effect of lauric acid on methanogens, pyrosequencing showed altered archaeal populations in the rumen of dairy cows, possibly due to their physical association with protozoa (Hristov et al. 2012). In a more recent study, nutritional intervention during the early life of ruminants with bromocholoromethane, an antimethanogen compound, showed shifts in the archaeal community inhabiting the rumen. Differences in methane emissions by the archaeal community in the rumen of neonates and adult goats indicated competition occurring in the developing rumen to occupy different niches that offer potential for rumen manipulation (Abecia et al. 2014). Hence, diversity analysis of methanogens through NGS offers the potential for intervention with methanogens that could help to develop a strategy to curb production of methane from ruminants.

24.2.1.3 Rumen Anaerobic Fungi

In the past, several molecular techniques including qPCR and DGGE were developed for estimating the predominance and diversity of different genera as well as species of anaerobic rumen fungi (ARF). Initially, 18S rRNA gene sequences were employed for comparing fungal diversity and quantification, but now 18S rRNA-based analyses of fungi have become obsolete because of the highly conserved nature of the 18S rRNA gene. Alternatively, internal transcribed spacer regions (ITS) of rRNA genes have been used to study the diversity of ARF. Due to the presence of extremely AT-rich ITS region and intraindividual variations in ITS regions, characterization of fungi using ITS has become a challenge (Nicholson et al. 2005;Tuckwell et al. 2005). In an attempt to target alternative genetic markers, Dagar et al. (2011) have reported the use of D1/D2 domains of 28S rDNA to differentiate *Orpinomyces* to species level.

The phylum *Neocallimastigomycota* represents the earliest-diverging lineage of the zoosporic ARF, and based on thallus morphology and zoospores, ARF have been classified into six genera (*Anaeromyces, Orpinomyces, Cyllamyces, Piromyces, Neocallimastix*, and *Caecomyces*) and 18 species (Griffith et al. 2009). Advances in NGS in recent years applied to the rumen microbiome have given additional insights into the community structure and diversity of ARF in the herbivorous gut, which enabled us to identify several putative novel lineages (potential candidate genera) within the phylum *Neocallimastigomycota*. Results from pyrosequencing-based approaches employed to explore the ARF in a large number of wild/zoo animals are in concurrence with the observation that *Piromyces* and *Caecomyces* comprise the predominant genera in hindgut fermenters, whereas among foregut fermenters, *Neocallimastix* and *Piromyces* showed a greater prevalence (Liggenstoffer et al. 2010). Similar findings were also reported in the foregut of sheep, deer and cows over different seasons (Kittelmann et al. 2013). However, Sirohi et al. (2013) showed an abundance of *Orpinomyces* compared to other ARF genera in the foregut of cattle. Therefore, NGS technology has enabled rumen microbiologists to better characterize ruminal fungi and enhanced our understanding of the global genus-level diversity that exists within the *Neocallimastigomycota*.

24.2.1.4 Rumen Protozoa

Ciliated protozoa are common inhabitants of the rumen and contribute nearly 50 % of the viable biomass (10^5–10^6 cells per mL) (Sylvester et al. 2004, 2009). Their existence in the rumen is important for cellulose digestion and a stable internal environment. Shifts in protozoal populations can occur due to changes in diet that can have a direct impact on microbial protein production and also influence microbial protein yield and the nitrogen utilization (Brown et al. 2006; Wang et al. 2009).

Hitherto, more than 42 genera of rumen protozoa have been described with *Entodinium* being the most abundant, accounting for 95 % in concentrate-rich diets. Other major protozoa in the rumen are *Dasytricha, Ostracondium, Diplodinium, Eudiplodinium, Diploplastron, Epidinium, Ophryoscolex*, and *Polypastron.* Although microscopy has been the most reliable method for protozoal enumeration, molecular approaches such as DGGE and T-RFLP are also being used for diversity analysis (Skillman et al. 2006;Tymensen et al. 2012). Validation methods to explore protozoan diversity in the rumen using NGS technology are yet to be described, but with rapid changes in computational tools, we anticipate that new studies on ruminal protozoans will emerge.

24.2.1.5 Rumen Plasmidome

Since the rumen function is supported by a complex bio-web of rumen microbes, insight into the role of plasmids as a mobile element for transferring genetic information is crucial to further enhance our knowledge of rumen microbial ecology and evolution. The heterogeneity of the rumen microbial environment and its potential role in animal physiology raise several interesting questions regarding gene transfer and the evolution of shared functions among its microbial members. Its confined niche-like nature, which makes it less prone to external forces, presents an opportunity to gain basic understanding related to mechanisms of lateral gene transfer (Kav et al. 2012; Mizrahi 2012). Rumen microbiologists/biotechnologists have primarily restricted their research to bacterial isolates and their plasmids; however, with the advent of NGS technology, plasmids from uncultivable bacteria can now be identified (Kav et al. 2012). Their study suggested that rumen plasmids facilitate performance of rumen functions, such as respiration and amino acid, protein, and carbohydrate metabolisms. Phylogenetic analysis of different genes carried by a plasmid contigs indicated that they can be mobilized across taxonomic levels, to as high as phylum level. Based on plasmid homology with rumen microbes isolated from different locations around the globe, they seem to have a central evolutionary role in the rumen microbial population (Mizrahi 2012). Elucidating the genomics of microbial interactions within the rumen gut is expected to have implications for future developments in food sustainability, renewable energy, and economics; therefore, further research in this direction is warranted.

24.3 Whole-Genome Sequencing

Sequencing technologies in combination with annotation pipelines have made genome sequencing of individual microorganisms inexpensive and easily accessible to the scientific community. Currently, 20 genome sequences from rumen microbes are available publicly, of which half are closed genomes (Morgavi et al. 2013). The North American Consortium for the Genomics of Fibrolytic Ruminal Bacteria (FibRumBa database: http://jcvi.org/rumenomics/), founded in 2000, has begun a collaborative project with the J. Craig Venter Institute of Genomic Research to sequence the whole genomes of major cellulose-degrading bacteria in the rumen, i.e., *Fibrobacter succinogenes* S85, *Ruminococcus albus*, and *Prevotella bryantii* (Sirohi et al. 2012; Morgavi et al. 2013). The genome sequence of *F. succinogenes* showed a many-fold increase in carbohydrate-hydrolyzing enzymes compared to previous studies on this organism. Recently, a draft genome sequence of *Oscillibacter ruminantium* strain GH1 from Korean native cattle (Hanwoo) has been published (Lee et al. 2012), indicating the presence of genes that code for many fibrolytic enzymes and butyric acid production. Species of *Butyrivibrio* and *Pseudobutyrivibrio* genera have now gained attention due to their strong xylan-degrading potential. The recently completed draft genome sequence of *Butyrivibrio proteoclasticus* B316 showed the presence of a large number of genes responsible for hemicellulose degradation, including esterases, pectate lyases, xylanases, xylosidases, etc. Considering the plant cell wall structure, *Butyrivibrio proteoclasticus* B316 seems to be the initial plant cell wall colonizer and degrader and, therefore, provides access to cellulose by the primary cellulose degrader (Morgavi et al. 2013). The bacterial genome sequences completed to date and the amount of information added regarding their gene sequences are tremendous. This knowledge is likely to contribute beneficially to aspects of rumen digestive processes and may also suggest how to increase functional properties such as conjugated linoleic acid (CLA) levels in milk and meat (Morgavi et al. 2013).

The Ag Research Ltd., a New Zealand-based research group led by Graeme Attwood, for the first time sequenced the whole genome of a rumen methanogen, *Methanobrevibacter ruminantium* (Leahy et al. 2010). Similarly, a draft genome sequence of *Methanobrevibacter boviskoreani* strain JH1 from Hanwoo cattle and strain AbM4 from sheep abomasa have been published (Leahy et al. 2013; Lee et al. 2013). A comparison of methanogenic AbM4 and JH1 was illustrated by Kumar et al. (2014a) which enhanced our knowledge of the key enzymes and proteins that can be targeted for methane inhibition. The completed *M. ruminantium* genome and draft sequences from other methanogenic species may pave the way for identification of the underlying cellular mechanisms that define these microbes, leading to a better understanding of their microecology within the rumen. Currently, this research is mainly at an exploratory stage, but several promising cues for chemogenomic targets are being investigated at critical intervention points for the inhibition of rumen methanogens. The analysis of more genome sequences of rumen methanogens would help to identify potential methane inhibitors. In the near future, with the development of "The Hungate 1000," a catalogue of 1,000 reference microbial genomes from the rumen (http://www.hungate1000.org.nz/), the genomic dataset of the rumen microbiome will be numerically higher; thereby more fibrolytic bacteria and more targets for anti-methanogenic strategies can be identified in addition to understanding rumen function and develop strategies to improve fermentation efficiency. Parallel to "the Hungate 1000" another intriguing project "the Fungi 1000" has recently started under the leadership of Prof. Joey Saptafora involving the US Department of Energy's Joint Genome Institute (JGI). The project will provide a great opportunity to fill in the phylogenetic knowledge gap and simultaneously get the unique support and expertise on fungi from the JGI's sequencing and annotation team. The JGI, using the high-throughput metagenomic approach, was the first to generate a draft anaerobic fungal genome of *Piromyces* sp. E2. Despite the *Piromyces* sp. E2 genome being available online through http://genome.jgi.doe.gov/PirE2_1/PirE2_1.home.html, it has hitherto not been described in the literature. Further, Youssef et al. (2013) reported the genome sequencing of *Orpinomyces* sp. strain C1A, in which they used a combined strategy that utilized both SMRT and Illumina sequencing approaches. Consequently, the sequencing data generated through these projects should provide a strong impetus to rumen microbial diversity studies and open new avenues for targeting genes that are crucial for methane production, the biofuel industry, and ruminant nutrition.

24.4 Transcriptomics/Metatranscriptomics

Sequence-based metagenomic results only reveal the presence or absence of a particular gene or gene family and do not provide any information on gene expression or the viability status of the organism. Therefore, results should be considered preliminary and not conclusive measures of the relative abundance. To sidestep the challenges of sequence-based analysis, transcriptome-based approaches have also recently been applied to examine the metabolic and functional capacity of the rumen microbes. Transcriptomics gives an unbiased perspective of gene transcription in situ, providing a true reflection of metabolic activities (Sorek and Cossart 2009). Wang et al. (2011) used a combination of transcriptomics and proteomics techniques to identify carbohydrate-active enzymes (CAZymes) that were expressed and secreted by *Neocallimastix patriciarum* W5. The ARF was cultured on a number of recalcitrant cellulosic substrates to induce expression of cellulases, and the transcriptome was then sequenced using a combination of Roche 454 and Illumina sequencing technologies. A total of 219 glycosyl hydrolase (GH) contigs from twenty-five different GH families were identified, with a number of these enzymes displaying novel cellulase activities. In a recent study, a metatranscriptomic approach was used to describe the functional diversity of the eukaryotic microorganisms within the rumen of muskoxen (*Ovibos moschatus*), with a focus on plant cell wall-degrading enzymes (Qi et al. 2011). Messenger RNA obtained from the rumen sample was sequenced on the Illumina Genome Analyzer and obtained 2.8 GB of sequences which were assembled into 59,129 contigs. The study showed many lignocellulose-degrading enzyme modules including GH, carbohydrate esterases (CE), and polysaccharide lyases (PL) and therefore identifies many candidate genes coding for potentially valuable plant cell wall-degrading enzymes.

24.5 Proteomics and Metabolomics

Proteomics and metabolomics are other remarkable approaches that are able to (a) characterize gene function and (b) recognize and categorize biological and regulatory cascades by developing functional linkages between proteins (Chaucheyras-Durand and Ossa 2014). Identification and characterization of proteins by mass spectrometry followed by various separation techniques has now made it possible to study the simultaneous expression of more than 1,000 proteins. Furthermore, through whole-metabolic profile monitoring in living biological systems, metabolomics may be of great use in identification of the biochemical consequences of disease and describing altered physiological states induced by genetic mutations, drugs, diet, or stress. In a recent study, employing a multiple metabolomic platform and a combination of proton nuclear magnetic resonance (NMR) spectroscopy, gas chromatography–mass spectrometry (GC-MS), and direct flow injection tandem mass spectroscopy, 93 metabolites were identified and quantified in the rumen samples of dairy cows having ketosis (Saleem et al. 2013). Further, increased concentrations of several toxic, inflammatory, and unnatural compounds, including putrescine, methylamines, ethanolamine, and short-chain fatty acids were also reported. Perturbations in several amino acids (phenylalanine, ornithine, lysine, leucine, arginine, valine, and phenylacetylglycine) were also evident (Saleem et al. 2013). Profiling of ^{1}H-NMR spectra in the urine and serum of a large cohort of sheep subjected to road transport revealed that the transported animals experienced altered gut and energy metabolism, muscle catabolism, and possibly a renal response (Li et al. 2011). In another metabolomics study, Klein et al. (2011) suggested that the glycerophosphocholine–phosphocholine ratio could be a prognostic biomarker for the risk of ketosis. These approaches permit a far more detailed understanding of metabolic causes and effects due to altered health conditions or diet and therefore can help in improving animal health and productivity.

24.6 Limitations of Nucleic Acid and Non-Nucleic Acid-Based Approaches

Sampling, handling, and downstream processing of rumen samples are critical when characterizing the composition of the gut microbiome under different ecological conditions. The sampling procedure for ruminal contents is not well defined and different sampling techniques have been adopted by different groups of researchers. Collection of rumen samples has been either from a fistulated animal (Brulc et al. 2009; Pitta et al. 2010) or obtained by stomach tube (Lodge-Ivey et al. 2009) resulting in numerous sample types, including whole rumen contents, liquid contents, solid contents, adherent fiber, and rumen epithelium. Such an array of sampling techniques and sample types accounts for a portion of the huge variation in bacterial diversity associated with the rumen microbiome (Kim et al. 2011a, b).

Further, there are no records on storage time and temperature for ruminal contents dedicated to molecular studies. Preservation of microbial DNA in published studies has shown variable temperatures and conditions such as at room temperature, at −20 C, at −80 C, under liquid nitrogen, or by freeze-drying (Chaucheyras-Durand and Ossa 2014). However, DNA exposed to room temperature for more than 24 h leads to its fragmentation (Cardona et al. 2012). In the case of metatranscriptomic analysis, integrity of total RNA (RIN) is a critical parameter. Degradation of RNA also compromises results of downstream applications, such as reverse-transcription qPCR or microarray studies (Chaucheyras-Durand and Ossa 2014). Though samples for RNA extraction can be stored at room or low temperature for <24 h, it is recommended that samples be archived at −20 °C for longer periods. Storage at −20 °C can also result in a loss of RNA yields (personal communication with Dr. Kasera). The use of RNase inhibitor therefore is necessary to avoid degradation over time, but efficiency of such preservation is dependent on the nature of the sample (Cardona et al. 2012).

Since most of the culture-independent techniques rely on DNA/RNA extraction, a major challenge that often can introduce a bias in microbial diversity comparisons is the genomic DNA extraction (Wu et al. 2010; Henderson et al. 2013). Ruminal bacteria in the microbiome are composed of both Gram-positive and Gram-negative bacteria (Whitford et al. 1998). The cell walls of Gram-positive bacteria are rigid and difficult to lyse and therefore release of nucleic acid contents is difficult. Traditional methods such as the phenol–chloroform extraction (Yu and Morrison 2004) as well as the cetyl-trimethyl ammonium bromide methods (Min et al. 2002) were standardized for genomic DNA extraction from ruminal contents which resulted in better DNA recovery and ultimately the recovery of both common and rare bacterial populations in a sample (Henderson et al. 2013). However, several reports compared the impact of different commercial DNA extraction kits on bacterial diversity patterns (Ariefdjohan et al. 2010; Wu et al. 2010; Maukonen et al. 2012; Claassen et al. 2013; Kennedy et al. 2014). When compared to most traditional methods, the PSP Stool Mini Kit was comparable to the hot phenol bead-beating method and was found to be effective in the recovery of *Firmicutes* (Wu et al. 2010), whose cell walls are hard to lyse and release nucleic acids. Although several commercial kits are routinely employed by researchers, it is imperative to do a baseline comparison to optimize the genomic DNA extraction protocols that provide better yields and quality (Yu and Morrison 2004) before making conclusions on bacterial diversity estimations. Further, most community-based studies of both rumen and gastrointestinal microbiomes have targeted different hypervariable regions of the 16S rRNA gene (Stahl et al. 1988;Yu and Morrison 2004) and the presence of variable copy numbers of 16S rRNA and sequence variation may account for the apparent incongruities between these reports (Větrovský and Baldrian 2013). The PCR conditions are also crucial including inaccurate primer/target annealing, or annealing temperatures that may lead to false-positive results. With the sequencing analysis, issues concerning different datasets/affiliation of sequences can arise and can lead to misinterpretation of the data. This can be due to the use of the various pipelines relating to sequence databases of different composition and quality (Chaucheyras-Durand and Ossa 2014). Although, next-generation sequencing platforms continue to make strides in achieving better read lengths, identification of the genetic element to the lowest level still needs improvements in precision.

24.7 The Rumen Microbiome: A Potentially Rich Genetic Resource

The ruminal microbiome is integral to ruminant production. Ruminants and the rumen microbiome have coevolved over time in a symbiotic relationship which is vital for the health and production of ruminants. The symbiotic relationship is altered due to various reasons such as pathogen invasion, antimicrobials, exposure to chemical and microbial pollutants, toxins, and other deleterious agents resulting in dysbiosis, a condition that can harm the host, other animals, and/or humans. With advances in genomic technology, our knowledge of the rumen microbiome has increased several-fold; however, the relevance to host metabolism and productivity still remains obscure.

24.7.1 Host Specificity Versus Core Ruminal Microbiome

The rumen microbiome has two types of microbial niches: autochthonous (native) and allochthonous (invaders) (Weimer 1998). The native populations are host specific and resilient to perturbations in the rumen environment, while the invaders will only survive and establish if there are unoccupied niches in the rumen microbiome (Weimer et al. 2010b). Variations in the rumen microbiome within the animal are minimal compared to variations between animals, even when all animals are fed the same diets (Welkie et al. 2010). Inter-animal variation is considered to be a major challenge to modulate rumen function, and therefore, it is imperative that we analyze the rumen microbiome from a large group of animals to account for inter-animal variation (Li et al. 2009; Jami and Mizrahi 2012a, b); however, maintaining a large group of fistulated animals is unethical and expensive. With the dawn of genomic technology, it has become plausible to increase the resolution of rumen microbiome analysis even across large groups of animals and also monitor shifts over prolonged intervals with greater accuracy. In the last 2 years, utilizing genomic technology, it has become apparent that a "core microbiome" exists in most ruminant animals that vary in abundance by diet, breed, age, and geographic location. As the information on the composition of the core microbiome is unraveled, the functional capability of the rumen microbiome can be of great relevance to improve animal productivity. Thus, genomic technology expands the scope of rumen microbiology research to a more global scale.

24.7.2 Enhanced Feed Efficiency

Although numerous efforts have been made to manipulate rumen function (Fig. 24.2) through nutritional interventions (Calsamiglia et al. 2007; Patra and Saxena 2009; Nevel and Demeyer 1996), cause-and-effect relationships have not yet been established (Krause et al. 2013; Chaucheyras-Durand and Ossa 2014;). Specifically, the current lack of knowledge of the rumen microbiome hinders effective enhancement of rumen function to improve ruminant production (Firkins and Yu 2006; Zhou and Hernandez-Sanabria 2009). Manipulation of the rumen microbiome has the potential to enhance beneficial processes (fiber digestion), alter inefficient processes (prevent degradation of protein in the rumen, reduce methane production), and enhance protective process (detoxification mechanisms) which are crucial for animal health and milk production (Weimer 1998; Nagaraja et al. 1997). Early nutritional interventions such as feed supplements, additives, and modifiers were used to manipulate the rumen microbiome (Chaucheyras-Durand et al. 2008; Martin et al. 2010). Indeed, the identification of "superbugs" that have the potential to efficiently convert a diverse spectrum of feeds to food products (Krause et al. 2013) had been the "holy grail" of numerous investigators over the past six decades (Hobson and Stewart 1997; Russell and Rychlik 2001). Although early studies focused on cultivable fiber-digesting bacteria, including *Butyrivibrio fibrisolvens, Ruminococcus albus, Ruminococcus flavefaciens*, and *Fibrobacter succinogenes* (Kobayashi 2006; Stevenson and Weimer 2007; Russell et al. 2009), their success was thwarted by a lack of molecular genetic

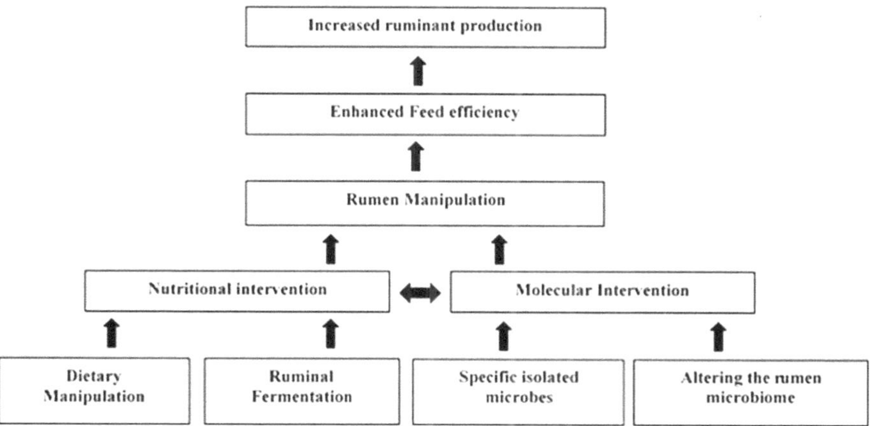

Fig. 24.2 Strategies for enhanced ruminant production

machinery to introduce these isolated bacterial species into the rumen (Teather et al. 1997). Similarly, isolation of amino acid-fermenting bacteria such as *Clostridium aminophilum*, *Clostridium sticklandii*, and *Peptostreptococcus anaerobius* (Russell et al. 1988; Chen and Russell 1989) and the subsequent identification of hyperammonia-producing bacteria (Attwood et al. 1998;Wallace et al. 2004) in the rumen illustrated their role in protein digestion; however, their significance in ruminant nutrition is limited (Krause et al. 2013). Finally, *Synergistes jonesii* which can detoxify 3,4-dihydoxypyridine (goitrogen) present in a legume *Leucaena* has the potential to protect ruminants from *Leucaena* toxicity (Allison et al. 1992). Increasing the proportion of concentrate in diets can lead to lactic acidosis in the rumen. The production of lactic acid in the rumen is linked to an abundance of *Lactobacillus* spp. and *Streptococcus bovis* (Plaizier et al. 2012; Krause et al. 2013). Shu et al. (1999, 2000) and Gill et al. (2000) developed a vaccine targeting these two organisms to prevent lactic acidosis. These are examples that illustrate the use of a few microbes to modulate rumen function. However, as the microbiome is colonized by millions of microbes, it is unlikely that such a limited assessment of cultivable bacteria from the rumen microbiome will lead to major advances in the field (Morgavi et al. 2013).

Moving forward, identification of the core microbiome and manipulating dietary and physiological aspects to optimize the core microbiome could be the most viable option for enhancing animal production. Databases provide enormous genetic information on the microbial ecology in the rumen, but this information when complemented with nutrition and metabolic aspects of ruminants can identify the "superbugs" that can provide greater insights to modulate diets and improve feed efficiency. This would allow for the selection of unconventional feeds that could support the optimized rumen microbiome and also improve animal production. In this post-genomics era, utilizing "omic" approaches offers the opportunity to manipulate the rumen microbiome for greater production efficiency of ruminants across the world.

24.7.3 Methane Production

Enteric emissions of greenhouse gases are of major concern, especially due to their role in climate change (Kumar et al. 2009). A negative implication of methane production by ruminants is a loss of 2–12 % of gross feed energy. These values can be affected by factors such as the type of feed, feed intake, and/or compounds affecting methanogenesis in the rumen (Johnson and Johnson 1995). Therefore, mitigation of methane emissions from ruminants can lower greenhouse gases and increase the efficiency of livestock production (Kumar et al. 2009). Different strategies, such as dietary manipulations (Beauchemin et al. 2008), use of chemical feed additives, halogenated methane analogues (Anderson et al. 2008), probiotics (Newbold and Rode 2006; Puniya et al. 2014), bacteriocins (Kumar et al. 2015), plant extracts (Patra et al. 2011) etc., were used to address this problem; however, these mitigation strategies are all associated with limitations. Using the chemogenomics approach, Wright et al. (2004) developed two vaccines, VF3 and VF7, that caused a 7.7 % methane reduction in sheep (g per kg of dry matter intake), despite targeting only a minority (20 %) of methanogens present within the host animals. The same group created a vaccine targeting five methanogens (*Methanobrevibacter* spp. strains 1Y and AK-87, *M. millerae* ZA-10, *Methanomicrobium mobile* BP, and *Methanosphaera stadtmanae* MCB-3) that was administered to sheep in three doses (Williams et al. 2009). Notably, immunization with the latter vaccine caused methane output to increase by 18 %, despite the fact that a larger proportion of the methanogenic population (52 %) was targeted. Thus, further work is needed to optimize the individual components of these vaccines such that the most potent methanogens are specifically targeted. Researchers believe that anti-methanogenic vaccines will only yield short-term reductions in methanogens and/or methanogenesis, due to the possible proteolytic degradation and lower persistence of host antibodies in the rumen (Li et al. 2007; Cook et al. 2008; Lascano and Cárdenas 2010). Vaccination of sheep with entodinial or mixed protozoa antigens reduced protozoa; the released IgG antibodies against rumen protozoa remained active and continued to bind the target cells for up to 8 h (Williams et al. 2008). Vaccines targeting single surface antigens may not be effective, as methanogenic archaea differ largely based on their host, diet, as well as geographical regions (Zhou and Hernandez-Sanabria 2009). A new vaccine has been developed using subcellular fractions (cytoplasmic and cell wall-derived protein) of *Methanobrevibacter ruminantium* M1 (Wedlock et al. 2010). Twenty sheep were vaccinated; then booster doses were given after 3 weeks, and the antisera were found to agglutinate and decrease the growth of archaeal methanogens and methane production in vitro. Overall, genome sequencing can be applied to investigate and better understand the complex interactions between methanogens and other microbes in the rumen suggesting development of novel methane mitigation strategies. The genomic technology has provided the platform for vaccine production against methanogens in the rumen. Further advances in genomic technology can help explore methanogen diversity as well as help in screening anti-methanogenic agents to reduce methane production from ruminants.

24.7.4 Biomass Degradation

As is evident from various studies (Hess et al. 2011; Sauer et al. 2012; Wang et al. 2012), the ruminal microbial community contains a unique source of enzymes that help in the rapid and efficient conversion of lignocellulosic biomass. However, due to the inherent recalcitrant nature of lignocellulosic materials and their short retention time in the rumen, about 50 % of the cellulosic biomass passes through the rumen undigested (Weimer et al. 2009). Further, conversion of cellulosic material to monosaccharides is also the critical step for cellulose-based biofuel production (Wyman 2007). The rumen microbiome is a repository of enzymes necessary for industrial biofuel production as well as a source of microbial cell factories. Therefore, investigation of the cow's rumen has recently attracted many industrial microbiologists/biotechnologists to identify novel enzymes involved with lignocellulose breakdown using "omic" tools and high-throughput bioinformatic pipelines. Modern biofuel production systems involve pretreatment of lignocellulosic compounds followed by enzymatic hydrolysis, simulating the rumen system. Enzymes of bacterial origin seem to be more important in lignocellulose hydrolysis when compared to fungal enzymes. Some classic examples are construction of engineered *E. coli* with enzymes for succinate production from hemicellulose and extraction of enzymes from *Fibrobacter succinogenes* and *Fusarium* spp. that are specific for xylan degradation (Zheng et al. 2012).

24.7.5 CAZymes

Complex carbohydrate present in the rumen is controlled by an array of enzymes involved in their assembly (glycosyl transferase) and their degradation (GH, PL and CE), collectively called Carbohydrate-Active enZymes (CAZymes). Therefore, understanding the CAZymes encoded by the genome of an organism (CAZome) provides information on the nature and extent of the metabolism of complex carbohydrates in the rumen. Several metagenomic studies have reported diversity of CAZymes (Park et al. 2010) from the rumen microbiome of Angus beef cattle, dairy cattle, reindeer, yak, and Indian buffalo (Brulc et al. 2009; Hess et al. 2011; Dai et al. 2012; Pope et al. 2012; Patel et al. 2014) using CAZYmes analysis. Pope et al. (2012) reported novel CAZymes and polysaccharide utilization loci (PUL)-like systems that contain genes encoding β-1,4-endoglucanases and β-1,4-endoxylanases in the foregut microbiome of tammar wallaby, native to South and Western Australia. Besides sequence-based metagenomic approaches, functional metagenomics approaches are also being applied to identify hydrolytic enzymes using specific substrates. The approach was initially used by Ferrer and coworkers (Ferrer et al. 2005) who identified nine endoglucanases,

12 esterases, and one cyclodextrinase from the rumen of a dairy cow. Wang et al. (2012) focused on sequencing and bioinformatic analysis of 120 fosmid clones that expressed hydrolase activities towards cellulose and xylan and reported the presence of novel CAZy genes. However, the goal of cataloguing genes from the key population of the rumen microbial community is yet to be achieved.

24.7.6 Rumen Microbial Cell Factories

Organic acids are an important metabolic intermediate in the rumen. Many rumen bacteria, such as *Anaerobiospirillum succiniciproducens, Actinobacillus succinogenes*, and *Mannheimia succiniciproducens*, derive energy by decarboxylating succinate and producing propionate, a nutrient source for ruminants (Sauer et al. 2012). Zou et al. (2011) reported use of the rumen environment for the production of succinic acid using *Actinobacillus succinogenes.* The rumen produces CO_2 and carbonate, which are essential substrates for a key enzyme, PEP carboxykinase, used in the biosynthesis of succinic acid. Likewise, other short-chain organic acids such as butyric, valeric, and caproic are produced by *Megasphaera elsdenii* (Sauer et al. 2012). Recently the genome sequence of *Oscillibacter ruminantium* identified the genes for production of butyric acid (Lee et al. 2012). Therefore, rumen bacteria can be promising candidates for use as microbial cell factories.

Other than as microbial cell factories and machinery for lignocellulose degradation, rumen microbes have also been screened for other bioactivities, including isoflavone bioconversion, CLA production, novel lipases (Liu et al. 2009; Bayer et al. 2010), and polyphenol oxidase, for degrading insecticides (Beloqui et al. 2006; Math et al. 2010). In addition, cysteine phytases have been isolated by screening ruminal genomic DNA from cows and goats using degenerate primer sets (Huang et al. 2011).

24.7.7 Food Safety

The rumen microbiome has a critical role in the overall health and production performance of ruminants. There have been several developments in the livestock sector across the world to improve milk and meat production from ruminants. The use of growth promoters to increase productivity was routinely practiced in most parts of the world. However, metagenomic evidence indicates that the gastrointestinal tracts in humans as well as animals have become large reservoirs of antimicrobial-resistant genes, collectively referred to as the resistome (Rolain 2013). The increased prevalence of these genes is attributed to the excessive use, misuse, and abuse of antibiotics in both human and veterinary interventions. It has been shown that AR genes that are prevalent in the gut microbiome of farm animals can be laterally transferred to other animals and humans via contact or through the food chain (Looft et al. 2012; Cantas et al. 2013). Earlier research was focused on the identification of zoonotic pathogens, but with the recent reports (Nordmann et al. 2011; Ransom-Jones et al. 2012) on international pandemics such as extended spectrum of beta-lactamases (CTX-M 14 and 15) and carbapenemase KPC-1 which demonstrate the rapid mobility of AR genes across a variety of hosts and countries thus heightening our concerns on AR as a major global concern of the twenty-first century. With a high risk associated with AR spreading from animals to humans, it is imperative to routinely screen different reservoirs for AR genes and identify those AR genes that can be of potential risk to human and animal health. Tremendous insights not only on the rumen microbiome but also on the associated resistome (genes that confer resistance to antibiotics of both pathogenic and nonpathogenic bacteria; Durso et al. 2011) as well as the virulome (genes that make the microbe pathogenic) in an ecosystem are gained using metagenomic approaches. This information has shed light on potential mechanisms and pathways of AR gene disseminations across different ecosystems. Further, comprehensive analysis of metagenomes indicated the extent of homology shared between different resistomes and assessed the potential for horizontal gene transfer through mobilomes (mobile genetic elements) of phylogenetically distant microbiomes as a mechanism for AR acquisition (Forsberg et al. 2012). To address this global problem, information on AR genes in their respective ecosystems alone is not sufficient but calls for a robust surveillance across different reservoirs. The combination of different disciplines with active involvement from clinicians, veterinarians, molecular biologists, agricultural scientists, engineers and stakeholders is pivotal to reduce the spread of AR and protect human and animal health.

24.7.8 Global Food Security

Can genomics help alleviate the food insecurities projected for 2050? It is estimated that the human population will reach 9.2 billion by 2050. In developing countries, meat consumption will increase by 45 % and milk consumption will increase by 46 % by 2030. How do we feed the burgeoning populations in 2050 with depleting agricultural resources, increased urbanization, and increased competition for food and land across the world? Developed countries have made great strides in increasing the production efficiency of food-producing animals and agricultural crops. However, to meet dairy and meat needs, there has been a constant increase in livestock numbers in developing countries to enhance food production. This increase in livestock numbers is not sustainable; therefore, the only option is to increase food production per unit basis by further improving production efficiency in developed and undeveloped countries. To achieve this, advanced concepts of science resolved to the nano-level, complemented with infrastructure and technology, should be augmented. At this juncture, genomics when applied to the rumen microbiome of different ruminant species can nail down the "superbugs" or core microbial consortium that is resilient and production-efficient and that can support enhanced ruminant production. Optimizing the rumen microbiome is only around the corner as the sequencing costs continue to decline over time and robust and sophisticated bioinformatic tools continue to develop which can ultimately contribute to the advancement of a sustainable and productive ruminant system.

24.8 Conclusion

The rumen microbiome is integral to ruminant production; however, much of it was a gray area until the advent of next-generation sequencing and "omic" technologies. Metagenomics is a powerful lens that provides an unprecedented view of the minute microbial populations in the rumen. Because of the vast diversity and complexity of the rumen ecosystem, "omic"-based studies will facilitate understanding the consortium of rumen microbes at species level and the roles they play in the ecosystem. The technologies can potentially expand our knowledge of host–microbe interactions and the rumen functions that may allow achieving significant improvements in animal health and productivity in addition to decreasing environmental pollutants and limiting contamination of the food chain.

References

Abecia L, Waddams KE, Martínez-Fernandez G et al (2014) An antimethanogenic nutritional intervention in early life of ruminants modifies ruminal colonization by archaea. Archaea 2014:1–12, Article ID 841463

Allison MJ, Mayberry WR, Mcsweeney CS et al (1992) *Synergistes jonesii*, gen. nov., sp. nov.: a rumen bacterium that degrades toxic pyridinediols. Syst Appl Microbiol 15(4):522–529

Anderson RC, Krueger NA, Stanton TB et al (2008) Effects of select nitrocompounds on *in vitro* ruminal fermentation during conditions of limiting or excess added reductant. Bioresour Technol 99(18):8655–8661

Ariefdjohan MW, Savaiano DA, Nakatsu CH (2010) Comparison of DNA extraction kits for PCR-DGGE analysis of human intestinal microbial communities from fecal specimens. Nutr J 9(1):23

Attwood GT, Klieve AV, Ouwerkerk D et al (1998) Ammonia-hyperproducing bacteria from New Zealand ruminants. Appl Environ Microbiol 64(5):1796–1804

Bayer S, Kunert A, Ballschmiter M et al (2010) Indication for a new lipolytic enzyme family: isolation and characterization of two esterases from a metagenomic library. J Mol Microbiol Biotechnol 18(3):181–187

Beauchemin K, Kreuzer M, O'mara F et al (2008) Nutritional management for enteric methane abatement: a review. Anim Prod Sci 48(2):21–27

Beloqui A, Pita M, Polaina J et al (2006) Novel polyphenol oxidase mined from a metagenome expression library of bovine rumen biochemical properties, structural analysis, and phylogenetic relationships. J Biol Chem 281(32):22933–22942

Branton D, Deamer DW, Marziali A et al (2008) The potential and challenges of nanopore sequencing. Nat Biotechnol 26(10):1146–1153

Brown M, Ponce C, Pulikanti R (2006) Adaptation of beef cattle to high-concentrate diets: performance and ruminal metabolism. J Anim Sci 84(13 suppl):E25–E33

Brulc JM, Antonopoulos DA, Miller ME et al (2009) Gene-centric metagenomics of the fiber-adherent bovine rumen microbiome reveals forage specific glycoside hydrolases. Proc Natl Acad Sci U S A 106(6):1948–1953

Callaway T, Dowd S, Edrington T et al (2010) Evaluation of bacterial diversity in the rumen and feces of cattle fed different levels of dried distillers grains plus solubles using bacterial tag-encoded FLX amplicon pyrosequencing. J Anim Sci 88(12):3977–3983

Calsamiglia S, Busquet M, Cardozo PW et al (2007) Invited review: essential oils as modifiers of rumen microbial fermentation. J Dairy Sci 90(6):2580–2595

Cantas L, Shah SQ, Cavaco LM et al (2013) A brief multidisciplinary review on antimicrobial resistance in medicine and its linkage to the global environmental microbiota. Front Microbiol 4:96

Cardona S, Eck A, Cassellas M et al (2012) Storage conditions of intestinal microbiota matter in metagenomic analysis. BMC Microbiol 12(1):158

Chaucheyras-Durand F, Ossa F (2014) Review: the rumen microbiome: composition, abundance, diversity, and new investigative tools. Prof Anim Sci 30(1):1–12

Chaucheyras-Durand F, Walker N, Bach A (2008) Effects of active dry yeasts on the rumen microbial ecosystem: past, present and future. Anim Feed Sci Technol 145(1):5–26

Chen G, Russell JB (1989) More monensin-sensitive, ammonia-producing bacteria from the rumen. Appl Environ Microbiol 55(5):1052–1057

Claassen S, du Toit E, Kaba M et al (2013) A comparison of the efficiency of five different commercial DNA extraction kits for extraction of DNA from faecal samples. J Microbiol Methods 94(2):103–110

Cook S, Maiti P, Chaves A et al (2008) Avian (IgY) antimethanogen antibodies for reducing ruminal methane production: *in vitro* assessment of their effects. Anim Prod Sci 48(2):260–264

Crosby LD, Criddle CS (2003) Understanding bias in microbial community analysis techniques due to rrn operon copy number heterogeneity. Biotechniques 34(4):790–803

Dagar SS, Kumar S, Mudgil P et al (2011) D1/D2 domain of large-subunit ribosomal DNA for differentiation of *Orpinomyces* spp. Appl Environ Microbiol 77(18): 6722–6725

Dahllöf I (2002) Molecular community analysis of microbial diversity. Curr Opin Biotechnol 13(3):213–217

Dai X, Zhu Y, Luo Y et al (2012) Metagenomic insights into the fibrolytic microbiome in yak rumen. PLoS One 7(7):e40430

Deng W, Xi D, Mao H et al (2008) The use of molecular techniques based on ribosomal RNA and DNA for rumen microbial ecosystem studies: a review. Mol Biol Rep 35(2):265–274

Denman SE, Tomkins NW, McSweeney CS (2007) Quantitation and diversity analysis of ruminal methanogenic populations in response to the antimethanogenic compound bromochloromethane. FEMS Microbiol Ecol 62(3):313–322

Durso LM, Harhay GP, Bono JL et al (2011) Virulence-associated and antibiotic resistance genes of microbial populations in cattle feces analyzed using a metagenomic approach. J Microbiol Methods 84(2):278–282

Evans PN, Hinds LA, Sly LI et al (2009) Community composition and density of methanogens in the foregut of the Tammar wallaby (*Macropus eugenii*). Appl Environ Microbiol 75(8):2598–2602

Ferrer M, Golyshina OV, Chernikova TN et al (2005) Novel hydrolase diversity retrieved from a metagenome library of bovine rumen microflora. Environ Microbiol 7(12):1996–2010

Firkins J, Yu Z (2006) Characterisation and quantification of the microbial populations in the rumen. Ruminant physiology, digestion, metabolism and impact of nutrition on gene expression, immunology and stress. Wageningen Academic Publishers, Wageningen, pp 19–54

Forsberg KJ, Reyes A, Wang B et al (2012) The shared antibiotic resistome of soil bacteria and human pathogens. Science 337(6098):1107–1111

Gill HS, Shu Q, Leng RA (2000) Immunization with *streptococcus bovis* protects against lactic acidosis in sheep. Vaccine 18(23):2541–2548

Godoy-Vitorino F, Goldfarb KC, Karaoz U et al (2012) Comparative analyses of foregut and hindgut bacterial communities in hoatzins and cows. Int Soc Microb Ecol J 6(3):531–541

Griffith G, Ozkose E, Theodorou M et al (2009) Diversity of anaerobic fungal populations in cattle revealed by selective enrichment culture using different carbon sources. Fungal Ecol 2(2):87–97

Henderson G, Cox F, Kittelmann S et al (2013) Effect of DNA extraction methods and sampling techniques on the apparent structure of cow and sheep rumen microbial communities. PLoS One 8(9):e74787

Hespell R, Akin D, Dehority B (1997) Bacteria, fungi, and protozoa of the rumen. In: Mackie RI, White BA,

Isaacson RE (eds) Gastrointestinal microbiology, vol 2. Gastrointestinal microbes and host interactions. Chapman and Hall, New York, pp 59–141

Hess M, Sczyrba A, Egan R et al (2011) Metagenomic discovery of biomass-degrading genes and genomes from cow rumen. Science 331(6016):463–467

Hobson PN, Stewart CS (1997) The rumen microbial ecosystem, 2nd edn. Springer Science and Bussiness Media, New York

Hristov A, Callaway T, Lee C et al (2012) Rumen bacterial, archaeal, and fungal diversity of dairy cows in response to ingestion of lauric or myristic acid. J Anim Sci 90(12):4449–4457

Huang H, Zhang R, Fu D et al (2011) Diversity, abundance and characterization of ruminal cysteine phytases suggest their important role in phytate degradation. Environ Microbiol 13(3):747–757

Jami E, Mizrahi I (2012a) Composition and similarity of bovine rumen microbiota across individual animals. PLoS One 7(3):e33306

Jami E, Mizrahi I (2012b) Similarity of the ruminal bacteria across individual lactating cows. Anaerobe 18(3):338–343

Jami E, Israel A, Kotser A et al (2013) Exploring the bovine rumen bacterial community from birth to adulthood. Int Soc Microb Ecol J 7(6):1069–1079

Jami E, White BA, Mizrahi I (2014) Potential role of the bovine rumen microbiome in modulating milk composition and feed efficiency. PLoS One 9(1):e85423

Janssen PH, Kirs M (2008) Structure of the archaeal community of the rumen. Appl Environ Microbiol 74(12):3619–3625

Johnson KA, Johnson DE (1995) Methane emissions from cattle. J Anim Sci 73(8):2483–2492

Kav AB, Sasson G, Jami E et al (2012) Insights into the bovine rumen plasmidome. Proc Natl Acad Sci U S A 109(14):5452–5457

Kennedy NA, Walker AW, Berry SH et al (2014) The impact of different DNA extraction kits and laboratories upon the assessment of human gut microbiota composition by 16 s rRNA gene sequencing. PLoS One 9(2):e88982

Khafipour E, Li S, Plaizier JC et al (2009) Rumen microbiome composition determined using two nutritional models of subacute ruminal acidosis. Appl Environ Microbiol 75(22):7115–7124

Kim M, Morrison M, Yu Z (2011a) Evaluation of different partial 16S rRNA gene sequence regions for phylogenetic analysis of microbiomes. J Microbiol Methods 84(1):81–87

Kim M, Morrison M, Yu Z (2011b) Status of the phylogenetic diversity census of ruminal microbiomes. FEMS Microbiol Ecol 76(1):49–63

Kittelmann S, Seedorf H, Walters WA et al (2013) Simultaneous amplicon sequencing to explore co-occurrence patterns of bacterial, archaeal and eukaryotic microorganisms in rumen microbial communities. PLoS One 8(2):e47879

Klein MS, Buttchereit N, Miemczyk SP et al (2011) NMR metabolomic analysis of dairy cows reveals milk glycerophosphocholine to phosphocholine ratio as prognostic biomarker for risk of ketosis. J Proteome Res 11(2):1373–1381

Klieve A, Hennessy D, Ouwerkerk D et al (2003) Establishing populations of *Megasphaera elsdenii* YE 34 and *Butyrivibrio fibrisolvens* YE 44 in the rumen of cattle fed high grain diets. J Appl Microbiol 95(3):621–630

Kobayashi Y (2006) Inclusion of novel bacteria in rumen microbiology: need for basic and applied science. Anim Sci J 77(4):375–385

Kocherginskaya SA, Aminov RI, White BA (2001) Analysis of the rumen bacterial diversity under two different diet conditions using denaturing gradient gel electrophoresis, random sequencing, and statistical ecology approaches. Anaerobe 7(3):119–134

Koike S, Kobayashi Y (2001) Development and use of competitive PCR assays for the rumen cellulolytic bacteria: *Fibrobacter succinogenes, Ruminococcus albus* and *Ruminococcus flavefaciens*. FEMS Microbiol Lett 204(2):361–366

Kong Y, Teather R, Forster R (2010) Composition, spatial distribution, and diversity of the bacterial communities in the rumen of cows fed different forages. FEMS Microbiol Ecol 74(3):612–622

Kong Y, Xia Y, Seviour R et al (2012) *In situ* identification of carboxymethyl cellulose–digesting bacteria in the rumen of cattle fed alfalfa or triticale. FEMS Microbiol Ecol 80(1):159–167

Krause D, Nagaraja T, Wright A et al (2013) Board-invited review: rumen microbiology: leading the way in microbial ecology. J Anim Sci 91(1):331–341

Kumar S, Puniya AK, Puniya M et al (2009) Factors affecting rumen methanogens and methane mitigation strategies. World J Microbiol Biotechnol 25(9):1557–1566

Kumar S, Dagar SS, Mohanty AK et al (2011) Enumeration of methanogens with a focus on fluorescence in situ hybridization. Naturwissenschaften 98(6):457–472

Kumar S, Dagar SS, Puniya AK (2012) Isolation and characterization of methanogens from rumen of Murrah buffalo. Ann Microbiol 62(1):345–350

Kumar S, Choudhury PK, Carro MD et al (2014) New aspects and strategies for methane mitigation from ruminants. Appl Microbiol Biotechnol 98(1):31–44

Kumar S, Dagar SS, Ebrahimi H et al (2015) Prospective use of bacteriocinogenic pediococcus pentosaceus as direct-fed microbial having methane reducing potential. J Integr Agric 14:561–566

Larue R, Yu Z, Parisi VA et al (2005) Novel microbial diversity adherent to plant biomass in the herbivore gastrointestinal tract, as revealed by ribosomal intergenic spacer analysis and rrs gene sequencing. Environ Microbiol 7(4):530–543

Lascano CE, Cárdenas E (2010) Alternatives for methane emission mitigation in livestock systems. Revista Brasileira de Zootecnia 39:175–182

Leahy SC, Kelly WJ, Altermann E et al (2010) The genome sequence of the rumen methanogen *Methanobrevibacter ruminantium* reveals new possibilities for controlling ruminant methane emissions. PLoS One 5(1):e8926

Leahy SC, Kelly WJ, Li D et al (2013) The complete genome sequence of *Methanobrevibacter* sp. AbM4. Stand Genomic Sci 8(2):215

Lee D-H, Zo Y-G, Kim S-J (1996) Nonradioactive method to study genetic profiles of natural bacterial communities by PCR-single-strand-conformation polymorphism. Appl Environ Microbiol 62(9):3112–3120

Lee G-H, Kumar S, Lee J-H et al (2012) Genome sequence of *Oscillibacter ruminantium* strain GH1, isolated from rumen of Korean native cattle. J Bacteriol 194(22):6362

Lee J-H, Rhee M-S, Kumar S et al (2013) Genome sequence of *Methanobrevibacter* sp. strain JH1, isolated from rumen of Korean native cattle. Genome Announc 1(1):e00002–e00013

Li C-J, Li RW, Y-h W et al (2007) Pathway analysis identifies perturbation of genetic networks induced by butyrate in a bovine kidney epithelial cell line. Funct Integr Genomics 7(3):193–205

Li M, Penner G, Hernandez-Sanabria E et al (2009) Effects of sampling location and time, and host animal on assessment of bacterial diversity and fermentation parameters in the bovine rumen. J Appl Microbiol 107(6):1924–1934

Li J, Wijffels G, Yu Y et al (2011) Altered fatty acid metabolism in long duration road transport: an NMR-based metabonomics study in sheep. J Proteome Res 10(3):1073–1087

Liggenstoffer AS, Youssef NH, Couger M et al (2010) Phylogenetic diversity and community structure of anaerobic gut fungi (phylum *Neocallimastigomycota*) in ruminant and non-ruminant herbivores. Int Soc Microb Ecol J 4(10):1225–1235

Liu K, Wang J, Bu D et al (2009) Isolation and biochemical characterization of two lipases from a metagenomic library of China Holstein cow rumen. Biochem Biophys Res Commun 385(4):605–611

Liu L, Li Y, Li S et al (2012) Comparison of next-generation sequencing systems. BioMed Res Int 1–11, Article ID 251364

Lodge-Ivey S, Browne-Silva J, Horvath M (2009) Technical note: bacterial diversity and fermentation end products in rumen fluid samples collected via oral lavage or rumen cannula. J Anim Sci 87(7):2333–2337

Looft T, Johnson TA, Allen HK et al (2012) In-feed antibiotic effects on the swine intestinal microbiome. Proc Natl Acad Sci U S A 109(5):1691–1696

Luton PE, Wayne JM, Sharp RJ et al (2002) The mcrA gene as an alternative to 16S rRNA in the phylogenetic analysis of methanogen populations in landfill. Microbiology 148(11):3521–3530

Mackie RI, Aminov RI, Hu W et al (2003) Ecology of uncultivated Oscillospira species in the rumen of cattle, sheep, and reindeer as assessed by microscopy and molecular approaches. Appl Environ Microbiol 69(11):6808–6815

Martin C, Morgavi D, Doreau M (2010) Methane mitigation in ruminants: from microbe to the farm scale. Anim Int J Anim Biosci 4(03):351–365

Math RK, Islam SMA, Cho KM et al (2010) Isolation of a novel gene encoding a 3, 5, 6-trichloro-2-pyridinol degrading enzyme from a cow rumen metagenomic library. Biodegradation 21(4):565–573

Maukonen J, Simões C, Saarela M (2012) The currently used commercial DNA-extraction methods give different results of clostridial and actinobacterial populations derived from human fecal samples. FEMS Microbiol Ecol 79(3):697–708

Min B, Attwood G, Reilly K et al (2002) *Lotus corniculatus* condensed tannins decrease in vivo populations of proteolytic bacteria and affect nitrogen metabolism in the rumen of sheep. Can J Microbiol 48(10):911–921

Mizrahi I (2012) The rumen plasmidome: a genetic communication hub for the rumen microbiome. Mob Genet Elements 2(3):152

Mohammed R, Stevenson D, Weimer P et al (2012) Individual animal variability in ruminal bacterial communities and ruminal acidosis in primiparous Holstein cows during the periparturient period. J Dairy Sci 95(11):6716–6730

Morgavi DP, Kelly WJ, Janssen PH et al (2013) Rumen microbial (meta)genomics and its application to ruminant production. Anim Int J Anim Biosci 7(Suppl 1):184–201

Morrison M, Daugherty SC, Nelson WC et al (2010) The FibRumBa database: a resource for biologists with interests in gastrointestinal microbial ecology, plant biomass degradation, and anaerobic microbiology. Microb Ecol 59(2):212–213

Mosoni P, Martin C, Forano E et al (2011) Long-term defaunation increases the abundance of cellulolytic ruminococci and methanogens but does not affect the bacterial and methanogen diversity in the rumen of sheep. J Anim Sci 89(3):783–791

Mrazek J, Tepšič K, Avguštin G et al (2006) Diet-dependent shifts in ruminal butyrate-producing bacteria. Folia Microbiol 51(4):294–298

Nagaraja TG, Newbold CJ, van Nevel CJ et al (1997) Manipulation of ruminal fermentation. In: Stewart CS, Hobson PJ (eds) Rumen microbial ecosystem. Blackie Academic & Professional, London, pp 523–632

Nevel CJ, Demeyer DI (1996) Control of rumen methanogenesis. Environ Monit Assess 42(1–2):73–97

Newbold CJ, Rode L (2006) Dietary additives to control methanogenesis in the rumen. Int Congr Ser 1293:138–147, Elsevier

Nicholson MJ, Theodorou MK, Brookman JL (2005) Molecular analysis of the anaerobic rumen fungus Orpinomyces–insights into an AT-rich genome. Microbiology 151(1):121–133

Nordmann P, Poirel L, Walsh TR et al (2011) The emerging NDM carbapenemases. Trends Microbiol 19(12):588–595

Ouwerkerk D, Klieve A, Forster R (2002) Enumeration of *Megasphaera elsdenii* in rumen contents by real-time Taq nuclease assay. J Appl Microbiol 92(4):753–758

Park BH, Karpinets TV, Syed MH et al (2010) CAZymes analysis toolkit (CAT): web service for searching and analyzing carbohydrate-active enzymes in a newly sequenced organism using CAZy database. Glycobiology 20(12):1574–1584

Patel DD, Patel AK, Parmar NR et al (2014) Microbial and carbohydrate active enzyme profile of buffalo rumen metagenome and their alteration in response to variation in the diet. Gene 545(1):88–94

Patra AK, Saxena J (2009) Dietary phytochemicals as rumen modifiers: a review of the effects on microbial populations. Antonie Van Leeuwenhoek 96(4): 363–375

Patra A, Kamra D, Bhar R et al (2011) Effect of *Terminalia chebula* and *Allium sativum* on in vivo methane emission by sheep. J Anim Physiol Anim Nutr 95(2):187–191

Pitta DW, Pinchak WE, Dowd SE et al (2010) Rumen bacterial diversity dynamics associated with changing from bermudagrass hay to grazed winter wheat diets. Microb Ecol 59(3):511–522

Pitta D, Kumar S, Veiccharelli B et al (2014a) Bacterial diversity associated with feeding dry forage at different dietary concentrations in the rumen contents of Mehshana buffalo *Bubalus bubalis* using 16S pyrotags. Anaerobe 25:31–41

Pitta D, Kumar S, Vecchiarelli B et al (2014b) Temporal dynamics in the rumen microbiome of dairy cows during the transition period. J Anim Sci 92:4014–4022

Plaizier J, Khafipour E, Li S et al (2012) Subacute ruminal acidosis (SARA), endotoxins and health consequences. Anim Feed Sci Technol 172(1):9–21

Pope PB, Mackenzie AK, Gregor I et al (2012) Metagenomics of the Svalbard reindeer rumen microbiome reveals abundance of polysaccharide utilization loci. PLoS One 7(6):e38571

Poulsen M, Schwab C, Jensen BB et al (2013) Methylotrophic methanogenic *Thermoplasmata* implicated in reduced methane emissions from bovine rumen. Nat Commun 4:1428

Puniya AK, Salem AZM, Kumar S et al (2014) Role of live microbial feed supplements with reference to anaerobic fungi in ruminant productivity. J Integr Agric. doi:10.1016/S2095-3119(14)60837-6

Qi M, Wang P, O'Toole N et al (2011) Snapshot of the eukaryotic gene expression in muskoxen rumen—a metatranscriptomic approach. PLoS One 6(5):e20521

Ransom-Jones E, Jones DL, McCarthy AJ et al (2012) The fibrobacteres: an important phylum of cellulose-degrading bacteria. Microb Ecol 63(2):267–281

Regensbogenova M, Kisidayova S, Michalowski T et al (2004) Rapid identification of rumen protozoa by restriction analysis of amplified 18S rRNA gene. Acta Protozool 43(3):219–224

Reilly K, Attwood G (1998) Detection of *Clostridium proteoclasticum* and closely related strains in the rumen by competitive PCR. Appl Environ Microbiol 64(3):907–913

Roesch LF, Fulthorpe RR, Riva A et al (2007) Pyrosequencing enumerates and contrasts soil microbial diversity. Int Soc Microb Ecol J 1(4):283–290

Rolain J-M (2013) Food and human gut as reservoirs of transferable antibiotic resistance encoding genes. Front Microbiol 4:173

Russell JB, Rychlik JL (2001) Factors that alter rumen microbial ecology. Science 292(5519):1119–1122

Russell J, Strobel H, Chen G (1988) Enrichment and isolation of a ruminal bacterium with a very high specific activity of ammonia production. Appl Environ Microbiol 54(4):872–877

Russell JB, Muck RE, Weimer PJ (2009) Quantitative analysis of cellulose degradation and growth of cellulolytic bacteria in the rumen. FEMS Microbiol Ecol 67(2):183–197

Saleem F, Bouatra S, Guo AC et al (2013) The bovine ruminal fluid metabolome. Metabolomics 9(2):360–378

Sauer M, Marx H, Mattanovich D (2012) From rumen to industry. Microb Cell Fact 11(1):121

Sekhavati MH, Mesgaran MD, Nassiri MR et al (2009) Development and use of quantitative competitive PCR assays for relative quantifying rumen anaerobic fungal populations in both *in vitro* and *in vivo* systems. Mycol Res 113(10):1146–1153

Shanks OC, Kelty CA, Archibeque S et al (2011) Community structures of fecal bacteria in cattle from different animal feeding operations. Appl Environ Microbiol 77(9):2992–3001

Shu Q, Gill H, Hennessy D et al (1999) Immunisation against lactic acidosis in cattle. Res Vet Sci 67(1):65–71

Shu Q, Gill H, Leng R et al (2000) Immunization with a *Streptococcus bovis* vaccine administered by different routes against lactic acidosis in sheep. Vet J 159(3):262–269

Sirohi SK, Singh N, Dagar SS et al (2012) Molecular tools for deciphering the microbial community structure and diversity in rumen ecosystem. Appl Microbiol Biotechnol 95(5):1135–1154

Sirohi SK, Choudhury PK, Puniya AK et al (2013) Ribosomal ITS1 sequence-based diversity analysis of anaerobic rumen fungi in cattle fed on high fiber diet. Ann Microbiol 63(4):1571–1577

Skillman LC, Toovey AF, Williams AJ et al (2006) Development and validation of a real-time PCR method to quantify rumen protozoa and examination of variability between Entodinium populations in sheep offered a hay-based diet. Appl Environ Microbiol 72(1):200–206

Soliva CR, Meile L, Cieślak A et al (2004) Rumen simulation technique study on the interactions of dietary lauric and myristic acid supplementation in suppressing ruminal methanogenesis. Br J Nutr 92(04):689–700

Sorek R, Cossart P (2009) Prokaryotic transcriptomics: a new view on regulation, physiology and pathogenicity. Nat Rev Genet 11(1):9–16

Stahl DA, Flesher B, Mansfield HR et al (1988) Use of phylogenetically based hybridization probes for studies of ruminal microbial ecology. Appl Environ Microbiol 54(5):1079–1084

Staley JT, Konopka A (1985) Measurement of in situ activities of nonphotosynthetic microorganisms in aquatic and terrestrial habitats. Ann Rev Microbiol 39(1):321–346

Stevens CE, Hume ID (1998) Contributions of microbes in vertebrate gastrointestinal tract to production and conservation of nutrients. Physiol Rev 78(2):393–427

Stevenson DM, Weimer PJ (2007) Dominance of *Prevotella* and low abundance of classical ruminal bacterial species in the bovine rumen revealed by relative quantification real-time PCR. Appl Microbiol Biotechnol 75(1):165–174

Sundset MA, Edwards JE, Cheng YF et al (2009) Molecular diversity of the rumen microbiome of Norwegian reindeer on natural summer pasture. Microb Ecol 57:335–348

Sylvester JT, Karnati SK, Yu Z et al (2004) Development of an assay to quantify rumen ciliate protozoal biomass in cows using real-time PCR. J Nutr 134(12):3378–3384

Sylvester J, Karnati S, Dehority B et al (2009) Rumen ciliated protozoa decrease generation time and adjust 18S ribosomal DNA copies to adapt to decreased transfer interval, starvation, and monensin. J Dairy Sci 92(1):256–269

Tajima K, Nagamine T, Matsui H et al (2001) Phylogenetic analysis of archaeal 16S rRNA libraries from the rumen suggests the existence of a novel group of archaea not associated with known methanogens. FEMS Microbiol Lett 200(1):67–72

Tatsuoka N, Mohammed N, Mitsumori M et al (2004) Phylogenetic analysis of methyl coenzyme-M reductase detected from the bovine rumen. Lett Appl Microbiol 39(3):257–260

Teather RM, Hefford HMA, Forster RJ (1997) Genetics of rumen bacteria. In: Hobson PN, Stewart CS (eds) Rumen microbial ecosystem. Blackie Academic & Professional, London, pp 427–456

Timp W, Mirsaidov UM, Wang D et al (2010) Nanopore sequencing: electrical measurements of the code of life. IEEE Trans Nanotechnol 9(3):281–294

Tuckwell DS, Nicholson MJ, McSweeney CS et al (2005) The rapid assignment of ruminal fungi to presumptive genera using ITS1 and ITS2 RNA secondary structures to produce group-specific fingerprints. Microbiology 151(5):1557–1567

Tymensen L, Barkley C, McAllister TA (2012) Relative diversity and community structure analysis of rumen protozoa according to T-RFLP and microscopic methods. J Microbiol Methods 88(1):1–6

Uyeno Y, Sekiguchi Y, Kamagata Y (2010) rRNA-based analysis to monitor succession of faecal bacterial communities in Holstein calves. Lett Appl Microbiol 51(5):570–577

Větrovský T, Baldrian P (2013) The variability of the 16S rRNA gene in bacterial genomes and its consequences for bacterial community analyses. PLoS One 8(2):e57923

Wallace RJ, Chaudhary LC, Miyagawa E et al (2004) Metabolic properties of *Eubacterium pyruvativorans*, a ruminal 'hyper-ammonia-producing' anaerobe with metabolic properties analogous to those of *Clostridium kluyveri*. Microbiology 150(Pt 9):2921–2930

Wang M, Wang H, Yu L (2009) Effects of NDF content on protozoal community and grazing rate in rumen. J Anim Vet Adv 8(9):1746–1752

Wang T-Y, Chen H-L, Lu M et al (2011) Functional characterization of cellulases identified from the cow rumen fungus Neocallimastix patriciarum W5 by transcriptomic and secretomic analyses. Biotechnol Biofuels 4:24

Wang X, Li X, Zhao C et al (2012) Correlation between composition of the bacterial community and concentration of volatile fatty acids in the rumen during the transition period and ketosis in dairy cows. Appl Environ Microbiol 78(7):2386–2392

Wedlock D, Pedersen G, Denis M et al (2010) Development of a vaccine to mitigate greenhouse gas emissions in agriculture: vaccination of sheep with methanogen fractions induces antibodies that block methane production in vitro. N Z Vet J 58(1):29–36

Weimer PJ (1998) Manipulating ruminal fermentation: a microbial ecological perspective. J Anim Sci 76(12):3114–3122

Weimer PJ, Russell JB, Muck RE (2009) Lessons from the cow: what the ruminant animal can teach us about consolidated bioprocessing of cellulosic biomass. Bioresour Technol 100(21):5323–5331

Weimer P, Stevenson D, Mertens D (2010a) Shifts in bacterial community composition in the rumen of lactating dairy cows under milk fat-depressing conditions. J Dairy Sci 93(1):265–278

Weimer PJ, Stevenson DM, Mantovani HC et al (2010b) Host specificity of the ruminal bacterial community in the dairy cow following near-total exchange of ruminal contents. J Dairy Sci 93(12):5902–5912

Welkie DG, Stevenson DM, Weimer PJ (2010) ARISA analysis of ruminal bacterial community dynamics in lactating dairy cows during the feeding cycle. Anaerobe 16(2):94–100

Whitford MF, Forster RJ, Beard CE et al (1998) Phylogenetic analysis of rumen bacteria by comparative sequence analysis of cloned 16S rRNA genes. Anaerobe 4(3):153–163

Whitford MF, Teather RM, Forster RJ (2001) Phylogenetic analysis of methanogens from the bovine rumen. BMC Microbiol 1(1):5

Williams YJ, Rea SM, Popovski S et al (2008) Reponses of sheep to a vaccination of entodinial or mixed rumen protozoal antigens to reduce rumen protozoal numbers. Br J Nutr 99(01):100–109

Williams YJ, Popovski S, Rea SM et al (2009) A vaccine against rumen methanogens can alter the composition of archaeal populations. Appl Environ Microbiol 75(7):1860–1866

Wright A, Kennedy P, O'Neill C et al (2004) Reducing methane emissions in sheep by immunization against rumen methanogens. Vaccine 22(29):3976–3985

Wu GD, Lewis JD, Hoffmann C et al (2010) Sampling and pyrosequencing methods for characterizing bacterial communities in the human gut using 16S sequence tags. BMC Microbiol 10(1):206

Wyman CE (2007) What is (and is not) vital to advancing cellulosic ethanol. Trends Biotechnol 25(4):153–157

Yanagita K, Kamagata Y, Kawaharasaki M et al (2000) Phylogenetic analysis of methanogens in sheep rumen ecosystem and detection of *Methanomicrobium mobile* by fluorescence *in situ* hybridization. Biosci Biotechnol Biochem 64(8):1737–1742

Youssef NH, Couger M, Struchtemeyer CG et al (2013) The genome of the anaerobic fungus *Orpinomyces* sp. strain C1A reveals the unique evolutionary history of a remarkable plant biomass degrader. Appl Environ Microbiol 79(15):4620–4634

Yu Z, Morrison M (2004) Improved extraction of PCR-quality community DNA from digesta and fecal samples. Biotechniques 36(5):808–813

Zheng Z, Chen T, Zhao M et al (2012) Engineering *Escherichia coli* for succinate production from hemicellulose via consolidated bioprocessing. Microb Cell Fact 11(1):37

Zhou M, Hernandez-Sanabria E (2009) Assessment of the microbial ecology of ruminal methanogens in cattle with different feed efficiencies. Appl Environ Microbiol 75(20):6524–6533

Zoetendal EG, Cheng B, Koike S et al (2004) Molecular microbial ecology of the gastrointestinal tract: from phylogeny to function. Curr Issues Intest Microbiol 5(2):31–48

Zou W, Zhu L-W, Li H-M et al (2011) Significance of CO_2 donor on the production of succinic acid by *Actinobacillus succinogenes* ATCC 55618. Microb Cell Fact 10:87